Inequalities for Differential and Integral Equations

This is Volume 197 in
MATHEMATICS IN SCIENCE AND ENGINEERING
Edited by William F. Ames, *Georgia Institute of Technology*

A list of recent titles in this series appears at the end of this volume.

Inequalities for Differential and Integral Equations

B.G. Pachpatte
*Marathwada University,
Aurangabad,
India*

ACADEMIC PRESS
San Diego • Boston • New York • London • Sydney • Tokyo • Toronto

This book is printed on acid-free paper.

Copyright © 1998 by ACADEMIC PRESS LIMITED

All Rights Reserved.
No part of this publication may be reproduced or transmitted in any form by photostat, microfilm, or any other means without written permission from the publishers

ACADEMIC PRESS INC
525B Street
Suite 1900
San Diego, CA 92101, USA
http://www.apnet.com

ACADEMIC PRESS LIMITED
24-28 Oval Road
LONDON
NW1 7DX, UK
http://www.hbuk.co.uk/ap/

A catalogue record for this book is available from the British Library

ISBN 0-12-543430-8

Typeset by Laser Words, Madras, India
Printed in Great Britain by MPG Books Ltd, Bodmin, Cornwall

98 99 00 01 02 03 EB 9 8 7 6 5 4 3 2 1

Contents

Preface ix

Introduction 1

1. **Linear Integral Inequalities** 9
   1.1 Introduction 9
   1.2 The Inequalities of Gronwall and Bellman 9
   1.3 Some Generalizations of the Gronwall–Bellman Inequality 12
   1.4 Volterra-Type Integral Inequalities 17
   1.5 The Inequalities of Gamidov and Rodrigues 25
   1.6 Simultaneous Inequalities 30
   1.7 Pachpatte's Inequalities 32
   1.8 Integro-differential Inequalities 44
   1.9 Inequalities with Several Iterated Integrals 52
   1.10 Inequalities Involving Product Integrals 62
   1.11 Applications 75
      1.11.1 Second-order Integro-differential Equations 75
      1.11.2 Perturbation of Volterra Integral Equations 78
      1.11.3 Higher Order Integro-differential Equations 81
      1.11.4 Integral Equation Involving Product Integrals 84
   1.12 Miscellaneous Inequalities 86
   1.13 Notes 96

2. **Nonlinear Integral Inequalities I** 99
   2.1 Introduction 99
   2.2 Inequalities Involving Comparison 100
   2.3 The Inequalities of Bihari and Langenhop 107
   2.4 Generalizations of Gronwall–Bellman–Bihari Inequalities 117

	2.5	Inequalities with Volterra-Type Kernels 126
	2.6	Inequalities with Nonlinearities in the Integral 135
	2.7	Pachpatte's Inequalities I 148
	2.8	Pachpatte's Inequalities II 158
	2.9	Integro-differential Inequalities 171
	2.10	Inequalities with Iterated Integrals 181
	2.11	Applications 186
		2.11.1 Second Order Nonlinear Differential Equations 187
		2.11.2 Perturbed Integro-differential Equations 192
		2.11.3 Higher Order Integro-differential Equations 198
		2.11.4 Estimates of the Solutions of Certain Differential Equations 203
	2.12	Miscellaneous Inequalities 206
	2.13	Notes 218
3.	**Nonlinear Integral Inequalities II** 221	
	3.1	Introduction 221
	3.2	Dragomir's Inequalities 221
	3.3	Pachpatte's Inequalities I 227
	3.4	The Inequalities of Ou-Iang and Dafermos 233
	3.5	Pachpatte's Inequalities II 236
	3.6	Pachpatte's Inequalities III 243
	3.7	Pachpatte's Inequalities IV 251
	3.8	The Inequalities of Haraux and Engler 267
	3.9	Pachpatte's Inequalities V 270
	3.10	Inequalities Involving Iterated Integrals 276
	3.11	Applications 291
		3.11.1 Volterra Integral Equations 291
		3.11.2 On Some Epidemic Models 294
		3.11.3 Certain Integral and Differential Equations 297
		3.11.4 Certain Integro-differential and Differential Equations 301
	3.12	Miscellaneous Inequalities 303
	3.13	Notes 320

CONTENTS

4. Multidimensional Linear Integral Inequalities 323
  4.1 Introduction 323
  4.2 Wendroff's Inequality 323
  4.3 Pachpatte's Inequalities I 327
  4.4 Pachpatte's Inequalities II 334
  4.5 Pachpatte's Inequalities III 343
  4.6 Snow's Inequalities 354
  4.7 Generalizations of Snow's Inequalities 364
  4.8 Pachpatte's Inequalities IV 374
  4.9 Inequalities in Several Variables 396
  4.10 Young's Inequality and its Generalizations .. 409
  4.11 Applications 417
     4.11.1 Hyperbolic Partial Integro-differential Equations 417
     4.11.2 Non-Self-Adjoint Hyperbolic Differential and Integro-differential Equations 424
     4.11.3 Perturbations of Hyperbolic Partial Differential Equations 428
     4.11.4 Simultaneous Integral Equations in Two Variables 437
  4.12 Miscellaneous Inequalities 439
  4.13 Notes 456

5. Multidimensional Nonlinear Integral Inequalities 459
  5.1 Introduction 459
  5.2 Generalizations of Wendroff's Inequality 460
  5.3 Wendroff-type Inequalities 468
  5.4 Generalizations of Pachpatte's Inequalities .. 479
  5.5 Pachpatte's Inequalities I 485
  5.6 Pachpatte's Inequalities II 495
  5.7 Inequalities in Many Independent Variables ... 508
  5.8 Pachpatte's Inequalities III 527
  5.9 Pachpatte's Inequalities IV 537
  5.10 Pachpatte's Inequalities V 544
  5.11 Applications 556
     5.11.1 Hyperbolic Partial Differential Equations 556

5.11.2 Hyperbolic Partial Integro-differential Equations 559
5.11.3 Higher Order Hyperbolic Partial Differential Equations ... 561
5.11.4 Multivariate Hyperbolic Partial Integro-differential Equations 562
5.12 Miscellaneous Inequalities 565
5.13 Notes 588

References 591

Index 609

Preface

Inequalities have played a dominant role in the development of all branches of mathematics, and they have a central place in the attention of many mathematicians. One reason for much of the successful mathematical development in the theory of ordinary and partial differential equations is the availability of some kinds of inequalities and variational principles involving functions and their derivatives. Differential and integral inequalities have become a major tool in the analysis of the differential and integral equations that occur in nature or are constructed by people. A good deal of information on this subject may be found in a number of monographs published during the last few years.

Most of the inequalities developed so far in the literature, which provide explicit known bounds on the functions appearing in differential, integral and other equations, perform quite well in practice and hence have found widespread acceptance in a variety of applications. Because of this, it is not surprising that numerous studies of new types of inequalities have been made in order to achieve many new developments in various branches of mathematical science and engineering practice. In the past few years, various investigators have discovered many useful and new integral inequalities through research into various branches of differential and integral equations, where inequalities are often the basis of important lemmas for proving various theorems or for approximating various functions. Many of the inequalities have only recently appeared in the literature, and cannot yet be found in books.

The aims of this monograph are to provide a systematic study of integral inequalities, which find numerous applications in the theory of various classes of differential and integral equations, to be a valuable reference for researchers into differential and integral equations and to serve as a graduate textbook. Taking into account the vast literature, the choice of material for a book devoted to integral inequalities is a difficult task. In this book an attempt has been made to present all the core material as well as those inequalities that have been recently discovered and have proven most useful in applications. In the author's view, this book is vital reading for mathematical analysts, pure and applied mathematicians, physicists, engineers, computer scientists and graduate students in those disciplines.

I wish to express my sincere thanks to Professor William F. Ames for his constant interest in this work and inviting me to write this monograph for his series *Mathematics in Science and Engineering.* It is a pleasure to record my gratitude to the members of the book production department of Academic Press for their helpful cooperation. I am also indebted to my family members for their encouragement, understanding and patience during the writing of the book.

B. G. Pachpatte
May 1997

Introduction

The importance of inequalities has long been recognized in the field of mathematics. The mathematical foundations of the theory of inequalities were established in part during the 18th and 19th century by mathematicians such as K. F. Gauss (1777–1855), A. L. Cauchy (1789–1857) and P. L. Chebyshev (1821–94). In the years thereafter the influence of inequalities has been immense, and the subject has attracted many distinguished mathematicians, including H. Poincaré (1854–1912), A. M. Lyapunov (1857–1918), O. Hölder (1859–1937) and J. Hadamard (1865–1963).

In a subject like inequalities, which has applications in every part of mathematics, historical and bibliographical questions are particularly troublesome. Often, apart from in particular applications, inequalities are regarded as auxiliary in character and are usually obtained by *ad hoc* methods rather than as a consequence of some underlying theory of inequalities. It is not easy to speak of the origin of a familiar inequality used in various applications. It is quite likely to occur first as an auxiliary proposition without explicit statement in the study of a certain branch of mathematics and then be rediscovered many years later by a number of authors. In the case of many inequalities the proofs provided by the original authors have been reproduced or a variety of proofs have been given.

Analysis has been the dominant branch of mathematics for the last three centuries, and inequalities are the heart of analysis. Although inequalities play a fundamental role in all branches of mathematics, the development of the subject as a branch of modern mathematics came in the 20th century through the pioneering work *Inequalities* by G. H. Hardy, J. E. Littlewood and G. Pólya, which appeared in 1934. This theoretical foundation, which has been developed by other mathematicians, has in turn led to the discovery of many new inequalities and interesting applications in various fields of mathematics.

This century has seen considerable and fruitful research in the field of inequalities and their applications in various branches of mathematics. The subject of inequalities is endlessly fascinating and remains a remarkably active and challenging area of current research, despite its relatively long history. Although various introductory and survey monographs on inequalities have appeared in the past few decades, the subject is sufficiently broad and active that there appears always to be room, indeed possibly always to be a need, for yet another one. In fact, it is fair to attribute the main driving force of the rapid development of this field to the users rather than to the inventors of inequalities. Its theory is beautiful, its techniques powerful, and its impact upon science and technology most profound.

Since 1934, when the key work of Hardy *et al.*, *Inequalities*, was published, several papers devoted to inequalities have been published which deal with new inequalities that are effective in many fields of applications of mathematics. It appears that in the theory of inequalities, the three fundamental inequalities, namely, the AM-GM inequality, the Hölder (in particular, Cauchy-Schwarz) inequality and the Minkowski inequality, have played dominant roles. A detailed discussion of these inequalities can be found in the book *Inequalities* by E. F. Beckenbach and R. Bellman, which appeared in 1961, and the book *Analytic Inequalities* by D. S. Mitrinović, published in 1970, see also Agarwal and Pang (1995), and Mitrinović *et al.* (1991, 1992).

In the theory of inequalities, an enormous amount of effort has been devoted to the sharpening of classical approaches and to the discovery of new ones. In the course of discovering a new approach to inequalities, one is naturally inspired by the classical one. In return, the classical approach may be used to establish more inequalities. The modern theory of inequalities, as well as the continuing and growing interest in this field, have deep classical roots which have retained great importance for several centuries at the core of mathematics.

Foremost among the mathematical challenges in modern science and technology is the field of nonlinear differential equations and the theory of such equations has challenged most of the world's greatest mathematicians, providing a tremendous stimulus for many of the modern developments of mathematics. The importance of differential equations for understanding many physical and mathematical problems was first recognized by Sir Isaac Newton in the 17th century, and he used them in his study of the motion of particles and planets. The development of the subject as a branch of modern mathematics came in the 19th and 20th centuries through the pioneering work of a number of mathematicians, notably, Birkhoff, Cauchy, Lyapunov, Picard, Poincaré, Riemann, Bendixion and many others.

Nonlinear differential equations arise in essentially every branch of modern science, engineering and mathematics. However, in only a very few special cases is it possible to obtain useful solutions to nonlinear equations via analytical calculations. In practice, a nonlinear equation is usually approximated by a linear equation which can be solved explicitly, and the solution of the linear equation is taken as an approximation to the solution of the nonlinear equation. It should be noted that not all nonlinear problems can be approximated suitably by linear problems. In recent years, several approaches and methods have been developed for the study of nonlinear systems which cannot be approximated sufficiently well by tractable linear systems.

The study of nonlinear differential equations is a fascinating question which is at the very heart of the understanding of many important problems of the natural sciences. It has already drawn a great deal of attention, not only among mathematicians themselves, but from various other disciplines as well. The mathematical description of dynamical processes encountered in physical, biological and applied sciences requires the investigation of ordinary, functional and partial differential equations. There exists a vast literature on the study of ordinary and partial differential equations. A good deal of information in this regard may be found in the monographs by Bellman (1953), Bellman and Cooke (1963), Coddington and Levinson (1955), Halanay (1966), Coppel (1965), Hale (1977), Hartman (1964), Hille

(1969), Lakshmikantham and Leela (1969), Reid (1971), Sansone and Conti (1964), Szarski (1965), Walter (1970), Courant and Hilbert (1962), Garbedian (1964), Sobolev (1964), Mikhailov (1978) and Sneddon (1957) to mention but a few.

Integral equations have been of considerable significance in the history of mathematics and have held a central place in the attention of mathematicians during the last few decades. The beginning of the theory of integral equations can be traced back to N. H. Abel who found an integral equation in 1812 starting from a problem in mechanics. In 1895, V. Volterra emphasized the significance of the theory of integral equations. In 1900, I. Fredholm made his famous contribution to integral equations, which led to a fascinating period in the development of mathematical analysis. During the past few years, integral equations have proved to be of immense use in applied fields, such as automatic control theory, network theory and the dynamics of nuclear reactors. The mathematical literature provides a good deal of information, and an excellent account of this subject may be found in the monographs by Burton (1983), Miller (1971), Corduneanu (1973, 1977, 1991), Gripenberg *et al.* (1990), Krasnoselskii (1964) and Tricomi (1957), to mention a few.

Many problems, arising in a wide variety of application areas, give rise to mathematical models which form boundary value problems for ordinary or partial differential equations. The foremost desire of an investigator is to solve the problem explicitly. If little theory is available and no explicit solution is readily obtainable, generally the ensuing line of attack is to identify circumstances under which the complexity of the problem may be reduced. In the past few years the growth of this theory has taken beautiful and unexpected paths and will continue with great vigour in the next few decades.

Integral inequalities that give explicit bounds on unknown functions provide a very useful and important device in the study of many qualitative as well as quantitative properties of solutions of nonlinear differential equations. One of the best known and widely used inequalities in the study of nonlinear differential equations can be stated as follows.

If u is a continuous function defined on the interval $J = [\alpha, \alpha + h]$ and

$$0 \leq u(t) \leq \int_{\alpha}^{t} [bu(s) + a] \, ds, \quad t \in J,$$

where a and b are nonnegative constants, then

$$0 \le u(t) \le ah\, e^{bh}, \quad t \in J.$$

This inequality was found by Gronwall (1919) while investigating the dependence of systems of differential equations with respect to a parameter. In fact the roots of such an inequality can be found in the work of Peano (1885–86), who proved explicitly the special case of the above inequality with $a = 0$. Its usefulness is felt from the very beginning when one encounters the most elementary theorem on the existence and uniqueness of solutions and their dependence on parameters and initial values.

After the discovery of the integral inequality resulting from Gronwall (1919), a number of mathematicians have shown their considerable interest to generalize the original form of this inequality. During the period 1919–75 a large number of papers appeared in the literature which were partly inspired by the Gronwall inequality and its applications. An extensive survey of integral inequalities of the Gronwall type which are adequate in many applications in the theory of differential and integral equations may be found in Beesack (1975).

The year 1973 marked a new beginning in the theory of such inequalities due to the discovery of the following remarkable inequality by Pachpatte (1973a).

If u, f, g are nonnegative continuous functions on $R_+ = [0, \infty)$, $u_0 \ge 0$ is a constant and

$$u(t) \le u_0 + \int_0^t f(s) \left(u(s) + \int_0^s g(\sigma) u(\sigma)\, d\sigma \right) ds, \quad t \in R_+,$$

then

$$u(t) \le u_0 \left[1 + \int_0^t f(s) \exp\left(\int_0^s [f(\sigma) + g(\sigma)]\, d\sigma \right) ds \right], \quad t \in R_+.$$

There are two interesting features related to this inequality. First, in the special case when $g = 0$, it reduces to the well-known generalization of Gronwall's inequality given by Bellman (1943); and second, it is applicable to the more general situations for which the earlier inequalities do not apply directly. In the years following the publication of the above inequality, Pachpatte made his fundamental contribution to the theory of such inequalities, providing ready and powerful tools in investigating various problems in the theory of differential and integral equations. Over the last twenty years the study of such inequalities has undergone formidable development, and various investigators have discovered many useful and new inequalities centred on the above inequality which will be an indispensable source for a long time to come.

One of the most useful inequalities in the development of the theory of differential equations is given in the following theorem.

If u, f are nonnegative continuous functions on $R_+ = [0, \infty)$, $u_0 \geq 0$ is a constant and

$$u^2(t) \leq u_0^2 + 2 \int_0^t f(s)u(s)\,ds, \quad t \in R_+,$$

then

$$u(t) \leq u_0 + \int_0^t f(s)\,ds, \quad t \in R_+.$$

$\square$

It appears that this inequality was first considered by Ou-Iang (1957), while investigating the boundedness of solutions of certain second-order differential equations. In the literature there are many papers which make use of this inequality to obtain global existence, uniqueness, stability, and other properties of the solutions of various nonlinear differential equations. The importance of this inequality stems from the fact that it is applicable in certain situations in which other available inequalities do not apply directly. Although stimulating research work related to differential and integral inequalities arising in the theory of differential and integral equations has been undertaken in the literature, it seems that the importance of various generalizations and extensions of the above inequality have been overlooked by the investigators. Motivated and inspired by the great and enduring success enjoyed by this inequality in various applications, in a series of recent papers Pachpatte (1994a–c; 1995a,b,e; 1996a–e; 1997; in press a–j) has established a number of new inequalities related to the above inequality, which can be used as handy tools in the study of certain new classes of differential, integral and integro-differential equations. The growth of the theory and applications of these inequalities are still under development and promise to open up new vistas for research in the near future.

In the study of global regular solutions for the dynamic antiplane shear problem in nonlinear viscoelasticity, Engler (1989) used the following slight variant of the inequality given by Haraux (1981).

If $a \in L^1(0, T, R_+)$, $u_0 > 0$ is a constant, and the function $u: [0, T] \to [1, \infty)$ satisfies the inequality

$$u(t) \leq u_0 \left(1 + \int_0^t a(s) u(s) \log u(s) \, ds \right), \quad 0 \leq t \leq T,$$

then

$$u(t) \leq u_0^{\exp\left(u_0 \int_0^t a(s) \, ds\right)}, \quad 0 \leq t \leq T.$$

The above inequality departs somewhat from the structure of the classical Gronwall inequality, since it contains an extra logarithmic factor in the integrand and the new bound obtained on the unknown function $u(t)$. In view of the important applications of the above inequality given in Engler (1989) and Haraux (1981), Pachpatte (1991, 1992, 1994d) has discovered a number of new generalizations and extensions of this inequality which can be used as powerful tools in the study of certain new classes of differential and integral equations. We believe that these inequalities will serve as a model for future investigations.

Since 1975 an enormous amount of effort has been devoted to the discovery of new types of inequalities and their applications in various branches of ordinary and partial differential and integral equations. Owing to the tremendous success enjoyed during the past few years, a large number of papers have appeared in the literature, which are partly inspired by the challenge of research in various branches of differential and integral equations, where inequalities are often the basis of important lemmas for proving various theorems or approximating various functions. A considerable number of isolated researchers have developed valuable work in the field of integral inequalities and their applications in the theory of differential and integral equations in the past few years. The literature on this theory and its applications is vast and rapidly growing larger. Part of this growth is due to the fact that the subject is genuinely rich and lends itself to many different approaches and applications.

INTRODUCTION

The present monograph deals with a detailed account of all core inequalities and the important advances made during the past twenty years or so in the theory of inequalities, which provides explicit bounds on the unknown functions and some of their important applications to differential and integral equations. This is a vast area with connections to many branches of ordinary and partial differential and integral equations, and to give a comprehensive account of this subject is thus impossible. For treatments of topics not discussed here, as well as additional perspectives and more complete references, readers are referred to the main sources mentioned in the references.

The monograph consists of five chapters and an extensive list of references. Chapter 1 contains a number of linear integral inequalities which find widespread applications in the theory of various classes of differential and integral equations. Chapters 2 and 3 deal with a variety of nonlinear integral inequalities which can be used to achieve great diversity in both the theory and applications of various types of nonlinear differential and integral equations. Chapters 4 and 5 are devoted respectively to the linear and nonlinear integral inequalities involving two or more independent variables which play the central role in the study of numerous problems in the theory of different classes of partial differential and integral equations. Each chapter contains some relevant applications. At the end of each chapter a section on miscellaneous inequalities is given, indicating further sources of information. If a source is not mentioned for any inequality this indicates that it is new to the literature.

Throughout the work, unless otherwise stated, all the functions which appear in the inequalities are assumed to be real-valued in their domains of definitions. In what follows, R denotes the set of real numbers and $R_+ = [0, \infty)$, $R_1 = [1, \infty)$, $I = [t_0, \infty)$, $t_0 \in R$ and $J = [\alpha, \beta]$ are the given subsets of R. The ordinary first-order derivative of a function $x(t)$ defined for $t \in R$ is denoted by x' or $\dot{x}$ or $\frac{dx}{dt}$ and the higher-order derivatives are denoted in the usual way. The first-order partial derivatives of a function $z(x, y)$ defined for $x, y \in R$, with respect to x and y are denoted by $z_x(x, y)$ and $z_y(x, y)$ or $\frac{\partial}{\partial x} z(x, y)$ and $\frac{\partial}{\partial y} z(x, y)$ or $D_1 z(x, y)$ and $D_2 z(x, y)$ respectively and the higher-order partial derivatives of $z(x, y)$ are denoted in the usual way. Most of the notations, definitions and symbols used throughout the work are standard and explained if necessary at appropriate places.

Chapter One

Linear Integral Inequalities

1.1 Introduction

The classical integral inequalities, which give explicit bounds for an unknown function, have played a fundamental role in establishing the foundations of the theory of differential and integral equations. Because of the importance of such inequalities in the theory of differential and integral equations, over the years various investigators have discovered many useful inequalities in order to achieve a diversity of desired goals. This Chapter presents some basic linear integral inequalities which find significant applications in the study of various classes of differential and integral equations. Some immediate applications of certain inequalities will also be presented in order to study the various properties of the solutions of some integral and integro-differential equations. Some useful inequalities established by various investigators are also given as miscellaneous inequalities.

1.2 The Inequalities of Gronwall and Bellman

In the study of the qualitative behaviour of solutions of differential and integral equations, certain integral inequalities often play a fundamental role. In this section the basic inequalities due to Gronwall (1919) and Bellman

(1943), which find numerous applications in the study of various classes of differential and integral equations, are presented.

Gronwall (1919, p. 293) established the following integral inequality.

Theorem 1.2.1 *Let u be a continuous function defined on the interval $J = [\alpha, \alpha + h]$ and*

$$0 \leq u(t) \leq \int_{\alpha}^{t} [bu(s) + a] \, ds, \quad t \in J, \tag{1.2.1}$$

where a and b are nonnegative constants. Then

$$0 \leq u(t) \leq ah\, e^{bh}, \quad t \in J. \tag{1.2.2}$$

$\square$

Proof: By analogy with the process of integrating a linear differential equation of first order, we take $u = z \exp(b(t - \alpha))$. Let the maximum of z on J occur at $t = t_1$. For this value of t, (1.2.1) gives

$$0 \leq z_{\max} \exp(b(t_1 - \alpha)) \leq \int_{\alpha}^{t_1} [bz(s) \exp(b(s - \alpha)) + a] \, ds,$$

whence by the mean value theorem

$$0 \leq z_{\max} \exp(b(t_1 - \alpha)) \leq z_{\max} \int_{\alpha}^{t_1} b \exp(b(s - \alpha)) \, ds + \int_{\alpha}^{t_1} a \, ds$$

$$= z_{\max} (\exp(b(t_1 - \alpha)) - 1) + a(t_1 - \alpha),$$

or finally

$$0 \leq z_{\max} \leq a(t_1 - \alpha) \leq ah,$$

from which (1.2.2) follows at once.

∎

It is interesting to note that the roots of the essential idea of such an inequality can be traced back at least to Peano (1885–86), which explicitly dealt with the special case of Theorem 1.2.1 having $a = 0$, and proved some quite general results concerning differential inequalities as well as maximal and minimal solutions of differential equations.

LINEAR INTEGRAL INEQUALITIES

In a paper published in 1943, Bellman proved the following inequality.

Theorem 1.2.2 *Let u and f be continuous and nonnegative functions defined on $J = [\alpha, \beta]$, and let c be a nonnegative constant. Then the inequality*

$$u(t) \leq c + \int_\alpha^t f(s)u(s)\,ds, \quad t \in J, \qquad (1.2.3)$$

implies that

$$u(t) \leq c \exp\left(\int_\alpha^t f(s)\,ds\right), \quad t \in J. \qquad (1.2.4)$$

$\square$

Proof: Define a function $z(t)$ by the right side of (1.2.3); then we observe that $z(\alpha) = c$, $u(t) \leq z(t)$ and

$$z'(t) = f(t)u(t) \leq f(t)z(t), \quad t \in J. \qquad (1.2.5)$$

We multiply (1.2.5) by $\exp(-\int_\alpha^t f(s)\,ds)$ and apply the identity

$$z'(t)\exp\left(-\int_\alpha^t f(s)\,ds\right) - z(t)f(t)\exp\left(-\int_\alpha^t f(s)\,ds\right)$$

$$= \frac{d}{dt}\left[z(t)\exp\left(-\int_\alpha^t f(s)\,ds\right)\right],$$

to obtain

$$\frac{d}{dt}\left[z(t)\exp\left(-\int_\alpha^t f(s)\,ds\right)\right] \leq 0. \qquad (1.2.6)$$

Integration of (1.2.6) from α to t gives

$$z(t)\exp\left(-\int_\alpha^t f(s)\,ds\right) - z(\alpha) \leq 0. \qquad (1.2.7)$$

Using $z(\alpha) = c$ and $u(t) \leq z(t)$ in (1.2.7) we get the desired inequality in (1.2.4).

■

It is clear that Bellman's result includes that of Gronwall, since $\int_\alpha^t a\,ds \leq ah$ for $\alpha \leq t \leq \alpha + h$. Bellman's inequality, given in Theorem 1.2.2, exerted tremendous influence in the subsequent years, and the study of such inequalities has grown into a substantial field, with many important applications in various branches of differential and integral equations.

Kamke (1945, p. 93) proved and made use of an inequality which was, in fact, that of Gronwall. Among the early users of inequalities of the above type in the theory of ordinary differential equations was Reid (1930), who employed a slightly more general form of Bellman's inequality to study the properties of solutions of infinite systems of ordinary linear differential equations. After the discovery of the above inequality by Bellman, inequalities of this type are known in the literature as 'Bellman's lemma or inequality', the 'Gronwall–Bellman inequality', the 'Bellman–Gronwall inequality' or 'Gronwall's inequality'. In view of the repeated mention of the inequality given in Theorem 1.2.2, it will be referred to here as the 'Gronwall–Bellman inequality'.

1.3 Some Generalizations of the Gronwall–Bellman Inequality

On the basis of various motivations, the Gronwall–Bellman inequality has been extended and used considerably in various contexts. This section gives some useful generalizations and variants of the Gronwall–Bellman inequality.

Bellman (1958) proved and made use of the following variant of the inequality given by himself in Theorem 1.2.2 to study the asymptotic behaviour of the solutions of linear differential-difference equations.

Theorem 1.3.1 *Let u and f be continuous and nonnegative functions defined on $J = [\alpha, \beta]$, and let $n(t)$ be a continuous, positive and nondecreasing function defined on J; then*

$$u(t) \le n(t) + \int_\alpha^t f(s)u(s)\,ds, \quad t \in J, \qquad (1.3.1)$$

implies that

$$u(t) \le n(t) \exp\left(\int_\alpha^t f(s)\,ds\right), \quad t \in J. \qquad (1.3.2)$$

□

LINEAR INTEGRAL INEQUALITIES

Proof: Let $u(t) = n(t)w(t)$. Then

$$w(t) \leq 1 + \int_\alpha^t f(s) \frac{n(s)w(s)}{n(t)} ds$$

$$\leq 1 + \int_\alpha^t f(s)w(s) ds.$$

Now an application of Theorem 1.2.2 yields

$$w(t) \leq \exp\left(\int_\alpha^t f(s) ds\right).$$

Using this bound on $w(t)$ in $u(t) = n(t)w(t)$, we get the required inequality in (1.3.2).

∎

Gollwitzer (1969) gave the following generalization of the Gronwall-Bellman inequality.

Theorem 1.3.2 *Let u, f, g and h be nonnegative continuous functions defined on $J = [\alpha, \beta]$, and*

$$u(t) \leq f(t) + g(t) \int_\alpha^t h(s)u(s) ds, \quad t \in J. \tag{1.3.3}$$

Then

$$u(t) \leq f(t) + g(t) \int_\alpha^t h(s)f(s) \exp\left(\int_s^t h(\sigma)g(\sigma) d\sigma\right) ds, \quad t \in J. \tag{1.3.4}$$

□

Proof: Define a function $z(t)$ by

$$z(t) = \int_\alpha^t h(s)u(s) ds, \tag{1.3.5}$$

then $z(\alpha) = 0$, $u(t) \leq f(t) + g(t)z(t)$ and

$$z'(t) = h(t)u(t) \leq h(t)f(t) + h(t)g(t)z(t). \tag{1.3.6}$$

Multiplying (1.3.6) by the integrating factor $\exp(-\int_\alpha^t h(\sigma)g(\sigma)\,d\sigma)$ we have

$$\frac{d}{dt}\left[z(t)\exp\left(-\int_\alpha^t h(\sigma)g(\sigma)\,d\sigma\right)\right] \leq h(t)f(t)\exp\left(-\int_\alpha^t h(\sigma)g(\sigma)\,d\sigma\right). \tag{1.3.7}$$

By setting $t = s$ in (1.3.7) and integrating it with respect to s from α to t we get

$$z(t)\exp\left(-\int_\alpha^t h(\sigma)g(\sigma)\,d\sigma\right) \leq \int_\alpha^t h(s)f(s)\exp\left(-\int_\alpha^s h(\sigma)g(\sigma)\,d\sigma\right) ds. \tag{1.3.8}$$

Using the bound on $z(t)$ from (1.3.8) in $u(t) \leq f(t) + g(t)z(t)$, we get the required inequality in (1.3.4).

∎

Observe that the inequality given in Theorem 1.3.2 with $g(t) = 1$ was also established by Jones (1964). In addition, several generalizations of Theorem 1.3.2 when $g(t) = 1$, including subsequent extensions to discrete and discontinuous functional equations are contained in Jones (1964).

Remark 1.3.1 Note that in Theorem 1.3.2 equality holds in (1.3.4) for a subinterval $J_1 = [\alpha, \beta_1]$ of J if equality holds in (1.3.3) for $t \in J_1$. The result remains valid if $\leq$ is replaced by $\geq$ in (1.3.3) and (1.3.4). Both (1.3.3) and (1.3.4) with $\leq$ therein replaced by $\geq$ remain valid if $\int_\alpha^t$ is replaced by $\int_t^\beta$ and $\int_s^t$ by $\int_t^s$ throughout.

△

As pointed out by Beesack (1975), if the integrals in Theorem 1.3.2 are Lebesgue integrals the hypotheses can be relaxed to: u, f, g, h are measurable functions such that hu, hf, $hg \in L(J)$. The equality and inequality conditions are then to be understood to hold almost everywhere, and the stated condition for equality is necessary as well as sufficient. Similar remarks apply to all of our subsequent theorems, which we shall mostly state for the continuous case.

Pachpatte (1975a) employed the following variant of the inequality given in Theorem 1.3.2 in obtaining various generalizations of Bellman's inequality given in Theorem 1.2.2.

Theorem 1.3.3 *Let u, g and h be nonnegative continuous functions defined on* $J = [\alpha, \beta]$, $n(t)$ *be a continuous, positive and nondecreasing function defined on J and*

$$u(t) \leq n(t) + g(t) \int_\alpha^t h(s)u(s)\,ds, \quad t \in J; \tag{1.3.9}$$

then

$$u(t) \leq n(t) \left[1 + g(t) \int_\alpha^t h(s) \exp\left(\int_s^t h(\sigma)g(\sigma)\,d\sigma \right) ds \right], \quad t \in J. \tag{1.3.10}$$

□

The proof follows by the same argument as in the proof of Theorem 1.3.1, applying the inequality given in Theorem 1.3.2 with J replaced by R_+.

A fairly general version of Theorem 1.2.2 is given in the following theorem.

Theorem 1.3.4 *Let u, p, q, f and g be nonnegative continuous functions defined on* $J = [\alpha, \beta]$, *and*

$$u(t) \leq p(t) + q(t) \int_\alpha^t [f(s)u(s) + g(s)]\,ds, \quad t \in J. \tag{1.3.11}$$

Then

$$u(t) \leq p(t) + q(t) \int_\alpha^t [f(s)p(s) + g(s)] \exp\left(\int_s^t f(\sigma)q(\sigma)\,d\sigma \right) ds, \quad t \in J. \tag{1.3.12}$$

□

Proof: Define a function $z(t)$ by

$$z(t) = \int_\alpha^t [f(s)u(s) + g(s)]\,ds,$$

Now by following the proof of Theorem 1.3.2, we get the desired inequality in (1.3.12).

■

Remark 1.3.2 By setting $q(t) = 1$ in Theorem 1.3.4 we arrive at the inequality given by Chandirov (1956). If we take $g(t) = 0$ in Theorem 1.3.4, we get the inequality given in Theorem 1.3.2.

△

The following variant of the Gronwall–Bellman inequality is given by Gollwitzer (1969).

Theorem 1.3.5 Let u, v, h and k be nonnegative continuous functions defined on $J = [\alpha, \beta]$, and

$$u(t) \geq v(x) - k(t) \int_x^t h(s)v(s)\,ds, \quad \alpha \leq x \leq t \leq \beta. \tag{1.3.13}$$

Then

$$u(t) \geq v(x) \exp\left(-k(t) \int_x^t h(s)\,ds\right), \quad \alpha \leq x \leq t \leq \beta. \tag{1.3.14}$$

□

Proof: Let

$$z(x) = u(t) + k(t) \int_x^t h(s)v(s)\,ds, \quad \alpha \leq x \leq t. \tag{1.3.15}$$

From (1.3.15) we have

$$z'(x) = -h(x)v(x)k(t) \geq -h(x)z(x)k(t), \quad \alpha \leq x \leq t,$$

since $z(x) \geq v(x)$. The derivative $z'(x)$ at the end points is taken to be the limit from the interior of $[\alpha, t]$. Then, using the integrating factor $r(x) = \exp(-k(t) \int_x^t h(s)\,ds)$, we have $(rz)'(x) \geq 0$ and so $(rz)(t) \geq (rz)(x)$ on $[\alpha, t]$. Now $z(x) \geq v(x)$ and $z(t) = u(t)$ and so we have the result given in (1.3.14). This result is best possible in the sense that if equality holds in (1.3.13) on $[\alpha, t]$, then equality holds in (1.3.14) on $[\alpha, t]$.

■

Remark 1.3.3 This theorem is similar to a special case of the Langenhop inequality (Langenhop, 1960), and an estimate for u which is independent of x is obtained by setting $x = \alpha$.

△

1.4 Volterra-Type Integral Inequalities

Integral inequalities which satisfies a Volterra-type integral inequality have wide applications in the theory of differential and integral equations. In this section we consider some Volterra-type integral inequalities which are more adequate in certain situations.

Chu and Metcalf (1967) proved the following linear generalization of the Gronwall–Bellman inequality.

Theorem 1.4.1 *Let u and f be continuous functions defined on $J = [\alpha, \beta]$ and let the function $k(t, s)$ be continuous and nonnegative on the triangle $\Delta: \alpha \leq s \leq t \leq \beta$. If*

$$u(t) \leq f(t) + \int_\alpha^t k(t, s) u(s) \, ds, \quad t \in J, \tag{1.4.1}$$

then

$$u(t) \leq f(t) + \int_\alpha^t H(t, s) f(s) \, ds, \quad t \in J, \tag{1.4.2}$$

where

$$H(t, s) = \sum_{i=1}^\infty k_i(t, s), \quad (t, s) \in \Delta$$

is the resolvent kernel of k, and the k_i are the iterated kernels of k. □

Proof: From (1.4.1), one has

$$u(t) \leq f(t) + \int_\alpha^t k(t, s) f(s) \, ds + \int_\alpha^t k(t, s) \int_\alpha^s k(s, \sigma) u(\sigma) \, d\sigma \, ds$$

$$= f(t) + \int_\alpha^t k_1(t, s) f(s) \, ds + \int_\alpha^t k_2(t, s) u(s) \, ds,$$

for $t \in J$. The remainder of the proof is by induction and a standard estimation procedure showing the resulting series to be uniformly convergent.

∎

Note that the earlier results, in which an explicit upper bound for u was obtained, are merely those cases for which the resolvent kernel H can be summed in closed form. For example, if $k(t, s) = g(t)h(s) \geq 0$, $\alpha \leq s \leq t \leq \beta$, then

$$H(t, s) = \sum_{i=1}^{\infty} \frac{g(t)h(s)}{(i-1)!} \left(\int_s^t g(\sigma)h(\sigma) \, d\sigma \right)^{i-1}$$

$$= g(t)h(s) \exp\left(\int_s^t g(\sigma)h(\sigma) \, d\sigma \right),$$

since one can show by induction that each k_i, $i = 1, 2, \ldots$, is given by the appropriate term in the sum of H. As noted in Chu and Metcalf (1967), the sharpest bound for solutions of an integral inequality is provided by the maximal solution or the exact solution of the corresponding integral equation, provided such a solution exists. Thus the result given in Theorem 1.4.1, which involves Neumann series, is sharp, as the bound is the exact solution of the corresponding integral equation.

Remark 1.4.1 Note that Beesack (1969) extended Theorem 1.4.1 to the case that u, $f \in L^2(J)$ and $k \in L^2(\Delta)$ and note that the results remain valid if $\leq$ is replaced by $\geq$ in both (1.4.1) and (1.4.2). The inequality given in Theorem 1.4.1 includes as a special case the inequality given in Theorem 1.2.2. Further we note that Willett (1965) considered the inequality in Theorem 1.4.1 and obtained the upper bound for u under the assumption that either $k(t, s)$ or $(\partial/\partial t)k(t, s)$ is degenerate or directly separable in the following sense:

$$k(t, s) \leq \sum_{i=1}^{n} h_i(t)g_i(s),$$

or a similar relation holds for $(\partial/\partial t)k(t, s)$.

Δ

The following theorem presents slight variants of the inequality given by Norbury and Stuart (1987) which are sometimes applicable more conveniently.

Theorem 1.4.2 *Let u and $k(t, s)$ be as in Theorem 1.4.1 and $k(t, s)$ be nondecreasing in t for each $s \in J$.*

(i) If

$$u(t) \leq c + \int_\alpha^t k(t,s) u(s) \, ds, \quad t \in J, \tag{1.4.3}$$

where $c \geq 0$ is a constant. Then

$$u(t) \leq c \exp\left(\int_\alpha^t k(t,s) \, ds \right), \quad t \in J. \tag{1.4.4}$$

(ii) Let $n(t)$ be a positive continuous and nondecreasing function for $t \in J$. If

$$u(t) \leq n(t) + \int_\alpha^t k(t,s) u(s) \, ds, \quad t \in J, \tag{1.4.5}$$

then

$$u(t) \leq n(t) \exp\left(\int_\alpha^t k(t,s) \, ds \right), \quad t \in J. \tag{1.4.6}$$

$\square$

Proof: (i) Fix any T, $\alpha \leq T \leq \beta$. Then, for $\alpha \leq t \leq T$, we have

$$u(t) \leq c + \int_\alpha^t k(T,s) u(s) \, ds, \tag{1.4.7}$$

Define a function $z(t)$ by the right side of (1.4.7), then $z(\alpha) = c$, $u(t) \leq z(t)$ for $\alpha \leq t \leq T$ and

$$z'(t) = k(T,t) u(t) \leq k(T,t) z(t), \quad \alpha \leq t \leq T. \tag{1.4.8}$$

By setting $t = s$ in (1.4.8) and integrating it with respect to s from α to t we get

$$z(T) \leq c \exp\left(\int_\alpha^T k(T,s) \, ds \right). \tag{1.4.9}$$

Since T is arbitrary, from (1.4.9) with T replaced by t and $u(t) \leq z(t)$ we get the inequality in (1.4.4).

(ii) Since $n(t)$ is a positive continuous and nondecreasing function for $t \in J$, from (1.4.5) we observe that

$$\frac{u(t)}{n(t)} \leq 1 + \int_\alpha^t k(t, s) \frac{u(s)}{n(s)} \, ds, \quad t \in J.$$

Now an application of the inequality given in (i) yields the desired result in (1.4.6).

∎

Remark 1.4.2 Note that the inequality given in part (i) was obtained in Norbury and Stuart (1987) under the assumptions of the existence and nonnegativity of $(\partial/\partial t)k(t, s)$.

△

Pachpatte (in press k) proved the following inequality, which in turn is a further generalization of the inequality given by Norbury and Stuart (1987).

Theorem 1.4.3 *Let u, p, q, r and f be nonnegative continuous functions defined on $J = [\alpha, \beta]$. Let $k(t, s)$ and its partial derivative $(\partial/\partial t)k(t, s)$ be nonnegative continuous functions for $\alpha \leq s \leq t \leq \beta$, and*

$$u(t) \leq p(t) + q(t) \int_\alpha^t k(t, s)[r(s)u(s) + f(s)] \, ds, \quad t \in J. \quad (1.4.10)$$

Then

$$u(t) \leq p(t) + q(t) \int_\alpha^t B(\sigma) \exp\left(\int_\sigma^t A(\tau) \, d\tau \right) d\sigma, \quad t \in J, \quad (1.4.11)$$

where

$$A(t) = k(t, t)r(t)q(t) + \int_\alpha^t \frac{\partial}{\partial t} k(t, s) r(s) q(s) \, ds, \quad t \in J, \quad (1.4.12)$$

$$B(t) = k(t, t)[r(t)p(t) + f(t)]$$

$$+ \int_\alpha^t \frac{\partial}{\partial t} k(t, s)[r(s)p(s) + f(s)] \, ds, \quad t \in J. \quad (1.4.13)$$

□

Proof: Define a function $z(t)$ by

$$z(t) = \int_\alpha^t k(t,s)[r(s)u(s) + f(s)]\,ds. \tag{1.4.14}$$

Differentiating (1.4.14) and using $u(t) \leq p(t) + q(t)z(t)$ and the fact that $z(t)$ is monotonic nondecreasing in t and (1.4.12), (1.4.13) we observe that

$$\begin{aligned}
z'(t) &= k(t,t)[r(t)u(t) + f(t)] + \int_\alpha^t \frac{\partial}{\partial t} k(t,s)[r(s)u(s) + f(s)]\,ds \\
&\leq k(t,t)[r(t)(p(t) + q(t)z(t)) + f(t)] \\
&\quad + \int_\alpha^t \frac{\partial}{\partial t} k(t,s)[r(s)(p(s) + q(s)z(s)) + f(s)]\,ds \\
&\leq z(t)[k(t,t)r(t)q(t) + \int_\alpha^t \frac{\partial}{\partial t} k(t,s) r(s) q(s)\,ds] \\
&\quad + k(t,t)[r(t)p(t) + f(t)] + \int_\alpha^t \frac{\partial}{\partial t} k(t,s)[r(s)p(s) + f(s)]\,ds \\
&= A(t)z(t) + B(t). \tag{1.4.15}
\end{aligned}$$

The inequality (1.4.15) implies the estimate

$$z(t) \leq \int_\alpha^t B(\sigma) \exp\left(\int_\sigma^t A(\tau)\,d\tau\right) d\sigma. \tag{1.4.16}$$

Using (1.4.16) in $u(t) \leq p(t) + q(t)z(t)$, we get the required inequality in (1.4.11).

∎

Remark 1.4.3 Note that the special version of the above inequality with $r(t) = 1$ and $f(t) = 0$ was obtained by Movlyankulov and Filatov (1972) under the assumptions that u is a continuous function on J and that $p(t)$, $q(t)$ and $k(t,s)$ are nonnegative continuous functions on the respective domains of their definitions, and proved that if the inequality

$$u(t) \leq p(t) + q(t) \int_\alpha^t k(t,s)u(s)\,ds, \quad t \in J,$$

holds, then

$$u(t) \leq P(t) \exp\left(Q(t) \int_\alpha^t K(t,s)\,ds\right), \quad t \in J,$$

where

$$P(t) = \sup_{s \in [\alpha,\,t]} p(s), \quad Q(t) = \sup_{s \in [\alpha,\,t]} q(s), \quad K(t,s) = \sup_{\tau \in [s,\,t]} k(\tau,s)$$

△

Thompson (1971) gave some inequalities related to the positive solutions of a Volterra equation of the second kind. The equation considered is

$$u(t) = f(t) + \int_0^t k(t,s)u(s)\,ds, \tag{1.4.17}$$

where $k(t,s)$ is a Volterra-type kernel, that is $k(t,s) = 0$ for $s > t$. Unless otherwise stated, we will assume that $f \in L_2^+(I)$ and $k \in L_2^+(I \times I)$, where $I = \{t: 0 \leq t < \infty\}$ and $L_2^+(I) = \{h: h \in L_2(I) \text{ and } h(t) \geq 0 \text{ for } t \in I\}$.
Define $k_1(t,s) = k(t,s)$ and for $n \geq 2$,

$$k_n(t,s) = \int_s^t k(t,\sigma)k_{n-1}(\sigma,s)\,d\sigma.$$

By the *resolvent kernel* we mean

$$H(t,s) = \sum_{i=1}^n k_i(t,s).$$

This series converges in $L_2(I \times I)$ norm.

For completeness we state without proof the principal theorem for Volterra equations.

Theorem 1.4.4 *Let $f \in L_2(I)$, $k \in L_2(I \times I)$; then the equation (1.4.17) has a unique L_2 solution given by*

$$u(t) = f(t) + \int_0^t H(t,s) f(s)\,ds, \quad \text{a.e. on } I.$$

□

For the proof of this theorem the reader is referred to Tricomi (1957).

We will need the following Lemmas proved in Beesack (1969).

Lemma 1.4.1 *Let $f \in L_2^+(I)$ and $k \in L_2^+(I \times I)$ in equation (1.4.17), then $u(t) \geq 0$ a.e. on I.*

△

Lemma 1.4.2 *Let $v \in L_2^+(I)$ and let*

$$v(t) \leq f(t) + \int_0^t k(t,s) v(s)\,ds, \quad \text{a.e. on } I.$$

If $u(t)$ is the unique L_2 solution of (1.4.17), then $v(t) \leq u(t)$ a.e. on I.

△

Thompson (1971) gave the following inequality.

Theorem 1.4.5 *Let $f_i \in L_2^+(I)$ and let $k_i \in L_2^+(I \times I)$, and let u_i be the unique L_2 solution of*

$$u_i(t) = f_i(t) + \int_0^t k_i(t,s) u_i(s)\,ds,$$

where $i = 1, 2, \ldots, n$. If

$$F(t) = \sum_{i=1}^n f_i(t) \quad \text{for } t \in I$$

and

$$k(t,s) = \sum_{i=1}^n k_i(t,s) \quad \text{for } (t,s) \in I \times I,$$

then the unique L_2 solution u of

$$u(t) = F(t) + \int_0^t k(t,s) u(s)\,ds,$$

exists, and
$$\sum_{i=1}^{n} u_i(t) \leq u(t) \quad \text{a.e. on } I.$$

□

Proof: Since $f_i \in L_2^+(I)$ and $k_i \in L_2^+(I \times I)$, then $F \in L_2(I)$ and $k \in L_2(I \times I)$, so from Theorem 1.4.4, $u(t)$ exists. The proof that $\sum_{i=1}^{n} u_i(t) \leq u(t)$ a.e. on I is by induction on n.

The theorem is obvious for $n = 1$. Assume its truth for $i = 1, 2, \ldots, n - 1$. Let
$$v(t) = \sum_{i=1}^{n-1} f_i(t) + \int_0^t \sum_{i=1}^{n-1} k_i(t, s) v(s) \, ds.$$

Therefore
$$u_n(t) + v(t) = f_n(t) + \sum_{i=1}^{n-1} f_i(t) + \int_0^t k_n(t, s) u_n(s) \, ds$$
$$+ \int_0^t \sum_{i=1}^{n-1} k_i(t, s) v(s) \, ds, \quad \text{a.e. on } I.$$

From Lemma 1.4.1, $u_i(t) \geq 0$ a.e. on I for $i = 1, 2, \ldots, n$. Therefore
$$u_n(t) + v(t) \leq \sum_{i=1}^{n} f_i(t) + \int_0^t \sum_{i=1}^{n} k_i(t, s)[u_n(s) + v(s)] \, ds,$$

or
$$u_n(t) + v(t) \leq F(t) + \int_0^t k(t, s)[u_n(s) + v(s)] \, ds, \quad \text{a.e. on } I.$$

Therefore, from Lemma 1.4.2 we have $u_n(t) + v(t) \leq u(t)$. But by our induction assumption
$$\sum_{i=1}^{n-1} u_i(t) \leq v(t), \quad \text{a.e. on } I.$$

Therefore
$$\sum_{i=1}^{n} u_i(t) \leq u(t), \quad \text{a.e. on } I.$$

This completes the proof.

■

Remark 1.4.4 Note that the literature concerning the Volterra-type inequalities is particularly rich; the reader is referred to the books by Lakshmikantham and Leela (1969), Martinjuk and Gutowski (1979) and Walter (1970) and some of the references cited therein.

△

1.5 The Inequalities of Gamidov and Rodrigues

Integral inequalities of the Gronwall–Bellman type are also encountered frequently in the study of certain boundary value problems and retarded differential equations. This section presents some such inequalities which emphasize their importance in various fields of differential equations.

Gamidov (1969) proved the following inequalities and employed them to obtain bounds for the solutions of certain boundary value problems.

Theorem 1.5.1 Let u, f, g_i, h_i, $(i = 1, 2, \ldots, n)$ be continuous functions defined on J, let g_i and h_i be nonnegative in J, and

$$u(t) \leq f(t) + \sum_{i=1}^{n} g_i(t) \int_{\alpha}^{t} h_i(s) u(s) \, ds, \quad t \in J. \tag{1.5.1}$$

Then for $t \in J$,

$$u(t) \leq f(t) + g(t) \int_{\alpha}^{t} f(s) \sum_{i=1}^{n} h_i(s) \exp\left(\int_{s}^{t} g(\sigma) \sum_{i=1}^{n} h_i(\sigma) \, d\sigma\right) ds, \tag{1.5.2}$$

where $g(t) = \sup_i \{g_i(t)\}$.

□

Proof: From (1.5.1) we observe that

$$u(t) \leq f(t) + g(t) \int_{\alpha}^{t} \left(\sum_{i=1}^{n} h_i(s)\right) u(s) \, ds.$$

Now an application of Theorem 1.3.2 gives the required inequality in (1.5.2).

■

Theorem 1.5.2 Let u, f, g_1, g_2, h_i $(i = 1, 2, \ldots, n)$ be nonnegative continuous functions defined on J, and

$$u(t) \leq f(t) + g_1(t) \int_{t_1}^{t} h_1(s)u(s)\, ds + g_2(t) \sum_{i=2}^{n} c_i \int_{t_1}^{t_i} h_i(s)u(s)\, ds, \quad (1.5.3)$$

where $\alpha = t_1 \leq t_2 \leq \ldots \leq t_n = \beta$, and c_i are constants, and

$$\sum_{i=2}^{n} c_i \int_{t_1}^{t_i} h_i(s) \left[g_2(s) + g_1(s) \int_{t_1}^{s} h_1(\tau)g_2(\tau) \right.$$

$$\left. \times \exp\left(\int_{\tau}^{s} g_1(\sigma)h_1(\sigma)\, d\sigma\right) d\tau \right] ds < 1, \quad (1.5.4)$$

then

$$u(t) \leq p_1(t) + M p_2(t), \quad (1.5.5)$$

where

$$p_1(t) = f(t) + g_1(t) \int_{t_1}^{t} h_1(s)f(s) \exp\left(\int_{s}^{t} g_1(\sigma)h_1(\sigma)\, d\sigma\right) ds,$$

$$p_2(t) = g_2(t) + g_1(t) \int_{t_1}^{t} h_1(s)g_2(s) \exp\left(\int_{s}^{t} g_1(\sigma)h_1(\sigma)\, d\sigma\right) ds,$$

$$M = \left(\sum_{i=2}^{n} c_i \int_{t_1}^{t_i} h_i(s)p_1(s)\, ds\right)\left(1 - \sum_{i=2}^{n} c_i \int_{t_1}^{t_i} h_i(s)p_2(s)\, ds\right)^{-1}.$$

□

Proof: We put

$$m_i = c_i \int_{t_1}^{t_i} h_i(s)u(s)\, ds.$$

Then (1.5.3) can be restated as

$$u(t) \leq \left(f(t) + g_2(t) \sum_{i=2}^{n} m_i \right) + g_1(t) \int_{t_1}^{t} h_1(s)u(s)\, ds.$$

LINEAR INTEGRAL INEQUALITIES

Now an application of Theorem 1.3.2 yields

$$u(t) \leq \left(f(t) + g_2(t)\sum_{i=2}^{n} m_i\right) + g_1(t)\int_{t_1}^{t}\left(f(s) + g_2(s)\sum_{i=2}^{n} m_i\right)$$

$$\times h_1(s)\exp\left(\int_{s}^{t} h_1(\sigma)g_1(\sigma)\,d\sigma\right)ds$$

$$= p_1(t) + \sum_{i=2}^{n} m_i p_2(t). \tag{1.5.6}$$

Since

$$\sum_{i=2}^{n} m_i = \sum_{i=2}^{n} c_i \int_{t_1}^{t_i} h_i(s)u(s)\,ds$$

$$\leq \sum_{i=2}^{n} c_i \int_{t_1}^{t_i} h_i(s)\left[p_1(s) + \sum_{i=2}^{n} m_i p_2(s)\right]ds$$

$$= \sum_{i=2}^{n} c_i \int_{t_1}^{t_i} h_i(s)p_1(s)\,ds + \sum_{i=2}^{n} c_i \int_{t_1}^{t_i} h_i(s)\left(\sum_{i=2}^{n} m_i\right)p_2(s)\,ds,$$

we have

$$\sum_{i=2}^{n} m_i \left(1 - \sum_{i=2}^{n} c_i \int_{t_1}^{t_i} h_i(s)p_2(s)\,ds\right) \leq \sum_{i=2}^{n} c_i \int_{t_1}^{t_i} h_i(s)p_1(s)\,ds,$$

and hence

$$\sum_{i=2}^{n} m_i \leq \left(\sum_{i=2}^{n} c_i \int_{t_1}^{t_i} h_i(s)p_1(s)\,ds\right)\left(1 - \sum_{i=2}^{n} c_i \int_{t_1}^{t_i} h_i(s)p_2(s)\,ds\right)^{-1} = M. \tag{1.5.7}$$

The required inequality in (1.5.5) follows from (1.5.6) and (1.5.7). ∎

Remark 1.5.1 Note that in addition to the above inequalities Gamidov (1969) gave some such inequalities and used them to obtain the bounds on the solutions of various types of boundary value problems. For some other results analogous to the inequality given in Theorem 1.5.1, see the results given by Willett (1965) and Kong and Zhang (1989).

△

In order to establish the next result, the following inequality is needed, which in turn is a slight variant of Bellman's inequality given in Theorem 1.3.1.

Theorem 1.5.3 *Let f be a nonnegative continuous function defined for $t \in R_+$, such that $\int_0^\infty f(s)\,ds < \infty$ and $n(t) > 0$ be a continuous and decreasing function defined for $t \in R_+$. If $u(t) \geq 0$ is a bounded continuous function defined for $t \in R_+$ and satisfies*

$$u(t) \leq n(t) + \int_t^\infty f(s)u(s)\,ds, \quad t \in R_+, \tag{1.5.8}$$

then

$$u(t) \leq n(t)\exp\left(\int_t^\infty f(s)\,ds\right), \quad t \in R_+. \tag{1.5.9}$$

□

The proof of this theorem is straightforward and it is similar to the proof of Theorem 1.3.1 with suitable modifications.

Rodrigues (1980) proved, and made use of, the following inequality to study the growth and decay of solutions of perturbed retarded linear equations.

Theorem 1.5.4 *Let f, g be nonnegative continuous functions defined for $t \in R_+$. Let $\gamma(t) > 0$ be a decreasing continuous function, for $t \geq \sigma$ and σ sufficiently large, in such a way that $\beta = \int_\sigma^\infty g(s)\,ds + \int_\sigma^\infty f(s)\,ds < 1$. Suppose that u is a nonnegative continuous function such that γu is bounded and*

$$u(t) \leq c + \int_\sigma^t f(s)u(s)\,ds + \frac{1}{\gamma(t)}\int_t^\infty \gamma(s)g(s)u(s)\,ds, \quad t \in R_+, \tag{1.5.10}$$

for $t \geq \sigma$, where $c \geq 0$ is a constant. Then, for $t \in R_+$,

$$u(t) \leq (c/(1-\beta))\exp\left((1/(1-\beta))\int_t^\infty g(s)\,ds\right). \tag{1.5.11}$$

□

Proof: Let
$$v(t) = \max_{\sigma \leq s \leq t} u(s).$$
Then $v(t)$ is an increasing continuous function, such that $u(t) \leq v(t)$ and $\gamma(t)v(t)$ is bounded for $t \in R_+$. For a given $t \geq \sigma$, there exists $t_1 \in [\sigma, t]$ satisfying $v(t) = u(t_1)$. This implies,
$$v(t) \leq c + \int_\sigma^{t_1} f(s)v(s)\,ds + \frac{1}{\gamma(t_1)} \int_{t_1}^\infty \gamma(s)g(s)v(s)\,ds.$$
But
$$\int_{t_1}^\infty \gamma(s)g(s)v(s)\,ds = \int_{t_1}^t \gamma(s)g(s)v(s)\,ds + \int_t^\infty \gamma(s)g(s)v(s)\,ds$$
$$\leq \gamma(t_1)v(t) \int_\sigma^\infty g(s)\,ds + \int_t^\infty \gamma(s)g(s)v(s)\,ds.$$

Combining the above inequalities we get
$$v(t) \leq c + v(t)\left[\int_\sigma^\infty f(s)\,ds + \int_\sigma^\infty g(s)\,ds\right] + \frac{1}{\gamma(t)} \int_t^\infty \gamma(s)g(s)v(s)\,ds.$$
Then
$$\gamma(t)v(t) \leq (1/(1-\beta))\left[c\gamma(t) + \int_t^\infty \gamma(s)g(s)v(s)\,ds\right].$$
Using Theorem 1.5.3 we get
$$\gamma(t)v(t) \leq (c/(1-\beta))\gamma(t) \exp\left((1/(1-\beta)) \int_t^\infty g(s)\,ds\right),$$
and this completes the proof.

∎

Remark 1.5.2 Note that in Rodrigues (1980) the above inequality is very effectively used to study the asymptotic behaviour of the solutions of perturbed retarded differential equations.

△

1.6 Simultaneous Inequalities

In analysing the dynamics of a physical system governed by certain differential and integral equations, one often needs some new kinds of inequalities. Greene (1977) proved the following interesting inequality, which can be used in the analysis of various problems in the theory of certain systems of simultaneous differential and integral equations.

Theorem 1.6.1 Let k_1, k_2 and μ be nonnegative constants and let f, g and h_i, $(i = 1, 2, 3, 4)$ be nonnegative continuous functions defined for $t \in R_+$ with h_i bounded such that

$$f(t) \leq k_1 + \int_0^t h_1(s)f(s)\,ds + \int_0^t e^{\mu s}h_2(s)g(s)\,ds, \qquad (1.6.1)$$

$$g(t) \leq k_2 + \int_0^t e^{-\mu s}h_3(s)f(s)\,ds + \int_0^t h_4(s)g(s)\,ds, \qquad (1.6.2)$$

for all $t \in R_+$. Then there exist constants c_1, c_2 and M_1, M_2 such that

$$f(t) \leq M_1 e^{c_1 t}, \quad g(t) \leq M_2 e^{c_2 t}, \qquad (1.6.3)$$

for all $t \in R_+$.

$\square$

Proof: This result of Greene has received successively simpler proofs by Wang (1978) and Das (1979). Here we present the proof given by Das in (1979).

We note that (1.6.1) implies

$$e^{-\mu t}f(t) \leq k_1 + \int_0^t e^{-\mu s}h_1(s)f(s)\,ds + \int_0^t h_2(s)g(s)\,ds. \qquad (1.6.4)$$

Now define

$$F(t) = e^{-\mu t}f(t) + g(t). \qquad (1.6.5)$$

From (1.6.4), (1.6.2) and (1.6.5) we observe that

$$F(t) \leq M + \int_0^t h(s)F(s)\,ds,$$

where $M = k_1 + k_2$ and h is defined by

$$h(t) = \max\{[h_1(t) + h_3(t)], \quad [h_2(t) + h_4(t)]\}.$$

Now an application of Theorem 1.2.2 yields

$$F(t) \leq M \exp\left(\int_0^t h(s)\,ds\right). \tag{1.6.6}$$

Using (1.6.6) in (1.6.5) and splitting we get

$$f(t) \leq M \exp\left(\mu t + \int_0^t h(s)\,ds\right), \quad g(t) \leq M \exp\left(\int_0^t h(s)\,ds\right).$$

It is immediate that the bounds in (1.6.3) follow in view of the additional assumption of boundedness on the h_i.

∎

Pachpatte (1984) obtained the following useful generalization of Greene's inequality given in Theorem 1.6.1.

Theorem 1.6.2 *Let u, v, a, b, p, h_i ($i = 1, 2, 3, 4$) be nonnegative continuous functions defined on R_+ and*

$$u(t) \leq a(t) + p(t)\left[\int_0^t h_1(s)u(s)\,ds + \int_0^t e^{\mu s}h_2(s)v(s)\,ds\right], \tag{1.6.7}$$

$$v(t) \leq b(t) + p(t)\left[\int_0^t e^{-\mu s}h_3(s)u(s)\,ds + \int_0^t h_4(s)v(s)\,ds\right], \tag{1.6.8}$$

for all $t \in R_+$, where μ is a nonnegative constant. then

$$u(t) \leq e^{-\mu t}Q(t), \quad v(t) \leq Q(t), \tag{1.6.9}$$

for all $t \in R_+$, where

$$Q(t) = f(t) + p(t)\left[\int_0^t h(s)f(s)\exp\left(\int_s^t h(\sigma)p(\sigma)\,d\sigma\right)ds\right], \tag{1.6.10}$$

in which $f(t) = a(t) + b(t)$ and

$$h(t) = \max\{[h_1(t) + h_3(t)], [h_2(t) + h_4(t)]\}, \qquad (1.6.11)$$

for all $t \in R_+$.

□

Proof: From (1.6.7) we observe that

$$e^{-\mu t} u(t) \leq a(t) + p(t) \left[\int_0^t e^{-\mu s} h_1(s) u(s) \, ds + \int_0^t h_2(s) v(s) \, ds \right]. \qquad (1.6.12)$$

Define

$$w(t) = e^{-\mu t} u(t) + v(t). \qquad (1.6.13)$$

From (1.6.12), (1.6.8), (1.6.13) and (1.6.11) we observe that

$$w(t) \leq f(t) + p(t) \int_0^t h(s) w(s) \, ds.$$

The bounds in (1.6.9) follow from an application of Theorem 1.3.2 and splitting.

∎

Remark 1.6.1 Note that in the special case when $a(t)$ and $b(t)$ are nonnegative constants, $p(t) = 1$ and the functions h_i are bounded, then Theorem 1.6.2 reduces to the inequality established by Greene in Theorem 1.6.1. Furthermore, note that the result given in Theorem 1.6.2 can be very easily extended to systems of more than two inequalities of the forms (1.6.7) and (1.6.8).

△

1.7 Pachpatte's Inequalities

Pachpatte (1973a) proved a very useful generalization of the well known Gronwall–Bellman inequality. The importance of the inequality given in Pachpatte (1973a, Theorem 1) stems from the fact that it is applicable in more general situations in which the other available inequalities do not apply directly. This section gives the inequality established in Pachpatte (1973a) and its extended and generalized versions given by Pachpatte (1974a,

1975b–d, 1977a) which serve great purpose in the development of the theory of integral and integro-differential equations of the more general type.

A fairly general linear version of the Gronwall–Bellman inequality established by Pachpatte (1973a) is given in the following theorem.

Theorem 1.7.1 *Let u, f and g be nonnegative continuous functions defined on R_+, for which the inequality*

$$u(t) \leq u_0 + \int_0^t f(s)u(s)\,ds + \int_0^t f(s)\left(\int_0^s g(\sigma)u(\sigma)\,d\sigma\right)ds, \quad t \in R_+,$$
(1.7.1)

holds, where u_0 is a nonnegative constant. Then

$$u(t) \leq u_0\left[1 + \int_0^t f(s)\exp\left(\int_0^s [f(\sigma) + g(\sigma)]\,d\sigma\right)ds\right], \quad t \in R_+. \quad (1.7.2)$$

□

Proof: Define a function $v(t)$ by the right member of (1.7.1). Then $v(0) = u_0$, $u(t) \leq v(t)$ and

$$v'(t) = f(t)u(t) + f(t)\int_0^t g(\sigma)u(\sigma)\,d\sigma$$

$$\leq f(t)\left(v(t) + \int_0^t g(\sigma)v(\sigma)\,d\sigma\right). \quad (1.7.3)$$

Define a function $m(t)$ by

$$m(t) = v(t) + \int_0^t g(\sigma)v(\sigma)\,d\sigma, \quad (1.7.4)$$

then $m(0) = v(0) = u_0$, $v'(t) \leq f(t)m(t)$, $v(t) \leq m(t)$ and

$$m'(t) = v'(t) + g(t)v(t)$$

$$\leq [f(t) + g(t)]m(t). \quad (1.7.5)$$

The inequality (1.7.5) implies the estimation for $m(t)$ such that

$$m(t) \leq u_0 \exp\left(\int_0^t [f(\sigma) + g(\sigma)]\,d\sigma\right). \quad (1.7.6)$$

Using (1.7.6) in (1.7.3) we have

$$v'(t) \leq u_0 f(t) \exp\left(\int_0^t [f(\sigma) + g(\sigma)] d\sigma\right). \quad (1.7.7)$$

Now by setting $t = s$ in (1.7.7) and integrating it from 0 to t and substituting the bound on $v(t)$ in $u(t) \leq v(t)$ we get the desired inequality in (1.7.2).

∎

Remark 1.7.1 It is interesting to note that, in the special case when $g(t) = 0$, the above inequality reduces to Bellman's inequality, given in Theorem 1.2.2. However, the bound obtained in (1.7.2) is not sharp, as explained in Section 1.4 below Theorem 1.4.1, but it serves an important purpose in applications. The main reason for not getting the sharp bound in Theorem 1.7.1 is the wastage involved while using the fact that $v(t) \leq m(t)$ from (1.7.4) to obtain the inequality in (1.7.5) in proving the much more general result given in Theorem 1.7.1.

△

In the following two theorems we present some useful generalizations of Theorem 1.7.1, given by Pachpatte (1974a, 1975d, 1977a).

Theorem 1.7.2 Let u, f, g, h and p be nonnegative continuous functions defined on R_+, and u_0 be a nonnegative constant.

(i) If

$$u(t) \leq u_0 + \int_0^t [f(s)u(s) + p(s)] ds + \int_0^t f(s) \left(\int_0^s g(\sigma)u(\sigma) d\sigma\right) ds, \quad (1.7.8)$$

for $t \in R_+$, then

$$u(t) \leq u_0 + \int_0^t \left[p(s) + f(s) \left\{ u_0 \exp\left(\int_0^s [f(\sigma) + g(\sigma)] d\sigma\right) \right.\right.$$
$$\left.\left. + \int_0^s p(\sigma) \exp\left(\int_\sigma^s [f(\tau) + g(\tau)] d\tau\right) d\sigma \right\} \right] ds, \quad (1.7.9)$$

for $t \in R_+$.

(ii) If
$$u(t) \leq u_0 + \int_0^t f(s)u(s)\,ds + \int_0^t f(s)\left(\int_0^s [g(\sigma)u(\sigma) + p(\sigma)]\,d\sigma\right)ds, \tag{1.7.10}$$

for $t \in R_+$, then

$$u(t) \leq u_0 + \int_0^t f(s)\left[u_0 \exp\left(\int_0^s [f(\sigma) + g(\sigma)]\,d\sigma\right)\right.$$

$$\left. + \int_0^t p(\sigma)\exp\left(\int_\sigma^s [f(\tau) + g(\tau)]\,d\tau\right)d\sigma\right]ds, \tag{1.7.11}$$

for $t \in R_+$.

(iii) If

$$u(t) \leq u_0 + \int_0^t f(s)u(s)\,ds$$

$$+ \int_0^t g(s)\left(u(s) + \int_0^s h(\sigma)u(\sigma)\,d\sigma\right)ds, \tag{1.7.12}$$

for $t \in R_+$, then

$$u(t) \leq u_0 \left[\exp\left(\int_0^t f(\sigma)\,d\sigma\right) + \int_0^t g(s)\exp\left(\int_0^s [f(\sigma) + g(\sigma) + h(\sigma)]\,d\sigma\right)\right.$$

$$\left. \times \exp\left(\int_s^t f(\sigma)\,d\sigma\right)ds\right], \tag{1.7.13}$$

for $t \in R_+$.

(iv) If

$$u(t) \leq h(t) + p(t)\left[\int_0^t f(s)u(s)\,ds\right.$$

$$\left. + \int_0^t f(s)p(s)\left(\int_0^s g(\sigma)u(\sigma)\,d\sigma\right)ds\right], \tag{1.7.14}$$

for $t \in R_+$, then

$$u(t) \leq h(t) + p(t) \left[\int_0^t f(s) \left\{ h(s) + p(s) \int_0^s h(\sigma)[f(\sigma) + g(\sigma)] \right. \right.$$

$$\left. \left. \times \exp\left(\int_\sigma^s p(\tau)[f(\tau) + g(\tau)] \, d\tau \right) d\sigma \right\} ds \right], \quad (1.7.15)$$

for $t \in R_+$.

□

Proof: Since the proofs resemble one another, the details for (i) and (iv) only are given; the proofs of (ii) and (iii) can be completed by following the proofs of (i) and (iv).

(i) Define a function $z(t)$ by the right side of (1.7.8). Then $z(0) = u_0$, $u(t) \leq z(t)$ and

$$z'(t) = [f(t)u(t) + p(t)] + f(t) \int_0^t g(\sigma)u(\sigma) \, d\sigma$$

$$\leq p(t) + f(t) \left[z(t) + \int_0^t g(\sigma)z(\sigma) \, d\sigma \right]. \quad (1.7.16)$$

Define a function $v(t)$ by

$$v(t) = z(t) + \int_0^t g(\sigma)z(\sigma) \, d\sigma, \quad (1.7.17)$$

then $v(0) = z(0) = u_0$, $z'(t) \leq p(t) + f(t)v(t)$ from (1.7.16), and from (1.7.17) $z(t) \leq v(t)$ and

$$v'(t) = z'(t) + g(t)z(t)$$

$$\leq p(t) + [f(t) + g(t)]v(t). \quad (1.7.18)$$

The inequality (1.7.18) implies the estimate

$$v(t) \leq u_0 \exp\left(\int_0^t [f(\tau) + g(\tau)] \, d\tau \right)$$

$$+ \int_0^t p(\sigma) \exp\left(\int_\sigma^t [f(\tau) + g(\tau)] \, d\tau \right) d\sigma. \quad (1.7.19)$$

Using (1.7.19) in (1.7.16) we get

$$z'(t) \le p(t) + f(t) \left[u_0 \exp\left(\int_0^t [f(\tau) + g(\tau)] \, d\tau \right) \right.$$

$$\left. + \int_0^t p(\sigma) \exp\left(\int_\sigma^t [f(\tau) + g(\tau)] \, d\tau \right) d\sigma \right]. \quad (1.7.20)$$

Now by setting $t = s$ in (1.7.20) and integrating it from 0 to t and substituting the bound on $z(t)$ in $u(t) \le z(t)$ we get the desired inequality in (1.7.9).

(iv) Define a function $v(t)$ by

$$v(t) = \int_0^t f(s) u(s) \, ds + \int_0^t f(s) p(s) \left(\int_0^s g(\sigma) u(\sigma) \, d\sigma \right) ds,$$

then $v(0) = 0$, $u(t) \le h(t) + p(t) v(t)$ and

$$v'(t) = f(t) u(t) + f(t) p(t) \int_0^t g(\sigma) u(\sigma) \, d\sigma$$

$$\le f(t) \left[h(t) + p(t) \left\{ v(t) + \int_0^t g(\sigma) [h(\sigma) + p(\sigma) v(\sigma)] \, d\sigma \right\} \right]. \quad (1.7.21)$$

If we put

$$m(t) = v(t) + \int_0^t g(\sigma) [h(\sigma) + p(\sigma) v(\sigma)] \, d\sigma, \quad (1.7.22)$$

then $m(0) = v(0) = 0$, $v'(t) \le f(t)[h(t) + p(t) m(t)]$ from (1.7.21) and from (1.7.22) $v(t) \le m(t)$ and

$$m'(t) = v'(t) + g(t)[h(t) + p(t) v(t)]$$
$$\le h(t)[f(t) + g(t)] + p(t)[f(t) + g(t)] m(t). \quad (1.7.23)$$

The inequality (1.7.23) implies the estimate

$$m(t) \leq \int_0^t h(\sigma)[f(\sigma) + g(\sigma)] \exp\left(\int_\sigma^t p(\tau)[f(\tau) + g(\tau)] d\tau\right) d\sigma. \quad (1.7.24)$$

Using (1.7.24) in (1.7.21) we have

$$v'(t) \leq f(t) \left[h(t) + p(t) \int_0^t h(\sigma)[f(\sigma) + g(\sigma)] \right.$$

$$\left. \times \exp\left(\int_\sigma^t p(\tau)[f(\tau) + g(\tau)] d\tau\right) d\sigma \right]. \quad (1.7.25)$$

Now by setting $t = s$ in (1.7.25) and integrating it from 0 to t and substituting the bound on $v(t)$ in $u(t) \leq h(t) + p(t)v(t)$ we get the required inequality in (1.7.15).

■

Theorem 1.7.3 *Let u, k, p, f, g and h be nonnegative continuous functions defined on R_+ and u_0 is a nonnegative constant.*
(i) If

$$u(t) \leq u_0 + \int_0^t f(s)u(s) ds + \int_0^t f(s) \left(\int_0^s g(\sigma)u(\sigma) d\sigma\right) ds$$

$$+ \int_0^t f(s) \left[\int_0^s g(\sigma) \left(\int_0^\sigma h(\tau)u(\tau) d\tau\right) d\sigma\right] ds, \quad (1.7.26)$$

for $t \in R_+$, then

$$u(t) \leq u_0 \left[1 + \int_0^t f(s) \exp\left(\int_0^s f(\sigma) d\sigma\right) \right.$$

$$\left. \times \left[1 + \int_0^s g(\sigma) \exp\left(\int_0^\sigma [g(\tau) + h(\tau)] d\tau\right) d\sigma\right] ds\right], (1.7.27)$$

for $t \in R_+$.

(ii) If

$$u(t) \leq k(t) + p(t)\left[\int_0^t f(s)u(s)\,ds + \int_0^t f(s)p(s)\left(\int_0^s g(\tau)u(\tau)\,d\tau\right)ds\right.$$

$$\left. + \int_0^t f(s)p(s)\left[\int_0^s g(\tau)p(\tau)\left(\int_0^\tau h(\sigma)u(\sigma)\,d\sigma\right)d\tau\right]ds\right], \quad (1.7.28)$$

for $t \in R_+$, then

$$u(t) \leq k(t) + p(t)\left[\int_0^t f(s)\left[k(s) + p(s)\left\{\int_0^s \exp\left(\int_\tau^s f(\sigma)p(\sigma)\,d\sigma\right)\right.\right.\right.$$

$$\times \left(k(\tau)[f(\tau) + g(\tau)] + g(\tau)p(\tau)\int_0^\tau k(\sigma)[f(\sigma) + g(\sigma) + h(\sigma)]\right.$$

$$\left.\left.\left.\times \exp\left(\int_\sigma^\tau p(\xi)[f(\xi) + g(\xi) + h(\xi)]\,d\xi\right)d\sigma\right)d\tau\right\}\right]ds\right], \quad (1.7.29)$$

for $t \in R_+$.

□

The proof of this theorem follows by the same arguments as in the proof of inequalities (i) and (iv) in Theorem 1.7.2, with suitable modifications.

We shall now state and prove some useful variants of Theorem 1.7.1 given by Pachpatte in (1975b, c).

Theorem 1.7.4 *Let u, f and g be nonnegative continuous functions defined on R_+, and $n(t)$ be a positive and nondecreasing continuous function defined on R_+ for which the inequality*

$$u(t) \leq n(t) + \int_0^t f(s)u(s)\,ds + \int_0^t f(s)\left(\int_0^s g(\sigma)u(\sigma)\,d\sigma\right)ds, \quad (1.7.30)$$

holds for $t \in R_+$. Then

$$u(t) \leq n(t)\left[1 + \int_0^t f(s)\exp\left(\int_0^s [f(\sigma) + g(\sigma)]\,d\sigma\right)ds\right], \quad (1.7.31)$$

for $t \in R_+$.

□

Proof: Since $n(t)$ is positive and nondecreasing we observe from (1.7.30) that

$$\frac{u(t)}{n(t)} \leq 1 + \int_0^t f(s) \frac{u(s)}{n(s)} \, ds + \int_0^t f(s) \left(\int_0^s g(\sigma) \frac{u(\sigma)}{n(\sigma)} \, d\sigma \right) ds.$$

Now an application of Theorem 1.7.1 yields the desired inequality in (1.7.31).

∎

Theorem 1.7.5 *Let u, f, h, g and p be nonnegative continuous functions defined on R_+; f and g are positive and sufficiently smooth on R_+ and $u_0 \geq 0$ is a constant.*

(i) *If*

$$u(t) \leq u_0 + \int_0^t f(s) \left(h(s) + \int_0^s p(\sigma) u(\sigma) \, d\sigma \right) ds, \quad (1.7.32)$$

for $t \in R_+$, then

$$u(t) \leq \left(u_0 + \int_0^t f(s) h(s) \, ds \right) \exp \left[\int_0^t f(s) \left(\int_0^s p(\sigma) \, d\sigma \right) ds \right], \quad (1.7.33)$$

for $t \in R_+$.

(ii) *If*

$$u(t) \leq u_0 + \int_0^t f(s) \left[h(s) + \int_0^s g(\tau) \left(\int_0^\tau p(\sigma) u(\sigma) \, d\sigma \right) d\tau \right] ds, \quad (1.7.34)$$

for $t \in R_+$, then

$$u(t) \leq \left(u_0 + \int_0^t f(s) h(s) \, ds \right)$$

$$\times \exp \left\{ \int_0^t f(s) \left[\int_0^s g(\tau) \left(\int_0^\tau p(\sigma) \, d\sigma \right) d\tau \right] ds \right\}, \quad (1.7.35)$$

for $t \in R_+$.

□

LINEAR INTEGRAL INEQUALITIES

Proof: We shall give the details of the proof of (ii) only; the proof of (i) can be completed similarly.

(ii) First we assume that u_0 is positive. From (1.7.34) we have

$$u(t) \leq n(t) + \int_0^t f(s) \left[\int_0^s g(\tau) \left(\int_0^\tau p(\sigma) u(\sigma) \, d\sigma \right) d\tau \right] ds, \quad (1.7.36)$$

where

$$n(t) = u_0 + \int_0^t f(s) h(s) \, ds. \quad (1.7.37)$$

Clearly $n(t)$ is a positive and nondecreasing function for $t \in R_+$, and hence from (1.7.36) we observe that

$$\frac{u(t)}{n(t)} \leq 1 + \int_0^t f(s) \left[\int_0^s g(\tau) \left(\int_0^\tau p(\sigma) \frac{u(\sigma)}{n(\sigma)} \, d\sigma \right) d\tau \right] ds. \quad (1.7.38)$$

Define a function $z(t)$ by the right side of (1.7.38); then $z(0) = 1$ and

$$z'(t) = f(t) \left[\int_0^t g(\tau) \left(\int_0^\tau p(\sigma) \frac{u(\sigma)}{n(\sigma)} \, d\sigma \right) d\tau \right]. \quad (1.7.39)$$

From (1.7.39), it is easy to observe that

$$\left[\frac{1}{g(t)} \left(\frac{z'(t)}{f(t)} \right)' \right]' = p(t) \frac{u(t)}{n(t)}. \quad (1.7.40)$$

Now using the fact that $u(t)/n(t) \leq z(t)$ from (1.7.38) in (1.7.40), we have

$$\frac{\left[\frac{1}{g(t)} \left(\frac{z'(t)}{f(t)} \right)' \right]'}{z(t)} \leq p(t). \quad (1.7.41)$$

Since

$$\frac{1}{g(t)} \left(\frac{z'(t)}{f(t)} \right)' \geq 0,$$

$z'(t) \geq 0$ and $z(t) > 0$, from (1.7.41) we observe that

$$\frac{\left[\frac{1}{g(t)} \left(\frac{z'(t)}{f(t)} \right)' \right]'}{z(t)} \leq p(t) + \frac{\frac{1}{g(t)} \left(\frac{z'(t)}{f(t)} \right)' z'(t)}{z^2(t)},$$

i.e.

$$\left(\frac{\frac{1}{g(t)}\left(\frac{z'(t)}{f(t)}\right)'}{z(t)}\right)' \leq p(t). \tag{1.7.42}$$

By taking $t = \sigma$ in (1.7.42) and integrating it with respect to σ from 0 to t we have

$$\frac{(z'(t)/f(t))'}{z(t)} \leq g(t) \int_0^t p(\sigma)\,d\sigma. \tag{1.7.43}$$

Since $z'(t)/f(t) \geq 0$, $z'(t) \geq 0$ and $z(t) > 0$, as above from (1.7.43) we observe that

$$\left(\frac{z'(t)/f(t)}{z(t)}\right)' \leq g(t) \int_0^t p(\sigma)\,d\sigma. \tag{1.7.44}$$

By taking $t = \tau$ in (1.7.14) and integrating from 0 to t we have

$$\frac{z'(t)}{z(t)} \leq f(t) \int_0^t g(\tau)\left(\int_0^\tau p(\sigma)\,d\sigma\right) d\tau. \tag{1.7.45}$$

The inequality (1.7.45) implies the estimate

$$z(t) \leq \exp\left\{\int_0^t f(s)\left[\int_0^s g(\tau)\left(\int_0^\tau p(\sigma)\,d\sigma\right) d\tau\right] ds\right\}. \tag{1.7.46}$$

Using (1.7.46) in (1.7.38) we get the required inequality in (1.7.35).

If u_0 is nonnegative, we carry out the above procedure with $u_0 + \epsilon$ instead of u_0, where $\epsilon > 0$ is an arbitrary small constant, and subsequently pass to the limit as $\epsilon \to 0$ to obtain (1.7.35).

∎

Theorem 1.7.6 Let u, v, p, f, g and h be nonnegative continuous functions defined on J.

(i) If

$$u(t) \geq v(s) - p(t)\left[\int_s^t f(\sigma)v(\sigma)\,d\sigma + \int_s^t f(\sigma)\left(\int_\sigma^t g(\tau)v(\tau)\,d\tau\right) d\sigma\right], \tag{1.7.47}$$

LINEAR INTEGRAL INEQUALITIES

for $\alpha \leq s \leq t \leq \beta$, then

$$u(t) \geq v(s) \left[1 + p(t) \int_s^t f(\sigma) \exp\left(\int_\sigma^t [f(\tau)p(t) + g(\tau)] \, d\tau \right) d\sigma \right]^{-1}, \tag{1.7.48}$$

for $\alpha \leq s \leq t \leq \beta$.

(ii) If

$$u(t) \geq v(s) - p(t) \left[\int_s^t f(\sigma)v(\sigma) \, d\sigma + \int_s^t f(\sigma) \left(\int_\sigma^t g(\tau)v(\tau) \, d\tau \right) d\sigma \right.$$

$$\left. + \int_s^t f(\sigma) \left[\int_\sigma^t g(\tau) \left(\int_\tau^t h(\xi)v(\xi) \, d\xi \right) d\tau \right] d\sigma \right], \tag{1.7.49}$$

for $\alpha \leq s \leq t \leq \beta$, then

$$u(t) \geq v(s) \left\{ 1 + p(t) \int_s^t f(\tau) \exp\left(\int_\tau^t p(t)f(\sigma) \, d\sigma \right) \right.$$

$$\left. \times \left[1 + \int_\tau^t g(\sigma) \exp\left(\int_\sigma^t [h(\xi) + g(\xi)] \, d\xi \right) d\sigma \right] d\tau \right\}^{-1}, \tag{1.7.50}$$

for $\alpha \leq s \leq t \leq \beta$.

$\square$

Proof: (i) For fixed t in the interval J we define for $\alpha \leq s \leq t$

$$m(s) = u(t) + p(t) \left[\int_s^t f(\sigma)v(\sigma) \, d\sigma + \int_s^t f(\sigma) \left(\int_\sigma^t g(\tau)v(\tau) \, d\tau \right) d\sigma \right], \tag{1.7.51}$$

From (1.7.51) we have

$$m'(s) = -p(t)f(s) \left(v(s) + \int_s^t g(\tau)v(\tau) \, d\tau \right),$$

which in view of $v(s) \leq m(s)$ implies

$$m'(s) \geq -p(t)f(s) \left(m(s) + \int_s^t g(\tau)m(\tau) \, d\tau \right). \tag{1.7.52}$$

If we put

$$r(s) = m(s) + \int_s^t g(\tau)m(\tau)\,d\tau, \qquad (1.7.53)$$

it follows from (1.7.52), (1.7.53) and the fact that $m(s) \leq r(s)$ that the inequality

$$r'(s) + [f(s)p(t) + g(s)]r(s) \geq 0,$$

is satisfied, which implies the estimation for $r(s)$ such that

$$r(s) \leq u(t)\exp\left(\int_s^t [f(\tau)p(t) + g(\tau)]\,d\tau\right), \qquad (1.7.54)$$

since $r(t) = u(t)$. Using (1.7.54) in (1.7.52) we have

$$m'(s) \geq -p(t)u(t)f(s)\exp\left(\int_s^t [f(\tau)p(t) + g(\tau)]\,d\tau\right). \qquad (1.7.55)$$

Now by taking $s = \sigma$ in (1.7.55) and integrating it from s to t and substituting the bound on $m(s)$ in $v(s) \leq m(s)$ we get the desired inequality in (1.7.48).

The proof of (ii) can be completed by following the same arguments as in the proof of inequality in (i) with suitable modifications. For details, see Pachpatte (1977a).

∎

1.8 Integro-differential Inequalities

Integral inequalities involving functions and their derivatives have played a significant role in the developments of various branches of analysis. Pachpatte (1977b, 1978a, 1982a) has given some integral inequalities of the Gronwall–Bellman type involving functions and their derivatives which are useful in certain applications in the theory of differential and integro-differential equations. This section gives some of the inequalities obtained in Pachpatte (1977b, 1978a, 1982a) and the inequalities that are slight variants of those given therein.

Theorem 1.8.1 *Let u, u', a, b and c be nonnegative continuous functions defined on R_+.*

(i) If $b \geq 1$ and

$$u'(t) \leq a(t) + b(t) \int_0^t c(s)(u(s) + u'(s)) \, ds, \quad t \in R_+, \qquad (1.8.1)$$

then

$$u'(t) \leq a(t) + b(t) \int_0^t c(\sigma)[A(\sigma) + b(\sigma)B(\sigma)] \, d\sigma, \quad t \in R_+, \qquad (1.8.2)$$

where

$$A(t) = u(0) + a(t) + \int_0^t a(s) \, ds, \quad t \in R_+, \qquad (1.8.3)$$

$$B(t) = \int_0^t c(s) A(s) \exp\left(\int_s^t b(\tau)[c(\tau) + 1] \, d\tau \right) ds, \quad t \in R_+. \qquad (1.8.4)$$

(ii) If

$$u'(t) \leq a(t) + b(t) \left(u(t) + \int_0^t c(s)[u(s) + u'(s)] \, ds \right), \quad t \in R_+, \qquad (1.8.5)$$

then

$$u'(t) \leq a(t) + b(t) \left[u(0) \exp\left(\int_0^t [b(s) + c(s) + b(s)c(s)] \, ds \right) \right.$$

$$+ \int_0^t a(s)(1 + c(s))$$

$$\left. \times \exp\left(\int_s^t [b(\tau) + c(\tau) + b(\tau)c(\tau)] \, d\tau \right) ds \right], \quad t \in R_+. \qquad (1.8.6)$$

$\square$

Proof: (i) Define a function $m(t)$ by

$$m(t) = \int_0^t c(s)(u(s) + u'(s)) \, ds, \qquad (1.8.7)$$

then (1.8.1) can be restated as

$$u'(t) \leq a(t) + b(t)m(t). \tag{1.8.8}$$

Differentiating (1.8.7) and using (1.8.8) we have

$$m'(t) \leq c(t)(u(t) + a(t) + b(t)m(t)). \tag{1.8.9}$$

By taking $t = s$ in (1.8.8) and integrating it from 0 to t we have

$$u(t) \leq u(0) + \int_0^t a(s)\,ds + \int_0^t b(s)m(s)\,ds. \tag{1.8.10}$$

Now, using (1.8.10) in (1.8.9) and the fact that $b(t) \geq 1$ and (1.8.3) we have

$$m'(t) \leq c(t)\left[A(t) + b(t)\left(m(t) + \int_0^t b(s)m(s)\,ds\right)\right]. \tag{1.8.11}$$

Define a function $r(t)$ by

$$r(t) = m(t) + \int_0^t b(s)m(s)\,ds. \tag{1.8.12}$$

Differentiating (1.8.12) and using $m'(t) \leq c(t)[A(t) + b(t)r(t)]$ from (1.8.11) and the fact that $m(t) \leq r(t)$ we get

$$r'(t) = m'(t) + b(t)m(t)$$

$$\leq b(t)[c(t) + 1]r(t) + c(t)A(t). \tag{1.8.13}$$

The inequality (1.8.13) implies the estimate

$$r(t) \leq B(t), \tag{1.8.14}$$

where $B(t)$ is given by (1.8.4). Using (1.8.14) in (1.8.11) we have

$$m'(t) \leq c(t)[A(t) + b(t)B(t)],$$

which implies the estimate for $m(t)$ such that

$$m(t) \leq \int_0^t c(\sigma)[A(\sigma) + b(\sigma)B(\sigma)]\,d\sigma. \tag{1.8.15}$$

Using (1.8.15) in (1.8.8) we get the desired inequality in (1.8.2).

(ii) Define a function $z(t)$ by

$$z(t) = u(t) + \int_0^t c(s)(u(s) + u'(s))\,ds. \qquad (1.8.16)$$

Differentiating (1.8.16) and using the facts that $u'(t) \leq a(t) + b(t)z(t)$ and $u(t) \leq z(t)$ we have

$$z'(t) = u'(t) + c(t)(u(t) + u'(t))$$
$$\leq a(t)(1 + c(t)) + (b(t) + c(t) + b(t)c(t))z(t). \qquad (1.8.17)$$

The inequality (1.8.17) implies the estimate

$$z(t) \leq u(0) \exp\left(\int_0^t [b(s) + c(s) + b(s)c(s)]\,ds\right)$$
$$+ \int_0^t a(s)(1 + c(s)) \exp\left(\int_s^t [b(\tau) + c(\tau) + b(\tau)c(\tau)]\,d\tau\right) ds.$$
$$(1.8.18)$$

Using (1.8.18) in $u'(t) \leq a(t) + b(t)z(t)$ we get the desired inequality in (1.8.6).

∎

Remark 1.8.1 Note that the hypothesis $u'(t) \geq 0$ on the unknown function $u(t)$ in the above inequalities is somewhat stronger, and it will restrict the class of functions for which these inequalities are applicable. However, the importance of such inequalities lies in their fruitful utilization in situations for which the other available inequalities do not apply. Furthermore, note that the inequalities in Theorem 1.8.1 give the estimates on $u'(t)$ and consequently on $u(t)$ after integration. Similar remarks apply to all of the inequalities given in this section.

△

Theorem 1.8.2 *Let u, u', a and b be nonnegative continuous functions defined on R_+.*

(i) If

$$u'(t) \leq u(0) + \int_0^t a(s)(u(s) + u'(s))\,ds$$

$$+ \int_0^t a(s)\left(\int_0^s b(\sigma)u'(\sigma)\,d\sigma\right)ds, \quad (1.8.19)$$

for $t \in R_+$, then

$$u'(t) \leq u(0)\left[1 + \int_0^t 2a(s)\exp\left(\int_0^s [1 + a(\sigma) + b(\sigma)]\,d\sigma\right)ds\right], \quad (1.8.20)$$

for $t \in R_+$.

(ii) If

$$u'(t) \leq u(0) + \int_0^t a(s)(u(s) + u'(s))\,ds$$

$$+ \int_0^t a(s)\left(\int_0^s b(\sigma)(u(\sigma) + u'(\sigma))\,d\sigma\right)ds, \quad (1.8.21)$$

for $t \in R_+$, then

$$u'(t) \leq u(0)\left\{1 + \int_0^t a(s)\left[2\exp\left(\int_0^s [1 + a(\sigma)]\,d\sigma\right)\right.\right.$$

$$+ \int_0^s 3b(\tau)\exp\left(\int_\tau^s [1 + a(\sigma)]\,d\sigma\right)$$

$$\left.\left.\times \exp\left(\int_0^\tau [2 + a(\sigma) + b(\sigma)]\,d\sigma\right)d\tau\right]ds\right\}, \quad (1.8.22)$$

for $t \in R_+$.

□

Proof: (i) Define a function $m(t)$ by the right-hand side of (1.8.19). Then (1.8.19) can be restated as

$$u'(t) \leq m(t). \quad (1.8.23)$$

LINEAR INTEGRAL INEQUALITIES

Differentiating the function $m(t)$ and using (1.8.23) we have

$$m'(t) = a(t)(u(t) + u'(t)) + a(t) \int_0^t b(\sigma)u'(\sigma)\,d\sigma$$

$$\leq a(t)\left(u(t) + m(t) + \int_0^t b(\sigma)m(\sigma)\,d\sigma\right). \quad (1.8.24)$$

By taking $t = s$ in (1.8.23) and integrating from 0 to t we have

$$u(t) \leq u(0) + \int_0^t m(s)\,ds. \quad (1.8.25)$$

Using (1.8.25) in (1.8.24) we obtain

$$m'(t) \leq a(t)\left[u(0) + m(t) + \int_0^t m(s)\,ds + \int_0^t b(\sigma)m(\sigma)\,d\sigma\right]. \quad (1.8.26)$$

If we put

$$v(t) = u(0) + m(t) + \int_0^t m(s)\,ds + \int_0^t b(\sigma)m(\sigma)\,d\sigma, \quad (1.8.27)$$

then it follows by differentiating (1.8.27) and using the facts that $m'(t) \leq a(t)v(t)$ from (1.8.26) and $m(t) \leq v(t)$ from (1.8.27) that the inequality

$$v'(t) \leq [1 + a(t) + b(t)]v(t)$$

is satisfied, which implies the estimation for $v(t)$ such that

$$v(t) \leq 2u(0)\exp\left(\int_0^t [1 + a(s) + b(s)]\,ds\right), \quad (1.8.28)$$

since $v(0) = 2u(0)$. Using (1.8.28) in (1.8.26) we have

$$m'(t) \leq 2u(0)a(t)\exp\left(\int_0^t [1 + a(\sigma) + b(\sigma)]\,d\sigma\right),$$

which implies the estimate for $m(t)$ such that

$$m(t) \leq u(0)\left[1 + \int_0^t 2a(s)\exp\left(\int_0^s [1 + a(\sigma) + b(\sigma)]\,d\sigma\right)\,ds\right]. \quad (1.8.29)$$

Using (1.8.29) in (1.8.23) we obtain the desired bound in (1.8.20).

The proof of (ii) can be completed by following similar arguments as in the proof of inequality (i) given above, with suitable modifications. ∎

Theorem 1.8.3 *Let u, u', u'', f, h and g be nonnegative continuous functions defined on R_+.*

(i) If

$$u''(t) \leq f(t) + h(t) \int_0^t g(s)(u(s) + u'(s)) \, ds, \quad t \in R_+, \qquad (1.8.30)$$

then

$$u''(t) \leq f(t) + h(t) \int_0^t p(s) \exp\left(\int_s^t q(\sigma) \, d\sigma\right) ds, \quad t \in R_+, \qquad (1.8.31)$$

where

$$p(t) = g(t) \left[u(0) + (1+t)u'(0) + \int_0^t f(\tau) \, d\tau + \int_0^t (t-\tau) f(\tau) \, d\tau \right], \qquad (1.8.32)$$

$$q(t) = g(t) \left(\int_0^t h(\tau) \, d\tau + \int_0^t (t-\tau) h(\tau) \, d\tau \right). \qquad (1.8.33)$$

(ii) If

$$u''(t) \leq f(t) + h(t) \left(u'(t) + \int_0^t g(s)(u(s) + u'(s)) \, ds \right), \quad t \in R_+, \qquad (1.8.34)$$

then

$$u''(t) \leq f(t) + h(t) \left[u'(0) \exp\left(\int_0^t Q(s) \, ds\right) \right. $$
$$\left. + \int_0^t P(s) \exp\left(\int_s^t Q(\sigma) \, d\sigma\right) ds \right], \quad t \in R_+, \qquad (1.8.35)$$

where

$$P(t) = f(t) + g(t)\left[u(0) + (1+t)u'(0)\right.$$
$$\left. + \int_0^t f(\tau)\,d\tau + \int_0^t (t-\tau)f(\tau)\,d\tau\right], \quad (1.8.36)$$

$$Q(t) = h(t) + g(t)\left[\int_0^t h(\tau)\,d\tau + \int_0^t (t-\tau)h(\tau)\,d\tau\right]. \quad (1.8.37)$$

□

Proof: (i) Define a function $z(t)$ by

$$z(t) = \int_0^t g(s)(u(s) + u'(s))\,ds. \quad (1.8.38)$$

Then (1.8.30) can be restated as

$$u''(t) \le f(t) + h(t)z(t). \quad (1.8.39)$$

By setting $t = \sigma$ in (1.8.39) and integrating it with respect to σ from 0 to t we have

$$u'(t) \le u'(0) + \int_0^t f(\sigma)\,d\sigma + \int_0^t h(\sigma)z(\sigma)\,d\sigma. \quad (1.8.40)$$

By taking $t = \tau$ in (1.8.40) and integrating with respect to τ from 0 to t we have

$$u(t) \le u(0) + tu'(0) + \int_0^t (t-\tau)f(\tau)\,d\tau + \int_0^t (t-\tau)h(\tau)z(\tau)\,d\tau. \quad (1.8.41)$$

From (1.8.38), (1.8.40), (1.8.41) and the fact that $z(t)$ is nondecreasing in t for $t \in R_+$, we observe that

$$z'(t) = g(t)(u(t) + u'(t))$$

$$\leq g(t)\left(u(0) + tu'(0) + \int_0^t (t-\tau)f(\tau)\,d\tau + \int_0^t (t-\tau)h(\tau)z(\tau)\,d\tau\right.$$

$$\left. + u'(0) + \int_0^t f(\tau)\,d\tau + \int_0^t h(\tau)z(\tau)\,d\tau\right)$$

$$\leq p(t) + q(t)z(t). \tag{1.8.42}$$

The inequality (1.8.42) implies the estimate

$$z(t) \leq \int_0^t p(s)\exp\left(\int_s^t q(\sigma)\,d\sigma\right)\,ds. \tag{1.8.43}$$

Using (1.8.43) in (1.8.39) we get the desired inequality in (1.8.31).

(ii) Define a function $z(t)$ by

$$z(t) = u'(t) + \int_0^t g(s)(u(s) + u'(s))\,ds.$$

Now by following similar arguments as in the proof of (i) given above we get the desired inequality in (1.8.35).

∎

1.9 Inequalities with Several Iterated Integrals

In the study of qualitative properties of the solutions of certain higher order differential and integro-differential equations, the results given in earlier sections do not apply directly. Inspired by this fact, Pachpatte (1988a, 1995c, in press e, j) has given a number of inequalities which play a fundamental role in the study of certain higher order differential and integro-differential equations. This section gives some basic inequalities established in these references.

First of all, the following definitions and notations used to simplify the details of presentation are given. We define the differential operators L_i, $0 \leq i \leq n$, by

$$L_0 u(t) = u(t), \quad L_i u(t) = \frac{1}{r_i(t)}\frac{d}{dt}L_{i-1}u(t), \quad 1 \leq i \leq n,$$

LINEAR INTEGRAL INEQUALITIES

with $r_n(t) = 1$, where $u(t)$ and $r_i(t) > 0$ are some functions defined for $t \in R_+$. For $t \in R_+$, $t_0 = t$ and some functions $h(t) \geq 0$, $g(t) > 0$ and $r_j(t) > 0$, $j = 1, 2, \ldots, n-1$, we set

$$M[t, r, h(t_n)] = M[t, r_1, \ldots, r_{n-1}, h(t_n)]$$
$$= \int_0^t r_1(t_1) \int_0^{t_1} r_2(t_2) \ldots \int_0^{t_{n-2}} r_{n-1}(t_{n-1})$$
$$\times \int_0^{t_{n-1}} h(t_n) \, dt_n \, dt_{n-1} \ldots dt_2 \, dt_1,$$

$$A[t, g, h(t_n)] = \int_0^t \int_0^{t_1} \ldots \int_0^{t_{n-2}} \frac{1}{g(t_{n-1})} \int_0^{t_{n-1}} h(t_n) \, dt_n \, dt_{n-1} \ldots dt_1,$$

$$B[t, g, h(t_n)] = \int_0^t \frac{1}{g(t_1)} \int_0^{t_1} \int_0^{t_2} \ldots \int_0^{t_{n-1}} h(t_n) \, dt_n \, dt_{n-1} \ldots dt_2 \, dt_1,$$

$$E[t, g, h(t_n)] = \int_0^t \int_0^{t_1} \ldots \int_0^{t_{n-1}} \frac{1}{g(t_n)} \int_0^{t_n} \int_0^{\tau_1} \ldots \int_0^{\tau_{n-1}} h(\tau_n)$$
$$\times d\tau_n \, d\tau_{n-1} \ldots d\tau_1 \, dt_n \, dt_{n-1} \ldots dt_1.$$

For $t \in R_+$, $s_0 = t$ and some functions $f_i(t) \geq 0$, $p_i(t) > 0$, $i = 1, 2, \ldots, n$, we set

$$\sum_{k=1}^n \left(\int_0^t p_1(s_1) \int_0^{s_1} p_2(s_2) \ldots \int_0^{s_{k-2}} p_{k-1}(s_{k-1}) \int_0^{s_{k-1}} p_k(s_k) f_k(s_k) \right.$$
$$\left. \times ds_k \, ds_{k-1} \ldots ds_2 \, ds_1 \right)$$
$$= \int_0^t p_1(s_1) f_1(s_1) \, ds_1 + \int_0^t p_1(s_1) \int_0^{s_1} p_2(s_2) f_2(s_2) \, ds_2 \, ds_1 + \cdots$$
$$+ \int_0^t p_1(s_1) \int_0^{s_1} p_2(s_2) \ldots \int_0^{s_{n-2}} p_{n-1}(s_{n-1})$$
$$\times \int_0^{s_{n-1}} p_n(s_n) \, ds_n \, ds_{n-1} \ldots ds_2 \, ds_1.$$

In the following three theorems some basic inequalities given by Pachpatte (1988a, in press j) are presented.

Theorem 1.9.1 Let $u \geq 0$, $h \geq 0$, $r_j(t) > 0$, $j = 1, 2, \ldots, n-1$, $g(t) > 0$ be continuous functions defined on R_+ and $u_0 \geq 0$ be a constant.

(a_1) If
$$u(t) \leq u_0 + M[t, r, h(t_n)u(t_n)], \tag{1.9.1}$$

for $t \in R_+$, then
$$u(t) \leq u_0 \exp(M[t, r, h(t_n)]), \tag{1.9.2}$$

for $t \in R_+$.

(a_2) If
$$u(t) \leq u_0 + A[t, g, h(t_n)u(t_n)], \tag{1.9.3}$$

for $t \in R_+$, then
$$u(t) \leq u_0 \exp(A[t, g, h(t_n)]), \tag{1.9.4}$$

for $t \in R_+$.

(a_3) If
$$u(t) \leq u_0 + B[t, g, h(t_n)u(t_n)], \tag{1.9.5}$$

for $t \in R_+$, then
$$u(t) \leq u_0 \exp(B[t, g, h(t_n)]), \tag{1.9.6}$$

for $t \in R_+$.

(a_4) If
$$u(t) \leq u_0 + E[t, g, h(t_n)u(t_n)], \tag{1.9.7}$$

for $t \in R_+$, then
$$u(t) \leq u_0 \exp(E[t, g, h(t_n)]), \tag{1.9.8}$$

for $t \in R_+$.
$\square$

Theorem 1.9.2 Let $u \geq 0$, $p \geq 0$, $q \geq 0$, $r_j > 0$, $j = 1, 2, \ldots, n-1$, $g > 0$ be continuous functions defined on R_+ and $u_0 \geq 0$ be a constant.

(b_1) If
$$u(t) \leq u_0 + M[t, r, p(t_n)(u(t_n) + M[t_n, r, q(s_n)u(s_n)])], \tag{1.9.9}$$

for $t \in R_+$, then
$$u(t) \leq u_0[1 + M[t, r, p(t_n) \exp(M[t_n, r, [p(s_n) + q(s_n)]])]], \tag{1.9.10}$$

for $t \in R_+$.

(b_2) If
$$u(t) \leq u_0 + A[t, g, p(t_n)(u(t_n) + A[t_n, g, q(s_n)u(s_n)])], \quad (1.9.11)$$
for $t \in R_+$, then
$$u(t) \leq u_0[1 + A[t, g, p(t_n)\exp(A[t_n, g, [p(s_n) + q(s_n)]])]], \quad (1.9.12)$$
for $t \in R_+$.

(b_3) If
$$u(t) \leq u_0 + B[t, g, p(t_n)(u(t_n) + B[t_n, g, q(s_n)u(s_n)])], \quad (1.9.13)$$
for $t \in R_+$, then
$$u(t) \leq u_0[1 + B[t, g, p(t_n)\exp(B[t_n, g, [p(s_n) + q(s_n)]])]], \quad (1.9.14)$$
for $t \in R_+$.

(b_4) If
$$u(t) \leq u_0 + E[t, g, p(\tau_n)(u(\tau_n) + E[\tau_n, g, q(\sigma_n)u(\sigma_n)])], \quad (1.9.15)$$
for $t \in R_+$, then
$$u(t) \leq u_0[1 + E[t, g, p(\tau_n)\exp(E[\tau_n, g, [p(\sigma_n) + q(\sigma_n)]])]], \quad (1.9.16)$$
for $t \in R_+$.

$\square$

Theorem 1.9.3 Let u, h, r_j and g be as defined in Theorem 1.9.1. Let f and q be nonnegative continuous functions defined on R_+.

(c_1) If
$$u(t) \leq f(t) + q(t)M[t, r, h(t_n)u(t_n)], \quad (1.9.17)$$
for $t \in R_+$, then
$$u(t) \leq f(t) + q(t)M[t, r, f(t_n)h(t_n)]\exp(M[t, r, q(t_n)h(t_n)]), \quad (1.9.18)$$
for $t \in R_+$.

(c_2) If
$$u(t) \leq f(t) + q(t)A[t, g, h(t_n)u(t_n)], \quad (1.9.19)$$
for $t \in R_+$, then
$$u(t) \leq f(t) + q(t)A[t, g, f(t_n)h(t_n)]\exp(A[t, g, q(t_n)h(t_n)]), \quad (1.9.20)$$

for $t \in R_+$.

(c_3) If
$$u(t) \leq f(t) + q(t)B[t, g, h(t_n)u(t_n)], \qquad (1.9.21)$$
for $t \in R_+$, then
$$u(t) \leq f(t) + q(t)B[t, g, f(t_n)h(t_n)]\exp(B[t, g, q(t_n)h(t_n)]), \qquad (1.9.22)$$
for $t \in R_+$.

(c_4) If
$$u(t) \leq f(t) + q(t)E[t, g, h(t_n)u(t_n)], \qquad (1.9.23)$$
for $t \in R_+$, then
$$u(t) \leq f(t) + q(t)E[t, g, f(t_n)h(t_n)]\exp(E[t, g, q(t_n)h(t_n)]), \qquad (1.9.24)$$
for $t \in R_+$.
□

Proofs of Theorems 1.9.1–1.9.3: Since the proofs of the inequalities in Theorems 1.9.1–1.9.3 resemble one another, we give the details of inequalities (b_1) in Theorem 1.9.2 and (c_1) in Theorem 1.9.3 only. The proofs of all other inequalities can be carried out following the same ideas as in the proofs of the inequalities (b_1) and (c_1).

(b_1) It is sufficient to assume that u_0 is positive, since the standard limiting argument can be used to treat the remaining case. Define a function $z(t)$ by the right side of (1.9.9). Then from the definition of $z(t)$ and using the fact that $u(t) \leq z(t)$, it is easy to observe that
$$L_n z(t) \leq p(t)(z(t) + M[t, r, q(s_n)z(s_n)]). \qquad (1.9.25)$$
Define a function $v(t)$ by
$$v(t) = z(t) + M[t, r, q(s_n)z(s_n)]. \qquad (1.9.26)$$
From (1.9.26) and using (1.9.25) and the fact that $z(t) \leq v(t)$, we observe that
$$L_n v(t) = L_n z(t) + q(t)z(t)$$
$$\leq [p(t) + q(t)]v(t). \qquad (1.9.27)$$
From (1.9.27) and using the facts that $v(t) > 0$, $(d/dt)v(t) \geq 0$, $L_{i-1}v(t) \geq 0$ for $i = 2, 3, \ldots, n$, for $t \in R_+$, we observe that
$$\frac{L_n v(t)}{v(t)} \leq [p(t) + q(t)] + \frac{((d/dt)v(t))L_{n-1}v(t)}{v^2(t)},$$

i.e.
$$\frac{d}{dt}\left(\frac{L_{n-1}v(t)}{v(t)}\right) \leq [p(t)+q(t)]. \tag{1.9.28}$$

By setting $t = s_n$ in (1.9.28) and integrating it from 0 to t and using the fact that $L_i v(0) = 0$, for $i = 1, 2, \ldots, n-1$, we obtain

$$\frac{(d/dt)L_{n-2}v(t)}{v(t)} \leq r_{n-1}(t)\int_0^t [p(s_n)+q(s_n)]\,ds_n. \tag{1.9.29}$$

Again as above from (1.9.29) we observe that

$$\frac{d}{dt}\left(\frac{L_{n-2}v(t)}{v(t)}\right) \leq r_{n-1}(t)\int_0^t [p(s_n)+q(s_n)]\,ds_n. \tag{1.9.30}$$

By setting $t = s_{n-1}$ in (1.9.30) and integrating it from 0 to t and using the fact that $L_i v(0) = 0$, $i = 1, 2, \ldots, n-2$, we get

$$\frac{L_{n-2}v(t)}{v(t)} \leq \int_0^t r_{n-1}(s_{n-1})\int_0^{s_{n-1}} [p(s_n)+q(s_n)]\,ds_n\,ds_{n-1}.$$

Continuing in this way, we obtain

$$\frac{(d/dt)v(t)}{v(t)} \leq r_1(t)\int_0^t r_2(s_2)\int_0^{s_2} r_3(s_3)\ldots \int_0^{s_{n-2}} r_{n-1}(s_{n-1})$$

$$\times \int_0^{s_{n-1}} [p(s_n)+q(s_n)]\,ds_n\,ds_{n-1}\ldots ds_3\,ds_2. \tag{1.9.31}$$

By setting $t = s_1$ in (1.9.31) and integrating it from 0 to t and using the fact that $v(0) = u_0$, we get

$$v(t) \leq u_0 \exp(M[t, r, [p(s_n)+q(s_n)]]). \tag{1.9.32}$$

Using (1.9.32) in (1.9.25) we have

$$L_n z(t) \leq u_0 p(t) \exp(M[t, r, [p(s_n)+q(s_n)]]). \tag{1.9.33}$$

It is easy to observe that the inequality (1.9.33) implies the estimate

$$z(t) \leq u_0[1 + M[t, r, p(t_n)\exp(M[t_n, r, [p(s_n)+q(s_n)]])]]. \tag{1.9.34}$$

Now by using (1.9.34) in $u(t) \leq z(t)$ we get the required inequality in (1.9.10).

(c_1) Define a function $z(t)$ by

$$z(t) = M[t, r, h(t_n)u(t_n)]. \qquad (1.9.35)$$

Using the fact that $u(t) \leq f(t) + q(t)z(t)$ in (1.9.35) we get

$$z(t) \leq M[t, r, h(t_n)(f(t_n) + q(t_n)z(t_n))]$$
$$= a(t) + M[t, r, h(t_n)q(t_n)z(t_n)], \qquad (1.9.36)$$

where $a(t) = M[t, r, h(t_n)f(t_n)]$. Clearly $a(t)$ is nonnegative for all $t \in R_+$. It is sufficient to assume that $a(t)$ is positive, since the standard limiting argument can be used to treat the remaining case. Since $a(t)$ is positive, continuous and nondecreasing for $t \in R_+$, from (1.9.36) we observe that

$$\frac{z(t)}{a(t)} \leq 1 + M\left[t, r, h(t_n)q(t_n)\frac{z(t_n)}{a(t_n)}\right].$$

Now, by an application of Theorem 1.9.1, part (a_1), we obtain

$$z(t) \leq a(t)\exp(M[t, r, h(t_n)q(t_n)]). \qquad (1.9.37)$$

Now using (1.9.37) in $u(t) \leq f(t) + q(t)z(t)$, we get the required inequality in (1.9.18).

∎

Another interesting and useful generalization of the inequality (a_1) in Theorem 1.9.1 is given in the following theorem.

Theorem 1.9.4 *Let $u \geq 0$, $f_i \geq 0$, $p_i > 0$ for $i = 1, 2, \ldots, n$, be continuous functions defined on R_+ and $u_0 \geq 0$ is a constant.*

(d_1) *If*

$$u(t) \leq u_0 + \sum_{k=1}^{n}\left(\int_0^t p_1(s_1)\int_0^{s_1} p_2(s_2)\ldots \int_0^{s_{k-2}} p_{k-1}(s_{k-1})\right.$$
$$\left.\times \int_0^{s_{k-1}} p_k(s_k)f_k(s_k)u(s_k)\,ds_k\,ds_{k-1}\ldots ds_2\,ds_1\right), \qquad (1.9.38)$$

for $t \in R_+$, then

LINEAR INTEGRAL INEQUALITIES

$$u(t) \leq u_0 \exp\left[\sum_{k=1}^{n}\left(\int_0^t p_1(s_1) \int_0^{s_1} p_2(s_2) \ldots \int_0^{s_{k-2}} p_{k-1}(s_{k-1})\right.\right.$$
$$\left.\left.\times \int_0^{s_{k-1}} p_k(s_k) f_k(s_k)\, ds_k\, ds_{k-1} \ldots ds_2\, ds_1\right)\right], \quad (1.9.39)$$

for $t \in R_+$.

(d_2) Let $b(t)$ be a positive continuous and nondecreasing function defined for $t \in R_+$. If

$$u(t) \leq b(t) + \sum_{k=1}^{n}\left(\int_0^t p_1(s_1) \int_0^{s_1} p_2(s_2) \ldots \int_0^{s_{k-2}} p_{k-1}(s_{k-1})\right.$$
$$\left.\times \int_0^{s_{k-1}} p_k(s_k) f_k(s_k) u(s_k)\, ds_k\, ds_{k-1} \ldots ds_2\, ds_1\right), \quad (1.9.40)$$

for $t \in R_+$, then

$$u(t) \leq b(t) \exp\left[\sum_{k=1}^{n}\left(\int_0^t p_1(s_1) \int_0^{s_1} p_2(s_2) \ldots \int_0^{s_{k-2}} p_{k-1}(s_{k-1})\right.\right.$$
$$\left.\left.\times \int_0^{s_{k-1}} p_k(s_k) f_k(s_k)\, ds_k\, ds_{k-1} \ldots ds_2\, ds_1\right)\right], \quad (1.9.41)$$

for $t \in R_+$. □

Proof: (d_1) It is sufficient to assume that u_0 is positive, since the limiting argument can be used to treat the remaining case. Define a function $z(t)$ by the right-hand side of (1.9.38). Then differentiating $z(t)$ and rewriting we have

$$\frac{z'(t)}{p_1(t)} - f_1(t) u(t) = z_1(t), \quad (1.9.42)$$

where

$$z_1(t) = \sum_{k=2}^{n}\left(\int_0^t p_2(s_2) \int_0^{s_2} p_3(s_3) \ldots \int_0^{s_{k-2}} p_{k-1}(s_{k-1})\right.$$
$$\left.\times \int_0^{s_{k-1}} p_k(s_k) f_k(s_k) u(s_k)\, ds_k\, ds_{k-1} \ldots ds_3\, ds_2\right),$$

with $s_1 = t$. From the definition of $z_1(t)$ we observe that

$$\frac{z_1'(t)}{p_2(t)} - f_2(t)u(t) = z_2(t), \tag{1.9.43}$$

where

$$z_2(t) = \sum_{k=3}^{n} \left(\int_0^t p_3(s_3) \int_0^{s_3} p_4(s_4) \ldots \int_0^{s_{k-2}} p_{k-1}(s_{k-1}) \right.$$

$$\left. \times \int_0^{s_{k-1}} p_k(s_k) f_k(s_k) u(s_k) \, ds_k \, ds_{k-1} \ldots ds_4 \, ds_3 \right),$$

with $s_2 = t$. Continuing in this way, we obtain

$$\frac{z_{n-2}'(t)}{p_{n-1}(t)} - f_{n-1}(t)u(t) = z_{n-1}(t), \tag{1.9.44}$$

where

$$z_{n-1}(t) = \int_0^t p_n(s_n) f_n(s_n) u(s_n) \, ds_n,$$

from the definition of $z_{n-1}(t)$ and using the fact that $u(t) \leq z(t)$ it is easy to observe that

$$\frac{z_{n-1}'(t)}{z(t)} \leq p_n(t) f_n(t). \tag{1.9.45}$$

If $H(t)$ is any C^1-function defined on R_+, $H(t) \geq 0$ for all $t \in R_+$ and $H(0) = 0$ and the function $z(t)$ is defined as above, then it is easy to observe that (Medveď, 1993)

$$\int_0^t \frac{H'(s)}{z(s)} \, ds \geq \frac{H(t)}{z(t)}, \tag{$*$}$$

for $t \in R_+$. Indeed, integrating by parts the left-hand side of $(*)$ we obtain

$$\int_0^t \frac{H'(s)}{z(s)} \, ds = \frac{H(t)}{z(t)} + \int_0^t \frac{H(s) z'(s)}{z^2(s)} \, ds \geq \frac{H(t)}{z(t)},$$

for $t \in R_+$.

From $(*)$ and (1.9.45) it is easy to observe that

$$\frac{z_{n-1}(t)}{z(t)} \leq \int_0^t \frac{z_{n-1}'(s_n)}{z(s_n)} \, ds_n$$

LINEAR INTEGRAL INEQUALITIES

$$\leq \int_0^t p_n(s_n) f_n(s_n) \, ds_n. \tag{1.9.46}$$

Using (∗), (1.9.44) and (1.9.46) and the fact that $u(t) \leq z(t)$ we observe that

$$\frac{z_{n-2}(t)}{z(t)} \leq \int_0^t \frac{z'_{n-2}(s_{n-1})}{z(s_{n-1})} \, ds_{n-1}$$

$$= \int_0^t \frac{p_{n-1}(s_{n-1}) f_{n-1}(s_{n-1}) u(s_{n-1}) + p_{n-1}(s_{n-1}) z_{n-1}(s_{n-1})}{z(s_{n-1})} \, ds_{n-1}$$

$$\leq \int_0^t p_{n-1}(s_{n-1}) f_{n-1}(s_{n-1}) \, ds_{n-1} + \int_0^t p_{n-1}(s_{n-1}) \frac{z_{n-1}(s_{n-1})}{z(s_{n-1})} \, ds_{n-1}$$

$$\leq \int_0^t p_{n-1}(s_{n-1}) f_{n-1}(s_{n-1}) \, ds_{n-1}$$

$$+ \int_0^t p_{n-1}(s_{n-1}) \int_0^{s_{n-1}} p_n(s_n) f_n(s_n) \, ds_n \, ds_{n-1}. \tag{1.9.47}$$

Proceeding in this way we get

$$\frac{z_1(t)}{z(t)} \leq \sum_{k=2}^n \left(\int_0^t p_2(s_2) \int_0^{s_2} p_3(s_3) \ldots \int_0^{s_{k-2}} p_{k-1}(s_{k-1}) \right.$$

$$\left. \times \int_0^{s_{k-1}} p_k(s_k) f_k(s_k) \, ds_k \, ds_{k-1} \ldots ds_3 \, ds_2 \right), \tag{1.9.48}$$

From (1.9.48), (1.9.42) and using the fact that $u(t) \leq z(t)$, we observe that

$$\frac{z'(t)}{p_1(t)} - f_1(t) u(t) \leq z(t) \sum_{k=2}^n \left(\int_0^t p_2(s_2) \int_0^{s_2} p_3(s_3) \ldots \int_0^{s_{k-2}} p_{k-1}(s_{k-1}) \right.$$

$$\left. \times \int_0^{s_{k-1}} p_k(s_k) f_k(s_k) \, ds_k \, ds_{k-1} \ldots ds_3 \, ds_2 \right),$$

i.e.

$$\frac{z'(t)}{z(t)} \leq p_1(t)f_1(t)\frac{u(t)}{z(t)} + p_1(t) \sum_{k=2}^{n} \left(\int_0^t p_2(s_2) \int_0^{s_2} p_3(s_3) \ldots \right.$$

$$\left. \times \int_0^{s_{k-2}} p_{k-1}(s_{k-1}) \int_0^{s_{k-1}} p_k(s_k) f_k(s_k) \, ds_k \, ds_{k-1} \ldots ds_3 \, ds_2 \right)$$

$$\leq p_1(t)f_1(t) + p_1(t) \sum_{k=2}^{n} \left(\int_0^t p_2(s_2) \int_0^{s_2} p_3(s_3) \ldots \right.$$

$$\left. \times \int_0^{s_{k-2}} p_{k-1}(s_{k-1}) \int_0^{s_{k-1}} p_k(s_k) f_k(s_k) \, ds_k \, ds_{k-1} \ldots ds_3 \, ds_2 \right). \quad (1.9.49)$$

By taking $t = s_1$ in (1.9.49) and integrating it from 0 to t we get

$$z(t) \leq u_0 \exp \left[\sum_{k=1}^{n} \left(\int_0^t p_1(s_1) \int_0^{s_1} p_2(s_2) \ldots \int_0^{s_{k-2}} p_{k-1}(s_{k-1}) \right. \right.$$

$$\left. \left. \times \int_0^{s_{k-1}} p_k(s_k) f_k(s_k) \, ds_k \, ds_{k-1} \ldots ds_2 \, ds_1 \right) \right]. \quad (1.9.50)$$

Using (1.9.50) in $u(t) \leq z(t)$ we get the required inequality in (1.9.39).

The proof of (d_2) can be completed by using the idea of the proof of Theorem 1.7.4 and using the inequality given in (d_1).

∎

Remark 1.9.1 We note that the inequalities given in Theorem 1.9.4 are inspired by the inequalities given by Medveď (1993). For some other results analogous to the above, see the results given by Pachpatte (in press e, j).

△

1.10 Inequalities Involving Product Integrals

Helton (1977) gave two inequalities involving product integrals, which in turn are further generalizations of the inequality established by Pachpatte

(1973a, Theorem 1). In this section we shall give the inequalities established by Helton (1977).

Following Helton (1977), for completeness, we began with some definitions and notations used in this section. Definitions and integrals are of the subdivision-refinement type, and functions are from R to R or $R \times R$ to R, where R denotes the set of real numbers. Further, interval functions are assumed to be defined only for elements $\{x, y\}$ of $R \times R$ such that $x < y$. Lower-case letters are used to denote functions defined on R, and upper-case letters are used to denote functions defined on $R \times R$. If h and G are functions defined on R and $R \times R$, respectively, and $\{x_i\}_{i=0}^n$ is a subdivision of some interval $[a, b]$, then $h_i = h(x_i)$ for $i = 1, 2, \ldots, n$ and $G_i = G(x_{i-1}, x_i)$ for $i = 1, 2, \ldots, n$.

The statement that $\int_a^b G$ exits means there exists a number L such that, if $\epsilon > 0$, then there exists a subdivision D of $[a, b]$ such that, if $\{x_i\}_{i=0}^n$ is a refinement of D, then

$$\left| L - \sum_{i=1}^n G_i \right| < \epsilon.$$

Similarly, $\prod_a^b (1 + G)$ exists if there exists a number L such that, if $\epsilon > 0$, then there exists a subdivision D of $[a, b]$ such that, if $\{x_i\}_{i=0}^n$ is a refinement of D, then

$$\left| L - \prod_{i=1}^n (1 + G_i) \right| < \epsilon.$$

Also, G has bounded variation on $[a, b]$ if there exist a subdivision D of $[a, b]$ and a number B such that, if $\{x_i\}_{i=0}^n$ is a refinement of D, then

$$\sum_{i=1}^n |G_i| < B.$$

If G has bounded variation on $[a, b]$, then $\int_a^b G$ exists if and only if $\prod_x^y (1 + G)$ exists for $a \leq x \leq y \leq b$.

For convenience in notation, we adopt the conventions that

$$\prod_{i=p+1}^p (1 + G_i) = 1 \quad \text{and} \quad \sum_{i=p+1}^p G_i = 0.$$

These conventions simplify the representation of certain expressions that occur in this section.

Right and left integrals arise in this section. These are denoted by $\int_a^b G(u, v)h(v)$ and $\int_a^b h(u)G(u, v)$, respectively. Suppose $\{x_i\}_{i=0}^n$ denotes a subdivision of some interval $[a, b]$. Then, the preceding right and left integrals have approximating sums of the form

$$\sum_{i=1}^n G_i h_i \quad \text{and} \quad \sum_{i=1}^n h_{i-1} G_i,$$

respectively. Through the section, several different functions are involved in right or left integrals. For examples, integrals of the form

$$\int_a^b \left[\int_a^u h(r)F(r, s)\right] G(u, v) \quad \text{and} \quad \int_a^b G(u, v)\left[\prod_v^b (1 + F + G)\right]$$

arise. Here, the approximating sums are of the form

$$\sum_{i=1}^n \left[\int_a^{x_{i-1}} h(r)F(r, s)\right] G_i \quad \text{and} \quad \sum_{i=1}^n G_i \left[\prod_{x_i}^b (1 + F + G)\right],$$

respectively. Representations involving right and left integrals are necessary due to possible discontinuities of the functions involved.

If $\int_a^b G$ exists, then $\int_a^b |G(u, v) - \int_u^v G|$ exists and is zero. This result is of use in switching between difference inequalities and integral inequalities. Additional background on product integration can be obtained from the references given in Helton (1977).

In order to establish our main results, we need the following two lemmas given in Helton (1977).

Lemma 1.10.1 *Suppose c is a positive constant, h is a bounded function from R to R, each of F and G is a nonnegative function from $R \times R$ to R, each of $\int_a^b F$ and $\int_a^b G$ exists, each of*

$$\int_a^b h(u)G(u, v) \quad \text{and} \quad \int_a^b \left[\int_a^u h(r)F(r, s)\right] G(u, v)$$

exists and

$$h(t) \leq c + \int_a^t h(u)G(u, v) + \int_a^t \left[\int_a^u h(r)F(r, s)\right] G(u, v),$$

LINEAR INTEGRAL INEQUALITIES

for $a \leq t \leq b$. Then, if $a < t \leq b$ and $\{x_i\}_{i=0}^{n}$ is a subdivision of $[a, t]$, the following inequality holds:

$$h(t) \leq c \left[1 + \sum_{i=1}^{n} \left(\prod_{j=1}^{i-1} (1 + F_j + G_j) \right) G_i \right] + \sum_{i=1}^{n} (c_i + d_i) H(i, n),$$

where $H(n, n) = 1$ and

$$c_i = \int_{x_{i-1}}^{x_i} h(u) G(u, v) - h_{i-1} G_i \quad \text{for} \quad i = 1, 2, \ldots, n,$$

$$d_i = \int_{x_{i-1}}^{x_i} \left(\int_{a}^{u} h(r) F(r, s) \right) G(u, v) - \left(\sum_{j=1}^{i-1} h_{j-1} F_j \right) G_i$$

$$\text{for} \quad i = 1, 2, \ldots, n,$$

and

$$H(i, n) = H(i, n - 1) + H(i, n - 1) G_n + \left(\sum_{j=i}^{n-2} H(i, j) F_{j+1} \right) G_n,$$

$$\text{for} \quad i = 1, 2, \ldots, n - 1.$$

$\triangle$

Proof: This lemma is established by induction. We initially note that $h(a) \leq c$. Our induction argument begins with $n = 1$. Suppose $a < t \leq b$ and $\{x_i\}_{i=0}^{1}$ is a subdivision of $[a, t]$. Then,

$$h(t) \leq c + \int_{a}^{t} h(u) G(u, v) + \int_{a}^{t} \left(\int_{a}^{u} h(r) F(r, s) \right) G(u, v)$$

$$= c + h_0 G_1 + c_1 + d_1$$

$$\leq c + c G_1 + c_1 + d_1$$

$$= c \left[1 + \sum_{i=1}^{1} \left(\prod_{j=1}^{i-1} (1 + F_j + G_j) \right) G_i \right] + \sum_{i=1}^{1} (c_i + d_i) H(i, 1).$$

Therefore, the result is true for $n = 1$.

The result is now assumed to be true for all positive integers less than or equal to n. That is, if $a < t \leq b$, $1 \leq m \leq n$ and $\{x_i\}_{i=0}^{m}$ is a subdivision of $[a, t]$, then

$$h(t) \leq c + \int_a^t h(u)G(u, v) + \int_a^t \left(\int_a^u h(r)F(r, s) \right) G(u, v)$$

$$\leq c \left[1 + \sum_{i=1}^m \left(\prod_{j=1}^{i-1}(1 + F_j + G_j) \right) G_i \right] + \sum_{i=1}^m (c_i + d_i)H(i, m).$$

(1.10.1)

The desired inequality is next established for $n + 1$.

Suppose $a < t \leq b$ and $\{x_i\}_{i=0}^{n+1}$ is a subdivision of $[a, t]$. In order to simplify the proof of the final result, several intermediate results are established.

If $\{A_i\}_{i=1}^m$ and $\{B_i\}_{i=1}^m$ are nonnegative sequences, then

$$\prod_{j=1}^m (1 + A_j + B_j) = 1 + \sum_{i=1}^m \left(\prod_{j=1}^{i-1}(1 + A_j + B_j) \right) [A_i + B_i],$$

(1.10.2)

and

$$1 + \sum_{i=1}^m \left(\prod_{j=1}^{i-1}(1 + A_j + B_j) \right) B_i \leq \prod_{j=1}^m (1 + A_j + B_j).$$

(1.10.3)

These relations can be established by induction. By employing the two preceding relations, we have that

$$c \left[1 + \sum_{i=1}^n \left(\prod_{j=1}^{i-1}(1 + F_j + G_j) \right) G_i \right] G_{n+1}$$

$$+ \sum_{i=1}^n c \left[1 + \sum_{j=1}^{i-1} \left(\prod_{k=1}^{j-1}(1 + F_k + G_k) \right) G_j \right] F_i G_{n+1}$$

$$\leq c \left[1 + \sum_{i=1}^n \left(\prod_{j=1}^{i-1}(1 + F_j + G_j) \right) G_i \right] G_{n+1}$$

$$+ \sum_{i=1}^n c \left(\prod_{j=1}^{i-1}(1 + F_j + G_j) \right) F_i G_{n+1} \qquad \text{[From (1.10.3)]}$$

LINEAR INTEGRAL INEQUALITIES

$$= c\left[1 + \sum_{i=1}^{n}\left(\prod_{j=1}^{i-1}(1+F_j+G_j)\right)[F_i+G_i]\right]G_{n+1}$$

$$= c\left(\prod_{j=1}^{n}(1+F_j+G_j)\right)G_{n+1}. \quad \text{[From (1.10.2)]} \quad (1.10.4)$$

This relation is used in the next paragraph.

By using (1.10.1) and the relation from the preceding paragraph, we have that

$$\int_{x_n}^{x_{n+1}} h(u)G(u,v) + \int_{x_n}^{x_{n+1}}\left(\int_a^u h(r)F(r,s)\right)G(u,v)$$

$$= h_n G_{n+1} + c_{n+1} + \left(\sum_{i=1}^{n} h_{i-1}F_i\right)G_{n+1} + d_{n+1}$$

[From definitions of c_{n+1}, d_{n+1}]

$$\leq \left\{c\left[1 + \sum_{i=1}^{n}\left(\prod_{j=1}^{i-1}(1+F_j+G_j)\right)G_i\right]\right.$$

$$\left. + \sum_{i=1}^{n}(c_i+d_i)H(i,n)\right\}G_{n+1} + c_{n+1}$$

$$+ \left(\sum_{i=1}^{n}\left\{c\left[1 + \sum_{j=1}^{i-1}\left(\prod_{k=1}^{j-1}(1+F_k+G_k)\right)G_j\right]\right.\right.$$

$$\left.\left. + \sum_{j=1}^{i-1}(c_j+d_j)H(j,i-1)\right\}F_i\right)G_{n+1} + d_{n+1} \quad \text{[From (1.10.1)]}$$

$$\leq c\left(\prod_{j=1}^{n}(1+F_j+G_j)\right)G_{n+1} + \left(\sum_{i=1}^{n}(c_i+d_i)H(i,n)\right)G_{n+1} + c_{n+1}$$

$$+ \sum_{i=1}^{n}\left(\sum_{j=1}^{i-1}(c_j+d_j)H(j,i-1)\right)F_i G_{n+1} + d_{n+1}.$$

[From (1.10.4)] (1.10.5)

It can be established by induction that

$$\sum_{i=1}^{n}\left(\sum_{j=1}^{i-1}(c_j+d_j)H(j,i-1)\right)F_i = \sum_{i=1}^{n-1}(c_i+d_i)\sum_{j=1}^{n-1}H(i,j)F_{j+1}.$$

(1.10.6)

Now, by using the preceding relation, we have that

$$\sum_{i=1}^{n}(c_i+d_i)H(i,n) + \left(\sum_{i=1}^{n}(c_i+d_i)H(i,n)\right)G_{n+1} + c_{n+1}$$

$$+ \sum_{i=1}^{n}\left(\sum_{j=1}^{i-1}(c_j+d_j)H(j,i-1)\right)F_i G_{n+1} + d_{n+1}$$

$$= \sum_{i=1}^{n}(c_i+d_i)H(i,n) + \left(\sum_{i=1}^{n}(c_i+d_i)H(i,n)\right)G_{n+1} + c_{n+1}$$

$$+ \left(\sum_{i=1}^{n-1}(c_i+d_i)\sum_{j=1}^{n-1}H(i,j)F_{j+1}\right)G_{n+1} + d_{n+1}$$

[From (1.10.6)]

$$= \sum_{i=1}^{n}(c_i+d_i)\left(H(i,n) + H(i,n)G_{n+1}\right.$$

$$\left. + \sum_{j=1}^{n-1}H(i,j)F_{j+1}G_{n+1}\right) + [c_{n+1} + d_{n+1}]$$

$$= \sum_{i=1}^{n+1}(c_i+d_i)H(i,n+1). \qquad \text{[From Definition of } H(i,n+1)\text{]}$$

(1.10.7)

We are now prepared to complete the argument for $n+1$. By using the results of the two preceding paragraphs, we have that

$$h(t) \le c + \int_a^{x_{n+1}} h(u)G(u,v) + \int_a^{x_{n+1}}\left(\int_a^u h(r)F(r,s)\right)G(u,v)$$

$$\le c \left\{ 1 + \sum_{i=1}^{n} \left(\prod_{j=1}^{i-1} (1 + F_j + G_j) \right) G_i \right\}$$

$$+ \sum_{i=1}^{n} (c_i + d_i) H(i, n) + \int_{x_n}^{x_{n+1}} h(u) G(u, v)$$

$$+ \int_{x_n}^{x_{n+1}} \left(\int_{a}^{u} h(r) F(r, s) \right) G(u, v) \qquad \text{[From (1.10.1)]}$$

$$\le c \left[1 + \sum_{i=1}^{n} \left(\prod_{j=1}^{i-1} (1 + F_j + G_j) \right) G_i \right] + \sum_{i=1}^{n} (c_i + d_i) H(i, n)$$

$$+ c \left(\prod_{j=1}^{n} (1 + F_j + G_j) \right) G_{n+1}$$

$$+ \left(\sum_{i=1}^{n} (c_i + d_i) H(i, n) \right) G_{n+1} + c_{n+1}$$

$$+ \sum_{i=1}^{n} \left(\sum_{j=1}^{i-1} (c_j + d_j) H(j, i-1) \right) F_i G_{n+1} + d_{n+1}$$

$$\text{[From (1.10.5)]}$$

$$= c \left[1 + \sum_{i=1}^{n+1} \left(\prod_{j=1}^{i-1} (1 + F_j + G_j) \right) G_i \right]$$

$$+ \sum_{i=1}^{n+1} (c_i + d_i) H(i, n+1). \qquad \text{[From (1.10.7)]}$$

Therefore the result is true for $n + 1$. Hence the proof of the lemma is completed.

∎

Lemma 1.10.2 *Suppose c is a positive constant, h is a bounded function from R to R, each of F and G is a nonnegative function from $R \times R$ to R, each of $\int_a^b F$ and $\int_a^b G$ exists, each of*

$$\int_a^b h(u)G(u,v) \quad \text{and} \quad \int_a^b \left[\int_a^u h(r)G(r,s)\right] F(u,v)$$

exist and

$$h(t) \leq c + \int_a^t h(u)G(u,v) + \int_a^t \left(\int_a^u h(r)G(r,s)\right) F(u,v),$$

for $a \leq t \leq b$. Then, if $a < t \leq b$ and $\{x_i\}_{i=0}^n$ is a subdivision of $[a, t]$, the following inequality holds:

$$h(t) \leq c\left[1 + \sum_{i=1}^n G_i \left(\prod_{j=i+1}^n (1 + F_j + G_j)\right)\right] + \sum_{i=1}^n (c_i + d_i)H(i, n),$$

where $H(n, n) = 1$ and

$$c_i = \int_{x_{i-1}}^{x_i} h(u)G(u,v) - h_{i-1}G_i, \quad \text{for } i = 1, 2, \ldots, n,$$

$$d_i = \int_{x_{i-1}}^{x_i} \left(\int_a^u h(r)G(r,s)\right) F(u,v) - \left(\sum_{j=1}^{i-1} h_{j-1}G_j\right) F_i,$$

$$\text{for } i = 1, 2, \ldots, n,$$

$$H(i, n) = H(i, n-1) + H(i, n-1)G_n + \left(\sum_{j=i}^{n-2} H(i, j)G_{j+1}\right) F_n,$$

$$\text{for } i = 1, 2, \ldots, n-1.$$

△

Proof: This lemma can be established by a proof similar to the proof used to establish Lemma 1.10.1.

■

Theorem 1.10.1 *Let c be a positive constant, h a bounded function from R to R, each of F and G be a nonnegative function from R × R to R, each of $\int_a^b F$ and $\int_a^b G$ exists, each of*

$$\int_a^b h(u)G(u,v) \quad \text{and} \quad \int_a^b \left(\int_a^u h(r)F(r,s)\right) G(u,v)$$

exists and

$$h(t) \leq c + \int_a^t h(u)G(u,v) + \int_a^t \left(\int_a^u h(r)F(r,s)\right) G(u,v)$$

for $a \leq t \leq b$, then

$$h(t) \leq c \left[1 + \int_a^t \left(\prod_a^u (1 + F + G)\right) G(u,v)\right]$$

for $a \leq t \leq b$.

$\square$

Proof: Suppose the conclusion is false. Then, there exists a number $t, a < t \leq b$, such that

$$h(t) > c \left[1 + \int_a^t \left(\prod_a^u (1 + F + G)\right) G(u,v)\right].$$

Let d denote the positive number such that

$$d = h(t) - c \left[1 + \int_a^t \left(\prod_a^u (1 + F + G)\right) G(u,v)\right].$$

We note that the existence of

$$\prod_a^u (1 + F + G) \quad \text{and} \quad \int_a^t \left[\prod_a^u (1 + F + G)\right] G(u,v)$$

can be established from the existence of $\int_a^b F$ and $\int_a^b G$.

It follows from the existence of the integrals involved that there exists a subdivision D_1 of $[a, t]$ such that, $\{x_i\}_{i=0}^n$ is a refinement of D_1; then

$$\left| \int_a^t \left(\prod_a^u (1 + F + G)\right) G(u,v) - \sum_{i=1}^n \left(\prod_{j=1}^{i-1} (1 + F_j + G_j)\right) G_i \right| < d(2c)^{-1}.$$

Let β represent a nonnegative function of bounded variation from $R \times R$ to R such that, if $\{x_i\}_{i=1}^n$ is a subdivision of $[a, t]$ and $1 \leq i \leq n$, then

$$2(|F_i| + |G_i|) < \beta_i.$$

There exists a subdivision D_2 of $[a, t]$ and a number B such that, if $\{x_i\}_{i=0}^n$ is a refinement of D_2, then

$$\prod_{i=1}^n (1 + \beta_i) < B.$$

Since $\int_a^t h(u)G(u, v)$ exists, it follows that

$$\int_a^t \left| \int_x^y h(u)G(u, v) - h(x)G(x, y) \right|$$

exists and is zero. Hence, there exists a subdivision D_3 of $[a, t]$ such that, if $\{x_i\}_{i=1}^n$ is a refinement of D_3, then

$$\sum_{i=1}^n \left| \int_{x_{i-1}}^{x_i} h(u)G(u, v) - h_{i-1}G_i \right| < d(4B)^{-1}.$$

Since $\int_a^b \left[\int_a^u h(r)F(r, s) \right] G(u, v)$ exists, it follows that

$$\int_a^t \left| \int_x^y \left(\int_a^u h(r)F(r, s) \right) G(u, v) - \left(\int_a^x h(r)F(r, s) \right) G(x, y) \right|$$

exists and is zero. From this, the existence of $\int_a^b h(r)F(r, s)$ and the bounded variation of G, it follows that there exists a subdivision D_4 of $[a, t]$ such that, if $\{x_i\}_{i=0}^n$ is a refinement of D_4, then

$$\sum_{i=1}^n \left| \int_{x_{i-1}}^{x_i} \left(\int_a^u h(r)F(r, s) \right) G(u, v) - \left(\sum_{j=1}^{i-1} h_{j-1}F_j \right) G_i \right| < d(4B)^{-1}.$$

Let D denote the subdivision $\cup_{i=1}^4 D_i$ of $[a, t]$. Suppose $\{x_i\}_{i=0}^n$ is a refinement of D. It follows from Lemma 1.10.1 that

$$h(t) \le c \left[1 + \sum_{i=1}^n \left(\prod_{j=1}^{i-1} (1 + F_j + G_j) \right) G_i \right] + \sum_{i=1}^n (c_i + d_i) H(i, n),$$

where c_i, d_i and $H(i, n)$ are defined in Lemma 1.10.1. It follows from the manner in which the $H(i, n)$ are defined that

$$H(i, n) \le \prod_{j=1}^n (1 + \beta_j) < B.$$

Thus,

$$h(t) \le c \left[1 + \int_a^t \left({}_a\prod^u (1 + F + G) \right) G(u, v) \right]$$

$$+ c \left| \sum_{i=1}^n \left(\prod_{j=1}^{i-1} (1 + F_j + G_j) \right) G_i - \int_a^t \left({}_a\prod^u (1 + F + G) \right) G(u, v) \right|$$

$$+ \sum_{i=1}^n |c_i| |H(i, n)| + \sum_{i=1}^n |d_i| |H(i, n)|$$

$$< [h(t) - d] + c[d(2c)^{-1}] + B \sum_{i=1}^n |c_i| + B \sum_{i=1}^n |d_i|$$

$$< [h(t) - d] + d/2 + B[d(4B)^{-1}] + B[d(4B)^{-1}]$$

$$= h(t).$$

This is a contradiction. Therefore, the desired inequality is established. This completes the proof of Theorem 1.10.1.

∎

Theorem 1.10.2 *If c is a positive constant, h is a bounded function from R to R, each of F and G is a nonnegative function from $R \times R$ to R, each of $\int_a^b F$ and $\int_a^b G$ exists, each of*

$$\int_a^b h(u) G(u, v) \quad \text{and} \quad \int_a^b \left(\int_a^u h(r) G(r, s) \right) F(u, v)$$

exists and

$$h(t) \le c + \int_a^t h(u) G(u, v) + \int_a^t \left[\int_a^u h(r) G(r, s) \right] F(u, v)$$

for $a \le t \le b$, then

$$h(t) \le c \left[1 + \int_a^t G(u, v) \left({}_v\prod^t (1 + F + G) \right) \right]$$

for $a \le t \le b$.

□

Proof: This theorem can be established by a proof similar to the proof used to establish Theorem 1.10.1 by using Lemma 1.10.2 in place of Lemma 1.10.1.

∎

Davis and Chatfield (1970, Theorem 3, p. 744] established that if $\int_a^b G$ exists and $\int_a^b G^2$ exists and is zero, then

$$\prod_a^b (1+G) = \exp \int_a^b G.$$

Thus, if the additional restriction that each of $\int_a^b F^2$ and $\int_a^b G^2$ exists and is zero is added to Theorems 1.10.1 and 1.10.2, then their conclusions are

$$h(t) \le c \left[1 + \int_a^t \left(\exp \int_a^u (F+G) \right) G(u,v) \right]$$

and

$$u(t) \le c \left[1 + \int_a^t G(u,v) \left(\exp \int_v^t (F+G) \right) \right],$$

respectively. When given in these forms, Theorems 1.10.1 and 1.10.2 more closely resemble the Gronwall–Bellman inequality.

It is interesting to note that, when f and g are continuous functions

$$G(u,v) = g(u)(v-u) \quad \text{and} \quad F(u,v) = f(u)(v-u),$$

the inequality established in Theorem 1.10.1 reduces to the inequality given by Pachpatte (1973a, Theorem 1).

Finally, we note that Mingarelli (1981) has given a Stieltjes version of the inequality given by Pachpatte (1973a, Theorem 1) which includes most of the Stieltjes formulations of Gronwall's inequality and a more general result given by Herod (1969). For a detailed discussion on inequalities involving Stieltjes and stochastic integrals, we refer interested readers to Erbe and Kong (1990), Helton (1969), Jones (1964), Mao (1989) and Zakai (1967) and the related references cited therein.

For various other inequalities relating to the topics discussed in the above sections, we refer interested readers to the references of Antosiewicz

(1962), Arino and Györi (1984), Asirov and Atdaev (1973), Azbelev and Tsalyuk (1962), Bellman (1967), Chandra and Fleishman (1970), Ráb (1979), Ved' (1965) and Wright *et al.* (1971), as well as other suitable references (Bellman, 1957; Dannan, 1985; Filatov and Sharova, 1976; Mamedov *et al.*, 1980; Mitrinović and Pečarić, 1988; Movlyankulov and Filatov, 1972; Schmaedeke and Sell, 1968; Schröder, 1980; Venckova, 1977).

1.11 Applications

The literature on the applications of the inequalities given in earlier sections is quite extensive. Here, applications of certain inequalities to the study of the basic problems related to some integral and integro-differential equations are presented

1.11.1 Second-order Integro-differential Equations

In this section we shall study the boundedness and asymptotic behaviour of solutions of second-order integro-differential equations of the form

$$(r(t)x')' + a(t)x = f\left(t, x, \int_{t_0}^{t} k(t, s, x)\,ds\right), \tag{A}$$

by means of comparison with the solutions of the second-order linear differential equation

$$(r(t)x')' + a(t)x = 0, \tag{B}$$

where $k: I \times I \times R \to R$, $f: I \times R \times R \to R$ are continuous functions, $r(t) > 0$ and $a(t)$ are continuous functions defined on I and $r(t)$ is continuously differentiable on I. Many authors have studied the behaviour of solutions of special versions of equations (A) and (B) with different viewpoints.

In the following theorems we present results on the boundedness and asymptotic behaviour of the solutions of equation (A) under some suitable conditions on the functions involved in (A) and on the solutions of (B). These results are slight variants of the main results given by Pachpatte (1975e).

LINEAR INTEGRAL INEQUALITIES

Theorem 1.11.1 *Suppose that the functions k and f in (A) satisfy the conditions*

$$|k(t, s, x)| \leq h(s)|x|, \tag{1.11.1}$$

$$|f(t, x, z)| \leq g(t)(|x| + |z|), \tag{1.11.2}$$

where h and g are real-valued continuous functions defined on I and $\int_{t_0}^{\infty} h(s)\,ds < \infty$, $\int_{t_0}^{\infty} g(s)\,ds < \infty$. If $x_1(t)$ and $x_2(t)$ are bounded solutions of equation (B), then the corresponding solution $x(t)$ of equation (A) is bounded on I.

□

Proof: Let $x_1(t)$ and $x_2(t)$ be solutions of equation (B) such that

$$r(t)[x_1(t)x_2'(t) - x_1'(t)x_2(t)] \equiv 1, \tag{1.11.3}$$

and suppose that $x(t)$ is any solution of (A). Then by using the variation of constants formula, the solution $x(t)$ of (A) is given by

$$x(t) = c_1 x_1(t) + c_2 x_2(t) + \int_{t_0}^{t} [x_1(s)x_2(t) - x_1(t)x_2(s)]$$

$$\times f\left(s, x(s), \int_{t_0}^{s} k(s, \sigma, x(\sigma))\,d\sigma\right) ds, \tag{1.11.4}$$

where c_1 and c_2 are constants. From (1.11.4) and using (1.11.1) and (1.11.2) we obtain

$$|x(t)| \leq c + \int_{t_0}^{t} Mg(s)|x(s)|\,ds + \int_{t_0}^{t} Mg(s)\left(\int_{t_0}^{s} h(\sigma)|x(\sigma)|\,d\sigma\right) ds,$$

where $c \geq 0$ and $M \geq 0$ are constants and the upper bounds for $|c_1 x_1(t) + c_2 x_2(t)|$ and $|x_1(s)x_2(t) - x_1(t)x_2(s)|$. Now an application of Theorem 1.7.1 yields

$$|x(t)| \leq c\left[1 + \int_{t_0}^{t} Mg(s)\exp\left(\int_{t_0}^{s} [Mg(\sigma) + h(\sigma)]\,d\sigma\right) ds\right].$$

The above estimation implies the boundedness of the solution $x(t)$ of (A) on I.

■

Theorem 1.11.2 *Suppose that the functions k and f in (A) satisfy the conditions*

$$|k(t, s, x)| \leq h(s)|x|, \tag{1.11.5}$$

$$|f(t, x, z)| \leq g(t)(|x| + \exp(-\alpha t)|z|), \tag{1.11.6}$$

where $\alpha > 0$ is a constant, h and g are real-valued continuous functions defined on I and $\int_{t_0}^{\infty} h(s)\exp(-\alpha s) < \infty$, $\int_{t_0}^{\infty} g(s)\exp(-2\alpha s)\,ds < \infty$. If $x_1(t)$ and $x_2(t)$ are the solutions of (B) such that

$$|x_i(t)| \leq M_i \exp(-\alpha t), \quad i = 1, 2, \tag{1.11.7}$$

where $M_i > 0$ for $i = 1, 2$ are constants, then the corresponding solution $x(t)$ of (A) approach zero as $t \to \infty$.

□

Proof: Let $x_1(t)$ and $x_2(t)$ be solutions of (B) satisfying the condition (1.11.3) and suppose $x(t)$ is any solution of (A). Then by using the variation of constants formula, any solution $x(t)$ of (A) is given by the integral equation (1.11.4). From (1.11.4) and (1.11.5)–(1.11.7), we obtain

$$|x(t)| \leq c\exp(-\alpha t) + \exp(-\alpha t)\int_{t_0}^{t} Mg(s)\exp(-\alpha s)|x(s)|\,ds$$

$$+ \exp(-\alpha t)\int_{t_0}^{t} M\exp(-2\alpha s)g(s)\left(\int_{t_0}^{s} h(\sigma)|x(\sigma)|\,d\sigma\right)ds.$$

(1.11.8)

From (1.11.8) we have

$$|x(t)|\exp(\alpha t) \leq c + \int_{t_0}^{t} Mg(s)\exp(-2\alpha s)|x(s)|\exp(\alpha s)\,ds$$

$$+ \int_{t_0}^{t} Mg(s)\exp(-2\alpha s)\left(\int_{t_0}^{s} h(\sigma)\exp(-\alpha\sigma)|x(\sigma)|\right.$$

$$\left.\times \exp(\alpha\sigma)\,d\sigma\right)ds.$$

Now an application of Theorem 1.7.1 yields

$$|x(t)| \exp(\alpha t) \leq c \left[1 + \int_{t_0}^{t} Mg(s) \exp(-2\alpha s) \right.$$

$$\left. \times \exp\left(\int_{t_0}^{s} [Mg(\sigma)\exp(-2\alpha\sigma) + h(\sigma)\exp(-\alpha\sigma)] \, d\sigma \right) ds \right].$$

Now, multiplying both sides of the above inequality by $\exp(-\alpha t)$ and then taking the limit as $t \to \infty$, we get the desired result, and hence the proof of the theorem is complete.

∎

1.11.2. Perturbation of Volterra Integral Equations

In this section we shall study the stability, exponential asymptotic stability and the growth of the solutions of a system of Volterra integral equations of the form

$$x(t) = f(t) + \int_0^t [k(t,s)x(s) + F(t,s,x(s))] \, ds, \tag{P}$$

under the assumption that the unperturbed linear system

$$y(t) = f(t) + \int_0^t k(t,s)y(s) \, ds, \tag{L}$$

has certain stability properties. Here x, y, f and F are in R^n, the n-dimensional Euclidean space with Euclidean norm $|\cdot|$, and k is an $n \times n$ matrix. We assume that f is continuously differentiable on R_+, and that $k(t,s)$ and $F(t,s,x)$ are continuous for $t, s \in R_+$, $|x| < \infty$, and $F(t,s,x)$ is continuously differentiable in t. For the study of existence of solutions of the above equations, see Burton (1983), Cordunneau (1973, 1977, 1991) and Miller (1971).

We begin with the stability definitions that we use later in this section (see Pachpatte (1976a)). The following definitions are stated for (P). Of course, we can apply them to (L) as well, with the same initial values $x(0) = y(0) = f(0) = x_0$.

Definitions (1) The system (P) is called *globally uniformly stable* on R_+ if there exists a constant $M > 0$ such that $|x(t)| \leq M$ for $|x_0| < \infty$.

(2) The system (P) is called *exponentially asymptotically stable* on R_+ if there exist positive constants M and α such that $|x(t)| \leq M|x_0|\exp(-\alpha t)$ for $|x_0|$ sufficiently small.

(3) The system (P) is called *uniformly slowly growing* on R_+ if and only if for every $\alpha > 0$ there exists a constant $M > 0$, possibly depending on α, such that $|x(t)| \leq M|x_0|\exp(\alpha t)$, for $|x_0| < \infty$.

$\triangle$

It is known (Bownds and Cushing, 1975) if $U(t, s)$ is an $n \times n$ matrix (called the resolvent kernel) which solves the matrix equation

$$U(t, s) = I + \int_s^t k(t, \tau) U(\tau, s) \, d\tau,$$

for $0 \leq s \leq t < \infty$, where I is the $n \times n$ identity matrix, then the solution of (P) is given by

$$x(t) = y(t) + \int_0^t U(t, s) \left(F(s, s, x(s)) + \int_0^s \frac{\partial}{\partial s} F(s, \tau, x(\tau)) \, d\tau \right) ds,$$

(1.11.9)

where

$$y(t) = U(t, 0) x_0 + \int_0^t U(t, s) f'(s) \, ds,$$

(1.11.10)

is a solution of the linear system (L).

Our approach and arguments lead us to the following assumptions on the perturbation function F:

(H$_1$) $|F(t, t, x(t))| \leq p(t)|x(t)|$,

(H$_2$) $\left| \dfrac{\partial}{\partial t} F(t, s, x(s)) \right| \leq p(t) q(s) |x(s)|$,

or

(H$_3$) $\left| \dfrac{\partial}{\partial t} F(t, s, x(s)) \right| \leq p(t) \exp(-\alpha(t - s)) q(s) |x(s)|$,

or

(H$_4$) $\left| \dfrac{\partial}{\partial t} F(t, s, x(s)) \right| \leq p(t) \exp(\alpha(t - s)) q(s) |x(s)|$, for $0 \leq s \leq t < \infty$, where α is a positive constant and p and q are continuous functions defined on R_+ such that

(H$_5$) $\quad \int_0^\infty p(s)\,ds < \infty, \quad \int_0^\infty q(s)\,ds < \infty.$

We now state and prove the following theorems given by Pachpatte (1976a), which deals with the preservation of stability of (P).

Theorem 1.11.3 *Let the resolvent kernel $U(t, s)$ satisfy the condition*

$$|U(t, s)| \leq N, \qquad (1.11.11)$$

for $0 \leq s \leq t < \infty$, where N is a positive constant. Suppose that the linear system (L) is globally uniformly stable on R_+ and that F satisfies (H_1), (H_2) together with (H_5). Then the perturbed system (P) is globally uniformly stable on R_+.

□

Proof: The solutions of (P) and (L) with the same initial values are related by the integral equation (1.11.9). Using (1.11.9), (1.11.11), (H_1) and (H_2) together with global uniform stability of (L) on R_+, we obtain

$$|x(t)| \leq M + \int_0^t Np(s)|x(s)|\,ds + \int_0^t Np(s)\left(\int_0^s q(\tau)|x(\tau)|\,d\tau\right)ds,$$

where M and N are positive constants. Now an application of Theorem 1.7.1 yields

$$|x(t)| \leq M\left[1 + \int_0^t Np(s)\exp\left(\int_0^s [Np(\tau) + q(\tau)]\,d\tau\right)ds\right].$$

The above estimation in view of the assumption (H_5) implies the global uniform stability of (P) on R_+.

■

The following theorems deals with the preservation of exponential asymptotic stability and the rate of growth of the solution of (P) on R_+.

Theorem 1.11.4 *Let the resolvent kernel $U(t, s)$ satisfy the condition*

$$|U(t, s)| \leq N\exp(-\alpha(t - s)), \qquad (1.11.12)$$

for $0 \leq s \leq t < \infty$, where N and α are positive constants. Suppose that the linear system (L) is exponentially asymptotically stable on R_+ and F satisfies (H_1), (H_3), together with (H_5). Then the perturbed system (P) is exponentially asymptotically stable on R_+.

□

Theorem 1.11.5 *Let the resolvent kernel $U(t, s)$ satisfy the condition*

$$|U(t, s)| \leq N \exp(\alpha(t-s)), \qquad (1.11.13)$$

for $0 \leq s \leq t < \infty$, where N and α are positive constants. Suppose that the linear system (L) is uniformly slowly growing on R_+ and F satisfies (H_1), (H_4), together with (H_5). Then the perturbed system (P) is uniformly slowly growing on R_+.

□

The proofs of Theorems 1.11.4 and 1.11.5 can be carried out by following the same lines as in the proof of Theorem 1.11.3 (see Pachpatte (1974b, 1975f, 1976a)).

1.11.3 Higher Order Integro-differential Equations

This section presents applications of the inequality given in Theorem 1.9.2, part (b_2), to study the boundedness, uniqueness and continuous dependence of the solutions of equations of the form

$$(r(t)x^{(n-1)}(t))' = F(t, x(t), A[t, r, f(s_n, x(s_n))]), \qquad (C)$$

with the given initial conditions

$$x(0) = x_0, \quad x^{(i-1)}(0) = 0, \quad i = 2, 3, \ldots, n, \qquad (D)$$

where $r(t)$ is a real-valued positive continuous function defined for $t \in R_+$, which is continuously differentiable on R_+, $f : R_+ \times R \to R$, $F : R_+ \times R \times R \to R$ are continuous functions, x_0 is a given constant and $A[t, r, f]$ is the notation set in Section 1.9. The results given here concerning the problem (C) and (D) are taken from Pachpatte (in press, j)(see, also Pachpatte (in press, e)).

The following result deals with the boundedness of the solutions of the problem (C) and (D).

Theorem 1.11.6 *Suppose that the functions F and f in (C) satisfy the conditions*

$$|F(t, x, z)| \leq p(t)(|x| + |z|), \tag{1.11.14}$$

$$|f(t, x)| \leq q(t)|x|, \tag{1.11.15}$$

where $p(t)$ and $q(t)$ are real-valued nonnegative continuous functions defined on R_+. If $x(t)$ is any solution of the problem (C) and (D), then

$$|x(t)| \leq |x_0|[1 + A[t, r, p(t_n)\exp(A[t_n, r, [p(s_n) + q(s_n)]])]], \tag{1.11.16}$$

for $t \in R_+$.

□

Proof: If $x(t)$ is any solution of the problem (C) and (D), then it is easy to observe that $x(t)$ is also a solution of the equivalent integral equation

$$x(t) = x_0 + A[t, r, F(t_n, x(t_n), A[t_n, r, f(s_n, x(s_n))])]. \tag{1.11.17}$$

From (1.11.17), (1.11.14) and (1.11.15), we obtain

$$|x(t)| \leq |x_0| + A[t, r, p(t_n)(|x(t_n)| + A[t_n, r, q(s_n)|x(s_n)|])].$$

Now an application of the inequality in Theorem 1.9.2 given in part (b_2) yields the inequality given in (1.11.16). Thus the right-hand side of (1.11.16) gives us the bound on the solution $x(t)$ of the problem (C) and (D) in terms of the known functions.

■

We shall now discuss the uniqueness of the solutions of the problem (C) and (D).

Theorem 1.11.7 *Suppose that the functions F and f in (C) satisfy the conditions*

$$|F(t, x, z) - F(t, \bar{x}, \bar{z})| \leq p(t)(|x - \bar{x}| + |z - \bar{z}|), \tag{1.11.18}$$

$$|f(t, x) - f(t, \bar{x})| \leq q(t)|x - \bar{x}|, \tag{1.11.19}$$

where $p(t)$ and $q(t)$ are as in Theorem 1.11.6. Then the problem (C)-(D) has at most one solution on R_+.

□

Proof: The problem (C)–(D) is equivalent to the equation (1.11.17). Let $x(t)$ and $y(t)$ be two solutions of (C)–(D) on R_+. From (1.11.17) and using

(1.11.18) and (1.11.19) we have

$$|x(t) - y(t)| \leq A[t, r, p(t_n)(|x(t_n) - y(t_n)| + A[t_n, r, |x(s_n) - y(s_n)|])].$$

Now an application of the inequality in Theorem 1.9.2 given in part (b_2) with $c = 0$ and $u(t) = |x(t) - y(t)|$ yields $x(t) = y(t)$, i.e. there is at most one solution of the problem (C)-(D).

∎

The next result deals with the continuous dependence of the solution on the equation and the initial data.

Consider the following initial value problem

$$(r(t)y^{(n-1)}(t))' = \overline{F}(t, y(t), A[t, r, \overline{f}(s_n, y(s_n))]), \quad (C')$$

with the given initial conditions

$$y(0) = y_0, \quad y^{(i-1)}(0) = 0, \quad i = 2, 3, \ldots, n, \quad (D')$$

where $r(t)$ is as in (C), y_0 is a given constant and $\overline{f} : R_+ \times R \to R$, $\overline{F} : R_+ \times R \times R \to R$ are continuous functions and $A[t, r, \overline{f}]$ is the notation set in Section 1.9.

Theorem 1.11.8 *Suppose that the functions F and f in (C) satisfy the conditions (1.11.18) and (1.11.19), and further assume that*

$$|x_0 - y_0| \leq \epsilon, \quad (1.11.20)$$

$$G(t) = A[t, r, |F(t_n, y(t_n), A[t_n, r, f(s_n, y(s_n))])$$
$$- \overline{F}(t_n, y(t_n), A[t_n, r, \overline{f}(s_n, y(s_n))])|] \leq \epsilon, \quad (1.11.21)$$

where $\epsilon > 0$ is an arbitrary constant. Then the solution of the given initial value problem (C)-(D) depends continuously on F and f and the initial data.

□

Proof: The equivalent integral equations corresponding to (C)-(D) and (C')-(D') are respectively (1.11.17) and

$$y(t) = y_0 + A[t, r, \overline{F}(t_n, y(t_n), A[t_n, r, \overline{f}(s_n, y(s_n))])]. \quad (1.11.22)$$

From (1.11.17), (1.11.22) and using (1.11.18)-(1.11.21) we obtain

$$|x(t) - y(t)| \leq |x_0 - y_0| + A[t, r, |F(t_n, x(t_n), A[t_n, r, f(s_n, x(s_n))])$$
$$- F(t_n, y(t_n), A[t_n, r, f(s_n, y(s_n))])|] + G(t)$$

$$\leq 2\epsilon + A[t, r, p(t_n)(|x(t_n) - y(t_n)|$$
$$+ A[t_n, r, q(s_n)|x(s_n) - y(s_n)|])].$$

Now an application of the inequality in Theorem 1.9.2 given in part (b_2) yields

$$|x(t) - y(t)| \leq 2\epsilon[1 + A[t, r, p(t_n)\exp(A[t_n, r, [p(s_n) + q(s_n)]])]]. \tag{1.11.23}$$

On the compact interval $I_0 \subset R_+$, the quantity in the square bracket on the right side in (1.11.23) is bounded by some positive constant, say K. Therefore $|x(t) - y(t)| \leq 2K\epsilon$ on I_0, so the solution of the given initial value problem (C)–(D) depends continuously on F and f and the initial data. If $\epsilon \to 0$, then $|x(t) - y(t)| \to 0$ on the set I_0.

∎

1.11.4 Integral Equation Involving Product Integrals

In this section we give an application of Theorem 1.10.1 to the determination of a bound for the solution of an integral equation given by Helton (1977). In the following, suppose each of f and g is a function from R to R, each of F and G is a function from $R \times R$ to R, f is quasi-continuous on $[a, b]$, each of $\int_a^b F$ and $\int_a^b G$ exists and each of F and G has bounded variation on $[a, b]$. We consider the integral equation

$$h(t) = f(t) + \int_a^t h(u)G(u, v) + \int_a^t \left[\int_a^u h(r)F(r, s) \right] G(u, v),$$

for $a \leq t \leq b$. The existence of a bounded solution for this integral equation can be established by constructing an appropriate sequence of Picard iterates. Further, the uniqueness of this solution can be established by using Theorem 1.10.1.

In the following, we show that in some instances the inequality given in Theorem 1.10.1 produces a better bound for the solution of the preceding integral equation than does the Gronwall inequality involving product integrals. We use the following form of the Gronwall inequality given by Helton (1969, Theorem 4, p. 495).

Theorem H *Let c be a positive constant, h a bounded function from R to R, G a nonnegative function from $R \times R$ to R, $\int_a^b G$ exist and*

$$h(t) \leq c + \int_a^t h(u)G(u, v),$$

for $a \leq t \leq b$; then

$$h(t) \leq c \prod_a^t (1 + G),$$

for $a \leq t \leq b$.

□

Initially, a bound for $h(t)$ is constructed by using Theorem 1.10.1. Since f is quasi-continuous on $[a, b]$, f is bounded on $[a, b]$. Let c denote a bound for f on $[a, b]$. Then,

$$|h(t)| \leq c + \int_a^t |h(u)||G(u, v)| + \int_a^t \left(\int_a^u |h(r)||F(r, s)| \right) |G(u, v)|,$$

and hence applying Theorem 1.10.1, we have that

$$|h(t)| \leq c \left[1 + \int_a^t \left(\prod_a^u (1 + |F| + |G|) \right) |G(u, v)| \right]$$

$$\leq c \left[1 + \int_a^t \left(\prod_a^u (1 + |F| + |G|) \right) [|G(u, v)| + |F(u, v)|] \right]$$

$$= c \prod_a^t (1 + |F| + |G|).$$

Thus $c \prod_a^t (1 + |F| + |G|)$ is bound for $h(t)$ for $a \leq t \leq b$.

Now, a bound for $h(t)$ is constructed by using the Gronwall inequality given in Theorem H. To do this, we first note that

$$\int_a^t \left(\int_a^u h(r)F(r, s) \right) G(u, v) = \int_a^t h(u)F(u, v) \int_v^t G,$$

for $a \leq t \leq b$. This relation can be established by using the identity

$$\sum_{i=1}^n \left(\sum_{j=1}^{i-1} h_{j-1} F_j \right) G_i = \sum_{i=1}^{n-1} h_{i-1} F_i \left(\sum_{j=i+1}^n G_j \right),$$

for subdivisions $\{x_i\}_{i=0}^{n}$ of $[a, t]$. Now by using the preceding integral identity, we have that

$$|h(t)| \leq c + \int_a^t |h(u)||G(u, v)| + \int_a^t |h(u)||F(u, v)| \left|\int_v^t G\right|$$

$$= c + \int_a^t |h(u)| \left[|G(u, v)| + |F(u, v)| \left|\int_v^t G\right|\right]$$

$$\leq c + \int_a^t |h(u)|[|G(u, v)| + B|F(u, v)|],$$

where B denotes the least number such that $\left|\int_u^v G\right| \leq B$ for $a \leq u < v \leq b$. Thus by applying the form of the Gronwall inequality given in Theorem H to the preceding integral inequality, we have that

$$|h(t)| \leq c \prod_a^t [1 + (|G| + B|F|)],$$

for $a \leq t \leq b$. Whenever B satisfies the inequality $B > 1$, this bound for $h(t)$ is not as good as the bound for $h(t)$ obtained previously by using Theorem 1.10.1.

With regard to the applications of the inequalities given above, the literature is particularly vast and it is difficult to give a detailed account of the various applications. For a number of interesting applications of the inequalities given by Pachpatte in Section 1.7, we refer the readers to the recent results given by Dhakne and Pachpatte (1988a,b, 1991, 1994), Grace and Lalli (1980a,b), Lalli (1984) and Murge and Pachpatte (1987, 1992).

1.12 Miscellaneous Inequalities

1.12.1 Sansone and Conti (1964)

Let λ, ϕ, u be given functions defined for $t \in R_+$ such that λ is summable and nonnegative, ϕ is absolutely continuous and u is continuous. If

$$u(t) \leq \phi(t) + \int_0^t \lambda(s)u(s)\,ds,$$

then

$$u(t) \leq \phi(0) \exp\left(\int_0^t \lambda(s)\,ds\right) + \int_0^t \phi'(s) \exp\left(\int_s^t \lambda(\sigma)\,d\sigma\right) ds.$$

1.12.2 Rudakov (1970)

Let u, b, c be continuous functions in J, let b be nonnegative in J, and suppose

$$u(t) \leq u(r) + \int_r^t [b(s)u(s) + c(s)]|ds|,$$

for $t \in J$. Then

$$\left[u(\alpha) - \int_\alpha^t c(s) \exp\left(\int_\alpha^s b(\sigma)\,d\sigma\right) ds\right] \exp\left(-\int_\alpha^t b(s)\,ds\right) \leq u(t),$$

$$u(t) \leq \left[u(\alpha) + \int_\alpha^t c(s) \exp\left(-\int_\alpha^s b(\sigma)\,d\sigma\right) ds\right] \exp\left(\int_\alpha^t b(s)\,ds\right),$$

for $t \in J$.

1.12.3 Sardarly (1965)

Let u, a, b and q be continuous functions in J, let $c(t,s)$ be a continuous function for $\alpha \leq s \leq t \leq \beta$, let b and q be nonnegative in J, and suppose

$$u(t) \leq a(t) + \int_\alpha^t [q(t)b(s)u(s) + c(t,s)]\,ds, \quad t \in J.$$

Then for $t \in J$,

$$u(t) \leq a(t) + \int_\alpha^t c(t,s)\,ds + q(t) \int_\alpha^t b(s) \left(a(s) + \int_\alpha^s c(s,\sigma)\,d\sigma\right)$$

$$\times \exp\left(\int_s^t b(\sigma)q(\sigma)\,d\sigma\right) ds.$$

1.12.4

Let u, f, p, q be nonnegative continuous functions defined for $t \in R_+$ and $c \geq 0$ is a constant. If

$$u(t) \leq c + \int_0^t f(s)[p(s)u(s) + q(s)]\,ds, \quad t \in R_+,$$

then for $t \in R_+$,

$$u(t) \leq \left(c + \int_0^t f(s)q(s)\,ds\right) \exp\left(\int_0^t f(s)p(s)\,ds\right).$$

1.12.5 Pachpatte (1974c)

Let u, f, g and h be nonnegative continuous functions defined for $t \in R_+$. If

$$u(t) \leq f(t) + g(t) \int_t^\infty h(s)u(s)\,ds, \quad t \in R_+,$$

and $\int^\infty f(s)h(s)\,ds < \infty$, $\int^\infty g(s)h(s)\,ds < \infty$, then for $t \in R_+$

$$u(t) \leq f(t) + g(t) \int_t^\infty f(s)h(s) \exp\left(\int_t^s g(\sigma)h(\sigma)\,d\sigma\right) ds.$$

1.12.6 Gamidov (1969)

Let u, a_1, a_2, a_3 and h be nonnegative continuous functions in J, and u is differentiable in J, and suppose

$$u'(t) \leq a_1(t) + a_2(t) \int_\alpha^\beta h(s)u(s)\,ds + a_3(t)u(t),$$

where

$$u(\alpha) = 0, \quad \int_\alpha^\beta h(s) \int_\alpha^s a_2(\tau) \exp\left(\int_\tau^s a_3(\sigma)\,d\sigma\right) d\tau\,ds = A < 1.$$

Then
$$u(t) \leq \int_\alpha^t [a_1(s) + Ka_2(s)] \exp\left(\int_s^t a_3(\sigma)\,d\sigma\right) ds,$$
where
$$K = (1/(1-A)) \int_\alpha^\beta h(s) \int_\alpha^s a_1(\tau) \exp\left(\int_\tau^s a_3(\sigma)\,d\sigma\right) d\tau\,ds.$$

1.12.7 Coppel (1965)

Let u be a bounded continuous function in $J_0 = [\alpha, \infty]$, and suppose
$$u(t) \leq a \exp[-\gamma(t-\alpha)] + \int_\alpha^\infty b \exp(-\gamma|t-s|) u(s)\,ds, \quad t \in J_0,$$
where $a \geq 0$, $b \geq 0$, and $\gamma > 0$ are constants and $b < \gamma/2$. Then
$$u(t) \leq (a/b)(\gamma - \delta) \exp[-\delta(t-\alpha)], \quad t \in J_0,$$
where $\delta = (\gamma^2 - 2b\gamma)^{1/2}$.

1.12.8 Coppel (1965)

Let u be a continuous function in J, and suppose
$$u(t) \leq a \exp[-\gamma(\beta - t)] + \int_\alpha^\beta b \exp(-\gamma|t-s|) u(s)\,ds, \quad t \in J,$$
where $a \geq 0$, $b \geq 0$, and $\gamma > 0$ are constants and $b < \gamma/2$. Then
$$u(t) \leq (a/b)(\gamma - \delta) \exp[-\delta(\beta - t)], \quad t \in J,$$
where $\delta = (\gamma^2 - 2b\gamma)^{1/2}$.

1.12.9

Let a, b, f, g be nonnegative continuous functions defined for $t \in R_+$. Let $\gamma(t) > 0$ be a decreasing continuous function, for $t \geq \sigma$ and σ sufficiently large, in such a way that

$$\beta = \int_\sigma^\infty a(s)\,ds + \int_\sigma^\infty f(s)\,ds < 1.$$

Suppose that u is a nonnegative continuous function such that γu is bounded and

$$u(t) \leq k + \int_\sigma^t [a(s)u(s) + b(s)]\,ds + [1/\gamma(t)]\int_t^\infty \gamma(s)[f(s)u(s) + g(s)]\,ds,$$

for $t \geq \sigma$, where $k \geq 0$ is a constant. Then, for $t \in R_+$,

$$u(t) \leq (M/(1-\beta))\exp\left(1/(1-\beta)\int_t^\infty f(s)\,ds\right),$$

where

$$M = k + \int_\sigma^\infty b(s)\,ds + \int_\sigma^\infty g(s)\,ds.$$

1.12.10 Mahmudov and Musaev (1969)

Let the function u be positive and continuous on $[0, T]$ and satisfy the inequality

$$u(t) \leq f(t) + \int_0^t [\phi_1(s)u(s) + \psi_1(t,s)]\,ds + \int_0^T [\phi_2(s)u(s) + \psi_2(t,s)]\,ds,$$

where $\phi_i(t) \geq 0$, $\psi_i(t,s) \geq 0$, $i = 1, 2$, and $f(t) \geq 0$ are continuous functions and

$$Q(t) = \left[1 + \exp\left(\int_0^T \phi_1(s)\,ds\right)\int_0^T \phi_1(s)\,ds\right]\int_0^T \phi_2(s)\,ds < 1.$$

Then

$$u(t) \leq (1/(1-Q(t)))\left[1 + \exp\left(\int_0^T \phi_1(s)\,ds\right)\int_0^T \phi_1(s)\,ds\right]$$
$$\times \int_0^T \phi_2(s)N(s)\,ds + N(t),$$

where

$$N(t) = f(t) + \int_0^T [\psi_1(t,s) + \psi_2(t,s)]\,ds$$

$$+ \exp\left(\int_0^T \phi_1(s)\,ds\right) \int_0^t \phi_1(\tau) f(\tau)\,d\tau$$

$$+ \exp\left(\int_0^T \phi_1(s)\,ds\right) \int_0^t \phi_1(\tau) \left(\int_0^\tau [\psi_1(\tau,s) + \psi_2(\tau,s)]\,ds\right) d\tau.$$

1.12.11

Let u, p and q be nonnegative continuous functions defined on J, $k(t,s)$ be a continuous and nonnegative function on the rectangle $\Box: \alpha \leq s \leq t \leq \beta$ and nondecreasing in t for each $s \in J$. If

$$u(t) \leq p(t) + q(t) \int_\alpha^t k(t,s) u(s)\,ds, \quad t \in J,$$

then

$$u(t) \leq p(t) + q(t) \int_\alpha^t k(t,s) p(s) \exp\left(\int_s^t k(t,\sigma) q(\sigma)\,d\sigma\right) ds, \quad t \in J.$$

1.12.12 Hastings (1966)

Suppose that u is an integrable function on $[0, T]$, and for t in this interval,

$$u(t) \leq A(t) + \int_0^t (t-s) B(s) u(s)\,ds,$$

where $B(t)$ is continuous and nonnegative and $A(t)$ is continuous. Let $z(t)$ be any solution to $z'' - B(t)z = 0$ which is positive for all $t \geq 0$. Then for $t \in [0, T]$,

$$u(t) \leq A(t) + z(t) \int_0^t \int_0^s A(\sigma) B(\sigma) z(\sigma)/z^2(s)\,d\sigma\,ds.$$

1.12.13 Ved' (1965)

Let u, b, σ and $k(t, s)$ be nonnegative continuous functions for $\alpha \leq s \leq t \leq \beta$, and suppose that

$$u(t) \leq a_1 + \sigma(t) \left(a_2 + \int_\alpha^t b(s)u(s)\,ds + \int_\alpha^t \left(\int_\alpha^s k(s, \tau)u(\tau)\,d\tau \right) ds \right),$$

for $t \in J$, where $a_1 \geq 0$, $a_2 \geq 0$ are constants. Then

$$u(t) \leq a_2 \exp\left(\int_\alpha^t B(s)\sigma(s)\,ds \right) + \int_\alpha^t a_1 B(s) \exp\left(\int_s^t B(\tau)\sigma(\tau)\,d\tau \right) ds,$$

for $t \in J$, where $B(t) = b(t) + \int_\alpha^t k(t, \tau)\,d\tau$.

1.12.14 Pachpatte (1977a)

Let u, f and g be nonnegative continuous functions defined on R_+, and $n(t)$ be a positive nondecreasing function defined on R_+, and suppose

$$u(t) \leq n(t) + \int_0^t f(s)u(s)\,ds + \int_0^t f(s) \left[\int_0^s f(\tau) \left(\int_0^\tau g(\sigma)u(\sigma)\,d\sigma \right) d\tau \right] ds,$$

for $t \in R_+$, then

$$u(t) \leq n(t) \left\{ 1 + \int_0^t f(s) \left[1 + \int_0^s f(\tau) \exp\left(\int_0^\tau [f(\sigma) + g(\sigma)]\,d\sigma \right) d\tau \right] ds \right\},$$

for $t \in R_+$.

1.12.15 Agarwal and Thandapani (1981)

Let u, p, q, f_{ri}, $i = 1, 2, \ldots, r$ be nonnegative continuous functions defined on R_+, for which the inequality

$$u(t) \leq p(t) + q(t) \sum_{r=1}^n E_r(t, u),$$

holds, where

$$E_r(t, u) = \int_0^t f_{r1}(t_1) \int_0^{t_1} f_{r2}(t_2) \ldots \int_0^{t_{r-1}} f_{rr}(t_r)u(t_r)\,dt_r\,dt_{r-1}\ldots dt_2\,dt_1.$$

Then

$$u(t) \leq p(t) + q(t)\int_0^t \sum_{r=1}^n E'_r(s, p)\exp\left(\int_s^t \sum_{r=1}^n E'_r(\tau, q)\,d\tau\right) ds,$$

where

$$E'_r(t, \phi) = f_{r1}(t)\int_0^t f_{r2}(t_2)\ldots\int_0^{t_{r-1}} f_{rr}(t_r)\phi(t_r)\,dt_r\,dt_{r-1}\ldots dt_2.$$

1.12.16 Pachpatte (1980a)

Let u, u', u'' and a be nonnegative continuous functions defined on R_+, for which the inequality

$$u''(t) \leq u(0) + u'(0) + \int_0^t a(s)(u(s) + u'(s) + u''(s))\,ds,$$

holds for $t \in R_+$. Then

$$u''(t) \leq u(0) + u'(0) + \int_0^t a(s)\left[2(u(0) + u'(0))\exp\left(\int_0^s a(\tau)\,d\tau\right)\right.$$

$$+ \int_0^s (u'(0) + 2[u(0) + u'(0)])\exp\left(\int_0^\tau [2 + a(\sigma)]\,d\sigma\right)$$

$$\left. \times \exp\left(\int_\tau^s a(\sigma)\,d\sigma\right) d\tau \right] ds,$$

for $t \in R_+$.

1.12.17 Thandapani and Agarwal (1980)

Let u, $u^{(k)}$, h_k, $k = 0, 1, 2, \ldots$, and p, q be nonnegative continuous functions defined on R_+. Suppose that the following inequality

$$u^{(k)}(t) \leq p(t) + q(t) \sum_{j=0}^{k} \int_0^t h_j(s) u^{(j)}(s) \, ds, \quad t \in R_+.$$

Then

$$u^{(k)}(t) \leq p(t) + q(t) \int_0^t \phi_1(s) \exp\left(\int_s^t \phi_2(\tau) \, d\tau\right) ds,$$

where

$$\phi_1(t) = p(t) h_k(t) + \sum_{j=0}^{k-1} \sum_{i=0}^{j} u^{(j)}(0) h_i(t) \frac{t^{j-i}}{(j-i)!}$$

$$+ \sum_{j=0}^{k-1} \frac{h_{k-j-1}(t)}{(j)!} \int_0^t (t-x)^j p(x) \, dx,$$

$$\phi_2(t) = q(t) h_k(t) + \sum_{j=0}^{k-1} \frac{h_{k-j-1}(t)}{(j)!} \int_0^t (t-x)^j q(x) \, dx.$$

1.12.18 Pachpatte (1981a)

Let u, u', $\ldots$, $u^{(n)}$, a, b and c be nonnegative continuous functions defined on R_+, for which the inequality

$$u^{(n)}(t) \leq a(t) + b(t) \left(\sum_{j=0}^{n-1} u^{(j)}(t) + \int_0^t c(s) \sum_{j=0}^{n-1} u^{(j)}(s) \, ds \right),$$

holds for $t \in R_+$. Then

$$u^{(n)}(t) \leq a(t) + b(t) \left[\left(\sum_{j=0}^{n-1} u^{(j)}(0) \right) \exp\left(\int_0^t [1 + b(s) + c(s)] \, ds \right) \right.$$

$$\left. + \int_0^t a(s) \exp\left(\int_s^t [1 + b(\tau) + c(\tau)] \, d\tau \right) ds \right],$$

for $t \in R_+$.

1.12.19 Young (1985)

Let u, b and a_j be nonnegative continuous functions for $0 \leq t \leq T < \infty$, and $a_j \not\equiv 0$, $j = 0, 1, 2, \ldots, n$. If u satisfies

$$u(t) \leq b(t) + \sum_{j=0}^{n} \int_0^t a_0(t_1) \int_0^{t_1} a_1(t_2) \ldots$$

$$\int_0^{t_j} a_j(t_{j+1}) u(t_{j+1}) \, dt_{j+1} \ldots dt_2 \, dt_1,$$

then

$$u(t) \leq b(t) + \int_0^t a_0(s)[b(s) + v_2(s)] \, ds,$$

where v_2 is determined recursively from

$$v_k(t) \leq w_{k-2}^{-1}(t) \int_0^t [(a_0 + \ldots + a_{k-1})b + a_{k-1} v_{k+1}] w_{k-2}(s) \, ds,$$

for $k = n, n - 1, \ldots, 2$, with

$$w_j(t) = \exp\left(-\int_0^t (a_0 + \ldots + a_j) \, ds\right), \quad j = 0, 1, 2, \ldots, n,$$

and

$$v_{n+1}(t) \leq w_n^{-1}(t) \int_0^t (a_0 + \ldots + a_n) b w_n \, ds.$$

1.12.20 Jones (1964)

Let u, a, b, k be real-valued functions defined on J, which are either continuous or of bounded variation on J. Let b, k be nonnegative and let η be a nondecreasing, continuous function on J. If

$$u(t) \leq a(t) + b(t) \int_\alpha^t k(s) u(s) \, d\eta(s), \quad t \in J,$$

then

$$u(t) \leq a(t) + b(t) \int_\alpha^t a(s)k(s) \exp\left(\int_s^t b(\tau)k(\tau)\,\mathrm{d}\eta(\tau)\right) \mathrm{d}\eta(s), \quad t \in J.$$

1.13 Notes

The results given in Section 1.2 centre on the basic inequalities used in the theory of differential and integral equations and can be found in many standard books on differential and integral equations. Theorem 1.2.1 and its proof are taken from the original paper of Gronwall (1919). The special version of Theorem 1.2.1 with $a = 0$ appeared first in the work of Peano (1885–86). Theorem 1.2.2 is due to Bellman (1943).

Section 1.3 contains some useful generalizations and variants of the results given in Section 1.2. Theorem 1.3.1 is due to Bellman (1958), given while studying the asymptotic behaviour of the solutions of certain linear differential-difference equations. The results given in Theorems 1.3.2 and 1.3.5 are taken from Gollwitzer (1969). Theorem 1.3.3 is given by Pachpatte (1975a) and Theorem 1.3.4 is a further generalization of the inequality given by Chandirov (1956).

Section 1.4 deals with the Volterra-type inequalities. Theorem 1.4.1 is due to Chu and Metcalf (1967). Theorem 1.4.2 is a slight variant of the inequality given by Norbury and Stuart (1987). Theorem 1.4.3 is adapted from a recent paper of Pachpatte (in press, k). Theorem 1.4.4 is a fundamental result used in the theory of linear Volterra integral equations and can be found in Tricomi (1957). Theorem 1.4.5 is taken from a paper of Thompson (1971).

The results given in Sections 1.5 and 1.6 are mainly developed in order to apply them in certain applications. Theorems 1.5.1 and 1.5.2 are taken from Gamidov (1969) and Theorems 1.5.3 and 1.5.4 are taken from Rodrigues (1980). Theorem 1.6.1 is first proved by Greene (1977). Theorem 1.6.2 is a further generalization of Theorem 1.6.1 and is given by Pachpatte (1984).

Section 1.7 is dedicated to the inequality first investigated by Pachpatte (1973a) and its subsequent generalizations and variants given by Pachpatte (1974a, 1975b–d, 1977a) which play a fundamental role in the study of various problems in the theory of general classes of integral and integro-differential equations.

Section 1.8 deals with the inequalities involving functions and their derivatives established by Pachpatte (1977b, 1978a, 1982a) which serve a great purpose in certain applications. Section 1.9 is devoted to the inequalities established by Pachpatte (1988a, 1995c, in press, e,j) which are basically motivated by the idea of applying such inequalities to study the behaviour of solutions of certain higher order differential and integro-differential equations.

The results in Section 1.10 are due to Helton (1977), which in turn are the further generalizations of Pachpatte's inequality given in Theorem 1.7.1, involving product integrals. Section 1.11 contains applications of some of the inequalities given in earlier sections. Section 1.12 deals with the miscellaneous inequalities due to various investigators which can also be used in certain applications.

Chapter Two

Nonlinear Integral Inequalities I

2.1 Introduction

Nonlinear differential equations whose solutions cannot be found explicitly arise in essentially every branch of modern science, engineering and mathematics. One of the most useful methods available for studying a nonlinear system of ordinary differential equations is to compare it with a single first-order equation derived naturally from an estimate on a system. However, the bounds provided by the comparison method are sometimes difficult or impossible to calculate explicitly. In fact, in many applications explicit bounds are more useful while studying the behaviour of solutions of such systems. Another basic tool, which is typical among investigations on this subject, is the use of nonlinear integral inequalities which provide explicit bounds on the unknown functions. Over the last 40 years several new nonlinear integral inequalities have been developed in order to study the behaviour of solutions of such systems. This chapter considers various nonlinear integral inequalities discovered in the literature, which are very effective in studying the behaviour of solutions of systems of differential, integral and integro-differential equations. Applications of some of the inequalities are also presented, and some miscellaneous inequalities which can be used in certain applications are given.

2.2 Inequalities Involving Comparison

This section gives some basic comparison results related to ordinary differential equations which may be found in many classical books. These results will show that the comparison method provides not only a powerful tool for obtaining bounds for solutions, but also a unified approach to many such problems. To begin with, some basic definitions and preliminary results from Bainov and Simeonov (1992) and Beesack (1975) are collected.

Consider the initial value problem

$$x' = F(t, x), \quad x(\alpha) = x_0, \qquad (2.2.1)$$

where F is a continuous function on an open set D containing the point (α, x_0). There exist $\beta > \alpha$ and $b > 0$ such that the rectangle $E = \{(t, x): \alpha \leq t \leq \beta, |x - x_0| \leq b\}$ is contained in D.

A differentiable function x defined in an interval $\alpha \leq t \leq \beta_x$ is said to be a *solution* of (2.2.1) if $(t, x(t)) \in D$ for $\alpha \leq t \leq \beta_x$ and $x'(t) = F(t, x(t))$, $\alpha \leq t \leq \beta_x$, $x(\alpha) = x_0$. A solution r is called a *maximal (minimal)* solution of (2.2.1) on the interval $[\alpha, \beta_r]$ if $x(t) \leq r(t)$ ($x(t) \geq r(t)$) for $\alpha \leq t \leq \min(\beta_x, \beta_r)$ holds for every solution x of (2.2.1).

We shall start with the following theorem, which deals with the existence of a maximal and minimal solution of equation (2.2.1).

Theorem 2.2.1 *If the function F is continuous in the rectangle E, then equation (2.2.1) has a maximal and minimal solution. If $|F(t, x)| \leq M$ for $(t, x) \in E$ and $M > 0$ is a constant, then these solutions, as well as all others, are defined at least on the interval $[\alpha, \beta_1]$, where $\beta_1 - \alpha = \min(\beta - \alpha, bM^{-1})$. If $[\alpha, \beta_r)$ is the largest domain of existence of the maximal (minimal) solution r of (2.2.1), then there exists $\epsilon_0 > 0$ such that, for $0 \leq \epsilon < \epsilon_0$, the maximal (minimal) solution of the system*

$$x' = F(t, x) \pm \epsilon, \quad x(\alpha) = x_0, \qquad (2.2.2)$$

also exists on $[\alpha, \beta_r)$, where $+$ or $-$ holds according as r is maximal or minimal.

□

This theorem has been proved by various authors, and the reader is referred to the standard references by Hille (1969, p. 46), Reid (1971,

Theorem 4.1), Szarski (1965, Theorem 8.1) and Lakshmikantham and Leela (1969, Theorem 1.3.1).

Before giving the main comparison theorem, the following lemmas, which are of independent interest are presented.

Lemma 2.2.1 *Let $F(t, x)$ be continuous in an open set D containing a point (α, x_0), and let r be any solution of (2.2.1) for $t \in [\alpha, \beta_1]$. If x satisfies the stronger condition*

$$x'(t) < F(t, x(t)), \qquad \alpha \leq t \leq \beta_1, \qquad x(\alpha) < x_0, \qquad (2.2.3)$$

then

$$x(t) < r(t), \quad \alpha \leq t < \beta_1. \qquad (2.2.4)$$

The result remains valid if $<$ is replaced by $>$ in both (2.2.3) and (2.2.4).

△

Lemma 2.2.2 *Let $F(t, x)$ be continuous in an open set D containing a point (α, x_0), and let $\{\epsilon_n\}$ be a strictly monotonic sequence with $\lim_{n \to \infty} \epsilon_n = 0$. Consider the initial value problems*

$$y' = F(t, y) + \epsilon_n, \quad y(\alpha) = x_0 + \epsilon_n. \qquad (2.2.5)$$

Let $y = r_n(t)$ be the maximal or the minimal solution of (2.2.5) on $[\alpha, \beta_1]$ according as $\epsilon_n > 0$ or $\epsilon_n < 0$. Then the sequence $\{r_n\}$ converges uniformly on $[\alpha, \beta_1]$ to a function r which is the maximal solution of (2.2.1) if $\epsilon_n > 0$, or the minimal solution of (2.2.1) if $\epsilon_n < 0$.

△

The basic comparison theorem used in the theory of differential equations can now be formulated.

Theorem 2.2.2 *Let $F(t, x)$ be continuous in an open set D containing a point (α, x_0), and suppose that the initial value problem*

$$r' = F(t, r), \quad r(\alpha) = x_0, \qquad (2.2.6)$$

has a maximal solution $r = r(t)$ with domain $\alpha \leq t \leq \beta_1$. If x is any differentiable function on $[\alpha, \beta_1]$ such that $(t, x(t)) \in D$ for $t \in [\alpha, \beta_1]$ and

$$x'(t) \leq F(t, x(t)), \qquad \alpha \leq t \leq \beta_1, \qquad x(\alpha) \leq x_0, \qquad (2.2.7)$$

then

$$x(t) \leq r(t), \quad \alpha \leq t \leq \beta_1. \qquad (2.2.8)$$

Moreover, the result remains valid if 'maximal' is replaced by 'minimal' and $\leq$ is replaced by $\geq$ in both (2.2.7) and (2.2.8).

$\square$

The proof of this theorem, which depends on the above two lemmas, may be found in Szarski (1965, Theorem 9.5, p. 27), Lakshmikantham and Leela (1969, Theorem 1.4.1, p. 15), Walter (1970, Theorem 10, p. 68), Beesack (1975, Theorem 6.2, p. 37) and Bainov and Simeonov (1992, Theorem 6.3, p. 57). Since the proofs of Lemmas 2.2.1 and 2.2.2 and Theorem 2.2.2 are essentially contained in the above references, the explicit proofs will not be given here.

Brauer (1963) obtained bounds for solutions of a system of ordinary differential equation of the form

$$x' = f(t, x), \tag{P}$$

where x and f are n-dimensional vectors and $0 \leq t < \infty$. The symbol $|\cdot|$ will be used to denote any convenient vector norm. It is assumed that $f(t, x)$ is continuous for $0 \leq t < \infty$, $|x| < \infty$, but no assumptions are required on f to ensure the uniqueness of solutions of (P), as the arguments do not require uniqueness.

The following comparison theorem is well known and used considerably in the literature (Bainov and Simeonov, 1992; Beesack, 1975; Brauer, 1963; Brauer and Wong, 1969; Lakshmikantham and Leela, 1969; Szarski, 1965; Walter, 1970).

Theorem 2.2.3 *Suppose that there exists a continuous nonnegative function $w(t, r)$ on $0 \leq t < \infty$, $0 \leq r < \infty$, such that*

$$|f(t, x)| \leq w(t, |x|), \qquad 0 \leq t < \infty, \qquad |x| < \infty. \tag{2.2.9}$$

(i) Let $x(t)$ be a solution of (P) and $r(t)$ be the maximum solution of the scalar equation

$$r' = w(t, r), \tag{2.2.10}$$

with $r(0) = |x(0)|$, then $x(t)$ can be continued to the right as far as $r(t)$ exists, and

$$|x(t)| \leq r(t), \tag{2.2.11}$$

for all such t.

(ii) Let $x(t)$ be a solution of (P) and let $u(t)$ be the minimum solution of

$$u' = -w(t, u), \qquad (2.2.12)$$

with $u(0) = |x(0)|$. Then, for all $t \geq 0$ such that $x(t)$ exists and $u(t) \geq 0$, we have

$$|x(t)| \geq u(t). \qquad (2.2.13)$$

□

Theorem 2.2.3 involves a comparison between solutions of the system (P) and the scalar equations (2.2.10) and (2.2.12). As noted in (Brauer, 1963, p. 36), the bounds given in (2.2.11) and (2.2.13) are sometimes difficult or impossible to calculate explicitly. However, the comparison method provides not only a powerful tool for obtaining bounds for solutions, but also a unified approach to many such problems.

Brauer (1963) gave explicit bounds when the function $w(t, r)$ has the form $\lambda(t)\phi(r)$. In this case the condition (2.2.9) is replaced by

$$|f(t, x)| \leq \lambda(t)\phi(|x|). \qquad (2.2.14)$$

Then the equations (2.2.10) and (2.2.12) lead us to consider the separable first order equations

$$r' = \lambda(t)\phi(r), \qquad (2.2.15)$$

and

$$u' = -\lambda(t)\phi(u), \qquad (2.2.16)$$

which can be solved explicitly. We define $J(r) = \int_0^r ds/\phi(s)$. If this integral diverges at zero, the lower limit of integration can be replaced by any fixed $\epsilon > 0$, but we will use the lower limit zero for convenience. If $\int_0^\infty ds/\phi(s) = R \leq \infty$, the function J maps the half line $[0, \infty)$ onto the interval $[0, R)$, with $J(\infty) = R$. The function J has a positive derivative $1/\phi(r)$, and is therefore monotonic increasing. Thus the inverse function J^{-1} exists and is monotonic increasing on $[0, R)$.

If $r(t)$ is a solution of (2.2.15) with $r(0) = r_0$, then

$$\int_0^t \lambda(s)\, ds = \int_{r_0}^r \frac{ds}{\phi(s)} = J(r) - J(r_0).$$

Thus

$$J(r) = J(r_0) + \int_0^t \lambda(s)\,ds,$$

and

$$r(t) = J^{-1}\left(J(r_0) + \int_0^t \lambda(s)\,ds\right). \tag{2.2.17}$$

This solution exists as long as $J(r_0) + \int_0^t \lambda(s)\,ds$ is in the domain $[0, R)$ of J^{-1}. This requires $J(r_0) + \int_0^t \lambda(s)\,ds < R$ or $\int_0^t \lambda(s)\,ds < \int_{r_0}^\infty ds/\phi(s)$. It follows that the solution $r(t)$ of (2.2.15) with $r(0) = r_0$ exists on the interval $[0, T)$, where T is defined by $\int_0^T \lambda(s)\,ds = \int_{r_0}^\infty ds/\phi(s)$. If $\int_0^\infty ds/\phi(s) = \infty$, the solution exists on $0 \le t < \infty$. If $\int_0^\infty \lambda(s)\,ds < \int_{r_0}^\infty ds/\phi(s) \le \infty$, then $J(r_0) + \int_0^t \lambda(s)\,ds < R$, which implies $r(t) < \infty$, and the solution $r(t)$ remains bounded on $0 \le t < \infty$. Since $r(t)$ is monotonic increasing, this implies that $r(t)$ tends to a limit as $t \to \infty$.

Equation (2.2.16) can be solved by the same approach. It is found that the solution $u(t)$ of (2.2.16) with $u(0) = r_0$ is given by

$$u(t) = J^{-1}\left[J(r_0) - \int_0^t \lambda(s)\,ds\right],$$

defined so long as $J(r_0) - \int_0^t \lambda(s)\,ds \ge 0$. Thus $u(t)$ exists on the interval $[0, \tau]$, where $\int_0^\tau \lambda(s)\,ds = \int_0^{r_0} ds/\phi(s)$. If $\int_0^\infty \lambda(s)\,ds \le \int_0^{r_0} ds/\phi(s)$, $u(t)$ exists on $0 \le t < \infty$, and if $\int_0^\infty \lambda(s)\,ds < \int_0^{r_0} ds/\phi(s)$, $u(t) > 0$ on $0 \le t < \infty$. To avoid misinterpretation, recall that the convergence of the integral $\int_0^{r_0} ds/\phi(s)$ has been assumed.

A great many papers have been written on the various generalizations and extensions of the above given comparison theorems. In the following theorem we present a comparison theorem given by Opial (1960).

Theorem 2.2.4 *Let $J = [\alpha, \beta)$, and $f: J \times R^n \to R^n$ be continuous and satisfy $f(t, x) \le f(t, y)$ for $t \in J$ and $x, y \in R^n$ with $x \le y$. If a continuous map $x: J \to R^n$ satisfies*

$$x(t) \le c + \int_\alpha^t f(s, x(s))\,ds, \quad t \in J, \tag{2.2.18}$$

for some $c \in R^n$, and if $\psi: [\alpha, \beta_1) = J_1 \to R^n$ (where $\beta_1 \leq \beta$) is the maximal solution of the system

$$\psi(t) = c + \int_0^t f(s, \psi(s)) \, ds, \qquad (2.2.19)$$

then $x(t) \leq \psi(t)$ for $t \in J_1$.

□

The proof of this theorem is given in Beesack (1975, Theorem 4.1); see also Lakshmikantham and Leela (1969), Olech (1967), Sato (1953) and Walter (1970). To avoid repetition it is omitted here.

Another interesting and useful comparison theorem established by Lasota et al. (1971) is given in the following theorem.

Theorem 2.2.5 *Let $y, z, g: [0, T] \to R^n$, $k: [0, T] \times R^n \to R^n$, where R^n denotes the n-dimensional vector space with any convenient vector norm denoted by $|\cdot|$, and suppose*

$$y(t) \leq g(t) + \int_0^t k(s, y(s)) \, ds,$$

$$z(t) \geq g(t) + \int_0^t k(s, z(s)) \, ds,$$

for all $0 \leq t \leq T$. Assume k satisfies

$$|k(t, y) - k(t, \bar{y})| \leq a(t)|y - \bar{y}|,$$

for all y and $\bar{y}$ in R^n, where a is integrable on $0 \leq t \leq T$. Assume further that $k(t, y)$ is monotonic increasing in y. Then $y(t) \leq z(t)$ for all $0 \leq t \leq T$.

□

Proof: First, note that, under the above assumptions, the equation

$$u(t) = g(t) + \int_0^t k(s, u(s)) \, ds, \qquad (2.2.20)$$

has exactly one continuous solution on $0 \leq t \leq T$, and the sequence of successive approximations, obtained from any continuous zero approximation, converges uniformly to $u(t)$. A simple proof of this well-known fact runs as follows.

The space of continuous functions $u(t)$ from $[0, T]$ to R^n equipped with the norm
$$|u|_0 = \max\{|u(t)|\exp(-A(t)): 0 \le t \le T\},$$
where $A(t) = \int_0^t a(s)\,ds$ is a Banach space B. The map G defined by
$$(Gu)(t) = g(t) + \int_0^t k(s, u(s))\,ds$$
maps B into itself and is a contraction. Indeed,
$$|(Gu)(t) - (Gv)(t)|\exp(-A(t))$$
$$\le \exp(-A(t))\int_0^t |k(s, u(s)) - k(s, v(s))|\,ds$$
$$\le \exp(-A(t))|u - v|_0 \int_0^t a(s)\exp(A(s))\,ds$$
$$\le |u - v|_0(1 - \exp(-A(t))).$$
Setting $q = 1 - \exp(-A(t)) < 1$, we obtain
$$|Gu - Gv|_0 \le q|u - v|_0,$$
and the assertions stated above follow from the contraction fixed point theorem (Miller, 1971).

Second, note that, owing to the monotonicity of $k(t, y)$ the operator G is monotonic increasing, i.e. $u(t) \le v(t)$ on $[0, T]$ implies $(Gu)(t) \le (Gv)(t)$ on $[0, T]$. If we solve (2.2.20) by successive approximations starting with the zero approximation $v_0(t) = y(t)$, then $v_1 = Gv_0 \ge v_0$. Furthermore, the sequence $\{v_n\}$ defined by $v_{n+1} = Gv_n$ is increasing, since $v_{n-1} \le v_n$ implies $Gv_{n-1} \le Gv_n$, i.e. $v_n \le v_{n+1}$. Thus $v_n \uparrow u$, the solution of (2.2.20). Similarly, the sequence $\{w_n\}$, defined by $w_0(t) = z(t)$ and $w_{n+1} = Gw_n$, satisfies $w_n \downarrow u$. Thus
$$y(t) = v_0(t) \le u(t) \le w_0(t) = z(t),$$
for $0 \le t \le T$. This completes the proof. ∎

The comparison method plays a vital role in the investigations of existence, uniqueness, stability, boundedness and other properties of the solutions of various classes of differential and integral equations. The interested reader is referred to the recent results given in Constantin (1991) and Sugiyama (1976) and to Lakshmikantham and Leela (1969), Szarski (1965) and Walter (1970), which contain many good references on these topics.

2.3 The Inequalities of Bihari and Langenhop

The inequalities which furnish explicit bounds on unknown functions are more useful in studying many qualitative as well as quantitative properties of solutions of differential and integral equations. This section deals with some useful nonlinear integral inequalities which provide explicit bounds on unknown functions.

Bihari (1956) proved the following useful nonlinear generalization of Gronwall–Bellman inequality.

Theorem 2.3.1 *Let u and f be nonnegative continuous functions defined on R_+. Let $w(u)$ be a continuous nondecreasing function defined on R_+ and $w(u) > 0$ on $(0, \infty)$. If*

$$u(t) \leq k + \int_0^t f(s) w(u(s)) \, ds, \qquad (2.3.1)$$

for $t \in R_+$, where k is a nonnegative constant, then for $0 \leq t \leq t_1$,

$$u(t) \leq G^{-1}\left(G(k) + \int_0^t f(s) \, ds \right), \qquad (2.3.2)$$

where

$$G(r) = \int_{r_0}^r \frac{ds}{w(s)}, \quad r > 0, r_0 > 0, \qquad (2.3.3)$$

and G^{-1} is the inverse function of G and $t_1 \in R_+$ is chosen so that

$$G(k) + \int_0^t f(s) \, ds \in \text{Dom}(G^{-1}),$$

for all $t \in R_+$ lying in the interval $0 \leq t \leq t_1$.

Proof: We first assume that $k > 0$ and define a function $z(t)$ by the right-hand side of (2.3.1). Then we have $z(0) = k$ and $u(t) \leq z(t)$ and

$$z'(t) = f(t)w(u(t)) \leq f(t)w(z(t)), \qquad (2.3.4)$$

From (2.3.3) and (2.3.4) we have

$$\frac{d}{dt}G(z(t)) = \frac{z'(t)}{w(z(t))} \leq f(t). \qquad (2.3.5)$$

Now by setting $t = s$ in (2.3.5) and integrating it from 0 to t we have

$$G(z(t)) \leq G(k) + \int_0^t f(s)\,ds. \qquad (2.3.6)$$

Since $G^{-1}(z)$ is increasing, from (2.3.6) we have

$$z(t) \leq G^{-1}\left(G(k) + \int_0^t f(s)\,ds\right). \qquad (2.3.7)$$

Using (2.3.7) in $u(t) \leq z(t)$ gives the required inequality in (2.3.2).

If k is nonnegative, we carry out the above procedure with $k + \epsilon$ instead of k, where $\epsilon > 0$ is an arbitrary small constant, and subsequently pass to the limit as $\epsilon \to 0$ to obtain (2.3.2). The subinterval $0 \leq t \leq t_1$ is obvious.

∎

The inequality given in Theorem 2.3.1 is now known in the literature as Bihari's inequality. However, the above inequality seems to have appeared first in the work of LaSalle (1949). The case $k = 0$ of Bihari's inequality was dealt with in LaSalle (1949). Both LaSalle (1949) and Bihari (1956) applied their results to uniqueness questions for differential equations.

Remark 2.3.1 The choice of the point $r_0 \in R_+$ does not affect the final result (2.3.2). If $\bar{r}_0 \in R_+$ and

$$\bar{G}(r) = \int_{\bar{r}_0}^r \frac{ds}{w(s)}, \qquad r \in R_+,$$

then $\overline{G}(r) = G(r) - G(\overline{r}_0)$, so $\overline{G}^{-1}(v) = G^{-1}(v + G(\overline{r}_0))$. Consequently,

$$\overline{G}^{-1}\left(\overline{G}(k) + \int_0^t f(s)\,ds\right) = G^{-1}\left(G(k) + \int_0^t f(s)\,ds\right),$$

so that the bound in (2.3.2) is independent of the choice of the point $r_0 \in R_+$.

△

Remark 2.3.2 In some cases we may have $w(0) = 0$ at the end-point of the interval, say $(0, \infty)$ with w continuous on $[0, \infty)$ and $w > 0$ on $(0, \infty)$. To be specific, suppose w is also nondecreasing in $(0, \infty)$ and $f \geq 0$ on R_+. Then Theorem 2.3.1 is valid for functions $u \geq 0$ continuous on R_+, even if $u(\bar{t}) = 0$ for some $\bar{t} \in R_+$. To see this, let $\epsilon > 0$ and $u_\epsilon(t) = u(t) + \epsilon > 0$ for all $t \in R_+$. Then

$$u_\epsilon(t) \leq (k + \epsilon) + \int_0^t f(s)w(u_\epsilon(s) - \epsilon)\,ds$$

$$\leq (k + \epsilon) + \int_0^t f(s)w(u_\epsilon(s))\,ds,$$

and it follows that

$$u_\epsilon(t) \leq G^{-1}\left[G(k + \epsilon) + \int_0^t f(s)\,ds\right].$$

Letting $\epsilon \to 0$, we obtain (2.3.2).

△

Remark 2.3.3 Theorem 2.3.1 remains valid if $\leq$ is replaced by $\geq$ in both (2.3.1) and (2.3.2). Taking $w(u) = u$, $G(u) = \log u$, we obtain Bellman's inequality, given in Theorem 1.2.2.

△

Langenhop (1960) proved the following inequality, which provides an explicit lower bound for an unknown function.

Theorem 2.3.2 *Let u, f and w be as defined in Theorem 2.3.1 and*

$$u(t) \geq u(\sigma) - \int_\sigma^t f(s)w(u(s))\,ds, \quad 0 \leq \sigma \leq t < \infty, \qquad (2.3.8)$$

then

$$u(t) \geq G^{-1}\left(G(u(\sigma)) - \int_\sigma^t f(s)\,ds\right), \quad 0 \leq \sigma \leq t \leq t_2, \quad (2.3.9)$$

where G, G^{-1} are as defined in Theorem 2.3.1 and $t_2 \in R_+$ is chosen so that

$$G(u(\sigma)) - \int_\sigma^t f(s)\,ds \in \mathrm{Dom}(G^{-1}),$$

for all $t \in R_+$ lying in $0 \leq \sigma \leq t \leq t_2$.

□

Proof: We write (2.3.8) in the form

$$u(\sigma) \leq u(t) + \int_\sigma^t f(s)w(u(s))\,ds, \quad 0 \leq \sigma \leq t < \infty,$$

and let $t \in R_+$ be fixed. Continuing with the function $u(\sigma)$, $\sigma \in R_+$ and proceeding as in the proof of Theorem 2.3.1, we get the desired inequality in (2.3.9). We omit the details.

■

We note that the above result is also valid if $\geq$ is replaced by $\leq$ in both (2.3.8) and (2.3.9). Moreover, as pointed out by Beesack (1975, p. 14), the inequality (2.3.8) implies (2.3.9) even when w is not monotonic nondecreasing on R_+. This follows from a general result given by Walter (1970, p. 44); see also Redheffer (1964).

Constantin (1990a,b) obtained the following interesting and useful inequalities, similar to those of Bihari and Langenhop, which in fact are motivated by the inequality established by Morro (1982).

Theorem 2.3.3 *Let g, h and k be continuous functions on the interval $[t_0, t_1]$, $t_0, t_1 \in R$, $t_1 > t_0$, $g > 0$, $kh > 0$ and $k, g \in C^1([t_0, t_1], R)$, $w: R_+ \to R_+$ a continuous, monotonic nondecreasing function so that there exists $L > 0$ with $w(x) \geq Lx$ for each $x \in R_+$ and $w(0) = 0$. If a continuous positive function u has the property that*

$$u(t) \leq g(t) + k(t) \int_{t_0}^t h(s)w(u(s))\,ds, \quad t \in [t_0, t_1], \quad (2.3.10)$$

then

$$u(t) \leq G^{-1}\left[G(g(t_0)) + \int_{t_0}^{t}\left[k(s)h(s) + \max\left\{0, \frac{k'(s)}{Lk(s)}\right\}\right.\right.$$

$$\left.\left. + \max\left\{0, \frac{g'(s)}{Lg(s)}\right\}\right]ds\right], \quad (2.3.11)$$

for $t \in [t_0, t']$, where G is defined by (2.3.3) and t' is defined so that the existence condition of the right part of (2.3.11) should be assured. □

Proof: We denote by y the function

$$y(t) = \int_{t_0}^{t} h(s)w(u(s))\,ds,$$

and from (2.3.10) we obtain that $u(t) \leq g(t) + k(t)y(t)$ so that

$$k(t)y'(t) = k(t)h(t)w(u(t))$$
$$\leq k(t)h(t)w(g(t) + k(t)y(t)),$$

which implies that

$$\frac{k(t)y'(t)}{w(g(t) + k(t)y(t))} \leq k(t)h(t), \quad (2.3.12)$$

since $kh > 0$ and w is monotonic nondecreasing.

We observe that because $kh > 0$ we have $ky > 0$ for $t > t_0$ and so

$$g'(t) \leq 0 \Rightarrow \frac{g'(t)}{w(g(t) + k(t)y(t))} \leq 0,$$

$$g'(t) > 0 \Rightarrow \frac{g'(t)}{w(g(t) + k(t)y(t))} \leq \frac{g'(t)}{L(g(t) + k(t)y(t))} \leq \frac{g'(t)}{Lg(t)}.$$

Hence we have that

$$\frac{g'(t)}{w(g(t) + k(t)y(t))} \leq \max\left\{0, \frac{g'(t)}{Lg(t)}\right\}. \quad (2.3.13)$$

It is now obvious that for $t > t_0$

$$k'(t) \leq 0, \quad y(t) > 0 \Rightarrow \frac{y(t)k'(t)}{w(g(t)+k(t)y(t))} \leq 0,$$

$$k'(t) > 0, \quad y(t) > 0 \Rightarrow \frac{y(t)k'(t)}{w(g(t)+k(t)y(t))} \leq \frac{y(t)k'(t)}{L(g(t)+k(t)y(t))}$$

$$\leq \frac{y(t)k'(t)}{Lk(t)y(t)} = \frac{k'(t)}{Lk(t)},$$

$$k'(t) > 0, \quad y(t) < 0 \Rightarrow \frac{y(t)k'(t)}{w(g(t)+k(t)y(t))} \leq 0,$$

$$k'(t) \leq 0, \quad y(t) < 0 \Rightarrow \frac{y(t)k'(t)}{w(g(t)+k(t)y(t))} \leq \frac{y(t)k'(t)}{L(g(t)+k(t)y(t))}$$

$$\leq \frac{y(t)k'(t)}{Lk(t)y(t)} = \frac{k'(t)}{Lk(t)}$$

and hence we obtain for $t > t_0$

$$\frac{y(t)k'(t)}{w(g(t)+k(t)y(t))} \leq \max\left\{0, \frac{k'(t)}{Lk(t)}\right\}. \tag{2.3.14}$$

From (2.3.12), (2.3.13), (2.3.14) we conclude that

$$\frac{g'(t) + y'(t)k(t) + k'(t)y(t)}{w(g(t)+k(t)y(t))} \leq k(t)h(t) + \max\left\{0, \frac{g'(t)}{Lg(t)}\right\}$$

$$+ \max\left\{0, \frac{k'(t)}{Lk(t)}\right\}.$$

By integration on $[t_0, t_1]$ this yields

$$\int_{g(t_0)+k(t_0)y(t_0)}^{g(t)+k(t)y(t)} \frac{ds}{w(s)} \leq \int_{t_0}^{t} \left[k(s)h(s) + \max\left\{0, \frac{g'(s)}{Lg(s)}\right\}\right.$$

$$\left. + \max\left\{0, \frac{k'(s)}{Lk(s)}\right\}\right] ds. \tag{2.3.15}$$

From (2.3.3) and (2.3.15) and using the fact that $y(t_0) = 0$, we deduce that

$$G(g(t) + k(t)y(t)) - G(g(t_0)) \leq \int_{t_0}^{t} \left[k(s)h(s) + \max\left\{0, \frac{g'(s)}{Lg(s)}\right\} \right. $$
$$\left. + \max\left\{0, \frac{k'(s)}{Lk(s)}\right\} \right] ds,$$

from which

$$g(t) + k(t)y(t) \leq G^{-1}\left[G(g(t_0)) + \int_{t_0}^{t} \left[k(s)h(s) + \max\left\{0, \frac{g'(s)}{Lg(s)}\right\} \right.\right.$$
$$\left.\left. + \max\left\{0, \frac{k'(s)}{Lk(s)}\right\} \right] ds \right], \quad (2.3.16)$$

and using (2.3.16) in $u(t) \leq g(t) + k(t)y(t)$ the theorem is proved.

∎

Remark 2.3.4 If we take $g(t) = k \geq 0$ is a constant, and the function $k(t) = 1$ in (2.3.10), then the estimate obtained in (2.3.11) coincides with (2.3.2).

△

Theorem 2.3.4 Let $x(t)$ be a solution of the equation

$$x' = f(t, x), \quad t \geq t_0 \geq 0, \quad (2.3.17)$$

where $f(t, x)$ is continuous for $t_0 \leq t < \infty$ and $|x| < \infty$ and $|f(t, x)| \leq a(t)w(|x|) + b(t)$, where a, b are continuous and positive and w is continuous, monotonic, nondecreasing and positive. Then we have that

$$|x(t)| \leq G^{-1}\left[G\left(|x(t_0)| + \int_{t_0}^{t} b(s)\,ds\right) + \int_{t_0}^{t} a(s)\,ds \right], \quad (2.3.18)$$

where G is defined by (2.3.3) and

$$|x(t)| \geq G^{-1}\left[G(|x(t_0)|) - \int_{t_0}^{t} a(s)\,ds \right] - \int_{t_0}^{t} b(s)\,ds, \quad (2.3.19)$$

for all t for which

$$G\left(|x(t_0)| + \int_{t_0}^t b(s)\,ds\right) + \int_{t_0}^t a(s)\,ds,$$

and

$$G(|x(t_0)|) - \int_{t_0}^t a(s)\,ds,$$

is in the domain of G^{-1}.

□

Proof: From (2.3.17) we have that

$$x(t) = x(t_0) + \int_{t_0}^t f(s, x(s))\,ds,$$

from where we obtain that

$$|x(t)| \leq |x(t_0)| + \int_{t_0}^t b(s)\,ds + \int_{t_0}^t a(s)w(|x(s)|)\,ds.$$

We denote by y, g the functions

$$y(t) = \int_{t_0}^t a(s)w(|x(s)|)\,ds, \quad t \geq t_0,$$

$$g(t) = |x(t_0)| + \int_{t_0}^t b(s)\,ds, \quad t \geq t_0,$$

and hence we have that $|x(t)| \leq g(t) + y(t)$ for $t \geq t_0$,

$$y'(t) = a(t)w(|x(t)|) \leq a(t)w(g(t) + y(t)), \quad t \geq t_0,$$

$$g'(t) = b(t), \quad t \geq t_0,$$

from where we deduce that

$$\frac{g'(t) + y'(t)}{w(g(t) + y(t))} \leq a(t) + \frac{b(t)}{w(g(t) + y(t))}$$

$$\leq a(t) + \frac{b(t)}{w(g(t))}, \quad t \geq t_0,$$

and an integration yields

$$\int_{g(t_0)+y(t_0)}^{g(t)+y(t)} \frac{ds}{w(s)} \leq \int_{t_0}^{t} a(s)\,ds + \int_{g(t_0)}^{g(t)} \frac{ds}{w(s)}, \quad t \geq t_0. \tag{2.3.20}$$

From (2.3.3) and (2.3.20), and using the facts that $y(t_0) = 0$, $g(t_0) = |x(t_0)|$, we obtain that

$$G(g(t) + y(t)) - G(|x(t_0)|) \leq \int_{t_0}^{t} a(s)\,ds + G(g(t)) - G(|x(t_0)|)$$

and so

$$G(g(t) + y(t)) \leq G\left(|x(t_0)| + \int_{t_0}^{t} b(s)\,ds\right) + \int_{t_0}^{t} a(s)\,ds. \tag{2.3.21}$$

From (2.3.21) and $|x(t)| \leq g(t) + y(t)$ we get the desired inequality in (2.3.18).

For the second part, from (2.3.17) we observe that for every $t \leq T, t > t_0$ we have that

$$x(T) = x(t) + \int_{t}^{T} f(s, x(s))\,ds,$$

and so we deduce

$$|x(T)| \geq |x(t)| - \int_{t}^{T} b(s)\,ds - \int_{t}^{T} a(s)w(|x(s)|)\,ds.$$

Denoting $|x(t)| = v(t)$ we have

$$v(T) \geq v(t) - \int_{t}^{T} b(s)\,ds - \int_{t}^{T} a(s)w(v(s))\,ds.$$

We define for $t_0 \leq t \leq T$,

$$F(t) = v(T) + \int_{t}^{T} b(s)\,ds + \int_{t}^{T} a(s)w(v(s))\,ds, \quad t_0 \leq t \leq T,$$

and obtain that $F(t) \geq v(t)$, $t_0 \leq t \leq T$ and

$$F'(t) = -b(t) - a(t)w(v(t)), \quad t_0 \leq t \leq T.$$

Because w is monotonic nondecreasing we deduce that

$$F'(t) + b(t) + a(t)w(F(t)) \geq 0, \quad t_0 \leq t \leq T,$$

and so

$$\frac{F'(t)}{w(F(t))} + a(t) \geq -\frac{b(t)}{w(F(t))}, \quad t_0 \leq t \leq T. \tag{2.3.22}$$

and $F(t) \geq v(T) + \int_t^T b(s)\,ds$ we have that

$$w(F(t)) \geq w\left(v(T) + \int_t^T b(s)\,ds\right), \quad t_0 \leq t \leq T. \tag{2.3.23}$$

Using (2.3.23) in (2.3.22) we have

$$\frac{F'(t)}{w(F(t))} + a(t) \geq -\frac{b(t)}{w\left(v(T) + \int_t^T b(s)\,ds\right)}, \quad t_0 \leq t \leq T,$$

and an integration of the above inequality on $[0, T]$ and the use of (2.3.3) yields

$$G(F(T)) - G(F(t)) + \int_t^T a(s)\,ds \geq G(v(T)) - G\left(v(T) + \int_t^T b(s)\,ds\right).$$

Since $v(T) = F(T)$, we obtain that

$$G\left(v(T) + \int_t^T b(s)\,ds\right) \geq G(F(t)) - \int_t^T a(s)\,ds, \quad t_0 \leq t \leq T.$$

Because $F(t) \geq v(t)$, taking $t = t_0$ and using the monotonicity of G we obtain

$$G\left(v(T) + \int_{t_0}^T b(s)\,ds\right) \geq G(v(t_0)) - \int_{t_0}^T a(s)\,ds, \quad t_0 < T,$$

from where

$$v(T) \geq G^{-1}\left[G(v(t_0)) - \int_{t_0}^{T} a(s)\,ds\right] - \int_{t_0}^{T} b(s)\,ds, \quad t_0 < T.$$

Since $v(t) = |x(t)|$ and T is arbitrary, we deduce the inequality (2.3.19).

∎

Remark 2.3.5 If we take $b(t) = 0$ we obtain the class of equations analysed by Langenhop (1960) and Brauer (1963) and the same bounds as given therein.

△

2.4 Generalizations of Gronwall–Bellman–Bihari Inequalities

In the past few years many authors have obtained various generalizations and extensions of Gronwall–Bellman–Bihari inequalities. These generalizations are motivated by specific applications to the theory of differential and integral equations. In this section some useful generalizations of Gronwall–Bellman–Bihari inequalities containing combinations of the inequalities by taking a sum of two integrals will be given, one containing the unknown function in a linear form and the other in a nonlinear form.

Pachpatte (1975a) proved the following integral inequality.

Theorem 2.4.1 *Let u, f, g and h be nonnegative continuous functions defined on R_+. Let $w(u)$ be a continuous nondecreasing and submultiplicative function defined on R_+ and $w(u) > 0$ on $(0, \infty)$. If*

$$u(t) \leq u_0 + g(t)\int_0^t f(s)u(s)\,ds + \int_0^t h(s)w(u(s))\,ds, \qquad (2.4.1)$$

for all $t \in R_+$, where u_0 is a positive constant, then for $0 \leq t \leq t_1$,

$$u(t) \leq a(t)G^{-1}\left[G(u_0) + \int_0^t h(s)w(a(s))\,ds\right], \qquad (2.4.2)$$

where

$$a(t) = 1 + g(t) \int_0^t f(s) \exp\left(\int_s^t g(\sigma) f(\sigma) \, d\sigma \right) ds, \qquad (2.4.3)$$

for $t \in R_+$ and

$$G(r) = \int_{r_0}^r \frac{ds}{w(s)}, \quad r > 0, r_0 > 0, \qquad (2.4.4)$$

and G^{-1} is the inverse function of G, and $t_1 \in R_+$ is chosen so that

$$G(u_0) + \int_0^t h(s) w(a(s)) \, ds \in \mathrm{Dom}(G^{-1}),$$

for all $t \in R_+$ lying in the interval $0 \le t \le t_1$.

□

Proof: Define a function $n(t)$ by

$$n(t) = u_0 + \int_0^t h(s) w(u(s)) \, ds, \qquad (2.4.5)$$

then (2.4.1) can be restated as

$$u(t) \le n(t) + g(t) \int_0^t f(s) u(s) \, ds.$$

Since $n(t)$ is positive monotonic nondecreasing on R_+, by applying Theorem 1.3.3 we have

$$u(t) \le a(t) n(t), \qquad (2.4.6)$$

where $a(t)$ is defined by (2.4.3). From (2.4.5) and (2.4.6) we have

$$\begin{aligned} n'(t) &= h(t) w(u(t)) \\ &\le h(t) w(a(t) n(t)) \\ &\le h(t) w(a(t)) w(n(t)). \end{aligned} \qquad (2.4.7)$$

From (2.4.4) and (2.4.7) we have

$$\frac{d}{dt}G(n(t)) = \frac{n'(t)}{w(n(t))} \leq h(t)w(a(t)). \tag{2.4.8}$$

By taking $t = s$ in (2.4.8) and integrating it from 0 to t, we obtain

$$G(n(t)) \leq G(u_0) + \int_0^t h(s)w(a(s))\,ds. \tag{2.4.9}$$

The desired bound in (2.4.2) follows from (2.4.6) and (2.4.9). The subinterval $0 \leq t \leq t_1$ is obvious.

∎

Pachpatte (1975a) also gave the following theorem, which can be used in more general situations.

Theorem 2.4.2 *Let u, f, g and h be nonnegative continuous functions defined on R_+. Let $w(u)$ be a continuous nondecreasing subadditive and submultiplicative function defined on R_+ and $w(u) > 0$ on $(0, \infty)$. Let $p(t) > 0$, $\phi(t) \geq 0$ be continuous and nondecreasing functions defined on R_+ and $\phi(0) = 0$. If*

$$u(t) \leq p(t) + g(t)\int_0^t f(s)u(s)\,ds + \phi\left(\int_0^t h(s)w(u(s))\,ds\right), \tag{2.4.10}$$

for all $t \in R_+$, then for $0 \leq t \leq t_2$,

$$u(t) \leq a(t)\left[p(t) + \phi\left(F^{-1}\left[F(A(t)) + \int_0^t h(s)w(a(s))\,ds\right]\right)\right], \tag{2.4.11}$$

where $a(t)$ is defined by (2.4.3) and

$$A(t) = \int_0^t h(s)w(a(s)p(s))\,ds,$$

$$F(r) = \int_{r_0}^r \frac{ds}{w(\phi(s))}, \quad r > 0, r_0 > 0, \tag{2.4.12}$$

F^{-1} is the inverse of F, and $t_2 \in R_+$ is chosen so that

$$F(A(t)) + \int_0^t h(s)w(a(s))\,ds \in \mathrm{Dom}(F^{-1}),$$

for all $t \in R_+$ lying in the interval $0 \le t \le t_2$.

□

Proof: Define a function $n(t)$ by

$$n(t) = p(t) + \phi\left(\int_0^t h(s)w(u(s))\,ds\right). \qquad (2.4.13)$$

Then (2.4.10) can be restated as

$$u(t) \le n(t) + g(t)\int_0^t f(s)u(s)\,ds.$$

Since $n(t)$ is positive monotonic nondecreasing on R_+, by applying Theorem 1.3.3 we have

$$u(t) \le a(t)n(t), \qquad (2.4.14)$$

where $a(t)$ is defined by (2.4.3). From (2.4.13) and (2.4.14) we observe that

$$n(t) - p(t) = \phi\left(\int_0^t h(s)w(u(s))\,ds\right)$$

$$\le \phi\left(\int_0^t h(s)w(a(s)n(s))\,ds\right)$$

$$= \phi\left(\int_0^t h(s)w(a(s)p(s) + a(s)(n(s) - p(s)))\,ds\right)$$

$$\le \phi\left(A(t) + \int_0^t h(s)w(a(s))w(n(s) - p(s))\,ds\right).$$

$$(2.4.15)$$

Let $T \in R_+$ be any arbitrary number. Define a function $v(t)$ by

$$v(t) = A(t) + \int_0^t h(s)w(a(s))w(n(s) - p(s)) \, ds. \qquad (2.4.16)$$

From (2.4.16) and (2.4.15) for $0 \le t \le T$, we have

$$v(t) \le A(T) + \int_0^t h(s)w(a(s))w(\phi(v(s))) \, ds. \qquad (2.4.17)$$

Define a function $y(t)$ by the right-hand side of (2.4.17). Then $y(0) = A(T)$, $v(t) \le y(t)$ for $0 \le t \le T$ and

$$y'(t) \le h(t)w(a(t))w(\phi(y(t))), \quad 0 \le t \le T. \qquad (2.4.18)$$

From (2.4.12) and (2.4.18) we have

$$\frac{d}{dt} F(y(t)) = \frac{y'(t)}{w(\phi(y(t)))} \le h(t)w(a(t)), \quad 0 \le t \le T. \qquad (2.4.19)$$

By taking $t = s$ in (2.4.19) and integrating it from 0 to T we get

$$F(y(T)) \le F(A(T)) + \int_0^T h(s)W(a(s)) \, ds, \quad 0 \le t \le T < t_2.$$

This, together with $v(t) \le y(t)$ and (2.4.15) for $t = T$, gives

$$n(T) - p(T) \le \phi \left(F^{-1} \left[F(A(T)) + \int_0^T h(s)w(a(s)) \, ds \right] \right),$$

$$0 \le t \le T < t_2. \qquad (2.4.20)$$

Since T is arbitrary, the conclusion of the theorem is now clear from (2.4.20) and (2.4.14). The subinterval $0 \le t \le t_2$ is obvious.

■

Another interesting and useful inequality established by Pachpatte (1975a) is given in the following theorem, which basically involves the comparison principle.

Theorem 2.4.3 Let u, f, g and h be nonnegative continuous functions defined on R_+. Let $w(t, u)$ be a nonnegative continuous monotonic nondecreasing function in $u \geq 0$, for each fixed $t \in R_+$. Let p and ϕ be as defined in Theorem 2.4.2. If

$$u(t) \leq p(t) + g(t) \int_0^t f(s)u(s)\,ds + \phi\left(\int_0^t h(s)w(s, u(s))\,ds\right), \quad (2.4.21)$$

for all $t \in R_+$, then

$$u(t) \leq a(t)[p(t) + \phi(r(t))], \quad t \in R_+, \quad (2.4.22)$$

where $a(t)$ is defined by (2.4.3) and $r(t)$ is the maximal solution of

$$r'(t) = h(t)w(t, a(t)[p(t) + \phi(r(t))]), \quad r(0) = 0, \quad (2.4.23)$$

existing on R_+. □

Proof: Define a function $n(t)$ by

$$n(t) = p(t) + \phi\left(\int_0^t h(s)w(s, u(s))\,ds\right). \quad (2.4.24)$$

Then (2.4.21) can be restated as

$$u(t) \leq n(t) + g(t) \int_0^t f(s)u(s)\,ds.$$

Since $n(t)$ is positive monotonic nondecreasing on R_+, by applying Theorem 1.3.3 we have

$$u(t) \leq a(t)n(t), \quad (2.4.25)$$

where $a(t)$ is defined by (2.4.3). Now from (2.4.24) and (2.4.25) we have

$$u(t) \leq a(t)[p(t) + \phi(v(t))], \quad (2.4.26)$$

where

$$v(t) = \int_0^t h(s)w(s, u(s))\,ds. \tag{2.4.27}$$

From (2.4.27) and (2.4.26), it follows that

$$v'(t) \leq h(t)w(t, a(t)[p(t) + \phi(v(t))]). \tag{2.4.28}$$

A suitable application of Theorem 2.2.2 to (2.4.28) and (2.4.23) yields

$$v(t) \leq r(t), \tag{2.4.29}$$

where $r(t)$ is the maximal solution of (2.4.23) such that $r(0) = v(0) = 0$. Now from (2.4.26) and (2.4.29) the desired bound in (2.4.22) follows.

∎

It should be noted that Theorems 2.4.1 and 2.4.2 are the further generalizations of the similar theorems given by Muldowney and Wong (1968), Deo and Murdeshwar (1971) and Dhongade and Deo (1973). Theorem 2.4.3, in the special case when the first integral term on the right-hand side in (2.4.21) is absent, $h(t) = 1$, $\phi(r) = r$ and $p(t)$ is constant, was first established by Viswanathan (1963). This theorem may easily be modified to include the case in which w depends on three arguments, t, s and u; see, for instance, Willett (1965). Moreover, one also obtains as a special case a useful generalization of the Gronwall–Bellman inequality due to Bihari (1956).

Constantin (1990b) has given the following inequality, which can be used in certain applications.

Theorem 2.4.4 *Suppose*

(i) $u, k, g, h_1, h_2 : R_+ \to (0, \infty)$ *and continuous,*
(ii) $k, g \in C^1(R_+, R_+)$,
(iii) $H(u)$ *is a nonnegative monotonic nondecreasing and continuous function for $u > 0$ with $H(0) = 0$.*

If

$$u(t) \leq g(t) + k(t)\int_0^t h_1(s)u(s)\,ds + k(t)\int_0^t h_2(s)H(u(s))\,ds, \tag{2.4.30}$$

for $t \in R_+$, then

$$u(t) \leq G^{-1}\left[G(g(0)) + \int_0^t \left[k(s)\max\{h_1(s), h_2(s)\}\right.\right.$$

$$\left.\left. + \max\left\{0, \frac{k'(s)}{k(s)}\right\} + \max\left\{0, \frac{g'(s)}{g(s)}\right\}\right]ds\right], \quad (2.4.31)$$

for $t \in [0, t']$, where

$$G(r) = \int_{r_0}^r \frac{ds}{s + H(s)}, \quad r > 0, r_0 > 0, \quad (2.4.32)$$

and t' is defined so that the existence condition of the right-hand part of the inequality (2.4.31) should be assured.

□

Proof: From (2.4.30) we have

$$u(t) \leq g(t) + k(t)\int_0^t \max\{h_1(s), h_2(s)\}[u(s) + H(u(s))]\,ds.$$

We consider $w: R_+ \to R_+$ defined by $w(x) = x + H(x)$ and obtain that $w(x) \geq x$, $x \in R_+$, w is continuous monotonic nondecreasing and $w(0) = 0$. Applying Theorem 2.3.3 we deduce the inequality (2.4.31).

∎

The following generalization of Bihari's inequality is used by Talpalaru (1973).

Theorem 2.4.5 *Let $u, f: I \to R_+$ and $k(t, s): I \times I \to R_+$, $t_0 \leq s \leq t$, be integrable functions. Let $w(r)$ be a positive, continuous and nondecreasing function defined for $r > 0$. If*

$$u(t) \leq c + \int_{t_0}^t \left[f(s)w(u(s)) + \int_{t_0}^s k(s, \sigma)w(u(\sigma))\,d\sigma\right]ds, \quad (2.4.33)$$

for all $t \geq t_0$, where c is a nonnegative constant, then

$$\int_c^{u(t)} \frac{ds}{w(s)} \leq \int_{t_0}^t \left[f(s) + \int_{t_0}^s k(s, \sigma) \, d\sigma \right] ds. \quad (2.4.34)$$

Proof: As noted in the proof of Theorem 2.3.1 we assume that c is positive. Define a function $z(t)$ by the right-hand side of (2.4.33). Then $z(0) = c$, $u(t) \leq z(t)$ and we observe that

$$\frac{z'(t)}{w(z(t))} = f(t) \frac{w(u(t))}{w(z(t))} + \int_{t_0}^t k(t, \sigma) \frac{w(u(\sigma))}{w(z(t))} \, d\sigma$$

$$\leq f(t) \frac{w(u(t))}{w(z(t))} + \int_{t_0}^t k(t, \sigma) \frac{w(u(\sigma))}{w(z(\sigma))} \, d\sigma$$

$$\leq f(t) + \int_{t_0}^t k(t, \sigma) \, d\sigma. \quad (2.4.35)$$

Integrating (2.4.35) from t_0 to t and using the fact that $u(t) \leq z(t)$ we obtain

$$\int_c^{u(t)} \frac{ds}{w(s)} \leq \int_{t_0}^t \left[f(s) + \int_{t_0}^s k(s, \sigma) \, d\sigma \right] ds,$$

which yields the desired conclusion (2.4.34). The case when $c \geq 0$ can be treated as mentioned in the proof of Theorem 2.3.1.

∎

If we denote

$$\Omega(u) = \int^u \frac{ds}{w(s)},$$

then the inequality (2.4.34) can be written as

$$\Omega(u(t)) \leq \Omega(c) + \int_{t_0}^t \left[f(s) + \int_{t_0}^s k(s, \sigma) \, d\sigma \right] ds. \quad (2.4.36)$$

Furthermore, when $k(t, s) = 0$, the inequality (2.4.34) or (2.4.36) reduces to Bihari's inequality given in Theorem 2.3.1. If $w(r) = r$, the inequality

(2.4.36) becomes

$$u(t) \leq c \exp\left(\int_{t_0}^{t}\left[f(s) + \int_{t_0}^{s} k(s,\sigma)\,d\sigma\right]ds\right). \qquad (2.4.37)$$

If we take $w(r) = r$ and $k(t,s) = 0$, then the bound obtained in (2.4.37) reduces to the bound obtained in Bellman's inequality given in Theorem 1.2.2.

2.5 Inequalities with Volterra-Type Kernels

In the study of ordinary differential equations and integral equations, there often occur inequalities of the form

$$u(t) \leq p(t) + \int_0^t g(t,s) w(u(s))\,ds, \qquad (2.5.1)$$

where p, u are continuous functions defined on R_+, $g(t,s) \geq 0$ is continuous for $0 \leq s \leq t < \infty$ and $w: R_+ \to R_+$ is continuous. In this section we wish to give some results which provides upper bounds on the inequalities of the form (2.5.1).

We first present the following inequality needed in the proof of our next theorem, which is given by Constantin (1992).

Theorem 2.5.1 *Let a, b be continuous and positive functions defined on R_+ and $w: R_+ \to R_+$ be a continuous monotonic nondecreasing function such that $w(0) = 0$ and $w(x) > 0$ for $x > 0$. If u is a positive differentiable function on R_+ that satisfies*

$$u'(t) \leq a(t) w(u(t)) + b(t), \quad t \in R_+, \qquad (2.5.2)$$

then we have

$$u(t) \leq G^{-1}\left[G\left(u(0) + \int_0^t b(s)\,ds\right) + \int_0^t a(s)\,ds\right], \qquad (2.5.3)$$

for the values of t, so that the existence of the right-hand part of (2.5.3) should be assured, where

$$G(r) = \int_{r_0}^{r} \frac{ds}{w(s)}, \quad r > 0, r_0 > 0. \tag{2.5.4}$$

□

Proof: From (2.5.2) we have

$$u(t) \leq u(0) + \int_0^t b(s)\,ds + \int_0^t a(s)w(u(s))\,ds, \quad t \in R_+.$$

We denote by

$$y(t) = \int_0^t a(s)w(u(s))\,ds, \quad g(t) = u(0) + \int_0^t b(s)\,ds, \quad t \in R_+,$$

and we obtain $u(t) \leq g(t) + y(t)$ for $t \in R_+$. On the other hand

$$y'(t) = a(t)w(u(t)) \leq a(t)w(g(t) + y(t)), \quad t \in R_+,$$
$$g'(t) = b(t), \quad t \in R_+,$$

from which we deduce that

$$\frac{g'(t) + y'(t)}{w(g(t) + y(t))} \leq a(t) + \frac{b(t)}{w(g(t) + y(t))}$$
$$\leq a(t) + \frac{g'(t)}{w(g(t))}, \quad t \in R_+,$$

and an integration yields

$$G(g(t) + y(t)) \leq G\left(u(0) + \int_0^t b(s)\,ds\right) + \int_0^t a(s)\,ds. \tag{2.5.5}$$

Here, in order to obtain (2.5.5) we have used (2.5.4). The required inequality in (2.5.3) now follows from $u(t) \leq g(t) + y(t)$ and (2.5.5).

■

Constantin (1992) gave the following inequality, which provides an estimate for $u(t)$ in (2.5.1) involving $g(t, s)$ and bounds for the partial derivative $g_t(t, s)$.

Theorem 2.5.2 *Let us suppose that the partial derivative $g_t(t, s)$ of $g(t, s)$ exists and that there are positive continuous functions α, β on R_+ such that $g_t(t, s) \le \alpha(t)\beta(s)$. Suppose that*

(i) *$w: R_+ \to R_+$ is continuous nondecreasing subadditive, $w(0) = 0$ and $w(x) > 0$ for $x > 0$,*

(ii) $\lim_{x \to 0} \int_x^1 \dfrac{ds}{w(s)} < \infty$ *and* $\lim_{x \to \infty} \int_1^x \dfrac{ds}{w(s)} = \infty.$

If $m(t) = g(t, t)w$ satisfies conditions (i) and (ii) and p, u are positive continuous functions on R_+, such that (2.5.1) is satisfied, then

$$u(t) \le p(t) + G^{-1}\left[G\left(\int_0^t b(s)\,ds\right) + \int_0^t a(s)\,ds\right], \quad t \in R_+, \quad (2.5.6)$$

where

$$G(r) = \int_0^r \frac{ds}{w(s)}, \quad \text{on } R_+, \quad (2.5.7)$$

and

$$a(t) = m(t) + \alpha(t)\int_0^t \beta(s)\,ds, \quad (2.5.8)$$

$$b(t) = m(t)w(p(t)) + \alpha(t)\int_0^t \beta(s)w(p(s))\,ds, \quad (2.5.9)$$

for $t \in R_+$.

□

Proof: Conditions (i) and (ii) ensure that the function G is well defined on R_+ and is a bijection of R_+ onto R_+. This allows us to conclude that in this case in (2.5.6) we do not have to impose restrictions on the interval where this inequality holds.

Let

$$y(t) = \int_0^t g(t, s)w(u(s))\,ds, \quad t \in R_+.$$

We have that y is differentiable and

$$y'(t) = m(t)w(u(t)) + \int_0^t g_t(t,s)w(u(s))\,ds. \tag{2.5.10}$$

Since $u(t) \leq y(t) + p(t)$, w is subadditive, we deduce from (2.5.10) that

$$y'(t) \leq m(t)w(y(t)) + m(t)w(p(t)) + \int_0^t g_t(t,s)w(y(s)+p(s))\,ds. \tag{2.5.11}$$

Furthermore, from (2.5.11) we obtain

$$y'(t) \leq m(t)w(y(t)) + m(t)w(p(t))$$
$$+ \alpha(t)\left(\int_0^t \beta(s)\,ds\right)w(y(t)) + \alpha(t)\int_0^t \beta(s)w(p(s))\,ds$$
$$= a(t)w(y(t)) + b(t).$$

Now an application of Theorem 2.5.1 we deduce that

$$y(t) \leq G^{-1}\left[G\left(\int_0^t b(s)\,ds\right) + \int_0^t a(s)\,ds\right]. \tag{2.5.12}$$

Using (2.5.12) in $u(t) \leq y(t) + p(t)$ we get the desired inequality in (2.5.6). ∎

Remark 2.5.1 Condition (ii) required on w guarantees that the function G is well defined and G^{-1} is defined on R_+, which enables us to avoid restrictions on the interval where the estimate (2.5.6) is true. If this condition is not satisfied, we have to say that (2.5.6) is true for those values of $t \in R_+$ so that the existence condition of the right-hand part of (2.5.6) should be assured.

△

In order to establish our further results concerning (2.5.1), we introduce some notation and definitions (Beesack, 1984a) which we use in our further discussion. Given a continuous function $p: R_+ \to R$, we write

$$\hat{p}(t) = \max\{p(s): 0 \leq s \leq t\}, \quad \bar{p}(t) = \max\{1, \hat{p}(t)\}, \qquad (2.5.13)$$

for $t \in R_+$. A function $\phi: R_+ \to R_+$ is said to belong to the class $\mathscr{F}$ if ϕ is continuous and nondecreasing with $\phi(u) > 0$ for $u > 0$, and satisfies

$$v^{-1}\phi(u) \leq \phi(v^{-1}u) \quad \text{for } v \geq 1, u \geq 0. \qquad (2.5.14)$$

The class $\mathscr{F}$ was introduced in Dhongade and Deo (1973), but with the condition that (2.5.14) holds for $v > 0$, $u \geq 0$. It was noted by Beesack (1975, p. 65) and Beesack (1977, p. 207) that $\mathscr{F}$ then becomes merely the trivial class of linear functions; $\phi(u) = \phi(1)u$. The definition (2.5.14) was adopted to avoid this triviality. For a further discussion concerning the class $\mathscr{F}$, see Beesack (1975, 1977, 1984a,b).

Beesack (1984a) proved and made use of the inequalities given in the following theorem.

Theorem 2.5.3 *Let u and p be nonnegative continuous functions defined on R_+, $g(t, s) \geq 0$ is continuous for $0 \leq s \leq t < b (\leq \infty)$, as well as nondecreasing in t for each s, and $w: R_+ \to R_+$ is continuous and nondecreasing with $w(u) > 0$ for $u > 0$ and such that (2.5.1) is satisfied. Fix $r_0 > 0$, and define G by*

$$G(r) = \int_{r_0}^{r} \frac{ds}{w(s)}, \quad r > 0. \qquad (2.5.15)$$

(In case $w(0) > 0$, or in case $w(0^+)$ is finite, one may take $r_0 = 0$). Let G^{-1} denote the inverse function of G. Then if $\hat{p}$ and $\bar{p}$ are defined by (2.5.13), we have

$$u(t) \leq G^{-1}\left[G(\hat{p}(t)) + \int_0^t g(t, s)\, ds\right], \quad 0 \leq t < b_1. \qquad (2.5.16)$$

In the case where w is submultiplicative on R_+, then

$$u(t) \leq \hat{p}(t) G^{-1}\left[G(1) + (w(\hat{p}(t))/\hat{p}(t))\int_0^t g(t, s)\, ds\right], \quad 0 \leq t < b_2. \qquad (2.5.17)$$

If w is subadditive on R_+, then

$$u(t) \leq p(t) + G^{-1}\left[G\left(\int_0^t g(t,s)w(p(s))\,ds\right) + \int_0^t g(t,s)\,ds\right],$$

$$0 \leq t < b_3. \quad (2.5.18)$$

Finally, if $w \in \mathscr{F}$ (satisfies (2.5.14)), then

$$u(t) \leq \overline{p}(t)G^{-1}\left[G(1) + \int_0^t g(t,s)\,ds\right], \quad 0 \leq t < b(\leq \infty). \quad (2.5.19)$$

In (2.5.16)–(2.5.18), the numbers b_1 to b_3 are chosen so that the quantity in square brackets is in the range of G for $0 \leq t < b$. $\square$

Proof: Fix any T, $t_0 < T < b$. Then, for $0 \leq t \leq T$, we have

$$u(t) \leq \hat{p}(T) + \int_0^t g(T,s)w(u(s))\,ds = y(t), \quad (2.5.20)$$

and hence

$$y'(t) = g(T,t)w(u(t)) \leq g(T,t)w(y(t)),$$

i.e.

$$\frac{y'(t)}{w(y(t))} \leq g(T,t). \quad (2.5.21)$$

Integrate (2.5.21) over $[0,t]$ and use $y(0) = \hat{p}(T)$ to obtain

$$G(y(t)) \leq G(\hat{p}(T)) + \int_0^t g(T,s)\,ds, \quad 0 \leq t \leq T.$$

Now set $t = T$ to obtain

$$y(T) \leq G^{-1}\left[G(\hat{p}(T)) + \int_0^T g(T,s)\,ds\right], \quad 0 \leq t \leq T < b_1.$$

By (2.5.20) for $t = T$ we now obtain (2.5.16) with t replaced by T.

Suppose now that w is submultiplicative on R_+. We rewrite (2.5.20) in the form

$$u(t) \le \hat{p}(T) + \int_0^t g(T,s) w(\hat{p}(T) u(s)/\hat{p}(T))\, ds,$$

whence

$$u(t)/\hat{p}(T) \le 1 + w(\hat{p}(T))/\hat{p}(T) \int_0^t g(T,s) w(u(s)/\hat{p}(T))\, ds,$$

for $0 \le t \le T < \infty$. We may now proceed as above (or simply use Bihari's result) to obtain

$$u(t) \le \hat{p}(T) G^{-1}\left[G(1) + (w(\hat{p}(T))/\hat{p}(T)) \int_0^t g(T,s)\, ds \right],$$

for $0 \le t \le T < b_2$. Again we obtain (2.5.17) with t replaced by T.

To prove (2.5.18) when w is subadditive on R_+, we proceed as in Pachpatte (1975g, Theorem 2) and denote the right-hand side of (2.5.1) by $n(t)$. Then $v(t) = n(t) - p(t) \ge 0$ and $u(t) \le n(t)$, so

$$v(t) = \int_0^t g(t,s) w(u(s))\, ds$$

$$\le \int_0^t g(t,s) w(n(s))\, ds = \int_0^t g(t,s) w(p(s) + v(s))\, ds,$$

and hence

$$v(t) \le \int_0^t g(t,s) w(p(s))\, ds + \int_0^t g(t,s) w(v(s))\, ds.$$

We may now apply (2.5.16) to this inequality, which is just (2.5.1) with $p(t)$ replaced by $P(t) = \int_0^t g(t,s) w(p(s))\, ds$ and u replaced by v. We obtain (2.5.18).

Finally, suppose $w \in \mathscr{F}$. By (2.5.20) and (2.5.13), we have for $0 \le t \le T < b$,

$$u(t) \le \overline{p}(T) + \int_0^t g(T, s)w(u(s))\,ds,$$

whence by (2.5.14)

$$u(t)/\overline{p}(T) \le 1 + \int_0^t g(T, s)w(u(s)/\overline{p}(T))\,ds,$$

since $\overline{p}(T) \ge 1$. As for (2.5.17), the result now follows by applying Bihari's result to the last inequality. The fact that (2.5.19) holds for $0 \le t < b$ follows from the inequality $w(r) \le w(1)r$ for $r \ge 1$, so that by (2.5.15), $G(u) \to \infty$ as $u \to \infty$.

∎

Beesack (1984a) also obtained the upper bound on the following generalization of (2.5.1), namely

$$u(t) \le p(t) + \sum_{j=1}^k \int_{t_0}^t g_j(t, s)w_j(u(s))\,ds, \quad t_0 \le t < b, \qquad (2.5.22)$$

under some suitable conditions on the functions involved in (2.5.22). For similar results, see also Pinto (1990). The precise formulation of such results is very close to that of the results given in Theorem 2.5.3, and this is left to the reader.

Beesack (1977) dealt with more general inequalities of the form

$$f(u(t)) \le a(t) + H\left(t, \int_\alpha^t w(t, s, u(s))\,ds\right), \quad t \in J, \qquad (2.5.23)$$

and its special cases. Here we only state the main results given in Beesack (1977), since these results are essentially contained in Bainov and Simeonov (1972) and Beesack (1975, 1977), and hence the details will not be given here.

In what follows we shall be dealing with monotonic functions of several variables. If $F_1(t, s, u)$ is defined on $I_1 \times J_1 \times K_1$ where I_1, J_1, K_1

are intervals, then by saying that $F_1(u)$ is monotonic (or nondecreasing, increasing etc.) we shall mean that for each $(t, s) \in I_1 \times J_1$, the function $g(u) \equiv F_1(t, s, u)$ is monotonic (or nondecreasing, increasing etc.) on K_1. Similarly, if $F_2(t, s, v)$ is another such function defined on $I_2 \times J_2 \times K_2$, then we say, for example that $F_1(u), F_2(t)$ have the same parity (or the opposite parity) if $F_1(u), F_2(t)$ are both nonincreasing or both nondecreasing (or that one is nonincreasing and the other is nondecreasing).

The main results established by Beesack (1977) (see also Bainov and Simeonov (1972) and Beesack (1975)) are given in the following theorems.

Theorem 2.5.4 *Let $f(u)$ be continuous and strictly monotonic on an interval I, and let $H(t, v)$ be continuous on $J \times K$, where $J = [\alpha, \beta]$ and K is an interval containing zero, with $H(v)$ monotonic. Let $T_1 = \{(t, s) : \alpha \le s \le t \le \beta\}$, and suppose that $w(t, s, u)$ is continuous and of one sign on $T_1 \times I$ with $w(t)$ monotonic on J and $w(u)$ monotonic on I. Suppose also that the functions u and a are continuous on J with $u(J) \subset I$ and*

$$a(t) + H(t, v) \in f(I) \quad \text{for } t \in J, \quad |v| \le b, \tag{2.5.24}$$

for some constant $b > 0$. Let

$$f(u(t)) \le a(t) + H\left(t, \int_\alpha^t w(t, s, u(s))\,ds\right), \quad t \in J, \tag{2.5.25}$$

and let $r = r(t, T, \alpha)$ be the maximal (minimal) solution of the system

$$\frac{dr}{dt} = w(T, t, f^{-1}[a(t) + H(t, r)]), \quad \alpha \le t \le T \le \beta_1 \ (\le \beta),$$
$$r(\alpha) = 0, \tag{2.5.26}$$

if $w(u), f$ have the same (opposite) parity, where $\beta_1 > \alpha$ is chosen so that the maximal (minimal) solution exists for $\alpha \le t \le \beta_1$. Then provided $w(t), H(v)$ have the same parity,

$$u(t) \stackrel{\le}{\ge} f^{-1}[a(t) + H(t, \bar{r}(t))], \quad \alpha \le t \le \beta_1, \tag{2.5.27}$$

where $\bar{r}(t) = r(t, t, \alpha)$ and the upper inequality sign holds if (i) f is increasing and $w(u), H(v)$ have the same parity, while the lower inequality sign holds if (ii) f is decreasing and $w(u), H(v)$ have the opposite parity.

□

Theorem 2.5.5 *Under the hypotheses of Theorem 2.5.4, suppose that*

$$f(u(t)) \geq a(t) + H\left(t, \int_\alpha^t w(t, s, u(s))\,ds\right), \quad t \in J,$$

and that $w(t)$, $H(v)$ *have the opposite parity. Let* $\bar{r}(t) = r(t, t, \alpha)$ *where* $r(t, T, \alpha)$ *is the maximal (minimal) solution of (2.5.26) according as* $w(u)$, f *have the opposite (the same) parity. Then*

$$u(t) \gtreqless f^{-1}[a(t) + H(t, \bar{r}(t))], \quad \alpha \leq t \leq \beta_1,$$

where $\leq$ *or* $\geq$ *holds according as case (i) or (ii) holds in Theorem 2.5.4.*

□

We note that the special versions of Theorems 2.5.4 and 2.5.5 can be used to obtain earlier results on Gronwall–Bihari-like inequalities in a uniform manner. For more detailed discussions of the various special versions and variants of these theorems, see Bainov and Simeonov (1992) and Beesack (1975, 1977), and also Constantin (1991) and Pachpatte (1973a).

2.6 Inequalities with Nonlinearities in the Integral

There exist various nonlinear generalizations of the Gronwall–Bellman inequality which play a vital role in the study of qualitative properties of solutions of differential and integral equations. This section deals with some useful inequalities containing nonlinearities in the integral and investigated by Willett and Wong (1965), Gollwitzer (1969) and Pachpatte (1975c,d,h, 1984), beginning with the following definitions used in our subsequent discussion.

A continuous function $k(t)$ defined on some interval I is said to be convex if $k(\alpha x + \beta y) \leq \alpha k(x) + \beta k(y)$, $x, y \in I$, $\alpha, \beta \geq 0$, $\alpha + \beta = 1$, $k(t)$ is said to be concave if $-k(t)$ is convex. The previous inequality is a special case of Jensen's inequality for convex functions. The function $G(u)$ is said to be submultiplicative if $G(uv) \leq G(u)G(v)$, $u, v \geq 0$.

We start with the following fundamental inequality given by Willett and Wong (1965).

Theorem 2.6.1 Let u, a, b and k be nonnegative continuous functions defined on J, let $p \geq 1$ be constant, and suppose

$$u(t) \leq a(t) + b(t) \left(\int_\alpha^t k(s) u^p(s) \, ds \right)^{1/p}, \quad t \in J. \tag{2.6.1}$$

Then

$$u(t) \leq a(t) + b(t) \frac{\left(\int_\alpha^t k(s) e(s) a^p(s) \, ds \right)^{1/p}}{1 - [1 - e(t)]^{1/p}}, \quad t \in J, \tag{2.6.2}$$

where

$$e(t) = \exp\left(-\int_\alpha^t k(s) b^p(s) \, ds \right). \tag{2.6.3}$$

$\square$

Proof: Define a function $v(t)$ on J by

$$v(t) = e(t) \int_\alpha^t k(s) u^p(s) \, ds. \tag{2.6.4}$$

It follows from (2.6.1) and (2.6.4) that

$$u(t) \leq a(t) + b(t) v^{1/p}(t) e^{-1/p}(t)$$

and since

$$e'(t) = -e(t) k(t) b^p(t),$$

we have

$$v'(t) = e(t) k(t) u^p(t) - k(t) v(t) b^p(t)$$
$$\leq k(t) [a(t) e^{1/p}(t) + b(t) v^{1/p}(t)]^p - k(t) v(t) b^p(t).$$

Integrating the above inequality from α to t we get

$$v(t) \leq \int_\alpha^t k(s) [a(s) e^{1/p}(s) + b(s) v^{1/p}(s)]^p \, ds - \int_\alpha^t k(s) v(s) b^p(s) \, ds,$$

or

$$\left(v(t) + \int_\alpha^t k(s)v(s)b^p(s)\,ds\right)^{1/p}$$

$$\leq \left(\int_\alpha^t [a(s)k^{1/p}(s)e^{1/p}(s) + b(s)k^{1/p}(s)v^{1/p}(s)]^p\,ds\right)^{1/p}$$

$$\leq \left(\int_\alpha^t a^p(s)k(s)e(s)\,ds\right)^{1/p} + \left(\int_\alpha^t b^p(s)k(s)v(s)\,ds\right)^{1/p} \quad (2.6.5)$$

Here we have used the Minkowski inequality to get the last inequality in (2.6.5). From (2.6.5) we have

$$\left(v(t) + \int_\alpha^t k(s)v(s)b^p(s)\,ds\right)^{1/p} - \left(\int_\alpha^t b^p(s)k(s)v(s)\,ds\right)^{1/p}$$

$$\leq \left(\int_\alpha^t a^p(s)k(s)e(s)\,ds\right)^{1/p}. \quad (2.6.6)$$

The left member of (2.6.6) is a function of the form $z(x) = (m+x)^{1/p} - x^{1/p}$. For $p \geq 1$ and $m \geq 0$, $z(x)$ is a nonincreasing function of x for $x \geq 0$. Thus we may replace $\int_\alpha^t b^p(s)k(s)v(s)\,ds$ in (2.6.6) by a larger quantity and still have a valid inequality. It is easy to see from the definition of $v(t)$ given in (2.6.4) that

$$\int_\alpha^t b^p(s)k(s)v(s)\,ds \leq \left(\int_\alpha^t b^p(s)k(s)e(s)\,ds\right)\left(\int_\alpha^t k(s)u^p(s)\,ds\right)$$

$$= (1 - e(t))\int_\alpha^t k(s)u^p(s)\,ds. \quad (2.6.7)$$

The conclusion (2.6.2) follows upon substituting (2.6.4) and (2.6.7) in (2.6.5).

■

Remark 2.6.1 A special form of this inequality with $k(t) = 1$ was proved first by Willett (1964). Moreover, the continuity hypothesis was replaced by the usual Lebesgue integrability in Willett and Wong (1965).

△

The following theorem, in which the nonlinear function u^p in Theorem 2.6.1 is replaced by a general function is given by Gollwitzer (1969).

Theorem 2.6.2 *Let u, f, g and h be nonnegative continuous functions on the interval $[a, b]$, $G(u)$ be a continuous, strictly increasing, convex and submultiplicative function for $u > 0$, $\lim_{u \to \infty} G(u) = \infty$, $\alpha(t)$, $\beta(t)$ be continuous functions on $[a, b]$, $\alpha(t), \beta(t) > 0$, $\alpha(t) + \beta(t) = 1$, and*

$$u(t) \leq f(t) + g(t) G^{-1} \left(\int_a^t h(s) G(u(s))\, ds \right), \quad a \leq t \leq b. \tag{2.6.8}$$

Then

$$u(t) \leq f(t) + g(t) G^{-1} \left(\int_a^t h(s)\alpha(s) G(f(s)\alpha^{-1}(s)) \right.$$

$$\left. \times \exp\left(\int_s^t h(x)\beta(x) G(g(x)\beta^{-1}(x))\, dx \right) ds \right), \quad a \leq t \leq b.$$

$$\tag{2.6.9}$$

Furthermore, if

$$u(t) \geq u(x) - g(t) G^{-1} \left(\int_x^t h(s) G(u(s))\, ds \right), \quad a \leq x \leq t \leq b, \tag{2.6.10}$$

then

$$u(t) \geq \alpha(t) G^{-1} \left(\alpha^{-1}(t) G(u(x)) \exp\left(-\beta(t) G(g(t)\beta^{-1}(t)) \int_x^t h(s)\, ds \right) \right),$$

$$a \leq x \leq t \leq b. \tag{2.6.11}$$

□

Proof: Let $\alpha(t), \beta(t) > 0$, $\alpha(t) + \beta(t) = 1$. Rewrite (2.6.8) as

$$u(t) \leq \alpha(t)(f(t)\alpha^{-1}(t)) + \beta(t)(g(t)\beta^{-1}(t))G^{-1}\left(\int_a^t h(s)G(u(s))\,ds\right).$$

Since G is convex, submultiplicative and monotonic, we have

$$G(u(t)) \leq \alpha(t)G(f(t)\alpha^{-1}(t)) + \beta(t)G(g(t)\beta^{-1}(t))\int_a^t h(s)G(u(s))\,ds,$$

$$a \leq t \leq b.$$

The estimate in (2.6.9) follows by first applying Theorem 1.3.2 and then applying G^{-1} to both sides of the resulting inequality.

From (2.6.10), we have

$$u(x) \leq \alpha(t)(u(t)\alpha^{-1}(t)) + \beta(t)(g(t)\beta^{-1}(t))G^{-1}\left(\int_x^t h(s)G(u(s))\,ds\right),$$

$$a \leq x \leq t \leq b.$$

Since G is convex, submultiplicative and monotonic, we have

$$\alpha(t)G(u(t)\alpha^{-1}(t)) \geq G(u(x)) - \beta(t)G(g(t)\beta^{-1}(t))\int_x^t h(s)G(u(s))\,ds,$$

and the result given in (2.6.11) follows from applying Theorem 1.3.5. ∎

Gollwitzer (1969) also gave the following theorem.

Theorem 2.6.3 *Let u and h be nonnegative continuous functions on the interval $[a, b]$, $G(u)$ be a continuous, concave function for $u \geq 0$, continuously differentiable for $u > 0$, $G'(u) > 0$ for $u > 0$, $G(0) = 0$, $\lim_{u \to \infty} G(u) = \infty$, $C \geq 0$ a constant, and*

$$u(t) \leq C + G^{-1}\left(\int_a^t h(s)G(u(s))\,ds\right), \quad a \leq t \leq b. \tag{2.6.12}$$

Then

$$u(t) \le G^{-1}\left[G(C)\exp\left(\int_a^t h(s)\,ds\right)\right], \quad a \le t \le b. \quad (2.6.13)$$

Furthermore, if

$$u(t) \ge u(x) - G^{-1}\left(\int_x^t h(s)G(u(s))\,ds\right), \quad a \le x \le t \le b, \quad (2.6.14)$$

then

$$u(t) \ge G^{-1}\left[G(u(x))\exp\left(-\int_x^t h(s)\,ds\right)\right], \quad a \le x \le t \le b. \quad (2.6.15)$$

$\square$

Proof: It is sufficient to assume that C is positive, since a standard limiting argument can be used to treat the remaining case. Consider (2.6.12) and define $\psi(t)$ as

$$\psi(t) = C + G^{-1}\left(\alpha + \int_a^t h(s)G(u(s))\,ds\right), \quad \alpha > 0, a \le t \le b. \quad (2.6.16)$$

We note that $\psi(t)$ majorizes the right member of (2.6.12), and hence $\psi(t) \ge u(t)$. Since $G(u)$ is concave, the derivative $G'(u)$ is nonincreasing for $u > 0$. Since $\psi(t) - C > 0$ (here is where we use the constant α),

$$G'(\psi(s)) \le G'(\psi(s) - C), \quad a \le t \le b. \quad (2.6.17)$$

Furthermore, by the fundamental theorem of integral calculus,

$$\ln G(\psi(t)) - \ln G(C + G^{-1}(\alpha)) = \int_a^t G'(\psi(s))G(\psi(s))^{-1}\psi'(s)\,ds, \quad (2.6.18)$$

where

$$\psi'(s) = G(u(s))h(s)[G'(\psi(s) - C)]^{-1}.$$

Since $\psi(t) \geq u(t)$ and (2.6.17) holds, the integrand in (2.6.18) is majorized by $h(s)$. Hence

$$\ln (G(\psi(t))/G(C + G^{-1}(\alpha))) \leq \int_a^t h(s)\,ds, \quad a \leq t \leq b,$$

or

$$G(u(t)) \leq G(C)\exp\left(\int_a^t h(s)\,ds\right), \quad a \leq t \leq b,$$

since the continuity of G permits us to let α approach zero. Thus the first assertion has been proven.

Consider (2.6.14) and define $\phi(x)$ as

$$\phi(x) = u(t) + G^{-1}\left(\int_x^t h(s)G(u(s))\,ds\right), \quad a \leq x \leq t. \qquad (2.6.19)$$

We note that $\phi(x) \geq u(x)$. Assume that $u(t)$ is positive on $[a, b]$. Since G is concave and continuously differentiable, we can use the techniques given in the first part of the proof to show that

$$\ln G(\phi(t)) - \ln G(\phi(x)) \geq - \int_x^t h(s)\,ds, \quad a \leq x \leq t \leq b,$$

and hence

$$G(u(t)) \geq G(u(x))\exp\left(-\int_x^t h(s)\,ds\right), \quad a \leq x \leq t \leq b. \qquad (2.6.20)$$

The estimate given in (2.6.15) is now clear. If $u(t)$ is not positive on $[a, b]$, we can replace $u(t)$ by $\{u(t) + \epsilon\}$ in (2.6.14), $\epsilon > 0$, and then let $\epsilon \to 0$ in (2.6.20) to complete the proof.

∎

Remark 2.6.2 We note that in Bainov and Simeonov (1992) and Beesack (1975, 1977) the authors have given some variants and further generalizations of Gollwitzer's inequalities given in Theorems 2.6.2 and 2.6.3. In order to avoid duplications, these results will not be discussed here.

△

Our next two results deal with the nonlinear generalizations of Greene's inequality (1977), established by Pachpatte (1984), which can be used in some applications.

Theorem 2.6.4 *Let u, v, a, b, p and h_i ($i = 1, 2, 3, 4$) be nonnegative continuous functions defined on R_+, $G(r)$ be a continuous, strictly increasing, convex and submultiplicative function for $r > 0$, with $G(r) \to \infty$ as $r \to \infty$, $\alpha(t)$, $\beta(t)$ be continuous functions on R_+, $\alpha(t)$, $\beta(t) > 0$, $\alpha(t) + \beta(t) = 1$, and*

$$u(t) \leq a(t) + p(t)G^{-1}\left[\int_0^t h_1(s)G(u(s))\,ds + \int_0^t e^{\mu s}h_2(s)G(v(s))\,ds\right], \tag{2.6.21}$$

$$v(t) \leq b(t) + p(t)G^{-1}\left[\int_0^t e^{-\mu s}h_3(s)G(u(s))\,ds + \int_0^t h_4(s)G(v(s))\,ds\right], \tag{2.6.22}$$

for all $t \in R_+$, where μ is a nonnegative constant. Then

$$u(t) \leq G^{-1}[e^{\mu t}Q_0(t)], \quad v(t) \leq G^{-1}[Q_0(t)], \tag{2.6.23}$$

for all $t \in R_+$, where

$$Q_0(t) = f_0(t) + \beta(t)G(p(t)\beta^{-1}(t))\int_0^t h(s)f_0(s)$$

$$\times \exp\left(\int_s^t h(\sigma)\beta(\sigma)G(p(\sigma)\beta^{-1}(\sigma))\,d\sigma\right)ds, \tag{2.6.24}$$

in which

$$f_0(t) = \alpha(t)[G(a(t)\alpha^{-1}(t)) + G(b(t)\alpha^{-1}(t))], \tag{2.6.25}$$

and
$$h(t) = \max\{[h_1(t) + h_3(t)], [h_2(t) + h_4(t)]\}, \quad (2.6.26)$$

for all $t \in R_+$.

Proof: Rewrite (2.6.21) and (2.6.22) as

$$u(t) \leq \alpha(t)(a(t)\alpha^{-1}(t)) + \beta(t)(p(t)\beta^{-1}(t))$$
$$\times G^{-1}\left[\int_0^t h_1(s)G(u(s))\,ds + \int_0^t e^{\mu s}h_2(s)G(v(s))\,ds\right], \quad (2.6.27)$$

$$v(t) \leq \alpha(t)(b(t)\alpha^{-1}(t)) + \beta(t)(p(t)\beta^{-1}(t))$$
$$\times G^{-1}\left[\int_0^t e^{-\mu s}h_3(s)G(u(s))\,ds + \int_0^t h_4(s)G(v(s))\,ds\right], \quad (2.6.28)$$

Since G is convex, submultiplicative and monotonic, from (2.6.27) and (2.6.28), we have

$$G(u(t)) \leq \alpha(t)G(a(t)\alpha^{-1}(t)) + \beta(t)G(p(t)\beta^{-1}(t))$$
$$\times \left[\int_0^t h_1(s)G(u(s))\,ds + \int_0^t e^{\mu s}h_2(s)G(v(s))\,ds\right].$$

$$G(v(t)) \leq \alpha(t)G(b(t)\alpha^{-1}(t)) + \beta(t)G(p(t)\beta^{-1}(t))$$
$$\times \left[\int_0^t e^{-\mu s}h_3(s)G(u(s))\,ds + \int_0^t h_4(s)G(v(s))\,ds\right].$$

The bounds in (2.6.23) follow by first applying Theorem 1.6.2 with u, v, a, b and p replaced by $G(u(t))$, $G(v(t))$, $\alpha(t)G(a(t)\alpha^{-1}(t))$, $\alpha(t)G(b(t)\alpha^{-1}(t))$ and $\beta(t)G(p(t)\beta^{-1}(t))$ respectively and then applying G^{-1} to both sides of the resulting inequality.

∎

Theorem 2.6.5 *Let u, v, a, b, p and h_i be as in Theorem 2.6.4. Let $H(r)$ be a positive, continuous, strictly increasing, subadditive and submultiplicative function for $r > 0$ with $\lim_{r \to \infty} H(r) = \infty$ and*

$$u(t) \leq a(t) + p(t)H^{-1}\left[\int_0^t h_1(s)H(u(s))\,ds + \int_0^t e^{\mu s}h_2(s)H(v(s))\,ds\right], \tag{2.6.29}$$

$$v(t) \leq b(t) + p(t)H^{-1}\left[\int_0^t e^{-\mu s}h_3(s)H(u(s))\,ds + \int_0^t h_4(s)H(v(s))\,ds\right], \tag{2.6.30}$$

for all $t \in R_+$, where μ is a nonnegative constant. Then

$$u(t) \leq H^{-1}[e^{\mu t}Q_1(t)], \quad v(t) \leq H^{-1}[Q_1(t)], \tag{2.6.31}$$

for all $t \in R_+$, where

$$Q_1(t) = f_1(t) + H(p(t))\left[\int_0^t h(s)f_1(s)\exp\left(\int_s^t h(\sigma)H(p(\sigma))\,d\sigma\right)ds\right], \tag{2.6.32}$$

in which $f_1(t) = H(a(t)) + H(b(t))$, and $h(t)$ is defined by (2.6.26). $\square$

Proof: Since H is subadditive, submultiplicative and monotonic, we have from (2.6.29) and (2.6.30),

$$H(u(t)) \leq H(a(t)) + H(p(t))\left[\int_0^t h_1(s)H(u(s))\,ds + \int_0^t e^{\mu s}h_2(s)H(v(s))\,ds\right],$$

$$H(v(t)) \leq H(b(t)) + H(p(t))\left[\int_0^t e^{-\mu s}h_3(s)H(u(s))\,ds + \int_0^t h_4(s)H(v(s))\,ds\right].$$

The desired bound in (2.6.31) follows by first applying Theorem 1.6.2 with u, v, a, b and p replaced by $H(u(t))$, $H(v(t))$, $H(a(t))$, $H(b(t))$ and $H(p(t))$ respectively and then applying H^{-1} to both sides of the resulting inequality.

∎

The following theorem deals with the inequalities given by Pachpatte (1975c,d), which can be used in certain applications.

Theorem 2.6.6 *Let u, v, f, g, h and k be nonnegative continuous functions defined on $[a, b]$. Let G, α, β be defined as in Theorem 2.6.2 and*

$$u(t) \leq f(t) + g(t)G^{-1}\left(\int_a^t h(s)G(u(s))\,ds + \int_a^t h(s)\beta(s)G(g(s)\beta^{-1}(s))\right.$$
$$\left.\times \left(\int_a^s k(\tau)G(u(\tau))\,d\tau\right)ds\right), \qquad (2.6.33)$$

for $t \in [a, b]$. Then

$$u(t) \leq f(t) + g(t)G^{-1}\left(\int_a^t h(s)\left\{\alpha(s)G(f(s)\alpha^{-1}(s)) + \beta(s)G(g(s)\beta^{-1}(s))\right.\right.$$
$$\times \int_a^s \alpha(\tau)G(f(\tau)\alpha^{-1}(\tau))[h(\tau) + k(\tau)]\exp\left(\int_\tau^s \beta(\sigma)G(g(\sigma)\beta^{-1}(\sigma))\right.$$
$$\left.\left.\left.\times [h(\sigma) + k(\sigma)]\,d\sigma\right)d\tau\right\}ds\right), \qquad (2.6.34)$$

for $t \in [a, b]$.

Furthermore, if

$$u(t) \geq v(s) - g(t)G^{-1}\left(\int_s^t h(\sigma)G(v(\sigma))\,d\sigma\right.$$
$$\left.+ \int_s^t h(\sigma)\left(\int_\sigma^t k(\tau)G(v(\tau))\,d\tau\right)d\sigma\right), \qquad (2.6.35)$$

for $a \leq s \leq t \leq b$. Then

$$u(t) \geq \alpha(t)G^{-1}\left(\alpha^{-1}(t)G(v(s))\left\{1 + \beta(t)G(g(t)\beta^{-1}(t))\right.\right.$$
$$\left.\left.\times \int_s^t h(\sigma)\exp\left(\int_\sigma^t [h(\tau)\beta(t)G(g(t)\beta^{-1}(t)) + k(\tau)]\,d\tau\right)d\sigma\right\}^{-1}\right),$$
$$(2.6.36)$$

for $a \leq s \leq t \leq b$.

□

Proof: Rewrite (2.6.33) as

$$u(t) \leq \alpha(t)(f(t)\alpha^{-1}(t)) + \beta(t)(g(t)\beta^{-1}(t))G^{-1}\left(\int_a^t h(s)G(u(s))\,ds\right.$$
$$\left. + \int_a^t h(s)\beta(s)G(g(s)\beta^{-1}(s))\left(\int_a^s k(\tau)G(u(\tau))\,d\tau\right)ds\right).$$

Since G is convex, submultiplicative and monotonic, we have

$$G(u(t)) \leq \alpha(t)G(f(t)\alpha^{-1}(t)) + \beta(t)G(g(t)\beta^{-1}(t))\left(\int_a^t h(s)G(u(s))\,ds\right.$$
$$\left. + \int_a^t h(s)\beta(s)G(g(s)\beta^{-1}(s))\left(\int_a^s k(\tau)G(u(\tau))\,d\tau\right)ds\right).$$

The estimate given in (2.6.34) follows by first applying Theorem 1.7.2 part (iv) and then applying G^{-1} to both sides of the resulting inequality.

Rewrite (2.6.35) as

$$v(s) \leq \alpha(t)(u(t)\alpha^{-1}(t)) + \beta(t)(g(t)\beta^{-1}(t))G^{-1}\left(\int_s^t h(\sigma)G(v(\sigma))\,d\sigma\right.$$
$$\left. + \int_s^t h(\sigma)\left(\int_\sigma^t k(\tau)G(v(\tau))\,d\tau\right)d\sigma\right), \quad a \leq s \leq t \leq b.$$

Since G is convex, submultiplicative and monotonic, we have

$$\alpha(t)G(u(t)\alpha^{-1}(t)) \geq G(v(s)) - \beta(t)G(g(t)\beta^{-1}(t))\left(\int_s^t h(\sigma)G(v(\sigma))\,d\sigma\right.$$
$$\left. + \int_s^t h(\sigma)\left(\int_\sigma^t k(\tau)G(v(\tau))\,d\tau\right)d\sigma\right).$$

The estimate given in (2.6.36) follows by first applying Theorem 1.7.6 part (i) and then applying G^{-1} to both sides of the resulting inequality. ∎

In the next theorem we shall present the inequalities established by Pachpatte in (1975h).

Theorem 2.6.7 *Let u, f, g, h and k be as in Theorem 2.6.6 and $H(r)$ be as in Theorem 2.6.5 and*

$$u(t) \leq f(t) + g(t)H^{-1}\left(\int_a^t h(s)H(u(s))\,ds + \int_a^t h(s)H(g(s))\right.$$
$$\left. \times \left(\int_a^s k(\tau)H(u(\tau))\,d\tau\right)ds\right), \qquad (2.6.37)$$

for $t \in [a, b]$. Then

$$u(t) \leq H^{-1}\left[H(f(t)) + H(g(t))\left(\int_a^t h(s)\left\{H(f(s)) + H(g(s))\right.\right.\right.$$
$$\times \int_a^s H(f(\tau))[h(\tau) + k(\tau)]$$
$$\left.\left.\left. \times \exp\left(\int_\tau^s H(g(\sigma))[h(\sigma) + k(\sigma)]\,d\sigma\right)d\tau\right\}ds\right)\right], \qquad (2.6.38)$$

for $t \in [a, b]$.

Furthermore, if

$$u(t) \geq u(s) - g(t)H^{-1}\left(\int_s^t h(\tau)H(u(\tau))\,d\tau\right.$$
$$\left. + \int_s^t h(\tau)\left(\int_\tau^t k(\sigma)H(u(\sigma))\,d\sigma\right)d\tau\right), \qquad (2.6.39)$$

for $a \leq s \leq t \leq b$, then

$$u(t) \geq H^{-1}\left(H(u(s))\left\{1 + H(g(t))\int_s^t h(\tau)\right.\right.$$
$$\left.\left. \times \exp\left(\int_\tau^t [h(\sigma)H(g(t)) + k(\sigma)]\,d\sigma\right)d\tau\right\}^{-1}\right), \qquad (2.6.40)$$

for $a \leq s \leq t \leq b$.

□

Proof: Since H is subadditive, submultiplicative and monotonic, from (2.6.37) we have

$$H(u(t)) \leq H(f(t)) + H(g(t)) \left(\int_a^t h(s) H(u(s)) \, ds \right.$$

$$\left. + \int_a^t h(s) H(g(s)) \left(\int_a^s k(\tau) H(u(\tau)) \, d\tau \right) ds \right).$$

The estimate (2.6.38) follows by first applying Theorem 1.7.2 part (iv) and then applying H^{-1} to both sides of the resulting inequality.

Rewrite (2.6.39) as

$$u(s) \leq u(t) + g(t) H^{-1} \left(\int_s^t h(\tau) H(u(\tau)) \, d\tau \right.$$

$$\left. + \int_s^t h(\tau) \left(\int_\tau^t k(\sigma) H(u(\sigma)) \, d\sigma \right) d\tau \right). \qquad (2.6.41)$$

Since H is subadditive, submultiplicative and monotonic, from (2.6.41) we have

$$H(u(t)) \geq H(u(s)) - H(g(t)) \left[\int_s^t h(\tau) H(u(\tau)) \, d\tau \right.$$

$$\left. + \int_s^t h(\tau) \left(\int_\tau^t k(\sigma) H(u(\sigma)) \, d\sigma \right) d\tau \right].$$

The estimate (2.6.40) follows by first applying Theorem 1.7.6 part (i) and then applying H^{-1} to both sides of the resulting inequality. ∎

2.7. Pachpatte's Inequalities I

In the study of qualitative properties of the solutions of certain differential, integral and integro-differential equations some specific nonlinear integral inequalities play a crucial role. This section is concerned with some fundamental integral inequalities established by Pachpatte (1973a, 1976b, 1977a,c,d, 1978b), which play a vital role in the study of various properties of the solutions of certain differential, integral and integro-differential equations. In fact, these results are formulated in their more general forms, so that they offer greater versatility in certain applications.

We start with the following fundamental inequality established by Pachpatte (1973a).

Theorem 2.7.1 *Let u, f and g be nonnegative continuous functions defined on R_+, and $c > 0$, $p > 0$ and $p \neq 1$ are constants and suppose*

$$u(t) \leq c + \int_0^t f(s)u(s)\,ds + \int_0^t f(s)\left(\int_0^s g(\sigma)u^p(\sigma)\,d\sigma\right)ds, \quad t \in R_+. \tag{2.7.1}$$

(i) If $0 < p < 1$ and $E_0(t)$ is defined by

$$E_0(t) = 1 + (1-p)c^{p-1}\int_0^t g(\tau)\exp\left(-(1-p)\int_0^\tau f(\sigma)\,d\sigma\right)d\tau,$$

then

$$u(t) \leq c\left[1 + \int_0^t f(s)\exp\left(\int_0^s f(\sigma)\,d\sigma\right)(E_0(s))^{1/(1-p)}\,ds\right], \tag{2.7.2}$$

for $t \in R_+$.

(ii) If $1 < p < \infty$, then

$$u(t) \leq c\left[1 + \int_0^t f(s)\exp\left(\int_0^s f(\sigma)\,d\sigma\right)(\overline{E}_0(s))^{-1/(p-1)}\,ds\right], \tag{2.7.3}$$

for $t \in [0, \alpha)$, where $\overline{E}_0(t)$ is defined by the right-hand side of the definition of $E_0(t)$ for $1 < p < \infty$ and $\alpha = \sup\{t \in R_+ : \overline{E}_0(t) > 0\}$. Moreover, if we assume that $\overline{E}_0(t) > 0$ for all $t \in R_+$, then the inequality (2.7.3) remains valid for all $t \in R_+$.

□

Pachpatte (1976b) established the following useful inequalities.

Theorem 2.7.2 *Let u, f, g and h be nonnegative continuous functions defined on R_+ and $c > 0$ is a constant and suppose*

$$u(t) \leq c + \int_0^t f(s)u(s)\,ds + \int_0^t g(s)u(s)\left(u(s) + \int_0^s h(\sigma)u(\sigma)\,d\sigma\right)ds,$$
(2.7.4)

for $t \in R_+$. Then

$$u(t) \leq c\exp\left(\int_0^t \left[f(s) + cg(s)(E(s))^{-1}\exp\left(\int_0^s [f(\sigma)+h(\sigma)]\,d\sigma\right)\right]ds\right),$$
(2.7.5)

for $t \in [0, \alpha_1)$, where

$$E(t) = 1 - c\int_0^t g(\tau)\exp\left(\int_0^\tau [f(\sigma)+h(\sigma)]\,d\sigma\right)d\tau,$$

and $\alpha_1 = \sup\{t \in R_+ : E(t) > 0\}$. Moreover, if we assume that $E(t) > 0$ for all $t \in R_+$, then the inequality (2.7.5) remains valid for all $t \in R_+$. □

Theorem 2.7.3 *Let u, f, g, h and c be as in Theorem 2.7.2 and $p > 0$, $p \neq 1$ is a constant and suppose*

$$u(t) \leq c + \int_0^t f(s)u(s)\,ds + \int_0^t g(s)u^p(s)\left(u(s) + \int_0^s h(\sigma)u(\sigma)\,d\sigma\right)ds,$$
$$t \in R_+.$$
(2.7.6)

(i) If $0 < p < 1$ and $E_1(t)$ is defined by

$$E_1(t) = 1 - pc^p\int_0^t g(\tau)\exp\left(p\int_0^\tau [f(\sigma)+h(\sigma)]\,d\sigma\right)d\tau,$$
(2.7.7)

then

$$u(t) \leq c\exp\left(\int_0^t f(s)\,ds\right)\left[1 + (1-p)c^p\int_0^t g(s)(E_1(s))^{-1/p}\right.$$
$$\left.\times \exp\left(\int_0^s [pf(\tau)+h(\tau)]\,d\tau\right)\right]^{1/(1-p)},$$
(2.7.8)

for $t \in [0, \alpha_2)$, where $\alpha_2 = \sup\{t \in R_+ : E_1(t) > 0\}$.

(ii) If $1 < p < \infty$, then

$$u(t) \leq c \exp\left(\int_0^t f(s)\,ds\right) \left[1 - (p-1)c^p \int_0^t g(s)(\overline{E}_1(s))^{-1/p}\right.$$

$$\left. \times \exp\left(\int_0^s (pf(\tau) + h(\tau))\,d\tau\right) ds \right]^{-1/(p-1)}, \qquad (2.7.9)$$

for $t \in [0, \alpha_3)$, where $\overline{E}_1(t)$ is defined by the right-hand side of (2.7.7) for $1 < p < \infty$, and α_3 is the suprimum of $t \in R_+$ for which both $\overline{E}_1(t)$ and the expression between $[\ldots]$ in (2.7.9) are positive. Moreover, if we assume respectively that (1) $E_1(t) > 0$ and (2) $\overline{E}_1(t) > 0$ and the expression between $[\ldots]$ in (2.7.9) is positive for all $t \in R_+$, then the inequalities (2.7.8) and (2.7.9) remains valid for all $t \in R_+$.

□

Another interesting and useful inequality established by Pachpatte (1978b) is embodied in the following theorem.

Theorem 2.7.4 *Let u, f, g, h, c and p be as in Theorem 2.7.3 and suppose*

$$u(t) \leq c + \int_0^t f(s)u(s)\,ds + \int_0^t g(s)u^p(s)\left(u(s) + \int_0^s h(\sigma)u^{p+1}(\sigma)\,d\sigma\right) ds,$$

$$t \in R_+. \quad (2.7.10)$$

(i) *If $0 < p < 1$ and $E_2(t)$ is defined by*

$$E_2(t) = 1 - pc^p \int_0^t [g(\tau) + h(\tau)] \exp\left(p \int_0^\tau f(\sigma)\,d\sigma\right) d\tau, \qquad (2.7.11)$$

then

$$u(t) \leq c \exp\left(\int_0^t f(s)\,ds\right) \left[1 + (1-p)c^p \int_0^t g(s)(E_2(s))^{-1/p}\right.$$

$$\left. \times \exp\left(p \int_0^s f(\tau)\,d\tau\right) ds \right]^{1/(1-p)}, \qquad (2.7.12)$$

for $t \in [0, \alpha_4)$, where $\alpha_4 = \sup\{t \in R_+ : E_2(t) > 0\}$.

(ii) If $1 < p < \infty$, then

$$u(t) \leq c \exp\left(\int_0^t f(s)\,ds\right) \left[1 - (p-1)c^p \int_0^t g(s)(\overline{E}_2(s))^{-1/p}\right.$$

$$\left. \times \exp\left(p \int_0^s f(\tau)\,d\tau\right) ds\right]^{-1/(p-1)}, \qquad (2.7.13)$$

for $t \in [0, \alpha_5)$, where $\overline{E}_2(t)$ is defined by the right-hand side of (2.7.11) for $1 < p < \infty$ and α_5 is the suprimum of $t \in R_+$ for which both $\overline{E}_2(t)$ and the expression between [...] in (2.7.13) are positive.

Moreover, if we assume respectively (1) $E_2(t) > 0$ and (2) $\overline{E}_2(t) > 0$ and the expression between [...] in (2.7.13) is positive for all $t \in R_+$, then the inequalities (2.7.12) and (2.7.13) remain valid for all $t \in R_+$.

□

Proofs of Theorems 2.7.1–2.7.4: Since the proofs resemble one another, we give the details for Theorem 2.7.4 only; the proofs of Theorems 2.7.1–2.7.3 can be completed by following the proof of Theorem 2.7.4.

(i) Let $0 < p < 1$, and define a function $z(t)$ by the right-hand side of (2.7.10). Then $z(0) = c$, $u(t) \leq z(t)$ and

$$z'(t) \leq f(t)z(t) + g(t)z^p(t)\left(z(t) + \int_0^t h(\sigma)z^{p+1}(\sigma)\,d\sigma\right). \qquad (2.7.14)$$

Define a function $v(t)$ by

$$v(t) = z(t) + \int_0^t h(\sigma)z^{p+1}(\sigma)\,d\sigma. \qquad (2.7.15)$$

From (2.7.15), (2.7.14) and the fact that $z(t) \leq v(t)$, it follows that

$$v'(t) = z'(t) + h(t)z^{p+1}(t)$$
$$\leq f(t)v(t) + (g(t) + h(t))v^{p+1}(t). \qquad (2.7.16)$$

The inequality (2.7.16) can be written as

$$v^{-(p+1)}(t)v'(t) \leq f(t)v^{-p}(t) + (g(t) + h(t)). \tag{2.7.17}$$

Put $m(t) = v^{-p}(t)$ and so $m(0) = c^{-p}$, $m'(t) = -pv^{-p-1}(t)v'(t)$; then from (2.7.17) we obtain

$$m'(t) \geq -pf(t)m(t) - p(g(t) + h(t)). \tag{2.7.18}$$

The inequality (2.7.18) and the fact that $m(t) = v^{-p}(t)$ implies the estimate for $v(t)$ such that

$$v(t) \leq c(E_2(t))^{-1/p} \exp\left(\int_0^t f(\tau)\,d\tau\right). \tag{2.7.19}$$

Using (2.7.19) in (2.7.14) we get

$$z'(t) \leq f(t)z(t) + cg(t)z^p(t)(E_2(t))^{-1/p} \exp\left(\int_0^t f(\tau)\,d\tau\right). \tag{2.7.20}$$

Again by following the similar arguments as above, the inequality (2.7.20) implies the estimate for $z(t)$ such that

$$z(t) \leq c \exp\left(\int_0^t f(s)\,ds\right) \left[1 + (1-p)c^p \int_0^t g(s)(E_2(s))^{-1/p}\right.$$

$$\left. \times \exp\left(p\int_0^s f(\tau)\,d\tau\right)ds\right]^{1/(1-p)}. \tag{2.7.21}$$

Using (2.7.21) in $u(t) \leq z(t)$ we get the required inequality in (2.7.12).

(ii) Let $1 < p < \infty$ and define a function $z(t)$ by the right-hand side of (2.7.10). Then by following the same steps as in the proof of case (i), we see that all the inequalities from (2.7.14) to (2.7.20) hold for $0 \leq t < \alpha_5$. Furthermore, for $1 < p < \infty$, from (2.7.20) we get the estimate for $z(t)$ for $0 \leq t < \alpha_5$ such that

$$z(t) \leq c \exp\left(\int_0^t f(s)\,ds\right) \left[1 - (p-1)c^p \int_0^t g(s)(\overline{E}_2(s))^{-1/p}\right.$$

$$\left. \times \exp\left(p\int_0^s f(\tau)\,d\tau\right)ds\right]^{-1/(p-1)}. \tag{2.7.22}$$

Using (2.7.22) in $u(t) \leq z(t)$ we get the required inequality in (2.7.13). The last conclusion is obvious from the proofs of (i) and (ii) given above. The proof of Theorem 2.7.4 is complete.

∎

The following inequalities established by Pachpatte (1977a,c,d) can be similarly proved.

Theorem 2.7.5 *Let u, f, g and h be nonnegative continuous functions defined on R_+ and $c > 0$, $p > 0$, $p \neq 1$ are constants and*

$$u(t) \leq c + \int_0^t f(s)u(s)\,ds + \int_0^t f(s) \left(\int_0^s g(\tau) \right.$$
$$\left. \times \left(u(\tau) + \int_0^\tau h(\sigma) u^p(\sigma) d\sigma \right) d\tau \right) ds, \quad t \in R_+. \quad (2.7.23)$$

(i) *If $0 < p < 1$ and $F_0(t)$ is defined by*

$$F_0(t) = 1 + (1-p)c^{p-1} \int_0^t h(\tau) \exp\left(-(1-p)\int_0^\tau [f(\sigma) + g(\sigma)]\,d\sigma\right) d\tau,$$
$$(2.7.24)$$

then

$$u(t) \leq c \left[1 + \int_0^t f(\xi) \exp\left(\int_0^\xi f(\tau)\,d\tau \right) \left[1 + \int_0^\xi g(s)(F_0(s))^{1/(1-p)} \right. \right.$$
$$\left. \left. \times \exp\left(\int_0^s g(\tau)\,d\tau \right) ds \right] d\xi \right], \quad (2.7.25)$$

for $t \in R_+$.

(ii) *If $1 < p < \infty$, then*

$$u(t) \leq c \left[1 + \int_0^t f(\xi) \exp\left(\int_0^\xi f(\tau)\,d\tau \right) \left[1 + \int_0^\xi g(s)(\overline{F}_0(s))^{-1/(p-1)} \right. \right.$$
$$\left. \left. \times \exp\left(\int_0^s g(\tau)\,d\tau \right) ds \right] d\xi \right], \quad (2.7.26)$$

for $t \in [0, \beta)$, where $\overline{F}_0(t)$ is defined by the right-hand side of (2.7.24) for $1 < p < \infty$ and $\beta = \sup\{t \in R_+ : \overline{F}_0(t) > 0\}$. Moreover, if we assume that $\overline{F}_0(t) > 0$ for all $t \in R_+$, then the inequality (2.7.26) remains valid for all $t \in R_+$.

□

Theorem 2.7.6 Let u, f, g, h and c be as in Theorem 2.7.5 and suppose

$$u(t) \leq c + \int_0^t f(s)u(s)\,ds + \int_0^t f(s)\left(\int_0^s g(\tau)u(\tau)\right.$$

$$\left. \times \left(u(\tau) + \int_0^\tau h(\sigma)u(\sigma)\,d\sigma\right) d\tau\right) ds \qquad (2.7.27)$$

for $t \in R_+$, then

$$u(t) \leq c\left[1 + \int_0^t f(\xi)\exp\left(\int_0^\xi [f(s) + cg(s)(F(s))^{-1}]\right.\right.$$

$$\left.\left. \times \exp\left(\int_0^s [f(\tau) + h(\tau)]\,d\tau\right) ds\right) d\xi\right], \qquad (2.7.28)$$

for $t \in [0, \beta_1)$, where

$$F(t) = 1 - c\int_0^t g(s)\exp\left(\int_0^s [f(\tau) + h(\tau)]\,d\tau\right) ds,$$

and $\beta_1 = \sup\{t \in R_+ : F(t) > 0\}$. Moreover, if we assume that $F(t) > 0$ for all $t \in R_+$, then the inequality (2.7.28) remains valid for all $t \in R_+$.

□

Theorem 2.7.7 Let u, f, g, h, c and p be as in Theorem 2.7.3 and suppose

$$u(t) \leq c + \int_0^t f(s)u(s)\,ds + \int_0^t f(s)\left(\int_0^s g(\tau)u^p(\tau)\right.$$

$$\left. \times \left(u(\tau) + \int_0^\tau h(\sigma)u(\sigma)\,d\sigma\right) d\tau\right) ds, \qquad (2.7.29)$$

for $t \in R_+$.

(i) If $0 < p < 1$ and $F_1(t)$ is defined by

$$F_1(t) = 1 - pc^p \int_0^t g(s) \exp\left(p \int_0^s [f(\tau) + h(\tau)] d\tau\right) ds, \quad (2.7.30)$$

then

$$u(t) \le c \left[1 + \int_0^t f(s) \exp\left(\int_0^s f(\tau) d\tau\right) \left\{1 + (1-p)c^p \right.\right.$$
$$\left.\left. \times \int_0^s g(\tau)(F_1(\tau))^{-1} \exp\left(\int_0^\tau [pf(\sigma) + h(\sigma)] d\sigma\right) d\tau\right\}^{1/(1-p)} ds\right],$$

$$(2.7.31)$$

for $t \in [0, \beta_2)$, where $\beta_2 = \sup\{t \in R_+ : F_1(t) > 0\}$.

(ii) If $1 < p < \infty$, then

$$u(t) \le c \left[1 + \int_0^t f(s) \exp\left(\int_0^s f(\tau) d\tau\right) \left\{1 - (p-1)c^p \right.\right.$$
$$\left.\left. \times \int_0^s g(\tau)(\overline{F}_1(\tau))^{-1} \exp\left(\int_0^\tau [pf(\sigma) + h(\sigma)] d\sigma\right) d\tau\right\}^{-1/(p-1)}\right],$$

$$(2.7.32)$$

for $t \in [0, \beta_3)$, where $\overline{F}_1(t)$ is defined by the right-hand side of (2.7.30) for $1 < p < \infty$, and β_3 is the suprimum of $t \in R_+$ for which both $\overline{F}_1(t)$ and the expression between $\{\ldots\}$ in (2.7.32) are positive.

Moreover, if we assume respectively that (1) $F_1(t) > 0$ and (2) $\overline{F}_1(t) > 0$ and the expression between $\{\ldots\}$ in (2.7.32) is positive for all $t \in R_+$, then the inequalities (2.7.31) and (2.7.32) remains valid for all $t \in R_+$.

$\square$

Theorem 2.7.8 *Let u, f, g, h, c and p be as in Theorem 2.7.4 and suppose*

$$u(t) \le c + \int_0^t f(s)u(s)\,ds + \int_0^t f(s)\left(\int_0^s g(\tau)u^p(\tau)\right.$$
$$\left.\times \left(u(\tau) + \int_0^\tau h(\sigma)u^{p+1}(\sigma)\,d\sigma\right)d\tau\right)ds, \qquad (2.7.33)$$

for $t \in R_+$.
(i) If $0 < p < 1$ and $F_2(t)$ is defined by

$$F_2(t) = 1 - pc^p \int_0^t [g(\tau) + h(\tau)] \exp\left(p \int_0^\tau f(\sigma)\,d\sigma\right)d\tau, \qquad (2.7.34)$$

then

$$u(t) \le c\left[1 + \int_0^t f(\xi)\exp\left(\int_0^\xi f(\tau)\,d\tau\right)\left\{1 + (1-p)c^p\right.\right.$$
$$\left.\left.\times \int_0^\xi g(s)(F_2(s))^{-1/p}\exp\left(p\int_0^s f(\tau)\,d\tau\right)ds\right\}^{1/(1-p)} d\xi\right], \quad (2.7.35)$$

for $t \in [0, \beta_4)$, where $\beta_4 = \sup\{t \in R_+ : F_2(t) > 0\}$.
(ii) If $1 < p < \infty$, then

$$u(t) \le c\left[1 + \int_0^t f(\xi)\exp\left(\int_0^\xi f(\tau)\,d\tau\right)\left\{1 - (p-1)c^p\right.\right.$$
$$\left.\left.\times \int_0^\xi g(s)(\overline{F}_2(s))^{-1/p}\exp\left(p\int_0^s f(\tau)\,d\tau\right)ds\right\}^{-1/(p-1)} d\xi\right], \quad (2.7.36)$$

for $t \in [0, \beta_5)$, where $\overline{F}_2(t)$ is defined by the right-hand side of (2.7.34) for $1 < p < \infty$ and β_5 is the suprimum of $t \in R_+$ for which both $\overline{F}_2(t)$ and the expression between $\{\ldots\}$ in (2.7.36) are positive.

Moreover, if we assume respectively (1) $F_2(t) > 0$ and (2) $\overline{F}_2(t) > 0$ and the expression between {...} in (2.7.36) is positive for all $t \in R_+$, then the inequalities (2.7.35) and (2.7.36) remain valid for all $t \in R_+$.

□

As noted above, the inequality in Theorem 2.7.5 is dealt with in Pachpatte (1977a) and the inequalities in Theorems 2.7.6 and 2.7.7 are dealt with in Pachpatte (1977c), while the inequality in Theorem 2.7.8 is motivated by the inequality given in Theorem 2.7.4. In fact, the inequalities in Theorems 2.7.1 to 2.7.8 can be considered as further generalizations of the Gronwall–Bellman inequality, and in the special cases when $f(t) = 0$ these inequalities reduce to various new inequalities which can be used very effectively in certain applications.

2.8 Pachpatte's Inequalities II

The nonlinear integral inequalities of Bihari (1956) and Langenhop (1960) are often used to study the various problems in the theory of differential and integral equations. Pachpatte (1974d) established some new Bihari-like integral inequalities which can be used as handy tools in the study of certain classes of general integral and integro-differential equations for which the inequalities of Bihari and Langenhop do not apply directly. Since then, in a series of papers (Pachpatte 1975a,b,d,h, 1976c), Pachpatte has obtained a number of new results related to those given in Pachpatte (1974d), which are very effective in various applications. This section deals with some basic inequalities given in the above-mentioned papers.

Pachpatte (1974d) gave the inequalities in the following two theorems.

Theorem 2.8.1 *Let u and g be nonnegative continuous functions defined on R_+. Let $H(u)$ be a continuous nondecreasing function defined on R_+ and $H(u) > 0$ on $(0, \infty)$. If*

$$u(t) \leq c + \int_0^t g(s) \left(u(s) + \int_0^s g(\sigma) H(u(\sigma)) \, d\sigma \right) ds, \qquad (2.8.1)$$

for $t \in R_+$, where $c \geq 0$ is a constant, then for $0 \leq t \leq t_1$,

$$u(t) \le c + \int_0^t g(s) G^{-1}\left[G(c) + \int_0^s g(\sigma)\,d\sigma\right] ds, \qquad (2.8.2)$$

where

$$G(r) = \int_{r_0}^{r} \frac{ds}{s + H(s)}, \qquad r > 0,\ r_0 > 0, \qquad (2.8.3)$$

and G^{-1} is the inverse of G, and $t_1 \in R_+$ is chosen so that

$$G(c) + \int_0^t g(\sigma)\,d\sigma \in \mathrm{Dom}(G^{-1}),$$

for all $t \in R_+$ lying in the interval $0 \le t \le t_1$. $\square$

Theorem 2.8.2 *Let u, f and g be nonnegative continuous functions defined on R_+. Let $H(u)$ be a continuous nondecreasing and subadditive function defined on R_+ and $H(u) > 0$ on $(0, \infty)$. If*

$$u(t) \le f(t) + \int_0^t g(s)\left(u(s) + \int_0^s g(\sigma) H(u(\sigma))\,d\sigma\right) ds, \qquad (2.8.4)$$

for $t \in R_+$, then for $0 \le t \le t_2$,

$$u(t) \le f(t) + A(t) + \int_0^t g(s) G^{-1}\left[G(A(t)) + \int_0^s g(\sigma)\,d\sigma\right] ds, \qquad (2.8.5)$$

where

$$A(t) = \int_0^t g(\tau)\left(f(\tau) + \int_0^\tau g(\sigma) H(f(\sigma))\,d\sigma\right) d\tau, \qquad (2.8.6)$$

G, G^{-1} are as defined in Theorem 2.8.1 and $t_2 \in R_+$ is chosen so that

$$G(A(t)) + \int_0^t g(\sigma)\,d\sigma \in \mathrm{Dom}(G^{-1}),$$

for all $t \in R_+$ lying in the interval $0 \leq t \leq t_2$.

□

The next two theorems deal with the slight variants of the inequalities given by Pachpatte (1976c) and (1975h) respectively.

Theorem 2.8.3 Let u, h and k be nonnegative continuous functions defined on R_+. Let $H(u)$ be a continuous nondecreasing function defined on R_+ and $H(u) > 0$ on $(0, \infty)$. If

$$u(t) \leq c + \int_0^t h(s) H\left(u(s) + \int_0^s k(\sigma) H(u(\sigma)) \, d\sigma\right) ds, \quad (2.8.7)$$

for $t \in R_+$, where $c \geq 0$ is a constant, then for $0 \leq t \leq t_3$,

$$u(t) \leq c + \int_0^t h(s) H\left(F^{-1}\left[F(c) + \int_0^s [h(\sigma) + k(\sigma)] \, d\sigma\right]\right) ds, \quad (2.8.8)$$

where

$$F(r) = \int_{r_0}^r \frac{ds}{H(s)}, \quad r > 0, r_0 > 0, \quad (2.8.9)$$

F^{-1} is the inverse of F and $t_3 \in R_+$ is chosen so that

$$F(c) + \int_0^t [h(\sigma) + k(\sigma)] \, d\sigma \in \text{Dom}(F^{-1}),$$

for all $t \in R_+$ lying in the interval $0 \leq t \leq t_3$.

□

Theorem 2.8.4 Let u, f, g, h and k be nonnegative continuous functions defined on R_+. Let $H(u)$ be a continuous nondecreasing subadditive and submultiplicative function defined on R_+ and $H(u) > 0$ on $(0, \infty)$. If

$$u(t) \leq f(t) + g(t) \int_0^t h(s) H\left(u(s) + g(s) \int_0^s k(\sigma) H(u(\sigma)) \, d\sigma\right) ds,$$

$$(2.8.10)$$

for $t \in R_+$, then for $0 \leq t \leq t_4$,

$$u(t) \leq f(t) + g(t) \left[B(t) + \int_0^t h(s) H(g(s)) \right.$$

$$\left. \times H \left(F^{-1} \left[F(B(t)) + \int_0^s H(g(\sigma))[h(\sigma) + k(\sigma)] \, d\sigma \right] \right) ds \right], \tag{2.8.11}$$

where

$$B(t) = \int_0^t h(s) H \left(f(s) + g(s) \int_0^s k(\sigma) H(f(\sigma)) \, d\sigma \right) ds, \tag{2.8.12}$$

F and F^{-1} are as defined in Theorem 2.8.3 and $t_4 \in R_+$ is chosen so that

$$F(B(t)) + \int_0^t H(g(\sigma))[h(\sigma) + k(\sigma)] \, d\sigma \in \text{Dom}(F^{-1}),$$

for all $t \in R_+$ lying in the interval $0 \leq t \leq t_4$. □

Proofs of Theorems 2.8.1–2.8.4:. We will give the details for the proofs of Theorems 2.8.1 and 2.8.4 only; the proofs of Theorems 2.8.2 and 2.8.3 can be completed similarly.

In order to prove Theorem 2.8.1, it is sufficient to assume that c is positive, since the standard limiting argument can be used to treat the remaining case. Define a function $v(t)$ by the right member of (2.8.1). Then $v(0) = c$, $u(t) \leq v(t)$ and

$$v'(t) \leq g(t) \left(v(t) + \int_0^t g(\sigma) H(v(\sigma)) \, d\sigma \right). \tag{2.8.13}$$

If we put

$$m(t) = v(t) + \int_0^t g(\sigma) H(v(\sigma)) \, d\sigma, \tag{2.8.14}$$

it follows from (2.8.14), (2.8.13) and the fact that $v(t) \leq m(t)$, we have

$$m'(t) \leq g(t)(m(t) + H(m(t))). \tag{2.8.15}$$

Dividing both sides of (2.8.15) by $m(t) + H(m(t))$, integrating it from 0 to t and using (2.8.3), we obtain

$$G(m(t)) \leq G(c) + \int_0^t g(\sigma) \, d\sigma. \tag{2.8.16}$$

From (2.8.16) and (2.8.13) we have

$$v'(t) \leq g(t) G^{-1} \left[g(c) + \int_0^t g(\sigma) \, d\sigma \right]. \tag{2.8.17}$$

Now by taking $t = s$ in (2.8.17) and integrating it from 0 to t and substituting the value of $v(t)$ in $u(t) \leq v(t)$ we obtain the desired bound in (2.8.2). The subinterval $0 \leq t \leq t_1$ is obvious.

In order to prove Theorem 2.8.4, we define a function $v(t)$ by

$$v(t) = \int_0^t h(s) H \left(u(s) + g(s) \int_0^s k(\sigma) H(u(\sigma)) \, d\sigma \right) ds. \tag{2.8.18}$$

From (2.8.18) and using $u(t) \leq f(t) + g(t)v(t)$ and the subadditivity and submultiplicativity of H we have

$$v(t) \leq \int_0^t h(s) H \left(f(s) + g(s)v(s) + g(s) \int_0^s k(\sigma) H(f(\sigma) + g(\sigma)v(\sigma)) \, d\sigma \right) ds$$

$$\leq B(t) + \int_0^t h(s) H(g(s)) H \left(v(s) + \int_0^s k(\sigma) H(g(\sigma)) H(v(\sigma)) \, d\sigma \right) ds, \tag{2.8.19}$$

where $B(t)$ is given by (2.8.12).

Fix $T \in R_+$, any arbitrary number. Now from (2.8.19) we have for all $0 \le t \le T$,

$$v(t) \le B(T) + \int_0^t h(s)H(g(s))H\left(v(s) + \int_0^s k(\sigma)H(g(\sigma))H(v(\sigma))\,d\sigma\right)ds.$$

Now an application of Theorem 2.8.3 yields

$$v(t) \le B(T) + \int_0^t h(s)H(g(s))H\left(F^{-1}\left[F(B(T)) + \int_0^s H(g(\sigma))\right.\right.$$
$$\left.\left.\times [h(\sigma) + k(\sigma)]\,d\sigma\right]\right)ds, \quad 0 \le t \le T. \quad (2.8.20)$$

Now by setting $t = T$ in (2.8.20) and using the fact that $u(t) \le f(t) + g(t)v(t)$ is true for $t = T$, we obtain

$$u(T) \le f(T) + g(T)\left[B(T) + \int_0^T h(s)H(g(s))H\left(F^{-1}\left[F(B(T))\right.\right.\right.$$
$$\left.\left.\left.+ \int_0^s H(g(\sigma))[h(\sigma) + k(\sigma)]\,d\sigma\right]\right)ds\right], \quad 0 \le t \le T \le t_4. \quad (2.8.21)$$

Since T is arbitrary, the conclusion (2.8.11) is clear from (2.8.21). ∎

Pachpatte (1975b) established the inequalities given in the following two theorems.

Theorem 2.8.5 *Let u, f, g and h be nonnegative continuous functions defined on R_+. Let $w(u)$ be a continuous nondecreasing and submultiplicative function defined on R_+ and $w(u) > 0$ on $(0, \infty)$. If*

$$u(t) \le c + \int_0^t f(s)u(s)\,ds + \int_0^t f(s)\left(\int_0^s g(\sigma)u(\sigma)\,d\sigma\right)ds$$
$$+ \int_0^t h(s)w(u(s))\,ds, \quad (2.8.22)$$

for $t \in R_+$, where c is a positive constant, then for $0 \leq t \leq t_1$,

$$u(t) \leq a(t)\Omega^{-1}\left[\Omega(c) + \int_0^t h(s)w(a(s))\,ds\right], \qquad (2.8.23)$$

where

$$a(t) = 1 + \int_0^t f(s)\exp\left(\int_0^s [f(\sigma) + g(\sigma)]\,d\sigma\right)ds, \qquad (2.8.24)$$

$$\Omega(r) = \int_{r_0}^r \frac{ds}{w(s)}, \quad r > 0, r_0 > 0, \qquad (2.8.25)$$

and Ω^{-1} is the inverse of Ω, and $t_1 \in R_+$ is chosen so that

$$\Omega(c) + \int_0^t h(s)w(a(s))\,ds \in \mathrm{Dom}(\Omega^{-1}),$$

for all $t \in R_+$ lying in the interval $0 \leq t \leq t_1$. $\square$

Theorem 2.8.6 *Let u, f, g and h be nonnegative continuous functions defined on R_+. Let $w(u)$ be a continuous nondecreasing subadditive and submultiplicative function defined on R_+ and $w(u) > 0$ on $(0, \infty)$. Let $p(t) > 0$, $\phi(t) \geq 0$ be continuous and nondecreasing functions defined on R_+, and $\phi(0) = 0$. If*

$$u(t) \leq p(t) + \int_0^t f(s)u(s)\,ds + \int_0^t f(s)\left(\int_0^s g(\sigma)u(\sigma)\,d\sigma\right)ds$$

$$+ \phi\left(\int_0^t h(s)w(u(s))\,ds\right), \qquad (2.8.26)$$

for $t \in R_+$, then for $0 \leq t \leq t_2$,

$$u(t) \leq a(t)\left[p(t) + \phi\left(\psi^{-1}\left[\psi(b(t)) + \int_0^t h(s)w(a(s))\,ds\right]\right)\right], \qquad (2.8.27)$$

where $a(t)$ is defined by (2.8.24),

$$b(t) = \int_0^t h(s)w(a(s)p(s))\,ds, \qquad (2.8.28)$$

$$\psi(r) = \int_{r_0}^r \frac{ds}{w(\phi(s))}, \quad r > 0,\ r_0 > 0, \qquad (2.8.29)$$

and ψ^{-1} is the inverse of ψ, and $t_2 \in R_+$ is chosen so that

$$\psi(b(t)) + \int_0^t h(s)w(a(s))\,ds \in \mathrm{Dom}(\psi^{-1}),$$

for all $t \in R_+$ lying in the interval $0 \le t \le t_2$. $\square$

The following theorem is motivated by the inequality given by Pachpatte (1975a, Theorem 4) which basically involves the comparison principle.

Theorem 2.8.7 *Let u, f, g, h, p and ϕ be as in Theorem 2.8.6. Let $w(t, u)$ be a nonnegative continuous and monotonic nondecreasing function in $u \ge 0$ for each fixed $t \in R_+$. If*

$$u(t) \le p(t) + \int_0^t f(s)u(s)\,ds + \int_0^t f(s)\left(\int_0^s g(\sigma)u(\sigma)\,d\sigma\right) ds$$

$$+ \phi\left(\int_0^t h(s)w(s, u(s))\,ds\right), \qquad (2.8.30)$$

for $t \in R_+$, then

$$u(t) \le a(t)[p(t) + \phi(r(t))], \qquad (2.8.31)$$

where $a(t)$ is defined by (2.8.24), and $r(t)$ is the maximal solution of

$$r'(t) = h(t)w(t, a(t)[p(t) + \phi(r(t))]), \quad r(0) = 0, \qquad (2.8.32)$$

existing on R_+.

$\square$

The proofs of Theorems 2.8.5–2.8.7 follow by the same arguments as in the proofs of Theorems 2.4.1–2.4.3 and by making use of Theorem 1.7.4. Here we omit the details.

We next establish some integral inequalities, by considering one linear term and one or two nonlinear terms on the right-hand side. In what follows, we use the class of functions $\mathscr{F}$ as defined earlier in Section 2.5.

The inequalities in the following three theorems were established by Pachpatte (1975b).

Theorem 2.8.8 *Let u and g be nonnegative continuous functions defined on R_+ and $n(t) \geq 1$ be a monotonic nondecreasing continuous function defined on R_+ and $H \in \mathscr{F}$. If*

$$u(t) \leq n(t) + \int_0^t g(s)\left(u(s) + \int_0^s g(\sigma)H(u(\sigma))\,d\sigma\right), \qquad (2.8.33)$$

for $t \in R_+$, then for $0 \leq t \leq t_1$,

$$u(t) \leq n(t)a(t), \qquad (2.8.34)$$

where

$$a(t) = 1 + \int_0^t g(s)G^{-1}\left[G(1) + \int_0^s g(\sigma)\,d\sigma\right]ds, \qquad (2.8.35)$$

$$G(r) = \int_{r_0}^r \frac{ds}{s + H(s)}, \quad r > 0, r_0 > 0, \qquad (2.8.36)$$

and G^{-1} is the inverse of G, and $t_1 \in R_+$ is chosen so that

$$G(1) + \int_0^t g(s)\,ds \in \mathrm{Dom}(G^{-1}),$$

for all $t \in R_+$ lying in the interval $0 \leq t \leq t_1$. □

Proof: Since $n(t) \geq 1$ is monotonic nondecreasing and $H \in \mathscr{F}$, we observe that

$$\frac{u(t)}{n(t)} \leq 1 + \int_0^t g(s) \left(\frac{u(s)}{n(s)} + \int_0^s g(\sigma) H\left(\frac{u(\sigma)}{n(\sigma)}\right) d\sigma \right) ds.$$

Now an application of Theorem 2.8.1 yields the desired inequality in (2.8.34).

∎

Theorem 2.8.9 *Let u, g and h be nonnegative continuous functions defined on R_+. Let $H \in \mathscr{F}$ and let w be the same function as defined in Theorem 2.8.5. If*

$$u(t) \leq c + \int_0^t g(s) \left(u(s) + \int_0^s g(\sigma) H(u(\sigma)) d\sigma \right) ds + \int_0^t h(s) w(u(s)) ds, \quad (2.8.37)$$

for $t \in R_+$, where $c \geq 1$ is a constant, then for $0 \leq t \leq t_2$,

$$u(t) \leq a(t) \Omega^{-1} \left[\Omega(c) + \int_0^t h(s) w(a(s)) ds \right], \quad (2.8.38)$$

where $a(t)$ is defined by (2.8.35) in which G, G^{-1} are as defined in Theorem 2.8.8, Ω, Ω^{-1} are as defined in Theorem 2.8.5, and $t_2 \in R_+$ is chosen so that

$$G(1) + \int_0^t g(s) ds \in \text{Dom}(G^{-1}),$$

$$\Omega(c) + \int_0^t h(s) w(a(s)) ds \in \text{Dom}(\Omega^{-1}),$$

for all $t \in R_+$ lying in the interval $0 \leq t \leq t_2$.

□

Proof: Define a function $n(t)$ by

$$n(t) = c + \int_0^t h(s)w(u(s))\,ds. \qquad (2.8.39)$$

Then (2.8.37) can be restated as

$$u(t) \le n(t) + \int_0^t g(s)\left(u(s) + \int_0^s g(\sigma)H(u(\sigma))\,d\sigma\right)ds.$$

Since $n(t) \ge 1$ is monotonic nondecreasing and $H \in \mathscr{F}$, from Theorem 2.8.8 we have

$$u(t) \le a(t)n(t),$$

where $a(t)$ is defined by (2.8.35).

The rest of the proof follows by the same arguments as in the proof of Theorem 2.4.1 with suitable changes.

∎

Theorem 2.8.10 *Let u, g, h and H be as in Theorem 2.8.9. Let w, p and ϕ be as in Theorem 2.8.6 except that here $p \ge 1$. If*

$$u(t) \le p(t) + \int_0^t g(s)\left(u(s) + \int_0^s g(\sigma)H(u(\sigma))\,d\sigma\right)ds$$

$$+ \phi\left(\int_0^t h(s)w(u(s))\,ds\right), \qquad (2.8.40)$$

for $t \in R_+$, then for $0 \le t \le t_3$,

$$u(t) \le a(t)\left[p(t) + \phi\left(\psi^{-1}\left[\psi\left(\int_0^t h(s)w(a(s)p(s))\,ds\right.\right.\right.\right.$$

$$\left.\left.\left.\left.+ \int_0^t h(s)w(a(s))\,ds\right]\right)\right)\right], \qquad (2.8.41)$$

where $a(t)$ is defined by (2.8.35) in which G, G^{-1} are as defined in Theorem 2.8.8, ψ, ψ^{-1} are as defined in Theorem 2.8.6, and $t_3 \in R_+$ is *chosen so that*

$$G(1) + \int_0^t g(s)\,ds \in \text{Dom}(G^{-1}),$$

$$\psi\left(\int_0^t h(s)w(a(s)p(s))\,ds\right) + \int_0^t h(s)w(a(s))\,ds \in \text{Dom}(\psi^{-1}),$$

for all $t \in R_+$ lying in the interval $0 \le t \le t_3$.

□

The proof of this theorem can be completed by following the proof of Theorem 2.4.2 and applying Theorem 2.8.8. Here the details are omitted.

The inequalities in the next three theorems are also established by Pachpatte (1976c).

Theorem 2.8.11 *Let u, f and g be nonnegative continuous functions defined on R_+, $n(t) \ge 1$ be monotonic nondecreasing continuous function defined on R_+, and $H \in \mathscr{F}$. If*

$$u(t) \le n(t) + \int_0^t f(s)H\left(u(s) + \int_0^s g(\sigma)H(u(\sigma))\,d\sigma\right)\,ds, \qquad (2.8.42)$$

for $t \in R_+$, then for $0 \le t \le t_1$,

$$u(t) \le b(t)n(t), \qquad (2.8.43)$$

where

$$b(t) = 1 + \int_0^t f(s)H\left(F^{-1}\left[F(1) + \int_0^s [f(\sigma) + g(\sigma)]\,d\sigma\right]\right)\,ds, \qquad (2.8.44)$$

$$F(r) = \int_{r_0}^r \frac{ds}{H(s)}, \quad r > 0, r_0 > 0, \qquad (2.8.45)$$

F^{-1} is the inverse of F, and $t_1 \in R_+$ is chosen so that

$$F(1) + \int_0^t [f(\sigma) + g(\sigma)]\,d\sigma \in \text{Dom}(F^{-1}),$$

for all $t \in R_+$ lying in the interval $0 \le t \le t_1$.

□

Theorem 2.8.12 Let u, f, g and h be nonnegative continuous functions defined on R_+, and $H \in \mathscr{F}$. Let w be the same function as defined in Theorem 2.8.5. If

$$u(t) \leq c + \int_0^t f(s) H\left(u(s) + \int_0^s g(\sigma) H(u(\sigma)) \, d\sigma\right) ds$$

$$+ \int_0^t h(s) w(u(s)) \, ds, \tag{2.8.46}$$

for $t \in R_+$, where $c \geq 1$ is a constant, then for $0 \leq t \leq t_2$,

$$u(t) \leq b(t) \Omega^{-1}\left[\Omega(c) + \int_0^t h(s) w(b(s)) \, ds\right], \tag{2.8.47}$$

where $b(t)$ is defined by (2.8.44), in which F, F^{-1} are as defined in Theorem 2.8.11, Ω, Ω^{-1} are as defined in Theorem 2.8.5, and $t_2 \in R_+$ is chosen so that

$$F(1) + \int_0^t [f(s) + g(s)] \, ds \in \mathrm{Dom}(F^{-1}),$$

$$\Omega(c) + \int_0^t h(s) w(b(s)) \, ds \in \mathrm{Dom}(\Omega^{-1}),$$

for all $t \in R_+$ lying in the interval $0 \leq t \leq t_2$.

□

Theorem 2.8.13 Let u, f, g, h and w be as in Theorem 2.8.6. Let H, p and ϕ be as in Theorem 2.8.10. If

$$u(t) \leq p(t) + \int_0^t f(s) H\left(u(s) + \int_0^s g(\sigma) H(u(\sigma)) \, d\sigma\right) ds$$

$$+ \phi\left(\int_0^t h(s) w(u(s)) \, ds\right), \tag{2.8.48}$$

for $t \in R_+$, then for $0 \leq t \leq t_3$,

$$u(t) \leq b(t) \left[p(t) + \phi \left(\psi^{-1} \left[\psi \left(\int_0^t h(s)w(b(s)p(s)) \, ds \right) \right. \right. \right.$$
$$\left. \left. \left. + \int_0^t h(s)w(b(s)) \, ds \right] \right) \right], \quad (2.8.49)$$

where $b(t)$ is defined by (2.8.44), in which F, F^{-1} are as defined in Theorem 2.8.11, ψ, ψ^{-1} are as defined in Theorem 2.8.6, and $t_3 \in R_+$ is chosen so that

$$F(1) + \int_0^t [f(s) + g(s)] \, ds \in \mathrm{Dom}(F^{-1}),$$

$$\psi \left(\int_0^t h(s)w(b(s)p(s)) \, ds \right) + \int_0^t h(s)w(b(s)) \, ds \in \mathrm{Dom}(\psi^{-1}),$$

for all $t \in R_+$ lying in the interval $0 \leq t \leq t_3$.

$\square$

The details of the proofs of Theorems 2.8.11–2.8.13 follows by the same arguments as in the proofs of Theorems 2.8.8–2.8.10. Here we omit the details.

2.9 Integro-differential Inequalities

In the theory of ordinary differential and integro-differential equations one often has to deal with certain differential and integro-differential inequalities. This section deals with some basic nonlinear integro-differential inequalities established by Pachpatte (1977b, 1981a) in the course of attempts to apply them in certain applications.

Pachpatte (1977b) gave the following useful integro-differential inequality.

Theorem 2.9.1 *Let u, u' and b be nonnegative continuous functions defined on R_+ and $a > 0$ is a constant, If*

$$u'(t) \leq a + \int_0^t b(s)u'(s)(u(s) + u'(s))\,ds, \tag{2.9.1}$$

for $t \in R_+$, and $E(t)$ is defined by

$$E(t) = 1 - [a + u(0)] \int_0^t e^\sigma b(\sigma)\,d\sigma, \tag{2.9.2}$$

then

$$u'(t) \leq a \exp\left([a + u(0)] \int_0^t e^s b(s)(E(s))^{-1}\,ds\right), \tag{2.9.3}$$

for $t \in [0, \beta)$, where $\beta = \sup\{t \in R_+ : E(t) > 0\}$. Moreover, if we assume that $E(t) > 0$ for all $t \in R_+$, then the inequality (2.9.3) remains valid for all $t \in R_+$.

□

Proof: Define a function $z(t)$ by the right-hand side of (2.9.1). Then $z(0) = a$, $u'(t) \leq z(t)$ and

$$z'(t) = b(t)u'(t)(u(t) + u'(t))$$
$$\leq b(t)z(t)(u(t) + z(t)). \tag{2.9.4}$$

Integrating $u'(t) \leq z(t)$ from 0 to t and using it into (2.9.4) we get

$$z'(t) \leq b(t)z(t)\left(u(0) + z(t) + \int_0^t z(s)\,ds\right), \tag{2.9.5}$$

Define a function $v(t)$ by

$$v(t) = u(0) + z(t) + \int_0^t z(s)\,ds. \tag{2.9.6}$$

Differentiating (2.9.6) and using the facts that $z'(t) \leq b(t)z(t)v(t)$ from (2.9.5) and $z(t) \leq v(t)$ from (2.9.6) we obtain

$$v'(t) \leq b(t)v^2(t) + v(t).$$

The above inequality implies the estimate for $v(t)$ such that

$$v(t) \leq [a + u(0)] e^t (E(t))^{-1}, \quad t \in [0, \beta), \qquad (2.9.7)$$

Using (2.9.7) in (2.9.5) we have

$$z'(t) \leq b(t)z(t)[a + u(0)] e^t (E(t))^{-1}, \quad t \in [0, \beta).$$

The above inequality implies the estimate

$$z(t) \leq a \exp\left([a + u(0)] \int_0^t e^s b(s)(E(s))^{-1} ds\right), \quad t \in [0, \beta). \qquad (2.9.8)$$

Now using (2.9.8) in $u(t) \leq z(t)$ we get the desired inequality in (2.9.3).

∎

In the following theorem, a slightly different version of Theorem 2.9.1 is given which is convenient in certain applications.

Theorem 2.9.2 *Let u, u' and c be nonnegative continuous functions defined on R_+ and $a > 0$, $b \geq 0$ be constants. If*

$$u'(t) \leq a + b\left(u(t) + \int_0^t c(s)u'(s)(u(s) + u'(s)) ds\right), \qquad (2.9.9)$$

for $t \in R_+$ and $E_0(t)$ is defined by

$$E_0(t) = 1 - b[a + (1+b)u(0)] \int_0^t c(s) \exp((1+b)s) ds, \qquad (2.9.10)$$

then

$$u'(t) \leq [a + bu(0)] \exp\left(b \int_0^t \left[1 + c(s)(E(s))^{-1}(a + (1+b)u(0))\right. \right.$$
$$\left.\left. \times \exp((1+b)s)\right] ds\right), \qquad (2.9.11)$$

for $t \in [0, \beta_1)$, where $\beta_1 = \sup\{t \in R_+ : E_0(t) > 0\}$. Moreover, if we assume that $E_0(t) > 0$ for all $t \in R_+$, then the inequality (2.9.11) remains valid for all $t \in R_+$.

□

Proof: Define a function $z(t)$ by the right-hand side of (2.9.9). Then $z(0) = a + bu(0)$, $u'(t) \leq z(t)$ and

$$z'(t) = b(u'(t) + c(t)u'(t)(u(t) + u'(t)))$$
$$\leq bz(t)[1 + c(t)(u(t) + z(t))].$$

The rest of the proof can be completed by following the proof of Theorem 2.9.1 given above with suitable changes.

■

Another interesting and useful integro-differential inequality of the Bihari type established by Pachpatte (1977b) is given in the following theorem.

Theorem 2.9.3 *Let u, u' and b be nonnegative continuous functions defined on R_+ and $b \geq 1$. Let $w(u)$ be continuous nondecreasing function defined on R_+ and $w(u) > 0$ on $(0, \infty)$ and $a \geq 0$ be a constant. If*

$$u'(t) \leq a + \int_0^t b(s)w(u(s) + u'(s))\,ds, \quad (2.9.12)$$

for $t \in R_+$, then for $0 \leq t \leq t_1$,

$$u'(t) \leq a + \int_0^t b(s)w\left(G^{-1}\left[G(a + u(0)) + \int_0^s b(\sigma)\,d\sigma\right]\right)ds, \quad (2.9.13)$$

where

$$G(r) = \int_{r_0}^r \frac{ds}{s + w(s)}, \quad r > 0, r_0 > 0, \quad (2.9.14)$$

and G^{-1} is the inverse of G, and $t_1 \in R_+$ is chosen so that

$$G(a + u(0)) + \int_0^t b(\sigma)\,d\sigma \in \mathrm{Dom}(G^{-1}),$$

for all $t \in R_+$ lying in the interval $0 \le t \le t_1$.

□

Proof: It is sufficient to assume that a is positive, since a standard limiting argument can be used to treat the remaining case. Define a function $z(t)$ by the right-hand side of (2.9.12). Then $z(0) = a$, $u'(t) \le z(t)$ and

$$z'(t) = b(t)w(u(t) + u'(t)) \le b(t)w(u(t) + z(t)). \tag{2.9.15}$$

Integrating $u'(t) \le z(t)$ from 0 to t and using it in (2.9.15) we obtain

$$z'(t) \le b(t)w\left(u(0) + z(t) + \int_0^t z(s)\,ds\right). \tag{2.9.16}$$

Define a function $v(t)$ by

$$v(t) = u(0) + z(t) + \int_0^t z(s)\,ds. \tag{2.9.17}$$

Differentiating (2.9.17) and using the facts that $z'(t) \le b(t)w(v(t))$, and $z(t) \le v(t)$ from (2.9.16) and (2.9.17) and $b(t) \ge 1$ we have

$$v'(t) \le b(t)(v(t) + w(v(t))). \tag{2.9.18}$$

Dividing both sides of (2.9.18) by $(v(t) + w(v(t)))$ and then integrating from 0 to t and using (2.9.14) we have

$$G(v(t)) \le G(a + u(0)) + \int_0^t b(\sigma)\,d\sigma. \tag{2.9.19}$$

Now substituting the value of $v(t)$ from (2.9.19) in (2.9.16) and then integrating from 0 to t we obtain

$$z(t) \le a + \int_0^t b(s)w\left(G^{-1}\left[G(a + u(0)) + \int_0^s b(\sigma)\,d\sigma\right]\right)ds. \tag{2.9.20}$$

Now substituting (2.9.20) in $u(t) \le z(t)$, we get the desired inequality in (2.9.13).

■

A slightly different version of Theorem 2.9.3 is given in the following theorem.

Theorem 2.9.4 Let u, u', b, w and a be as in Theorem 2.9.3 and $c \geq 0$ be a constant. If

$$u'(t) \leq a + c\left(u(t) + \int_0^t b(s)w(u(s) + u'(s))\,ds\right), \qquad (2.9.21)$$

for $t \in R_+$, then for $0 \leq t \leq t_2$,

$$u'(t) \leq \exp(ct)\left[a + cu(0) + c\int_0^t \exp(-cs)b(s)\right.$$

$$\times w\left(G^{-1}\left[G([a + (1+c)u(0)])\right.\right.$$

$$\left.\left.\left. + (1+c)\int_0^s b(\sigma)\,d\sigma\right]\right)ds\right], \qquad (2.9.22)$$

where G, G^{-1} are as defined in Theorem 2.9.3, and $t_2 \in R_+$ is chosen so that

$$G([a + (1+c)u(0)]) + (1+c)\int_0^t b(\sigma)\,d\sigma \in \text{Dom}(G^{-1}),$$

for all $t \in R_+$ lying in the interval $0 \leq t \leq t_2$. $\square$

The proof of this theorem can be completed by following the proofs of Theorems 2.9.2 and 2.9.3. Here we omit the details.

The following variants of Theorem 2.9.3 can also be useful in certain applications.

Theorem 2.9.5 Let u, u' and b be nonnegative continuous functions defined on R_+, $u(0) = 0$ and $a \geq 0$ be a constant. Let $w(u)$ be a continuous nondecreasing submultiplicative function defined on R_+ and $w(u) > 0$ on $(0, \infty)$. If

$$u'(t) \leq a + \int_0^t b(s)w(u(s) + u'(s))\,ds, \qquad (2.9.23)$$

for $t \in R_+$, then for $0 \leq t \leq t_3$,

$$u(t) + u'(t) \leq (1+t)\Omega^{-1}\left[\Omega(a) + \int_0^t b(s)w(1+s)\,ds\right], \quad (2.9.24)$$

where

$$\Omega(r) = \int_{r_0}^r \frac{ds}{w(s)}, \quad r > 0, r_0 > 0, \quad (2.9.25)$$

and Ω^{-1} is the inverse of Ω and $t_3 \in R_+$ is chosen so that

$$\Omega(a) + \int_0^t b(s)w(1+s)\,ds \in \text{Dom}(\Omega^{-1}),$$

for all $t \in R_+$ lying in the interval $0 \leq t \leq t_3$.

□

Proof: As noted in the proof of Theorem 2.9.3 we assume that a is positive. Integrating (2.9.23) from 0 to t we have

$$u(t) \leq at + \int_0^t (t-s)b(s)w(u(s) + u'(s))\,ds$$

$$\leq at + t\int_0^t b(s)w(u(s) + u'(s))\,ds. \quad (2.9.26)$$

From (2.9.23) and (2.9.26) we observe that

$$u(t) + u'(t) \leq a(1+t) + (1+t)\int_0^t b(s)w(u(s) + u'(s))\,ds. \quad (2.9.27)$$

From (2.9.27) we have

$$\left(\frac{u(t) + u'(t)}{1+t}\right) \leq a + \int_0^t b(s)w\left(\frac{u(s) + u'(s)}{1+s}(1+s)\right)ds$$

$$\leq a + \int_0^t b(s)w(1+s)w\left(\frac{u(s) + u'(s)}{1+s}\right)ds.$$

Now an application of Bihari's inequality given in Theorem 2.3.1 yields the desired inequality in (2.9.24).

∎

Theorem 2.9.6 *Let u, u', b and a be as in Theorem 2.9.5. Let $w(t, v)$ be a nonnegative continuous monotonic nondecreasing function in v for $v \geq 0$ for each fixed $t \in R_+$. If*

$$u'(t) \leq a + \int_0^t b(s)w(s, u(s) + u'(s))\,ds, \quad t \in R_+, \qquad (2.9.28)$$

then

$$u(t) + u'(t) \leq p(t) + q(t)r(t), \quad t \in R_+, \qquad (2.9.29)$$

where $p(t) = a(1+t)$, $q(t) = (1+t)$ and $r(t)$ is the maximal solution of

$$r'(t) = b(t)w(t, p(t) + q(t)r(t)), \quad r(0) = 0, \qquad (2.9.30)$$

existing for all $t \in R_+$.

□

Proof: By following the proof of Theorem 2.9.5 we have

$$u(t) + u'(t) \leq p(t) + q(t)\int_0^t b(s)w(s, u(s) + u'(s))\,ds. \qquad (2.9.31)$$

Define a function $z(t)$ by

$$z(t) = \int_0^t b(s)w(s, u(s) + u'(s))\,ds,$$

then $z(0) = 0$, $u(t) + u'(t) \leq p(t) + q(t)z(t)$ and

$$z'(t) = b(t)w(t, u(t) + u'(t))$$

$$\leq b(t)w(t, p(t) + q(t)z(t)). \qquad (2.9.32)$$

Now a suitable application of the comparison Theorem 2.2.2 to (2.9.32) and (2.9.30) yields

$$z(t) \leq r(t), \qquad (2.9.33)$$

where $r(t)$ is the maximal solution of (2.9.30). The estimate (2.9.29) follows by using (2.9.33) in $u(t) + u'(t) \le p(t) + q(t)z(t)$ and the proof is complete.

∎

Pachpatte (1981a) gave a number of inequalities of the above type involving functions and their higher order derivatives. In the following, we shall merely state the following two results in Pachpatte (1981a), leaving the proofs to the reader.

Theorem 2.9.7 *Let $u, u', \ldots, u^{(n)}$, a, b and c be nonnegative continuous functions defined on R_+. Let $w(u)$ be a continuous nondecreasing subadditive and submultiplicative function defined on R_+ and $w(u) > 0$ on $(0, \infty)$. If*

$$u^{(n)}(t) \le a(t) + b(t)\left(\sum_{i=0}^{n-1} u^{(i)}(t) + \int_0^t c(s) w\left(\sum_{i=0}^{n-1} u^{(i)}(s)\right) ds\right), \quad (2.9.34)$$

for $t \in R_+$, then for $0 \le t \le t_1$,

$$u^{(n)}(t) \le a(t) + b(t)\left[p(t) + q(t)\Omega^{-1}\left[\Omega\left(\int_0^t r(s) w(p(s)) ds\right)\right.\right.$$
$$\left.\left. + \int_0^t r(s) w(q(s)) ds\right]\right], \quad (2.9.35)$$

where

$$p(t) = \left(\sum_{i=0}^{n-1} u^{(i)}(0)\right) \exp\left(\int_0^t [1 + b(s)] ds\right)$$
$$+ \int_0^t a(s) \exp\left(\int_s^t [1 + b(\sigma)] d\sigma\right) ds, \quad (2.9.36)$$

$$q(t) = \exp\left(\int_0^t [1 + b(s)] ds\right), \quad (2.9.37)$$

$$r(t) = c(t) \exp\left(-\int_0^t [1 + b(s)] ds\right), \quad (2.9.38)$$

and

$$\Omega(v) = \int_{v_0}^{v} \frac{ds}{w(s)}, \quad v > 0, v_0 > 0,$$

Ω^{-1} is the inverse of Ω, and $t_1 \in R_+$ is chosen so that

$$\Omega\left(\int_0^t r(s)w(p(s))\,ds\right) + \int_0^t r(s)w(q(s))\,ds \in \mathrm{Dom}(\Omega^{-1}),$$

for all $t \in R_+$ lying in the interval $0 \le t \le t_1$. □

Theorem 2.9.8 Let $u, u', \ldots, u^{(n)}, a, b$ and c be as in Theorem 2.9.7. Let $w(t, v)$ be as defined in Theorem 2.9.6. If

$$u^{(n)}(t) \le a(t) + b(t)\left(\sum_{i=0}^{n-1} u^{(i)}(t) + \int_0^t c(s)w(s, u^{(n)}(s))\,ds\right), \quad (2.9.39)$$

for $t \in R_+$, then

$$u^{(n)}(t) \le a(t) + b(t)(p(t) + q(t)R(t)), \quad (2.9.40)$$

for $t \in R_+$, where $p(t), q(t)$ are as defined in Theorem 2.9.7 and $R(t)$ is the maximal solution of

$$R'(t) = r(t)w(t, a(t) + b(t)[p(t) + q(t)R(t)]), \quad R(0) = 0, \quad (2.9.41)$$

existing on R_+, in which $r(t)$ is defined by (2.9.38) in Theorem 2.9.7. □

We note that the usefulness of the inequalities in Theorems 2.9.7 and 2.9.8 becomes apparent if we consider that $u(0), u'(0), \ldots, u^{(n-1)}(0), a, b$ and c are known and $u, u', \ldots, u^{(n)}$ are unknown functions. These inequalities give bounds in terms of the known functions which majorizes $u^{(n)}(t)$ and consequently $u(t)$ after n times integration. The same is true about the other inequalities given in this section. For a number of other inequalities of the above type, see Li (1963), Lin (1981, 1990), Lin and Wu (1989) and Pachpatte (1977c, 1978a,c, 1980a, 1982a).

2.10 Inequalities with Iterated Integrals

For the study of various properties of the solutions of higher order differential and integro-differential equations certain integral inequalities with several iterated integrals are very important in a variety of situations. This section presents some inequalities investigated by Pachpatte (in press e,j) with several iterated integrals. In what follows, the definitions and notations used in Section 1.9 are used to simplify the details of presentation.

The following results established by Pachpatte (in press e,j) can be considered as further generalizations of some of the inequalities given in Section 1.9. These results are formulated only for the iterated integrals as defined by $M[t, r, h]$ in Section 1.9. The results corresponding to the iterated integrals defined by $A[t, g, h]$, $B[t, g, h]$ and $E[t, g, h]$ in Section 1.9 can be formulated similarly.

Pachpatte (in press e,j) established the following inequalities, which can be considered as generalizations of Bihari's inequality given in Theorem 2.3.1 and Pachpatte's inequality given in Theorem 2.8.1.

Theorem 2.10.1 Let $u \geq 0$, $h \geq 0$, $r_j > 0$, $j = 1, 2, \ldots, n-1$, be continuous functions defined on R_+ and $c \geq 0$ be a constant. Let $w(u)$ be a continuously differentiable function defined on R_+, $w(u) > 0$ on $(0, \infty)$ and $w'(u) \geq 0$ for $u \in R_+$. If

$$u(t) \leq c + M[t, r, h(t_n)w(u(t_n))], \qquad (2.10.1)$$

for $t \in R_+$, then for $0 \leq t \leq t_1$,

$$u(t) \leq \Omega^{-1}[\Omega(c) + M[t, r, h(t_n)]], \qquad (2.10.2)$$

where

$$\Omega(v) = \int_{v_0}^{v} \frac{ds}{w(s)}, \quad v > 0, v_0 > 0, \qquad (2.10.3)$$

Ω^{-1} is the inverse of Ω and $\bar{t} \in R_+$ is chosen so that

$$\Omega(c) + M[t, r, h(t_n)] \in \text{Dom}(\Omega^{-1}),$$

for all $t \in R_+$ lying in the interval $0 \leq t \leq \bar{t}$.

□

Theorem 2.10.2 *Let $u \geq 0$, $f \geq 0$, $h \geq 0$, $r_j > 0$, $j = 1, 2, \ldots, n-1$, be continuous functions defined on R_+, $c \geq 0$ is a constant and let $w(u)$ be as given in Theorem 2.10.1. If*

$$u(t) \leq c + M[t, r, f(t_n)(u(t_n) + M[t_n, r, h(s_n)w(u(s_n))])], \quad (2.10.4)$$

for $t \in R_+$, then for $0 \leq t \leq \bar{t}$,

$$u(t) \leq c + M[t, r, f(t_n)G^{-1}[G(c) + M[t_n, r, [f(s_n) + h(s_n)]]]], \quad (2.10.5)$$

where

$$G(v) = \int_{v_0}^{v} \frac{ds}{s + w(s)}, \quad v > 0, v_0 > 0, \quad (2.10.6)$$

G^{-1} is the inverse of G and $\bar{t} \in R_+$ is chosen so that

$$G(c) + M[t, r, [f(s_n) + h(s_n)]] \in \mathrm{Dom}(G^{-1}),$$

for all $t \in R_+$ lying in the interval $0 \leq t \leq \bar{t}$. □

Proofs of Theorems 2.10.1 and 2.10.2: We will give the details for the proof of Theorem 2.10.2 only; the proof of Theorem 2.10.1 can be completed similarly.

Assume that $c > 0$ and define a function $z(t)$ by the right-hand side of (2.10.4). From the hypotheses, it is easy to see that $w(u)$ is monotonically increasing on $(0, \infty)$. From the definition of $z(t)$ and using the fact that $u(t) \leq z(t)$ we observe that

$$L_n z(t) \leq f(t)(z(t) + M[t, r, h(s_n)w(z(s_n))]), \quad (2.10.7)$$

where L_i, $i = 0, 1, 2, \ldots, n$ are the operators as defined in Section 1.9. Define a function $m(t)$ by

$$m(t) = z(t) + M[t, r, h(s_n)w(z(s_n))]. \quad (2.10.8)$$

From (2.10.8), (2.10.7) and the fact that $z(t) \leq m(t)$ we observe that

$$L_n m(t) = L_n z(t) + h(t)w(z(t))$$
$$\leq f(t)m(t) + h(t)w(m(t))$$
$$\leq [f(t) + h(t)](m(t) + w(m(t))). \quad (2.10.9)$$

From (2.10.9) and using the facts that $m(t) > 0$, $L_{n-1}m(t) \geq 0$, $(d/dt)(m(t) + w(m(t))) \geq 0$ for $t \in R_+$, it is easy to observe that

$$\frac{L_n m(t)}{m(t) + w(m(t))} \leq [f(t) + h(t)] + \frac{(L_{n-1}m(t))(d/dt)(m(t) + w(m(t)))}{(m(t) + w(m(t)))^2},$$

i.e.

$$\frac{d}{dt}\left(\frac{L_{n-1}m(t)}{m(t) + w(m(t))}\right) \leq [f(t) + h(t)]. \tag{2.10.10}$$

By setting $t = s_n$ in (2.10.10) and integrating it from 0 to t and using the fact that $L_{i-1}m(0) = 0$ for $i = 2, 3, \ldots, n$, we obtain

$$\frac{(d/dt)L_{n-2}m(t)}{m(t) + w(m(t))} \leq r_{n-1}(t) \int_0^t [f(s_n) + h(s_n)]\,ds_n. \tag{2.10.11}$$

Again, as above, from (2.10.11) we observe that

$$\frac{d}{dt}\left(\frac{L_{n-2}m(t)}{m(t) + w(m(t))}\right) \leq r_{n-1}(t) \int_0^t [f(s_n) + h(s_n)]\,ds_n. \tag{2.10.12}$$

By setting $t = s_{n-1}$ in (2.10.12) and integrating it from 0 to t and using the fact that $L_{i-2}m(0) = 0$ for $i = 3, 4, \ldots, n$, we get

$$\frac{L_{n-2}m(t)}{m(t) + w(m(t))} \leq \int_0^t r_{n-1}(s_{n-1}) \int_0^{s_{n-1}} [f(s_n) + h(s_n)]\,ds_n\,ds_{n-1}.$$

Continuing in this way we obtain

$$\frac{(d/dt)m(t)}{m(t) + w(m(t))} \leq r_1(t) \int_0^t r_2(s_2) \int_0^{s_2} r_3(s_3) \ldots \int_0^{s_{n-2}} r_{n-1}(s_{n-1})$$

$$\times \int_0^{s_{n-1}} [f(s_n) + h(s_n)]\,ds_n\,ds_{n-1} \ldots ds_3\,ds_2. \tag{2.10.13}$$

From (2.10.6) and (2.10.13) we have

$$\frac{d}{dt}G(m(t)) \leq r_1(t) \int_0^t r_2(s_2) \int_0^{s_2} r_3(s_3) \ldots \int_0^{s_{n-2}} r_{n-1}(s_{n-1})$$

$$\times \int_0^{s_{n-1}} [f(s_n) + h(s_n)]\,ds_n\,ds_{n-1} \ldots ds_3\,ds_2. \tag{2.10.14}$$

By taking $t = s_1$ in (2.10.14) and integrating it from 0 to t and using the fact that $m(0) = c$, we have

$$G(m(t)) \leq G(c) + M[t, r, [f(s_n) + h(s_n)]]. \qquad (2.10.15)$$

Using the bound on $m(t)$ from (2.10.15) in (2.10.7) we get

$$L_n z(t) \leq f(t) G^{-1}[G(c) + M[t, r, [f(s_n) + h(s_n)]]]. \qquad (2.10.16)$$

The inequality (2.10.16) implies the estimate

$$z(t) \leq c + M[t, r, f(t_n) G^{-1}[G(c) + M[t_n, r, [f(s_n) + h(s_n)]]]]. \qquad (2.10.17)$$

Using (2.10.17) in $u(t) \leq z(t)$ we get the required inequality in (2.10.5). If c is nonnegative, then the standard limiting argument can be used to get the required inequality in (2.10.5). The subinterval $0 \leq t \leq \bar{t}$ is obvious.

∎

The next result deals with the further generalization of the inequality given in part (a_1) in Theorem 1.9.1, which may be useful in some applications.

Theorem 2.10.3 *Let $u \geq 0$, $f \geq 0$, $h \geq 0$, $r_j > 0$, $p_j > 0$ for $j = 1, 2, \ldots, n-1$, be continuous functions defined on R_+ and $c \geq 0$ is a constant. Let $w(u)$ be a continuously differentiable and submultiplicative function defined on R_+, $w(u) > 0$ on $(0, \infty)$ and $w'(u) \geq 0$ for $u \in R_+$. If*

$$u(t) \leq c + M[t, r, f(t_n) u(t_n)] + M[t, p, h(t_n) w(u(t_n))], \qquad (2.10.18)$$

for $t \in R_+$, then for $0 \leq t \leq \bar{t}$,

$$u(t) \leq b(t) \Omega^{-1}[\Omega(c) + M[t, p, h(t_n) w(b(t_n))]], \qquad (2.10.19)$$

where

$$b(t) = \exp(M[t, r, f(t_n)]), \qquad (2.10.20)$$

Ω, Ω^{-1} are as defined in Theorem 2.10.1 and $\bar{t} \in R_+$ is chosen so that

$$\Omega(c) + M[t, p, h(t_n) w(b(t_n))] \in \mathrm{Dom}(\Omega^{-1}),$$

for all $t \in R_+$ lying in the interval $0 \leq t \leq \bar{t}$.

□

Proof: Assume that $c > 0$ and define a function $m(t)$ by

$$m(t) = c + M[t, p, h(t_n)w(u(t_n))]. \tag{2.10.21}$$

Then (2.10.18) can be restated as

$$u(t) \leq m(t) + M[t, r, f(t_n)u(t_n)]. \tag{2.10.22}$$

Since $m(t)$ is positive continuous and nondecreasing in t, we have

$$\frac{u(t)}{m(t)} \leq 1 + M\left[t, r, f(t_n)\frac{u(t_n)}{m(t_n)}\right].$$

Now an application of Theorem 1.9.1 part (a_1) yields

$$u(t) \leq b(t)m(t), \tag{2.10.23}$$

where $b(t)$ is defined by (2.10.20). The rest of the proof follows by similar arguments to those in the proof of Theorem 2.4.1 and in view of the proof of Theorem 2.10.2 given above. Here we omit the further details.

∎

We next formulate the following result, which deals with a further generalization of the inequality (d_1) given in Theorem 1.9.4.

Theorem 2.10.4 *Let $u \geq 0$, $f_i \geq 0$, $p_i > 0$ for $i = 1, 2, \ldots, n$, be continuous functions defined on R_+ and $c \geq 0$ is a constant and let $w(u)$ be as in Theorem 2.10.1. If*

$$u(t) \leq c + \sum_{k=1}^{n} \left(\int_0^t p_1(s_1) \int_0^{s_1} p_2(s_2) \cdots \int_0^{s_{k-2}} p_{k-1}(s_{k-1}) \right.$$

$$\left. \times \int_0^{s_{k-1}} p_k(s_k) f_k(s_k) w(u(s_k)) \, ds_k \, ds_{k-1} \cdots ds_2 \, ds_1 \right),$$

for $t \in R_+$ with $s_0 = t$, then for $0 \leq t \leq \bar{t}$,

$$u(t) \leq \Omega^{-1}\left[\Omega(c) + \sum_{k=1}^{n} \left(\int_0^t p_1(s_1) \int_0^{s_1} p_2(s_2) \cdots \int_0^{s_{k-2}} p_{k-1}(s_{k-1}) \right. \right.$$

$$\left. \left. \times \int_0^{s_{k-1}} p_k(s_k) f_k(s_k) \, ds_k \, ds_{k-1} \cdots ds_2 \, ds_1 \right) \right],$$

where Ω is defined by (2.10.3) in Theorem 2.10.1 and $\bar{t} \in R_+$ be chosen so that

$$\Omega(c) + \sum_{k=1}^{n} \left(\int_0^t p_1(s_1) \int_0^{s_1} p_2(s_2) \ldots \int_0^{s_{k-2}} p_{k-1}(s_{k-1}) \right.$$
$$\left. \times \int_0^{s_{k-1}} p_k(s_k) f_k(s_k) \, ds_k \, ds_{k-1} \ldots ds_2 \, ds_1 \right) \in \text{Dom}(\Omega^{-1}),$$

for all $t \in R_+$ lying in the interval $0 \leq t \leq \bar{t}$.

□

The proof of this theorem can be completed by following the proof of Theorem 1.9.4 and closely looking at the proof of Theorem 2.10.2 given above. Here we omit the details. We note that a slightly different version of Theorem 2.10.4 is given by Medveď (1993).

In concluding this section, it should be emphasized that the methods employed to establish results given in this section can be used to formulate a number of inequalities similar to those given in Sections 2.8 and 2.9 involving iterated integrals. In order to avoid monotony, the formulation of such results is not undertaken here.

The literature concerning the nonlinear inequalities of the type discussed in earlier sections is particularly rich and we refer the reader to Babkin (1953), Bobrowski et al. (1978), Butler and Rogers (1971a), Chandra and Lakshmikantham (1972), Chaplygin (1954), Dhongade and Deo (1973), Distefano (1968), Lakshmikantham (1973), Li (1963), Losonczi (1973), Maroni (1968), Martynyuk and Gutowski (1979), Olech (1967), Pinto (1990), Popenda (1986), Stachurska (1971), Tsamatos and Ntouyas (1991), Turinici (1982, 1984, 1986), Werbowski (1975), Wong (1975), Yang (1979), Young (1982) and Ziebur (1968), which also contain a good many references on these topics.

2.11 Applications

Most of the inequalities developed in the earlier discussion are motivated by their need in some kinds of application in the theory of differential and integral equations. The literature on the applications of such inequalities is considerable. This section presents applications of some of the inequalities given in earlier sections that have been investigated during the last two decades.

2.11.1 Second-Order Nonlinear Differential Equations

This section considers the second-order nonlinear differential equation of the form

$$u'' + f(t, u, u') = 0, \quad \text{(A)}$$

and a particular case of (A)

$$u'' + f(t, u) = 0. \quad \text{(B)}$$

The asymptotic behaviour of the solutions of (B) has been discussed by Cohen (1967), Tong (1982), Trench (1963), Waltman (1964) and others by using the Bellman and Bihari inequalities. Here we give the sufficient conditions under which the solutions of a second-order differential equation (A) are asymptotic to $at + b$, where a, b are real constants.

Constantin (1993) proved the following result, which deals with the asymptotic behaviour of the solutions of equation (A).

Theorem 2.11.1 *Suppose that the following are true.*

(i) The function $f(t, u, v)$ is continuous on $D = \{(t, u, v) : t \geq 1, u, v \in R\}$.

(ii) There are functions h_1, h_2, $h_3 : R_+ \to R_+$ continuous and such that $\int_1^\infty h_i(s)\,ds < \infty$ for $i = 1, 2, 3$ and

$$|f(t, u, v)| \leq h_1(t) g\left(\frac{|u|}{t}\right) + h_2(t)|v| + h_3(t),$$

for $(t, u, v) \in D$, where $g : R_+ \to R_+$ is continuous, nondecreasing function such that $g(x) > 0$ for $x > 0$, and if we denote

$$G(x) = \int_1^x \frac{ds}{g(s)}, \quad x > 0,$$

then $\lim_{x \to \infty} G(x) = \infty$.

Then for every solution $u(t)$ of (A) we have that $u(t) = at + b + o(t)$ as $t \to \infty$, where a, b are real constants.

□

Proof: By (i) we have that equation (A) has solutions $u(t)$ corresponding to arbitrary given initial values $u(1) = c_1$, $u'(1) = c_2$. Integrating (A) twice from 1 to t we get

$$u'(t) = c_2 - \int_1^t f(s, u(s), u'(s))\, ds, \quad t \geq 1, \qquad (2.11.1)$$

$$u(t) = c_1 + c_2(t-1) - \int_1^t (t-s) f(s, u(s), u'(s))\, ds, \quad t \geq 1. \qquad (2.11.2)$$

By (ii) and (2.11.1), (2.11.2) we obtain that

$$|u'(t)| \leq |c_2| + \int_1^t h_1(s) g\left(\frac{|u(s)|}{s}\right) ds + \int_1^t h_2(s) |u'(s)|\, ds$$

$$+ \int_1^t h_3(s)\, ds, \quad t \geq 1, \qquad (2.11.3)$$

$$\frac{|u(t)|}{t} \leq |c_1| + |c_2| + \int_1^t h_1(s) g\left(\frac{|u(s)|}{s}\right) ds + \int_1^t h_2(s) |u'(s)|\, ds$$

$$+ \int_1^t h_3(s)\, ds, \quad t \geq 1. \qquad (2.11.4)$$

Define a function $x(t)$ by the right-hand side of (2.11.4). Then by (2.11.3) and (2.11.4) we obtain that

$$x(t) \leq a(t) + \int_1^t h_1(s) g(x(s))\, ds + \int_1^t h_2(s) x(s)\, ds, \quad t \geq 1, \qquad (2.11.5)$$

where

$$a(t) = 1 + |c_1| + |c_2| + \int_1^t h_3(s)\, ds, \quad t \geq 1.$$

Fix $T > 1$. By (2.11.5) we obtain that

$$x(t) \leq a(T) + \int_1^t h_1(s) g(x(s))\, ds + \int_1^t h_2(s) x(s)\, ds, \quad 1 \leq t \leq T. \qquad (2.11.6)$$

Define a function $y(t)$ by the right-hand side of (2.11.6). Then $y(1) = a(T)$, $x(t) \leq y(t)$ and

$$y'(t) \leq h_1(t) g(y(t)) + h_2(t) y(t), \quad 1 \leq t \leq T.$$

Now, putting

$$w(t) = y(t) \exp\left(-\int_1^t h_2(s)\, ds\right), \quad 1 \leq t \leq T,$$

we can write the last inequality as

$$w'(t) \leq h_1(t) \exp\left(-\int_1^t h_2(s)\, ds\right) g(y(t)), \quad 1 \leq t \leq T.$$

Integrating this from 1 to t we obtain

$$y(t) \leq a(T) \exp\left(\int_1^t h_2(s)\, ds\right) + \exp\left(\int_1^t h_2(s)\, ds\right)$$

$$\times \int_1^t h_1(s) \exp\left(-\int_1^s h_2(\sigma)\, d\sigma\right) g(y(s))\, ds, \quad 1 \leq t \leq T,$$

thus

$$y(t) \leq a(T) \exp\left(\int_1^T h_2(s)\, ds\right) + \exp\left(\int_1^T h_2(s)\, ds\right)$$

$$\times \int_1^t h_1(s) \exp\left(-\int_1^s h_2(\sigma)\, d\sigma\right) g(y(s))\, ds, \quad 1 \leq t \leq T.$$

Applying Bihari's inequality given in Theorem 2.3.1 we obtain

$$y(t) \leq G^{-1}\left[G\left(a(T) \exp\left(\int_1^T h_2(s)\, ds\right)\right) + \exp\left(\int_1^T h_2(s)\, ds\right)\right.$$

$$\left. \times \int_1^t h_1(s) \exp\left(-\int_1^s h_2(\sigma)\, d\sigma\right) ds\right], \quad 1 \leq t \leq T.$$

By (2.11.6) and since $T \geq 1$ was arbitrarily chosen, we deduce that

$$x(t) \leq G^{-1}\left[G\left(a(t)\exp\left(\int_1^t h_2(s)\,ds\right)\right) + \exp\left(\int_1^t h_2(s)\,ds\right)\right.$$
$$\left.\times \int_1^t h_1(s)\exp\left(-\int_1^s h_2(\sigma)\,d\sigma\right)ds\right], \quad t \geq 1.$$

Our hypotheses imply that the right-hand side in the preceding inequality is bounded and thus there exists an $M > 0$ such that $x(t) \leq M$ for $t \geq 1$. From (2.11.3) and (2.11.4) one obtains

$$|u'(t)| \leq M, \quad t \geq 1,$$

$$\frac{|u(t)|}{t} \leq M, \quad t \geq 1.$$

By (ii) we have

$$\int_1^t |f(s, u(s), u'(s))|\,ds \leq \int_1^t h_1(s)g\left(\frac{|u(s)|}{s}\right)ds + \int_1^t h_2(s)|u'(s)|\,ds$$
$$+ \int_1^t h_3(s)\,ds \leq x(t) \leq M, \quad t \geq 1.$$

Thus the integral $\int_1^t f(s, u(s), u'(s))\,ds$ is absolutely convergent and consequently

$$\lim_{t\to\infty} \int_1^t f(s, u(s), u'(s))\,ds < \infty.$$

By (2.11.1) we obtain that there exists $a \in R$ such that $\lim_{t\to\infty} u'(t) = a$. Applying L'Hospital's rule we deduce that

$$\lim_{t\to\infty} \frac{u(t)}{t} = \lim_{t\to\infty} u'(t) = a.$$

The proof is complete. ∎

The following variant of Theorem 2.11.1 given by Constantin (1993) can be proved similarly.

Theorem 2.11.2 *Suppose that the following are true.*

(i) The function $f(t, u, v)$ is continuous on $D = \{(t, u, v): t \geq 1, u, v \in R_+\}$.

(ii) There are functions $h_1, h_2, h_3: R_+ \to R_+$ continuous and such that $\int_1^\infty h_i(s)\,ds < \infty$, for $i = 1, 2, 3$ and

$$|f(t, u, v)| \leq h_1(t)\frac{|u|}{t} + h_2(t)g(|v|) + h_3(t),$$

for $(t, u, v) \in D$, where $g: R_+ \to R_+$ is a continuous nondecreasing function such that $g(x) > 0$ for $x > 0$, and if we denote

$$G(x) = \int_1^x \frac{ds}{g(s)}, \quad x > 0,$$

then $\lim_{x \to \infty} G(x) = \infty$. Then for every solution $u(t)$ of (A) we have that $u(t) = at + b + o(t)$ as $t \to \infty$ where a, b are real constants.

□

The following Corollary of Theorem 2.11.1 given in Tong (1982) is interesting in itself.

Corollary 2.11.1 Let $f(t, u)$ be continuous on $D = \{(t, u): t \geq 1, u \in R\}$. Suppose there exist ϕ, $u \in C(R_+, R_+)$, w nondecreasing on R_+, $w(x) > 0$ for $x > 0$, such that

$$|f(t, u)| \leq \phi(t)w\left(\frac{|u|}{t}\right), \quad t \geq 1, u \in R,$$

and

$$\int_1^\infty \phi(s)\,ds < \infty, \quad \int_1^\infty \frac{ds}{w(s)} = \infty.$$

Then if $u(t)$ is a solution of (B) we have that $u(t) = at + b + o(t)$ as $t \to \infty$ where a, b are constants.

△

2.11.2. Perturbed Integro-differential Equations

This section is devoted to the study of asymptotic behaviour of the solutions of nonlinear integro-differential systems of the form

$$y' = A(t)y + f(t, y) + \int_{t_0}^{t} k(t, s, y(s))\,ds, \qquad (P)$$

where y, f, k are real n-vectors, $A(t)$ is a continuous $n \times n$ matrix, whose real-valued components are defined and continuous for $t \in R_{t_0} = [t_0, \infty)$, and $f(t, y), k(t, s, y)$ are continuous functions defined for $t, s \in R_{t_0}$ and $y \in R^n$, R^n the n dimensional Euclidean space. The norm of a matrix or a vector is defined as the sum of the absolute value of each of the components and is denoted by $\|\cdot\|$. Many papers have been devoted to a discussion of the asymptotic relationships between the solutions of (P) when the integral term is absent and those of the linear system

$$x' = A(t)x, \quad t \in R_{t_0}. \qquad (L)$$

Excellent accounts of this subject may be found in a number of standard texts on ordinary differential equations (Bellman, 1953; Coddington and Levinson, 1955; Halanay, 1966; Hartman, 1964; Lakshmikantham and Leela, 1969; Sansone and Conti, 1964) and a large number of papers appeared in the literature (Bihari, 1957; Brauer, 1964; Brauer and Wong, 1969; Hallam, 1969, 1970; Morchalo, 1984; Pachpatte, 1973b, 1974b, 1975f; Strauss and Yorke, 1967; Yang, 1984).

We shall denote by $X(t)$ the fundamental matrix of (L) which satisfies the initial condition $X(t_0) = I$, where I is the identity matrix.

Let $\Delta(t)$ be an $n \times n$ matrix and $\alpha(t)$ a positive function. We say that the systems (L) and (P) are asymptotically equivalent if, corresponding to each solution $x(t)$ of system (L), there exists a solution $y(t)$ of (P) with the property that

$$\|\Delta(t)[x(t) - y(t)]\| = o(\alpha(t)) \quad (t \to \infty), \qquad (2.11.7)$$

and conversely, to each solution $y(t)$ of (P) there corresponds a solution $x(t)$ of (L) such that (2.11.7) holds.

We shall now give the following result established by Talpalaru (1973).

Theorem 2.11.3 *Suppose that the following conditions are satisfied:*
(i) *there exists a nonsingular continuous matrix $\Delta(t)$ for $t \in R_{t_0}$ such that*

$$\|\Delta(t)X(t)\| \leq \alpha(t), \qquad (2.11.8)$$

where $\alpha(t)$ is a positive and continuous function for $t \in R_{t_0}$,
(ii) *there exists a continuous nonnegative function $a(t)$ such that*

$$\|X^{-1}(t)f(t, y)\| \leq a(t)w\left(\frac{\|\Delta(t)y\|}{\alpha(t)}\right), \qquad (2.11.9)$$

for $t \in R_{t_0}$, $\|y\| < \infty$, where $w(r)$ is a positive continuous and nondecreasing function on $r > 0$ and

$$\Omega(u) = \int_{u_0}^{u} \frac{ds}{w(s)} \to \infty, \quad \text{for } u \to \infty, \qquad (2.11.10)$$

(iii) *there exists a continuous, nonnegative function $h(t, s)$ defined for $t_0 \leq s \leq t < \infty$ such that*

$$\|X^{-1}(t)k(t, s, y)\| \leq h(t, s)w\left(\frac{\|\Delta(s)y\|}{\alpha(s)}\right), \qquad (2.11.11)$$

for $t_0 \leq s \leq t < \infty$, $\|y\| < \infty$,
(iv) *the functions $a(t)$ and $h(t, \sigma)$ satisfy*

$$\int_{t_0}^{\infty}\left[a(s) + \int_{t_0}^{s} h(s, \sigma)\,d\sigma\right] ds < \infty, \qquad (2.11.12)$$

then to each solution $y(t)$ of (P) there exists a solution $x(t)$ of (L) such that (2.11.7) holds.
□

Proof: To prove the existence of a solution $x(t)$ which satisfies (2.11.7) is the same as to prove the existence of a constant vector c such that $x(t) = X(t)c$ satisfies (2.11.7). Let $y(t)$ be a solution of the system (P) satisfying the condition $y(t_0) = y_0$. Then $y(t)$ can be written as

$$y(t) = X(t)y_0 + X(t)\int_{t_0}^{t} X^{-1}(s)\left[f(s, y(s)) + \int_{t_0}^{s} k(s, \sigma, y(\sigma))\,d\sigma\right] ds.$$

$$(2.11.13)$$

Multiplying both sides of (2.11.13) by $\Delta(t)$ and using (2.11.8) and (2.11.9) we have

$$\|\Delta(t)y(t)\| \leq \alpha(t)\|y_0\| + \alpha(t) \int_{t_0}^{t} \left[a(s)w\left(\frac{\|\Delta(s)y(s)\|}{\alpha(s)}\right) \right.$$

$$\left. + \int_{t_0}^{s} h(s,\sigma)w\left(\frac{\|\Delta(\sigma)y(\sigma)\|}{\alpha(\sigma)}\right) d\sigma \right] ds,$$

or

$$\frac{\|\Delta(t)y(t)\|}{\alpha(t)} \leq \|y_0\| + \int_{t_0}^{t} \left[a(s)w\left(\frac{\|\Delta(s)y(s)\|}{\alpha(s)}\right) \right.$$

$$\left. + \int_{t_0}^{s} h(s,\sigma)w\left(\frac{\|\Delta(\sigma)y(\sigma)\|}{\alpha(\sigma)}\right) d\sigma \right] ds.$$

Now an application of the inequality given in Theorem 2.4.5 yields

$$\Omega\left(\frac{\|\Delta(t)y(t)\|}{\alpha(t)}\right) \leq \Omega(\|y_0\|) + \int_{t_0}^{t} \left[a(s) + \int_{t_0}^{s} h(s,\sigma) d\sigma \right] ds.$$

It follows from (2.11.12) that there exists a constant $M > 0$, so that

$$\frac{\|\Delta(t)y(t)\|}{\alpha(t)} \leq M, \quad t \in R_{t_0}.$$

Observe that

$$\left\| \int_{t_0}^{t} X^{-1}(s) \left[f(s, y(s)) + \int_{t_0}^{s} k(s, \sigma, y(\sigma)) d\sigma \right] ds \right\|$$

$$\leq \int_{t_0}^{t} \left[\|X^{-1}(s)f(s, y(s))\| + \int_{t_0}^{s} \|X^{-1}(s)k(s, \sigma, y(\sigma))\| d\sigma \right] ds$$

$$\leq \int_{t_0}^{t} \left[a(s)w\left(\frac{\|\Delta(s)y(s)\|}{\alpha(s)}\right) + \int_{t_0}^{s} h(s,\sigma)w\left(\frac{\|\Delta(\sigma)y(\sigma)\|}{\alpha(\sigma)}\right) d\sigma \right] ds$$

$$\leq w(M) \int_{t_0}^{t} \left[a(s) + \int_{t_0}^{s} h(s, \sigma) \, d\sigma \right] ds, \quad t \in R_{t_0}.$$

As a consequence of the Lebesgue-dominated convergence theorem,

$$c = \lim_{t \to \infty} \int_{t_0}^{t} X^{-1}(s) \left[f(s, y(s)) + \int_{t_0}^{s} k(s, \sigma, y(\sigma)) \, d\sigma \right] ds + y_0, \quad (2.11.14)$$

exists. Using (2.11.14), we may rewrite (2.11.13) as

$$\Delta(t)y(t) = \Delta(t)X(t)y_0 + \Delta(t)X(t) \int_{t_0}^{\infty} X^{-1}(s) \left[f(s, y(s)) \right.$$

$$\left. + \int_{t_0}^{s} k(s, \sigma, y(\sigma)) \, d\sigma \right] ds - \Delta(t)X(t) \int_{t}^{\infty} X^{-1}(s) \left[f(s, y(s)) \right.$$

$$\left. + \int_{t_0}^{s} k(s, \sigma, y(\sigma)) \, d\sigma \right] ds$$

$$= \Delta(t)X(t)c - \Delta(t)X(t) \int_{t}^{\infty} X^{-1}(s) \left[f(s, y(s)) \right.$$

$$\left. + \int_{t_0}^{s} k(s, \sigma, y(\sigma)) \, d\sigma \right] ds.$$

Thus it follows from (2.11.8) and (2.11.12) that

$$\|\Delta(t)[y(t) - X(t)c]\| \leq \alpha(t) \left| \int_{t}^{\infty} X^{-1}(s) \left[f(s, y(s)) \right.\right.$$

$$\left.\left. + \int_{t_0}^{s} k(s, \sigma, y(\sigma)) \, d\sigma \right] ds \right| = o(\alpha(t)).$$

The proof is complete.

The converse of the above theorem given by Talpalaru (1973) is embodied in the following theorem.

Theorem 2.11.4 *Let the hypotheses of Theorem 2.11.3 be satisfied. Then, given any solution $x(t) = X(t)c$ of (L) with $\|c\| < M$, where*

$$M = \int_{t_0}^{\infty} \left[a(s) + \int_{t_0}^{s} h(s, \sigma) \, d\sigma \right] ds,$$

there exists a solution $y(t)$ of (P) such that (2.11.7) holds.

□

The proof of this theorem is based on the application of the well known Schauder–Tychonoff fixed point theorem, and the details are left to the reader.

Talpalaru (1973) also gave the following variant of Theorem 2.11.3.

Theorem 2.11.5 *Suppose that the following conditions are satisfied:*

(i) there exists a nonsingular continuous matrix $\Delta(t)$ such that (2.11.8) holds,

(ii) there exist functions $a(t)$ and $w(r)$ satisfying the same conditions as in Theorem 2.11.3 and

$$\|X^{-1}(t)(f(t, y_1) - f(t, y_2))\| \leq a(t) w \left(\frac{\|\Delta(t)(y_1 - y_2)\|}{\alpha(t)} \right), \quad t \in R_{t_0},$$
(2.11.15)

(iii) there exists a continuous nonnegative function $h(t, \sigma)$ defined for $t_0 \leq \sigma \leq t < \infty$ so that

$$\|X^{-1}(t)(k(t, \sigma, y_1(\sigma)) - k(t, \sigma, y_2(\sigma)))\|$$
$$\leq h(t, \sigma) w \left(\frac{\|\Delta(\sigma)(y_1(\sigma) - y_2(\sigma))\|}{\alpha(\sigma)} \right), \quad (2.11.16)$$

for $t_0 \leq \sigma \leq t < \infty$,

(iv) the functions $a(t)$ and $h(t, \sigma)$ satisfy

$$\int_{t_0}^{\infty} \left[a(s) + \int_{t_0}^{s} h(s, \sigma) \, d\sigma \right] ds < \infty, \quad (2.11.17)$$

and
$$\int_{t_0}^{\infty} \|X^{-1}(s)\| \left[\|f(s,0)\| + \int_{t_0}^{s} \|k(s,\sigma,0)\| \, d\sigma \right] ds < \infty. \quad (2.11.18)$$

Then, to each solution y(t) of (P) there exists a solution x(t) of (L) such that (2.11.7) holds.

□

The proof closely resembles that of Theorem 2.11.3, so only a few differences will be indicated. As in the proof of Theorem 2.11.3, we have

$$\Delta(t)y(t) = \Delta(t)X(t)y_0 + \Delta(t)X(t) \int_{t_0}^{t} X^{-1}(s) \left[f(s, y(s)) \right.$$
$$\left. + \int_{t_0}^{s} k(s, \sigma, y(\sigma)) \, d\sigma \right] ds$$

from which it follows that

$$\|\Delta(t)y(t)\| \le \|\Delta(t)X(t)\| \|y_0\| + \|\Delta(t)X(t)\| \int_{t_0}^{t} \left[\|X^{-1}(s)f(s,0)\| \right.$$
$$+ \int_{t_0}^{s} \|X^{-1}(s)k(s,\sigma,0)\| \, d\sigma \Big] ds + \int_{t_0}^{t} \Big[a(s)w\left(\frac{\|\Delta(s)y(s)\|}{\alpha(s)}\right)$$
$$+ \int_{t_0}^{s} h(s,\sigma)w\left(\frac{\|\Delta(\sigma)y(\sigma)\|}{\alpha(\sigma)}\right) d\sigma \Big] ds.$$

From (i) and (iv) it follows that

$$\frac{\|\Delta(t)y(t)\|}{\alpha(t)} \le \|y_0\| + \int_{t_0}^{t} \|X^{-1}(s)\| \left[\|f(s,0)\| + \int_{t_0}^{s} \|k(s,\sigma,0)\| \, d\sigma \right] ds$$
$$+ \int_{t_0}^{t} \Big[a(s)w\left(\frac{\|\Delta(s)y(s)\|}{\alpha(s)}\right)$$
$$+ \int_{t_0}^{s} h(s,\sigma)w\left(\frac{\|\Delta(\sigma)y(\sigma)\|}{\alpha(\sigma)}\right) d\sigma \Big] ds.$$

$$\leq M_1 + \int_{t_0}^{t} \left[a(s)w\left(\frac{\|\Delta(s)y(s)\|}{\alpha(s)}\right) \right.$$
$$\left. + \int_{t_0}^{s} h(s,\sigma)w\left(\frac{\|\Delta(\sigma)y(\sigma)\|}{\alpha(\sigma)}\right) d\sigma \right] ds.$$

Now an application of the inequality given in Theorem 2.4.5 yields

$$\frac{\|\Delta(t)y(t)\|}{\alpha(t)} \leq M, \quad t \in R_{t_0}.$$

To prove (2.11.7) we can proceed in the same manner as in the proof of Theorem 2.11.3.

We note that Talpalaru (1973) has given a detailed discussion through various remarks concerning the special versions of equation (P) studied by various authors in the literature. Further generalizations of the above results to higher order integro-differential equations are also indicated in Talpalaru (1973).

2.11.3 Higher Order Integro-differential Equations

This section is concerned with the behaviour of solutions of the higher order integro-differential equations of the form

$$y^{(n)} + \sum_{i=1}^{n} a_i(t)y^{(n-i)} = y^\alpha f\left(t, y, \int_0^t k(t,s,y)\,ds\right), \quad \text{(H)}$$

by means of comparison with solutions of the differential equation

$$y^{(n)} + \sum_{i=1}^{n} a_i(t)y^{(n-i)} = 0, \quad \text{(Q)}$$

where $a_i: R_+ \to R$, $k: R_+ \times R_+ \times R \to R$, $f: R_+ \times R \times R \to R$ are continuous functions and $0 < \alpha < 1$ is a constant. A great deal of attention has been given to the study of various general and special versions of equation (H) with or without delay arguments and many results have appeared in the literature (Butler and Rogers, 1971b; Grace and Lalli, 1980a,b; Graef and

Spikes, 1987; Kusano and Onose, 1973; Ladas, 1971; Lalli; 1984; Morchalo, 1979; Pachpatte, 1976d,c; Waltman, 1965; Zettl, 1970). Pachpatte (1976e) studied the boundedness, asymptotic behaviour and rate of growth of the solutions of (H) under some suitable conditions on the functions involved in (H) and on the solutions of equation (Q). In the following theorems we will present the results given by Pachpatte (1976e), which are based on the application of the special version of his inequality given in Theorem 2.7.3. Here it will be assumed that every solution $y(t)$ of (H) under discussion exists on an interval R_+.

In what follows, $\phi_1, \ldots, \phi_n$ denote the n linearly independent solutions of (Q) and $\phi = c_1\phi_1 + \cdots + c_n\phi_n$, where $c_1, \ldots, c_n$ are constants. We define

$$\sigma_j(t) = \frac{W_j(\phi_1, \ldots, \phi_n)(t)}{W(\phi_1, \ldots, \phi_n)(t)}, \quad t \in R_+,$$

where $W_j(\phi_1, \ldots, \phi_n)$ is the determinant obtained from the Wornskian $W(\phi_1, \ldots, \phi_n)$ by replacing the jth column by $(0, 0, \ldots, 0, 1)$.

The following result deals with the boundedness of the solutions of (H) under some suitable conditions on the functions involved in (H) and on the solutions of equation (Q).

Theorem 2.11.6 *Suppose that*

$$|\phi(t)| \le c, \quad t \in R_+, \quad (2.11.19)$$

$$\sum_{i=1}^{n} |\phi_i(t)\sigma_i(s)| \le M, \quad (2.11.20)$$

for $0 \le s \le t < \infty$, where $c > 0$, $M > 0$ are constants. Suppose that the functions f and k in (H) satisfy

$$|f(t, y, z)| \le p(t)(|y| + |z|), \quad t \in R_+, \quad (2.11.21)$$

$$|k(t, s, y)| \le q(s)|y|, \quad 0 \le s \le t < \infty, \quad (2.11.22)$$

where $p, q: R_+ \to R_+$ are continuous functions. If

$$E_1(t) = 1 - c^\alpha \alpha M \int_0^t p(\tau) \exp\left(\alpha \int_0^\tau q(\xi)\,d\xi\right) d\tau > 0, \quad (2.11.23)$$

for all $t \in R_+$ and

$$\int_0^\infty p(s)(E_1(s))^{-1/\alpha} \exp\left(\int_0^s q(\tau)\,d\tau\right) ds < \infty, \qquad (2.11.24)$$

then all solutions of (H) are bounded on R_+.

$\square$

Proof: By using the variation of constants formula, the solutions of (H) and (Q) are related by the integral equation (Coddington and Levinson, 1955)

$$y(t) = \phi(t) + \sum_{i=1}^n \phi_i(t) \int_0^t \sigma_i(s) y^\alpha(s) f\left(s, y(s), \int_0^s k(s, \tau, y(\tau))\,d\tau\right) ds, \qquad (2.11.25)$$

where $\phi(t) = c_1\phi_1(t) + \cdots + c_n\phi_n(t)$, $c_1, \ldots, c_n$ are constants and $\phi_1, \ldots, \phi_n$ are linearly independent solutions of (Q). From (2.11.25) and (2.11.19)–(2.11.22) we obtain

$$|y(t)| \leq c + \int_0^t M p(s)|y(s)|^\alpha \left(|y(s)| + \int_0^s q(\tau)|y(\tau)|\,d\tau\right) ds.$$

Now an application of the special version of Theorem 2.7.3 due to Pachpatte (1976b) (i.e. when $f(t) = 0$) yields

$$|y(t)| \leq c\left[1 + (1-\alpha)c^\alpha M \int_0^t p(s)(E_1(s))^{-1/\alpha} \exp\left(\int_0^s q(\tau)\,d\tau\right) ds\right]^{1/1-\alpha}.$$

The above estimation in view of the assumption (2.11.24) implies the boundedness of all solutions of (H) on R_+ and the theorem is proved.

∎

The next result shows that under some suitable conditions on the functions involved in (H) and on the solutions of equation (Q), all the solutions of (H) approach zero as $t \to \infty$.

Theorem 2.11.7 *Suppose that*

$$|\phi(t)| \leq c\exp(-\beta t), \quad t \in R_+, \qquad (2.11.26)$$

$$\sum_{i=1}^{n} |\phi_i(t)\sigma_i(s)| \leq M\exp(-\beta(t-s)), \quad 0 \leq s \leq t < \infty, \quad (2.11.27)$$

where $c > 0$, $M > 0$, $\beta > 0$ are constants. Suppose that the functions f and k in (H) satisfy

$$|f(t, y, z)| \leq p(t)(|y| + |z|), \quad t \in R_+, \qquad (2.11.28)$$

$$|k(t, s, y)| \leq \exp(-\beta s)q(s)|y|, \quad 0 \leq s \leq t < \infty, \qquad (2.11.29)$$

where $p, q \colon R_+ \to R_+$ are continuous functions. If

$$E_2(t) = 1 - M\alpha c^\alpha \int_0^t \exp(-\alpha\beta\tau)p(\tau)\exp\left(\alpha\int_0^\tau \exp(-\beta\xi)q(\xi)\,d\xi\right)d\tau > 0,$$
$$(2.11.30)$$

for all $t \in R_+$ and

$$\int_0^\infty \exp(-\alpha\beta s)p(s)\exp\left(\int_0^s \exp(-\beta\tau)q(\tau)\,d\tau\right)(E_2(s))^{-1/\alpha}\,ds < \infty,$$
$$(2.11.31)$$

then all solutions of (H) approach zero as $t \to \infty$.

□

Proof: If $\phi_1, \ldots, \phi_n$ are linearly independent solutions of (Q), then the solutions of (H) and (Q) are related by the integral equation (2.11.25). Using (2.11.25) and (2.11.26)–(2.11.29) we obtain

$$|y(t)| \leq c\exp(-\beta t) + \int_0^t M\exp(-\beta(t-s))p(s)|y(s)|^\alpha$$

$$\times \left(|y(s)| + \exp(-\beta s)\int_0^s q(\tau)|y(\tau)|\,d\tau\right)ds.$$

The above inequality can be written as

$$|y(t)|\exp(\beta t) \leq c + \int_0^t Mp(s)\exp(-\alpha\beta s)(|y(s)|\exp(\beta s))^\alpha$$

$$\times \left(|y(s)|\exp(\beta s) + \int_0^s q(\tau)\exp(-\beta\tau)\right.$$

$$\left.\times |y(\tau)|\exp(\beta\tau)\,d\tau\right)ds.$$

Now applying the special version of Theorem 2.7.3 due to Pachpatte (1976b) (i.e. when $f(t) = 0$) with $u(t) = |y(t)|\exp(\beta t)$, and then multiplying the resulting inequality by $\exp(-\beta t)$, we obtain

$$|y(t)| \leq c\exp(-\beta t)\left[1 + (1-\alpha)c^\alpha \int_0^t Mp(s)\exp(-\alpha\beta s)\right.$$

$$\left.\times \exp\left(\int_0^s q(\tau)\exp(-\alpha\tau)\,d\tau\right)(E_2(s))^{-1/\alpha}\,ds\right]^{1/1-\alpha}.$$

The above estimation in view of the assumption (2.11.31) yields the desired result, and the proof of the theorem is complete.

∎

A continuous function $z(t)$ will be called slowly growing if and only if for every $\epsilon > 0$ there exists a constant N, which may depend on ϵ such that

$$|z(t)| \leq N\exp(\epsilon t), \quad t \in R_+.$$

The following result demonstrates that all the solutions of (H) grow more slowly than any positive exponential.

Theorem 2.11.8 *Suppose that*

$$|\phi(t)| \leq c\exp(\beta t), \quad t \in R_+, \qquad (2.11.32)$$

$$\sum_{i=1}^n |\phi_i(t)\sigma_i(s)| \leq M\exp(\beta(t-s)), \quad 0 \leq s \leq t < \infty, \qquad (2.11.33)$$

where c, M and β are positive constants. Suppose that the functions f and k in (H) satisfy

$$|f(t, y, z)| \leq p(t)(|y| + |z|), \quad t \in R_+, \quad (2.11.34)$$

$$|k(t, s, y)| \leq \exp(\beta t)q(s)|y|, \quad 0 \leq s \leq t < \infty, \quad (2.11.35)$$

where $p, q \colon R_+ \to R_+$ are continuous functions. If

$$E_3(t) = 1 - M\alpha c^\alpha \int_0^t \exp(\alpha\beta\tau)p(\tau)\exp\left(\alpha \int_0^\tau \exp(\beta\xi)q(\xi)\,d\xi\right) d\tau > 0,$$
$$(2.11.36)$$

for all $t \in R_+$ and

$$\int_0^\infty \exp(\alpha\beta s)p(s)\exp\left(\int_0^s \exp(\beta\tau)q(\tau)\,d\tau\right)(E_3(s))^{-1/\alpha}\,ds < \infty, \quad (2.11.37)$$

then all solutions of (H) are slowly growing. $\square$

The proof of this theorem follows by a similar argument to that in the proof of Theorem 2.11.7, with suitable modifications, and hence the details are omitted.

2.11.4 Estimates of the Solutions of Certain Differential Equations

Morro (1982) obtained some interesting and new results on the mathematical structure of the hidden variable model with applications to continuum thermodynamics. Constantin (1990a,b) obtained an interesting generalization of the inequality used by Morro (1982) and applied it to a certain class of differential equations and to an estimate of the solution of the evolution equation for hidden variables. In this section we present the results given by Constantin (1990a,b) on the estimate of solution of the evolution equation for the hidden variables.

By referring to Constantin (1990a,b) we recall some notions about materials with hidden variables. A material with hidden variables $\{y_0, z_0, a_0, U, V, \mathscr{C}, f\}$ on $Y \times Z \times A$ consists of a ground value (y_0, z_0, a_0) of the variables $(y, z, a) \in Y \times Z \times A$, a representing the set of hidden variables, together with a connected neighbourhood $U \times V$ of (y_0, z_0) and the maps $\mathscr{C} \in \mathscr{C}^2(U \times A, \phi)$, $f \in \mathscr{C}^2(U \times V \times A, A)$ where Y, Z, A and ϕ denote finite-dimensional real-normed vector spaces with $\dim A \leq \dim Y + \dim Z$. A path is a bounded and piecewise continuously differentiable map $\pi \colon R \to U \times V$. The hidden variables are functions on R(time).

Consider the evolution equation

$$\dot{a}(t) = f(\pi(t), a(t)), \quad t \geq t_0, a(t_0) = a^*,$$

where the evolution function f satisfies:

(I) there is a map $\Lambda \in \mathscr{L}(A, A)$ and $w \colon R_+ \to R_+$ is a continuous monotonic nondecreasing function so that there exists $L > 0$ with $w(x) \geq Lx$ for each $x \in R_+$ and $w(0) = 0$ and $\int_{u_0}^{\infty} ds/w(s) = \infty$, $u_0 > 0$ is a constant, such that

$$\|f(\pi, a+b) - f(\pi, a) - \Lambda b\| \leq w(\|b\|),$$

for $\pi \in U \times V$, $a, a+b \in A$, and each eigenvalue of $\Lambda + I_A$ has a negative real part, where $\|\cdot\|$ denotes a convenient vector norm of an element in A,

(II) there is a positive constant ϵ such that

$$\|f(\pi + \omega, a) - f(\pi, a)\| \leq \epsilon \|\omega\|,$$

for $\pi, \pi + \omega \in U \times V, a \in A$.

We observe that this class of evolution equations is larger than that considered by Morro (1982) because it contains all functions of the form $w(\|b\|) = \delta\|b\|$ and also others (for example $w(\|b\|) = \|b\| + \|b\|^2$).

Let the hidden variables, as in Morro (1982), be a, $a+b \in A$ corresponding to the paths $\pi, \pi + \omega$. We have

$$\dot{a}(t) = f(\pi(t), a(t)), \quad t \geq t_0, a(t_0) = a^*,$$
$$\dot{a}(t) + \dot{b}(t) = f(\pi(t) + \omega(t), a(t) + b(t)), \quad t \geq t_0, a(t_0) + b(t_0) = a^* + b^*.$$

Denoting

$$\gamma = f(\pi + \omega, a + b) - f(\pi, a + b),$$
$$r = f(\pi, a + b) - f(\pi, a) - \Lambda b,$$

we obtain that
$$\dot{b} = \gamma + r + \Lambda b.$$

We find that
$$\frac{d}{dt}[\exp(-t\Lambda)b(t)] = \exp(-t\Lambda)[\gamma(t) + r(t)],$$

and an integration yields
$$b(t) = b(t_0)\exp((t-t_0)\Lambda) + \int_{t_0}^{t} [\gamma(s) + r(s)]\exp((t-s)\Lambda)\,ds.$$

Considering (I) and (II) and denoting by $-m$ the real part of the eigenvalue of Λ with the greatest real part, it follows that

$$\|b(t)\| \leq \exp(-m(t-t_0))\|b(t_0)\| + \int_{t_0}^{t} \exp(-m(t-s))\|\gamma(s) + r(s)\|\,ds$$

$$\leq \exp(-m(t-t_0))\|b(t_0)\| + \epsilon \int_{t_0}^{t} \exp(-m(t-s))\|\omega(s)\|\,ds$$

$$+ \int_{t_0}^{t} \exp(-m(t-s))w(\|b(s)\|)\,ds.$$

Let us assume that
$$v(t) = \|b(t)\|, \quad k(t) = \exp(-mt), \quad h(t) = \exp(mt),$$
$$g(t) = \exp(-m(t-t_0))\|b(t_0)\| + \epsilon \int_{t_0}^{t} \exp(-m(t-s))\|\omega(s)\|\,ds.$$

Now an application of Theorem 2.3.3 due to Constantin (1990a,b) yields

$$v(t) \leq G^{-1}\left[G(\|b(t_0)\|) + t - t_0 + \int_{t_0}^{t} \max\left\{0, \frac{g'(s)}{Lg(s)}\right\}\,ds\right], \quad (2.11.38)$$

where G, G^{-1} are defined as in Theorem 2.3.3. The bound obtained in (2.11.38) can be used in studying the independence property of the hidden variables of the present value of the physical variables (Morro (1982)).

2.12 Miscellaneous Inequalities

2.12.1 Rogers (1974)

Let $k(t, s, z)$ be a continuous R^n-valued function defined on $[0, 1] \times [0, 1] \times R^n$, nondecreasing in z componentwise for fixed t, s, and is uniformly Lipschitz in the last variable, and let $g(t, r)$ be a continuous R^n-valued function on $[0, 1] \times R^n$ nondecreasing in r componentwise for fixed t. Suppose $z(t)$ is a continuous solution of the integral inequality

$$z(t) \leq g\left(t, \int_0^t k(t, s, z(s)) \, ds\right),$$

then there is α in $[0, 1]$ such that

$$z(t) \leq g(t, p(t)),$$

and $p(t)$ is the maximal solution on $[0, \alpha)$ of the integral equation

$$r(t) = \int_0^t L(t, s, r(s)) \, ds,$$

where $L(t, s, r) = k(t, s, g(s, r))$.

2.12.2 Pachpatte (1975i)

Let $w_1 \in C[I \times R_+, R_+]$, $w_2 \in C[I \times I \times R_+, R_+]$ and assume that $w_2(t, s, r)$ is nondecreasing in r for each $t, s \in I$, where $I = [t_0, \infty)$, $t_0 \geq 0$. Suppose that $r(t)$ is the solution of the integro-differential equation

$$r'(\tau) = w_1(t, r(t)) + \int_{t_0}^t w_2(t, s, r(s) + \psi(s) e^{\alpha s}) \, ds, \quad \alpha > 0, \, r(t_0) = r_0,$$

existing to the right of t_0, where $\psi(t) \geq 0$ is a continuous function defined on I. Let $m(t) \geq 0$ be a continuous function on I such that $m(t_0) \leq r_0$ and

$$m'(t) \leq w_1(t, m(t)) + \int_{t_0}^{t} w_2(t, s, m(s) + \psi(s) e^{\alpha s}) \, ds,$$

where $m'(t) = \lim_{h \to 0^+} \sup (1/h)[m(t+h) - m(t)]$. Then

$$m(t) \leq r(t), \quad t \in R_+.$$

2.12.3 Willett and Wong (1965)

Let u, f and g be nonnegative continuous functions defined on R_+, $c > 0$, $p > 0$, $p \neq 1$ be constants and

$$u(t) \leq c + \int_0^t f(s) u(s) \, ds + \int_0^t g(s) u^p(s) \, ds, \quad t \in R_+.$$

(i) If $0 < p < 1$, then

$$u(t) \leq \exp\left(\int_0^t f(s) \, ds\right) \left[c^{1-p} + (1-p) \int_0^t g(s) \right.$$
$$\left. \times \exp\left(-(1-p) \int_0^s f(\sigma) \, d\sigma\right) ds \right]^{1/(1-p)}, \quad t \in R_+.$$

(ii) If $1 < p < \infty$, then

$$u(t) \leq \exp\left(\int_0^t f(s) \, ds\right) \left[c^{1-p} - (p-1) \int_0^t g(s) \right.$$
$$\left. \times \exp\left((p-1) \int_0^s f(\sigma) \, d\sigma\right) ds \right]^{-1/(p-1)}, \quad (2.12.1)$$

for $t \in [0, \alpha)$, where α is the supremum of $t \in R_+$ for which the expression between [...] in (2.12.1) is positive. Moreover, if we assume that the expression between [...] in (2.12.1) is positive for all $t \in R_+$, then the inequality (2.12.1) remains valid for all $t \in R_+$.

2.12.4 Pachpatte (1974a)

Let u, f, g and h be nonnegative continuous functions defined on R_+ and $c > 0$, $0 < p < 1$ are constants and

$$u(t) \leq c + \int_0^t f(s)\left[u(s) + \int_0^s f(\tau)\left(\int_0^\tau [g(\sigma)u(\sigma) + h(\sigma)u^p(\sigma)]\,d\sigma\right)d\tau\right]ds,$$

for $t \in R_+$. Then

$$u(t) \leq c + \int_0^t f(s)\left[c + \int_0^s f(\tau)\exp\left(\int_0^\tau [f(\sigma) + g(\sigma)]\,d\sigma\right)\right.$$

$$\times \left[c^{1-p} + (1-p)\int_0^\tau h(\sigma)\right.$$

$$\left.\left.\times \exp\left(-(1-p)\int_0^\sigma [f(\xi) + g(\xi)]\,d\xi\right)d\sigma\right]^{1/(1-p)} d\tau\right] ds,$$

for $t \in R_+$.

2.12.5 Pachpatte (1976f)

Let u, f, g and h be nonnegative continuous functions defined on R_+ and $c > 0$ is a constant and

$$u(t) \leq c + \int_0^t g(s)u(s)\left(u(s) + f(s) + \int_0^s h(\sigma)u(\sigma)\,d\sigma\right)ds,$$

for $t \in R_+$. Then

$$u(t) \leq c\exp\left(\int_0^t g(s)\left[f(s) + c\exp\left(\int_0^s [g(\sigma)f(\sigma) + h(\sigma)]\,d\sigma\right)\right.\right.$$

$$\left.\left.\times (E(s))^{-1}\right]ds\right), \qquad (2.12.2)$$

for $t \in [0, \alpha)$, where

$$E(t) = 1 - c \int_0^t g(\tau) \exp\left(\int_0^\tau [g(\sigma)f(\sigma) + h(\sigma)] \, d\sigma\right) d\tau,$$

and $\alpha = \sup\{t \in R_+ : E(t) > 0\}$. Moreover, if we assume that $E(t) > 0$ for all $t \in R_+$, then the inequality (2.12.2) remains valid for all $t \in R_+$.

2.12.6 Pachpatte (1974d)

Let u, f, g and h be nonnegative continuous functions defined on R_+. Let $H(u)$ be a continuous nondecreasing subadditive function defined on R_+ and $H(u) > 0$ on $(0, \infty)$. If

$$u(t) \leq f(t) + \int_0^t g(s) \left(u(s) + \int_0^s h(\sigma) u(\sigma) \, d\sigma \right.$$

$$\left. + \int_0^s [g(\sigma) + h(\sigma)] H(u(\sigma)) \, d\sigma \right) ds,$$

for $t \in R_+$, then for $0 \leq t \leq \bar{t}$,

$$u(t) \leq f(t) + Q(t) + \int_0^t g(s) G^{-1}\left[G(Q(t)) + \int_0^s [g(\sigma) + h(\sigma)] \, d\sigma\right] ds,$$

where

$$Q(t) = \int_0^t g(s) \left(f(s) + \int_0^s h(\sigma) f(\sigma) \, d\sigma \right.$$

$$\left. + \int_0^s [g(\sigma) + h(\sigma)] H(f(\sigma)) \, d\sigma \right) ds,$$

$$G(r) = \int_{r_0}^r \frac{ds}{s + H(s)}, \quad r > 0, r_0 > 0, \tag{2.12.3}$$

G^{-1} is the inverse of G, and $\bar{t} \in R_+$ is chosen so that

$$G(Q(t)) + \int_0^t [g(\sigma) + h(\sigma)] \, d\sigma \in \text{Dom}(G^{-1}),$$

for all $t \in R_+$ lying in the interval $0 \le t \le \bar{t}$.

2.12.7 Pachpatte (1976c)

Let u and f be nonnegative continuous functions defined on R_+, $n(t) \ge 1$ is a monotonic nondecreasing continuous function defined on R_+, and $H \in \mathcal{F}$, where $\mathcal{F}$ is the class of functions defined in Section 2.5. If

$$u(t) \le n(t) + \int_0^t f(s) \left(u(s) + \int_0^s f(\tau) \left(\int_0^\tau f(\sigma) H(u(\sigma)) \, d\sigma \right) d\tau \right) ds,$$

for $t \in R_+$ then for $0 \le t \le \bar{t}$,

$$u(t) \le b(t) n(t),$$

where

$$b(t) = 1 + \int_0^t f(s) \left[1 + \int_0^s f(\tau) G^{-1} \left[G(1) + \int_0^\tau f(\sigma) \, d\sigma \right] d\tau \right] ds, \quad (2.12.4)$$

G is defined by (2.12.3) and G^{-1} is the inverse of G, and $\bar{t} \in R_+$ is chosen so that

$$G(1) + \int_0^t f(\sigma) \, d\sigma \in \text{Dom}(G^{-1}),$$

for all $t \in R_+$ lying in the interval $0 \le t \le \bar{t}$.

2.12.8 Pachpatte (1976c)

Let u, f and h be nonnegative continuous functions defined on R_+, $H \in \mathcal{F}$, where $\mathcal{F}$ is the class of functions defined in Section 2.5, $w(u)$ be a continuous nondecreasing subadditive and submultiplicative function defined on R_+ and $w(u) > 0$ on $(0, \infty)$. Let $p(t) \geq 1$, $\phi(t) \geq 0$ be continuous nondecreasing functions defined on R_+ and $\phi(0) = 0$. If

$$u(t) \leq p(t) + \int_0^t f(s) \left(u(s) + \int_0^s f(\tau) \left(\int_0^\tau f(\sigma) H(u(\sigma)) \, d\sigma \right) d\tau \right) ds$$

$$+ \phi \left(\int_0^t h(s) w(u(s)) \, ds \right),$$

for $t \in R_+$, then for $0 \leq t \leq \bar{t}$,

$$u(t) \leq b(t) \left[p(t) + \phi \left(\psi^{-1} \left[\psi \left(\int_0^t h(s) w(b(s) p(s)) \, ds \right) \right. \right. \right.$$

$$\left. \left. \left. + \int_0^t h(s) w(b(s)) \, ds \right] \right) \right],$$

where $b(t)$ is defined by (2.12.4) in which G and G^{-1} are as defined therein and

$$\psi(r) = \int_{r_0}^r \frac{ds}{w(\phi(s))}, \quad r > 0, r_0 > 0, \tag{2.12.5}$$

ψ^{-1} is the inverse of ψ and $\bar{t} \in R_+$ is chosen so that

$$G(1) + \int_0^t f(s) \, ds \in \text{Dom}(G^{-1}),$$

$$\psi \left(\int_0^t h(s) w(b(s) p(s)) \, ds \right) + \int_0^t h(s) w(b(s)) \, ds \in \text{Dom}(\psi^{-1}),$$

for all $t \in R_+$ lying in the interval $0 \leq t \leq \bar{t}$.

2.12.9 Pachpatte (1976c)

Let u, f and h be nonnegative continuous functions defined on R_+, $H \in \mathcal{F}$, where $\mathcal{F}$ is the class of functions defined in Section 2.5, $w(t, u)$ be a nonnegative continuous monotonic nondecreasing function in $u \geq 0$ for each fixed

$t \in R_+$, the functions $p(t) \geq 1$, $\phi(t) \geq 0$ be continuous and nondecreasing on R_+, $\phi(0) = 0$. If

$$u(t) \leq p(t) + \int_0^t f(s) \left(u(s) + \int_0^s f(\tau) \left(\int_0^\tau f(\sigma) H(u(\sigma)) \, d\sigma \right) d\tau \right) ds$$

$$+ \phi \left(\int_0^t h(s) w(s, u(s)) \, ds \right),$$

for $t \in R_+$, then for $t \in I \subset R_+$,

$$u(t) \leq b(t)[p(t) + \phi(r(t))], \tag{2.12.6}$$

where $b(t)$ is defined by (2.12.4) in which G and G^{-1} are as defined therein and I is the largest interval of R_+ on which the right-hand side of (2.12.6) exists, and $r(t)$ is the maximal solution of

$$r'(t) = h(t) w(t, b(t)[p(t) + \phi(r(t))]), \quad r(0) = 0,$$

existing on R_+.

2.12.10 Pachpatte (1976g)

Let u, f, g, h, w, p and ϕ be as in Theorem 2.8.6 and $\alpha \geq 1$ be a constant. If

$$u(t) \leq p(t) + g(t) \left(\int_0^t h(s) u^\alpha(s) \, ds \right)^{1/\alpha} + \phi \left(\int_0^t f(s) w(u(s)) \, ds \right),$$

for $t \in R_+$, then for $0 \leq t \leq \bar{t}$,

$$u(t) \leq Q(t) \left[p(t) + \phi \left(\psi^{-1} \left[\psi \left(\int_0^t f(s) w(Q(s) p(s)) \, ds \right) \right. \right. \right.$$

$$\left. \left. \left. + \int_0^t f(s) w(Q(s)) \, ds \right] \right) \right],$$

where

$$Q(t) = 1 + g(t) \left[\frac{(\int_0^t h(s) e(s) \, ds)^{1/\alpha}}{1 - (1 - e(t))^{1/\alpha}} \right], \tag{2.12.7}$$

in which
$$e(t) = \exp\left(-\int_0^t h(s)g^p(s)\,ds\right),$$

ψ, ψ^{-1} are as in Theorem 2.8.6 and $\bar{t} \in R_+$ is chosen so that

$$\psi\left(\int_0^t f(s)w(Q(s)p(s))\,ds\right) + \int_0^t f(s)w(Q(s))\,ds \in \text{Dom}(\psi^{-1}),$$

for all $t \in R_+$ lying in the interval $0 \le t \le \bar{t}$.

2.12.11 Pachpatte (1976g)

Let u, f, g, h, p, ϕ and $w(t, u)$ be as in Theorem 2.8.7 and $\alpha \ge 1$ be a constant. If

$$u(t) \le p(t) + g(t)\left(\int_0^t h(s)u^\alpha(s)\,ds\right)^{1/\alpha} + \phi\left(\int_0^t f(s)w(s, u(s))\,ds\right),$$

for $t \in R_+$, then
$$u(t) \le Q(t)[p(t) + \phi(r(t))],$$

where $Q(t)$ is defined by (2.12.7) and $r(t)$ is the maximal solution of

$$r'(t) = f(t)w(t, Q(t)[p(t) + \phi(r(t))]), \quad r(0) = 0,$$

existing on R_+.

2.12.12 Pachpatte (1975g)

Let u, f and g be nonnegative continuous functions defined on R_+, and $n(t)$ be a positive monotonic nondecreasing continuous function defined on R_+ and $0 < \alpha < 1$ be a constant. If

$$u(t) \le n(t) + \int_0^t f(s)u(s)\,ds + \int_0^t f(s)\left(\int_0^s g(\sigma)(n(\sigma))^{1-\alpha}u^\alpha(\sigma)\,d\sigma\right)ds,$$

for $t \in R_+$, then
$$u(t) \le Q_1(t)n(t), \quad t \in R_+,$$

where

$$Q_1(t) = 1 + \int_0^t f(s) \exp\left(\int_0^s f(\sigma)\,d\sigma\right) \left[1 + (1-\alpha)\int_0^s g(\tau)\right.$$
$$\left.\times \exp\left(-(1-\alpha)\int_0^\tau f(\sigma)\,d\sigma\right) d\tau\right]^{1/(1-\alpha)} ds, \qquad (2.12.8)$$

for $t \in R_+$.

2.12.13 Pachpatte (1975g)

Let u, f, g, h, w, p and ϕ be as in Theorem 2.8.6 and $0 < \alpha < 1$ be a constant. If

$$u(t) \leq A(t) + \int_0^t f(s)u(s)\,ds + \int_0^t f(s)\left(\int_0^s g(\sigma)(A(\sigma))^{1-\alpha}u^\alpha(\sigma)\,d\sigma\right) ds,$$

for $t \in R_+$, where

$$A(t) = p(t) + \phi\left(\int_0^t h(s)w(u(s))\,ds\right), \qquad (2.12.9)$$

then for $0 \leq t \leq \bar{t}$,

$$u(t) \leq Q_1(t) \left[p(t) + \phi\left(\psi^{-1}\left[\psi\left(\int_0^t h(s)w(Q_1(s)p(s))\,ds\right)\right.\right.\right.$$
$$\left.\left.\left.+ \int_0^t h(s)w(Q_1(s))\,ds\right]\right)\right],$$

where $Q_1(t)$ is defined by (2.12.8), ψ, ψ^{-1} are as defined in Theorem 2.8.6 and $\bar{t} \in R_+$ is chosen so that

$$\psi\left(\int_0^t h(s)w(Q_1(s)p(s))\,ds\right) + \int_0^t h(s)w(Q_1(s))\,ds \in \text{Dom}(\psi^{-1}),$$

for all $t \in R_+$ lying in the interval $0 \leq t \leq \bar{t}$.

2.12.14 Pachpatte (1975g)

Let u, f, g, h, p, ϕ and $w(t, u)$ be as in Theorem 2.8.7 and $0 < \alpha < 1$ be a constant. If

$$u(t) \leq B(t) + \int_0^t f(s)u(s)\,ds + \int_0^t f(s)\left(\int_0^s g(\sigma)(B(\sigma))^{1-\alpha}u^\alpha(\sigma)\,d\sigma\right)ds,$$

for $t \in R_+$, where

$$B(t) = p(t) + \phi\left(\int_0^t h(s)w(s, u(s))\,ds\right), \qquad (2.12.10)$$

then

$$u(t) \leq Q_1(t)[p(t) + \phi(r(t))], \quad t \in R_+,$$

where $Q_1(t)$ is defined by (2.12.8) and $r(t)$ is the maximal solution of

$$r'(t) = h(t)w(t, Q_1(t)[p(t) + \phi(r(t))]), \quad r(0) = 0,$$

existing on R_+.

2.12.15 Pachpatte (1976b)

Let u, f and g be nonnegative continuous functions defined on R_+, and $n(t)$ be a positive monotonic nondecreasing continuous function defined on R_+. If

$$u(t) \leq n(t) + (1/n(t)) \int_0^t f(s)u(s)\left(u(s) + \int_0^s g(\sigma)u(\sigma)\,d\sigma\right)ds,$$

for $t \in R_+$ and

$$E_1(t) = 1 - \int_0^t f(\tau)\exp\left(\int_0^\tau g(\sigma)\,d\sigma\right)d\tau > 0,$$

for all $t \in R_+$, then

$$u(t) \leq Q_2(t)n(t), \quad t \in R_+,$$

where

$$Q_2(t) = \exp\left(\int_0^t f(s)(E_1(s))^{-1} \exp\left(\int_0^s g(\sigma)\,d\sigma\right) ds\right), \qquad (2.12.11)$$

for $t \in R_+$.

2.12.16 Pachpatte (1976b)

Let u, f, g, h, w, p and ϕ be as in Theorem 2.8.6. If

$$u(t) \leq A(t) + (1/A(t)) \int_0^t f(s)u(s)\left(u(s) + \int_0^s g(\sigma)u(\sigma)\,d\sigma\right) ds,$$

for $t \in R_+$, where $A(t)$ is defined by (2.12.9), then for $0 \leq t \leq \bar{t}$,

$$u(t) \leq Q_2(t)\left[p(t) + \phi\left(\psi^{-1}\left[\psi\left(\int_0^t h(s)w(Q_2(s)p(s))\,ds\right)\right.\right.\right.$$
$$\left.\left.\left. + \int_0^t h(s)w(Q_2(s))\,ds\right]\right)\right],$$

where $Q_2(t)$ is defined by (2.12.11), ψ, ψ^{-1} are as defined in Theorem 2.8.6 and $\bar{t} \in R_+$ is chosen so that

$$\psi\left(\int_0^t h(s)w(Q_2(s)p(s))\,ds\right) + \int_0^t h(s)w(Q_2(s))\,ds \in \mathrm{Dom}(\psi^{-1}),$$

for all $t \in R_+$ lying in the interval $0 \leq t \leq \bar{t}$.

2.12.17 Pachpatte (1976b)

Let u, f, g, h, p, ϕ and $w(t, u)$ be as in Theorem 2.8.7. If

$$u(t) \leq B(t) + (1/B(t)) \int_0^t f(s)u(s)\left(u(s) + \int_0^s g(\sigma)u(\sigma)\,d\sigma\right) ds,$$

for $t \in R_+$, where $B(t)$ is defined by (2.12.10), then

$$u(t) \leq Q_2(t)[p(t) + \phi(r(t))], \quad t \in R_+,$$

where $Q_2(t)$ is defined by (2.12.11), and $r(t)$ is the maximal solution of

$$r'(t) = h(t)w(t, Q_2(t)[p(t) + \phi(r(t))]) \quad r(0) = 0,$$

existing on R_+.

2.12.18 Pachpatte (1977b)

Let u, f and g be nonnegative continuous functions defined on R_+, and $n(t)$ be a positive monotonic nondecreasing continuous function defined on R_+. If

$$u(t) \leq f(t)n(t) + (u(t)/n(t)) \int_0^t g(s)u(s)\,ds,$$

for $t \in R_+$ and

$$E_2(t) = 1 - 2\int_0^t f(s)g(s)\,ds > 0,$$

for $t \in R_+$, then

$$u(t) \leq Q_3(t)n(t), \quad t \in R_+,$$

where

$$Q_3(t) = f(t)(E(t))^{-1/2},$$

for $t \in R_+$.

2.12.19 Pachpatte (1978c)

Let u, u' and b be nonnegative continuous functions defined on R_+ and

$$u'(t) \leq a + \int_0^t b(s)u(s)(u(s) + u'(s))\,ds, \quad t \in R_+,$$

where a is a positive constant. Then

$$u'(t) \leq a + [a+u(0)]^2 \int_0^t e^{2s} b(s)(1/E_3(s))^2 \, ds, \qquad (2.12.12)$$

for $t \in [0, \beta)$, where

$$E_3(t) = 1 - [a+u(0)] \int_0^t e^s b(s) \, ds,$$

and $\beta = \sup\{t \in R_+ : E_3(t) > 0\}$. Moreover, if we assume that $E_3(t) > 0$ for all $t \in R_+$, then the inequality (2.12.12) remains valid for all $t \in R_+$.

2.12.20 Pachpatte (1978c)

Let u, u' and b be nonnegative continuous functions defined on R_+ and

$$u'(t) \leq a + \int_0^t b(s)(u'(s))^\alpha (u(s) + u'(s)) \, ds, \quad t \in R_+,$$

where $a > 0$, $0 < \alpha < 1$ are constants. Then

$$u'(t) \leq \left[a^{1-\alpha} + (1-\alpha)[a+u(0)] \int_0^t e^s b(s)(E_4(s))^{-1/\alpha} \, ds \right]^{1/(1-\alpha)}, \qquad (2.12.13)$$

for $t \in [0, \beta)$, where

$$E_4(t) = 1 - \alpha[a+u(0)]^\alpha \int_0^t e^{\alpha s} b(s) \, ds,$$

and $\beta = \sup\{t \in R_+ : E_4(t) > 0\}$. Moreover, if we assume that $E_4(t) > 0$ for all $t \in R_+$, then the inequality (2.12.13) remains valid for all $t \in R_+$.

2.13 Notes

The material included in Section 2.2 contains some basic comparison theorems used for studying the behaviour of solutions of a system of differential equations. Theorems 2.2.1 and 2.2.2 in their present forms are taken from Beesack (1975); see also Bainov and Simeonov (1992). Theorem 2.2.3 is taken from Brauer (1963). Theorem 2.2.4 is due to Opial (1960) and Theorem 2.2.5 is taken from Lasota *et al.* (1971). The literature concerning the comparison theorems and their applications is particularly rich, and the reader is referred to the standard references (Lakshmikantham and Leela, 1969; Szarski, 1965; Walter, 1970).

Section 2.3 contains the basic nonlinear integral inequalities which provides explicit bounds on the unknown functions, which are very effective in the development of the theory of systems of ordinary differential equations. Theorem 2.3.1 is due to Bihari (1956). The special version of Bihari's inequality was also obtained independently by LaSalle (1949). Theorem 2.3.2 is due to Langenhop (1960). Theorems 2.3.3 and 2.3.4 are taken from Constantin (1990b).

The results given in Sections 2.4 and 2.5 are the further generalizations of the Gronwall–Bellman and Bihari inequalities mainly developed in order to apply them in certain applications. Theorems 2.4.1–2.4.3 are taken from Pachpatte (1975a). Theorem 2.4.4 is due to Constantin (1990b) and Theorem 2.4.5 can be found in Talpalaru (1973). Theorems 2.5.1 and 2.5.2 are due to Constantin (1992). Theorem 2.5.3 is due to Beesack (1984a) and Theorems 2.5.4 and 2.5.5 are taken from Beesack (1975, 1977) which basically involves the comparison principle.

Section 2.6 contains some useful inequalities with nonlinearities in integrals. Theorem 2.6.1 is due to Willett and Wong (1965) which is more convenient in certain applications. Theorems 2.6.2 and 2.6.3 are due to Gollwitzer (1969). Theorems 2.6.4 and 2.6.5 are taken from Pachpatte (1984) which are further generalizations of the Greene's inequality (1977). Theorems 2.6.6 and 2.6.7 are due to Pachpatte (1975c,d,h).

Section 2.7 deals with some specific nonlinear integral inequalities established by Pachpatte (1973a, 1976b, 1977a,c,d, 1978b), which can be used as ready and powerful tools in certain applications. Section 2.8 is devoted to the Bihari-like integral inequalities developed by Pachpatte (1974d, 1975a,b,h, 1976c), which can be used as ready tools in the study of more general versions of certain integral and integro-differential equations.

The results in Section 2.9 deal with the integro-differential inequalities developed by Pachpatte (1977b, 1981a) which are very effective in the study of certain differential and integro-differential equations. Section 2.10 is devoted to the inequalities involving several iterated integrals and investigated by Pachpatte in (in press e,j) which can be used as powerful tools in the study of certain higher order differential and integro-differential equations.

Section 2.11 is devoted to the applications of certain inequalities in studying the qualitative properties of solutions of certain differential and integro-differential equations. Section 2.12 contains some miscellaneous inequalities which can also be used in some applications.

Chapter Three

Nonlinear Integral Inequalities II

3.1 Introduction

The fundamental role played by the integral inequalities in the development of the theory of differential and integral equations is well known. In the literature there are many papers on integral inequalities and their applications in the theory of differential and integral equations. Although stimulating research work related to integral inequalities used in the theory of differential and integral equations has been undertaken in the past few years, it appears that there are certain classes of differential and integral equations for which the earlier results on such inequalities do not apply directly. Motivated by this fact, various investigators have discovered some useful integral inequalities in order to achieve a diversity of desired goals. This chapter deals with some basic nonlinear integral inequalities which can be used as handy tools in the study of certain new classes of differential and integral equations. Some immediate applications and miscellaneous inequalities which can be used in some new applications are also presented.

3.2 Dragomir's Inequalities

The integral inequalities which provide explicit bounds on unknown functions have proved to be very useful in the study of qualitative properties of the solutions of differential and integral equations. Dragomir (1987a,b) obtained some generalizations of the well-known Gronwall–Bellman inequality which can be used as convenient tools in applications. In this section we shall give the inequalities established by Dragomir (1987a,b) and their subsequent generalizations obtained by Dragomir (1992).

The main result given by Dragomir (1987a,b) is embodied in the following theorem.

Theorem 3.2.1 *Let $u, a, b: I = [\alpha, \beta] \to R_+$ be continuous functions. Let $L: I \times R_+ \to R_+$ be a continuous function such that*

$$0 \leq L(t, x) - L(t, y) \leq M(t, y)(x - y), \tag{L}$$

for $t \in I$ and $x \geq y \geq 0$, where $M: I \times R_+ \to R_+$ be a continuous function. If

$$u(t) \leq a(t) + b(t) \int_\alpha^t L(s, u(s)) \, ds, \tag{3.2.1}$$

for $t \in I$, then

$$u(t) \leq a(t) + b(t) \int_\alpha^t L(s, a(s)) \exp\left(\int_s^t M(\sigma, a(\sigma)) b(\sigma) \, d\sigma \right) ds, \tag{3.2.2}$$

for $t \in I$.

□

Proof: Define a function $z(t)$ by

$$z(t) = \int_\alpha^t L(s, u(s)) \, ds, \quad t \in I. \tag{3.2.3}$$

From (3.2.3), the fact that $u(t) \leq a(t) + b(t) z(t)$ and the condition (L) it follows that

$$\begin{aligned}
z'(t) &= L(t, u(t)) \\
&\leq L(t, a(t) + b(t) z(t)) \\
&= L(t, a(t) + b(t) z(t)) - L(t, a(t)) + L(t, a(t)) \\
&\leq M(t, a(t)) b(t) z(t) + L(t, a(t)).
\end{aligned} \tag{3.2.4}$$

The inequality (3.2.4) implies the estimate

$$z(t) \leq \int_\alpha^t L(s, a(s)) \exp\left(\int_s^t M(\sigma, a(\sigma))b(\sigma)\,d\sigma\right) ds. \tag{3.2.5}$$

Using (3.2.5) in $u(t) \leq a(t) + b(t)z(t)$, we get the desired inequality in (3.2.2).

∎

The following two corollaries are evident by the above theorem.

Corollary 3.2.1 Let u, a and b be as defined in Theorem 3.2.1. Let $G: I \times R_+ \to R_+$ be a continuous function such that

$$0 \leq G(t, x) - G(t, y) \leq N(t)(x - y), \tag{G}$$

for $t \in I$ and $x \geq y \geq 0$, where $N: I \to R_+$ is a continuous function. If

$$u(t) \leq a(t) + b(t) \int_\alpha^t G(s, u(s))\,ds, \tag{3.2.6}$$

for $t \in I$, then

$$u(t) \leq a(t) + b(t) \int_\alpha^t G(s, a(s)) \exp\left(\int_s^t N(\sigma)b(\sigma)\,d\sigma\right) ds, \tag{3.2.7}$$

for $t \in I$.

△

Corollary 3.2.2 Let u, a, b, $c: I \to R_+$ be continuous functions. Let $H: R_+ \to R_+$ be a continuous function such that

$$0 \leq H(x) - H(y) \leq Q(x - y), \tag{H}$$

for $t \in I$ and $x \geq y \geq 0$, where Q is a nonnegative constant. If

$$u(t) \leq a(t) + b(t) \int_\alpha^t c(s)H(u(s))\,ds, \tag{3.2.8}$$

for $t \in I$, Then

$$u(t) \leq a(t) + b(t) \int_\alpha^t c(s) H(a(s)) \exp\left(Q \int_s^t c(\sigma) b(\sigma)\,d\sigma\right) ds, \quad (3.2.9)$$

for $t \in I$.

△

Remark 3.2.1 By taking $H(u) = u$ in Corollary 3.2.2, we get the inequality which in turn is a further generalization of the well known Gronwall–Bellman inequality.

△

Dragomir (1992) has given the following generalization of his earlier result in Theorem 3.2.1.

Theorem 3.2.2 *Let $u, a, b: I \to R_+$ be continuous functions. Let $L: I \times R_+ \to R_+$ be a continuous function and $\psi: R_+ \to R_+$ be a continuous and strictly increasing function with $\psi(0) = 0$ and*

$$0 \leq L(t, x) - L(t, y) \leq M(t, y) \psi^{-1}(x - y), \quad (L_\psi)$$

for $t \in I$ and $x \geq y \geq 0$, where $M: I \times R_+ \to R_+$ is a continuous function and ψ^{-1} is the inverse of ψ. If

$$u(t) \leq a(t) + \psi\left(b(t) \int_\alpha^t L(s, u(s))\,ds\right), \quad (3.2.10)$$

for $t \in I$, then

$$u(t) \leq a(t) + \psi\left(b(t) \int_\alpha^t L(s, a(s)) \exp\left(\int_s^t M(\sigma, a(\sigma)) b(\sigma)\,d\sigma\right) ds\right), \quad (3.2.11)$$

for $t \in I$.

□

Proof: Define a function $z(t)$ by

$$z(t) = \int_\alpha^t L(s, u(s))\,ds. \quad (3.2.12)$$

From (3.2.12), the fact that $u(t) \leq a(t) + \psi(b(t)z(t))$ and the condition (L_ψ) it follows that

$$\begin{aligned}
z'(t) &= L(t, u(t)) \\
&\leq L(t, a(t) + \psi(b(t)z(t))) \\
&= L(t, a(t) + \psi(b(t)z(t))) - L(t, a(t)) + L(t, a(t)) \\
&\leq M(t, a(t))\psi^{-1}(\psi(b(t)z(t))) + L(t, a(t)) \\
&= M(t, a(t))b(t)z(t) + L(t, a(t)).
\end{aligned} \qquad (3.2.13)$$

The inequality (3.2.13) implies the estimate

$$z(t) \leq \int_\alpha^t L(s, a(s)) \exp\left(\int_s^t M(\sigma, a(\sigma))b(\sigma)\, d\sigma\right) ds. \qquad (3.2.14)$$

Using (3.2.14) in $u(t) \leq a(t) + \psi(b(t)z(t))$ we get the desired inequality in (3.2.11).

∎

The following two corollaries follow obviously from the above theorem.

Corollary 3.2.3 Let u, a, b and ψ be as in Theorem 3.2.2. Let $G: I \times R_+ \to R_+$ be a continuous function such that

$$0 \leq G(t, x) - G(t, y) \leq N(t)\psi^{-1}(x - y), \qquad (G_\psi)$$

for $t \in I$ and $x \geq y \geq 0$ and $N: I \to R_+$ is a continuous function. If

$$u(t) \leq a(t) + \psi\left(b(t) \int_\alpha^t G(s, u(s))\, ds\right), \qquad (3.2.15)$$

for $t \in I$, then

$$u(t) \leq a(t) + \psi\left(b(t) \int_\alpha^t G(s, a(s)) \exp\left(\int_s^t N(\sigma)b(\sigma)\, d\sigma\right) ds\right), \qquad (3.2.16)$$

for $t \in I$.

△

Corollary 3.2.4 Let u, a, b and ψ be as defined in Theorem 3.2.2 and $c: I \to R_+$ is a continuous function. Let $H: R_+ \to R_+$ be a continuous function such that

$$0 \leq H(x) - H(y) \leq Q\psi^{-1}(x-y), \qquad (H_\psi)$$

for $x \geq y \geq 0$, where $Q \geq 0$ is a constant. If

$$u(t) \leq a(t) + \psi\left(b(t)\int_\alpha^t c(s)H(u(s))\,ds\right), \qquad (3.2.17)$$

for $t \in I$, then

$$u(t) \leq a(t) + \psi\left(b(t)\int_\alpha^t c(s)H(a(s))\exp\left(Q\int_s^t c(\sigma)b(\sigma)\,d\sigma\right)ds\right), \qquad (3.2.18)$$

for $t \in I$.

△

The second generalization of Theorem 3.2.1 established by Dragomir (1992) is given in the following theorem.

Theorem 3.2.3 Let u, a, b, L and ψ be as defined in Theorem 3.2.2 and suppose in addition that

$$\psi^{-1}(xy) \leq \psi^{-1}(x)\psi^{-1}(y),$$

for $x, y \in R_+$. If

$$u(t) \leq a(t) + b(t)\psi\left(\int_\alpha^t L(s, u(s))\,ds\right), \qquad (3.2.19)$$

for $t \in I$, then

$$u(t) \leq a(t) + b(t)\psi\left(\int_\alpha^t L(s, a(s))\exp\left(\int_s^t M(\sigma, a(\sigma))\psi^{-1}(b(\sigma))\,d\sigma\right)ds\right), \qquad (3.2.20)$$

for $t \in I$.

□

The proof of the theorem can be completed by following the same arguments as in the proof of Theorem 3.2.2 with suitable changes. Furthermore, one can very easily obtain from Theorem 3.2.3 the two corollaries similar to Corollaries 3.2.3 and 3.2.4 given above. The details are omitted here.

3.3 Pachpatte's Inequalities I

The inequality given in Theorem 3.2.1 is used very effectively by Dragomir (1987a,b, 1988) to study the various properties of the solutions of certain differential and integral equations. In view of the important applications of the above-mentioned inequality, Pachpatte (in press k, l) obtained some interesting generalizations of the inequality in Theorem 3.2.1 which can be used as convenient tools in certain applications. In this Section we shall deal with some basic inequalities established in the above-mentioned papers.

Pachpatte (in press l) has obtained the following generalization of the inequality given in Theorem 3.2.1.

Theorem 3.3.1 *Let u, a and b be nonnegative continuous functions defined on R_+. Let $F(u)$ be a continuous strictly increasing convex and submultiplicative function for $u > 0$, $\lim_{u \to \infty} F(u) = \infty$, F^{-1} denotes the inverse function of F, $\alpha(t)$, $\beta(t)$ be continuous and positive functions defined on R_+, $\alpha(t) + \beta(t) = 1$, and L, M be the functions defined as in Theorem 3.2.1 satisfying the condition (L). If*

$$u(t) \leq a(t) + b(t) F^{-1}\left(\int_0^t L(s, F(u(s))) \, ds \right), \tag{3.3.1}$$

for $t \in R_+$, then

$$u(t) \leq a(t) + b(t) F^{-1}\Bigg(\int_0^t L(s, \alpha(s) F(a(s)\alpha^{-1}(s)))$$

$$\times \exp\left(\int_0^t M(\sigma, \alpha(\sigma) F(a(\sigma)\alpha^{-1}(\sigma))) \beta(\sigma) F(b(\sigma)\beta^{-1}(\sigma)) \, d\sigma \right) ds \Bigg), \tag{3.3.2}$$

for $t \in R_+$.

□

Proof: Rewrite (3.3.1) as

$$u(t) \leq \alpha(t)(a(t)\alpha^{-1}(t)) + \beta(t)(b(t)\beta^{-1}(t))F^{-1}\left(\int_0^t L(s, F(u(s)))\,ds\right). \quad (3.3.3)$$

Since F is convex, submultiplicative and monotonic, from (3.3.3) we see that

$$F(u(t)) \leq \alpha(t)F(a(t)\alpha^{-1}(t)) + \beta(t)F(b(t)\beta^{-1}(t))\int_0^t L(s, F(u(s)))\,ds, \quad (3.3.4)$$

The estimate given in (3.3.2) follows by first applying Theorem 3.2.1 to (3.3.4) and then applying F^{-1} to both sides of the resulting inequality.

∎

Remark 3.3.1 We note that the inequality established in Theorem 3.3.1 can be considered as a further generalization of the inequality in Theorem 2.6.2 given by Gollwitzer (1969).

△

The following generalization of Theorem 3.2.1 is motivated by the inequalities given by Pachpatte (1975c,h).

Theorem 3.3.2 *Let u, a, b, L and M be as in Theorem 3.3.1. Let $H(r)$ be a positive continuous strictly increasing, subadditive and submultiplicative function for $r > 0$ with $\lim_{r \to \infty} H(r) = \infty$. If*

$$u(t) \leq a(t) + b(t)H^{-1}\left(\int_0^t L(s, H(u(s)))\,ds\right), \quad (3.3.5)$$

for $t \in R_+$, then

$$u(t) \leq H^{-1}\left[H(a(t)) + H(b(t))\int_0^t L(s, H(a(s))) \times \exp\left(\int_s^t M(\sigma, H(a(\sigma)))H(b(\sigma))\,d\sigma\right)\,ds\right], \quad (3.3.6)$$

for $t \in R_+$.

□

Proof: Since H is subadditive, submultiplicative and monotonic, from (3.3.5) we have

$$H(u(t)) \leq H(a(t)) + H(b(t)) \int_0^t L(s, H(u(s))) \, ds. \tag{3.3.7}$$

Now an application of Theorem 3.2.1 with u, a, b replaced by $H(u(t))$, $H(a(t))$, $H(b(t))$ respectively, we get

$$H(u(t)) \leq H(a(t)) + H(b(t)) \int_0^t L(s, H(a(s)))$$

$$\times \exp\left(\int_s^t M(\sigma, H(a(\sigma))) H(b(\sigma)) \, d\sigma\right) ds. \tag{3.3.8}$$

Since H^{-1} is increasing this implies the required inequality in (3.3.6) by applying H^{-1} to (3.3.8).

∎

Another interesting and useful generalization of Theorem 3.2.1 given by Pachpatte (in press l) is embodied in the following theorem.

Theorem 3.3.3 *Let u, a, b, L and M be as in Theorem 3.3.1. Let $G(u)$ be a continuous nondecreasing function defined on R_+ and $G(u) > 0$ for $u > 0$. Let $w(u)$ be a continuous, subadditive, submultiplicative and nondecreasing function defined on R_+ and $w(u) > 0$ for $u > 0$. If*

$$u(t) \leq a(t) + b(t) G\left(\int_0^t L(s, w(u(s))) \, ds\right), \tag{3.3.9}$$

for $t \in R_+$, then for $0 \leq t \leq \bar{t}$,

$$u(t) \leq a(t) + b(t) G\left(\Omega^{-1}\left[\Omega\left(\int_0^t L(s, w(a(s))) \, ds\right) + \int_0^t M(s, w(a(s))) w(b(s)) \, ds\right]\right), \tag{3.3.10}$$

where

$$\Omega(r) = \int_{r_0}^{r} \frac{ds}{w(G(s))}, \quad r > 0, r_0 > 0, \qquad (3.3.11)$$

Ω^{-1} is the inverse of Ω and $\bar{t} \in R_+$ is chosen so that

$$\Omega\left(\int_0^t L(s, w(a(s))) \, ds\right) + \int_0^t M(s, w(a(s)))w(b(s)) \, ds \in \text{Dom}(\Omega^{-1}),$$

for all $t \in R_+$ lying in the interval $0 \le t \le \bar{t}$.   □

Proof: Define a function $z(t)$ by

$$z(t) = \int_0^t L(s, w(u(s))) \, ds. \qquad (3.3.12)$$

From (3.3.12) and using the fact that $u(t) \le a(t) + b(t)G(z(t))$ and the conditions on the functions w and L we observe that

$$z(t) \le \int_0^t L(s, w(a(s) + b(s)G(z(s)))) \, ds$$

$$\le \int_0^t L(s, w(a(s)) + w(b(s))w(G(z(s)))) \, ds$$

$$= \int_0^t [L(s, w(a(s)) + w(b(s))w(G(z(s)))) - L(s, w(a(s)))$$

$$+ L(s, w(a(s)))] \, ds$$

$$\le \int_0^t L(s, w(a(s))) \, ds + \int_0^t M(s, w(a(s)))w(b(s))w(G(z(s))) \, ds. \quad (3.3.13)$$

For an arbitrary $T \in R_+$ it follows from (3.3.13) that

$$z(t) \le \int_0^T L(s, w(a(s))) \, ds + \int_0^t M(s, w(a(s)))w(b(s))w(G(z(s))) \, ds,$$

for $0 \le t \le T$.

Now by following the same arguments as in the proof of Theorem 2.4.2 with suitable modifications we get the desired inequality in (3.3.10).

∎

Remark 3.3.2 We note that the inequality obtained in Theorem 3.3.3 can be considered as a further generalization of Bihari's inequality given in Theorem 2.3.1, which in turn is a further generalization of the well-known Gronwall–Bellman inequality.

△

The following theorem established by Pachpatte (in press k) deals with another useful generalization of the inequality given in Theorem 3.2.1.

Theorem 3.3.4 Let u, a, b and c be nonnegative continuous functions defined on R_+. Let $k(t, s)$ and its partial derivative $(\partial/\partial t)k(t, s)$ be nonnegative continuous functions for $0 \leq s \leq t < \infty$, and L, M be as in Theorem 3.3.1. If

$$u(t) \leq a(t) + b(t) \int_0^t k(t, s)[c(s)u(s) + L(s, u(s))] \, ds, \qquad (3.3.14)$$

for $t \in R_+$, then

$$u(t) \leq a(t) + b(t) \int_0^t A(\sigma) \exp\left(\int_\sigma^t B(\tau) \, d\tau\right) d\sigma, \qquad (3.3.15)$$

for $t \in R_+$, where

$$A(t) = k(t, t)[c(t)a(t) + L(t, a(t))]$$

$$+ \int_0^t \frac{\partial}{\partial t} k(t, s)[c(s)a(s) + L(s, a(s))] \, ds, \qquad (3.3.16)$$

$$B(t) = k(t, t)[c(t) + M(t, a(t))]b(t)$$

$$+ \int_0^t \frac{\partial}{\partial t} k(t, s)[c(s) + M(s, a(s))]b(s) \, ds, \qquad (3.3.17)$$

for $t \in R_+$.

□

Proof: Define a function $z(t)$ by

$$z(t) = \int_0^t k(t, s)[c(s)u(s) + L(s, u(s))]\, ds. \tag{3.3.18}$$

Differentiating (3.3.18) and using $u(t) \leq a(t) + b(t)z(t)$, the condition (L), the fact that $z(t)$ is monotonic nondecreasing in t, and (3.3.16), (3.3.17) we observe that

$$\begin{aligned} z'(t) &= k(t, t)[c(t)u(t) + L(t, u(t))] \\ &\quad + \int_0^t \frac{\partial}{\partial t} k(t, s)[c(s)u(s) + L(s, u(s))]\, ds \\ &\leq k(t, t)[c(t)(a(t) + b(t)z(t)) + L(t, a(t)) \\ &\quad + L(t, a(t) + b(t)z(t)) - L(t, a(t))] \\ &\quad + \int_0^t \frac{\partial}{\partial t} k(t, s)[c(s)(a(s) + b(s)z(s)) + L(s, a(s)) \\ &\quad + L(s, a(s) + b(s)z(s)) - L(s, a(s))]\, ds \\ &\leq k(t, t)[c(t)a(t) + L(t, a(t)) + c(t)b(t)z(t) \\ &\quad + M(t, a(t))b(t)z(t)] \\ &\quad + \int_0^t \frac{\partial}{\partial t} k(t, s)[c(s)a(s) + L(s, a(s)) + c(s)b(s)z(s) \\ &\quad + M(s, a(s))b(s)z(s)]\, ds \\ &\leq A(t) + B(t)z(t). \end{aligned} \tag{3.3.19}$$

The inequality (3.3.19) implies the estimate

$$z(t) \leq \int_0^t A(\sigma) \exp\left(\int_\sigma^t B(\tau)\, d\tau\right) d\sigma. \tag{3.3.20}$$

Using (3.3.20) in $u(t) \leq a(t) + b(t)z(t)$, we get the required inequality in (3.3.15).

∎

Remark 3.3.3 We note that the inequality given in Theorem 3.3.4 is motivated by the inequality given earlier in Theorem 2.4.1 and can be used more effectively in certain general situations. In the special case when $c(t) = 0$ and $k(t, s) = 1$ the inequality in Theorem 3.3.4 reduces to the inequality given in Theorem 3.2.1.

△

3.4 The Inequalities of Ou-Iang and Dafermos

In the development of the theory of abstract differential and integral equations, and also the various classes of partial differential equations, some specific types of integral inequalities have played an important role. This section deals with some such basic inequalities which provide explicit bounds on unknown functions.

One of the most useful inequality in the development of the theory of differential equations is given in the following theorem.

Theorem 3.4.1 *Let u and f be nonnegative continuous functions defined for $t \in R_+$. If*

$$u^2(t) \leq c^2 + 2 \int_0^t f(s) u(s) \, ds, \qquad (3.4.1)$$

for $t \in R_+$, where $c \geq 0$ is a constant, then

$$u(t) \leq c + \int_0^t f(s) \, ds, \qquad (3.4.2)$$

for $t \in R_+$.

□

Proof: In order to establish the inequality (3.4.2) we first assume that $c > 0$ and define a function $z(t)$ by the right-hand side of (3.4.1). Then $z(0) = c^2$, $u(t) \leq \sqrt{z(t)}$ and

$$z'(t) = 2f(t)u(t) \leq 2f(t)\sqrt{z(t)}. \qquad (3.4.3)$$

From (3.4.3) we have

$$(z(t))^{-1/2} z'(t) \leq 2f(t). \qquad (3.4.4)$$

By setting $t = s$ in (3.4.4) and integrating it from 0 to t we get

$$\sqrt{z(t)} \leq c + \int_0^t f(s)\,ds. \qquad (3.4.5)$$

The desired inequality in (3.4.2) follows by using (3.4.5) in $u(t) \leq \sqrt{z(t)}$.

If $c \geq 0$, we carry out the above procedure with $c + \epsilon$ instead of c, where $\epsilon > 0$ is an arbitrary small constant, and subsequently pass to the limit as $\epsilon \to 0$ to obtain (3.4.2).

■

Remark 3.4.1 It appears that this inequality was first given by Ou-Iang (1957) while studying the boundedness of solutions of certain second-order differential equations. In the past few years this inequality has been applied with considerable success to study the global existence, uniqueness, stability and other properties of the solutions of various nonlinear differential equations (Barbu, 1985; Brezis, 1973; Corduneanu, 1991; Haraux, 1981; Ikehata and Okazawa, 1990; Olekhnik, 1972; Tsutsumi and Fukuda, 1980). An interesting feature of this inequality lies in its fruitful utilization to the situations for which the other available inequalities do not apply directly.

△

Dafermos (1979) used the following inequality while attempting to establish a different connection between stability and the second law of thermodynamics.

Theorem 3.4.2 *Assume that the nonnegative functions $u(t) \in L^\infty[0, s]$, $g(t) \in L^1[0, s]$ satisfy the inequality*

$$u^2(\sigma) \leq M^2 u^2(0) + \int_0^\sigma [2\alpha u^2(t) + 2Ng(t)u(t)]\,dt, \qquad (3.4.6)$$

for $\sigma \in [0, s]$, where α, M and N are nonnegative constants. Then

$$u(s) \leq M e^{\alpha s} u(0) + N e^{\alpha s} \int_0^s g(t)\,dt. \qquad (3.4.7)$$

□

The proof of this theorem is straightforward in view of the proof of Theorem 3.4.1 given above. The details are omitted here.

Dafermos (1979) also used the following more general version of the above inequality in the proof of one of his main results.

Theorem 3.4.3 *Assume that the nonnegative functions $u(t) \in L^\infty[0, s]$ and $g(t) \in L^1[0, s]$ satisfy the inequality*

$$u^2(\sigma) \leq M^2 u^2(0) + \int_0^\sigma [(2\gamma + 4\beta\sigma)u^2(t) + 2Ng(t)u(t)]\,dt, \qquad (3.4.8)$$

for $\sigma \in [0, s]$, where β, γ, M and N are nonnegative constants. Then

$$u(s) \leq M \exp(\alpha s + \beta s^2)u(0) + N \exp(\alpha s + \beta s^2) \int_0^s g(t)\,dt, \qquad (3.4.9)$$

where $\alpha = \gamma + \beta/\gamma$.

□

Proof: We define a nonnegative function $z(\sigma)$ by

$$z^2(\sigma) = M^2 u^2(0) + \int_0^\sigma [(2\gamma + 4\beta\sigma)u^2(t) + 2Ng(t)u(t)]\,dt, \qquad (3.4.10)$$

for $\sigma \in [0, s]$, and we note that

$$2z(\sigma)z'(\sigma) = (2\gamma + 4\beta\sigma)u^2(\sigma) + 2Ng(\sigma)u(\sigma) + 4\beta \int_0^\sigma u^2(t)\,dt$$

$$\leq (2\alpha + 4\beta\sigma)z^2(\sigma) + 2Ng(\sigma)z(\sigma). \qquad (3.4.11)$$

Hence

$$z'(\sigma) \leq (\alpha + 2\beta\sigma)z(\sigma) + Ng(\sigma). \qquad (3.4.12)$$

Integrating the differential inequality (3.4.12) under the initial condition $z(0) = Mu(0)$ we arrive at (3.4.9). The proof is complete.

■

Remark 3.4.2 We note that Dafermos (1979) gave very interesting results which establish different connections between stability and the second law of thermodynamics. The proofs of the main results given in Dafermos

(1979) are based on the applications of the inequalities given in Theorems 3.4.2 and 3.4.3.

△

3.5 Pachpatte's Inequalities II

In the past few years the inequality given in Theorem 3.4.1 has been used considerably in the study of qualitative properties of the solutions of certain abstract differential, integral and partial differential equations. Inspired by the important role played by the inequalities given in Section 3.4, Pachpatte (1994c, 1995a, in press h) investigated some useful generalizations and variants of the inequalities given in Theorems 3.4.1–3.4.3, which can offer greater versatility in certain applications. In this section we shall present some fundamental inequalities established by Pachpatte (1994c, 1995a, in press h).

Pachpatte (1995a) obtained the following generalization of the inequality given in Theorem 3.4.1.

Theorem 3.5.1 *Let u, f and g be nonnegative continuous functions defined on R_+ and c be a nonnegative constant. If*

$$u^2(t) \leq c^2 + 2 \int_0^t [f(s)u^2(s) + g(s)u(s)] \, ds, \qquad (3.5.1)$$

for $t \in R_+$, then

$$u(t) \leq p(t) \exp\left(\int_0^t f(s) \, ds \right), \qquad (3.5.2)$$

for $t \in R_+$, where

$$p(t) = c + \int_0^t g(s) \, ds, \qquad (3.5.3)$$

for $t \in R_+$.

□

Proof: Let $c > 0$ and define a function $z(t)$ by the right-hand side of (3.5.1). Then $z(0) = c^2$, $u(t) \leq \sqrt{z(t)}$ and

$$z'(t) = 2[f(t)u^2(t) + g(t)u(t)]$$

$$\leq 2\sqrt{z(t)}[f(t)\sqrt{z(t)} + g(t)]. \quad (3.5.4)$$

From (3.5.4) we have

$$\frac{z'(t)}{\sqrt{z(t)}} \leq 2[f(t)\sqrt{z(t)} + g(t)]. \quad (3.5.5)$$

By taking $t = s$ in (3.5.5) and integrating from 0 to t we have

$$\sqrt{z(t)} \leq p(t) + \int_0^t f(s)\sqrt{z(s)}\, ds, \quad (3.5.6)$$

where $p(t)$ is defined by (3.5.3). Since $p(t)$ is positive and monotonic nondecreasing in t, by applying Theorem 1.3.1, we have

$$\sqrt{z(t)} \leq p(t) \exp\left(\int_0^t f(s)\, ds\right). \quad (3.5.7)$$

Using (3.5.7) in $u(t) \leq \sqrt{z(t)}$, we get the desired inequality in (3.5.2). The proof of the case when $c \geq 0$ can be completed as mentioned in the proof of Theorem 3.4.1.

∎

Pachpatte (1995a) also established the following inequality.

Theorem 3.5.2 *Let u, f, g and c be as in Theorem 3.5.1. Let $w(u)$ be a continuous, nondecreasing function defined on R_+ and $w(u) > 0$ on $(0, \infty)$. If*

$$u^2(t) \leq c^2 + 2\int_0^t [f(s)u(s)w(u(s)) + g(s)u(s)]\, ds, \quad (3.5.8)$$

for $t \in R_+$, then for $0 \leq t \leq \bar{t}$,

$$u(t) \leq \Omega^{-1}\left[\Omega(p(t)) + \int_0^t f(s)\, ds\right], \quad (3.5.9)$$

where $p(t)$ is defined by (3.5.3) and

$$\Omega(r) = \int_{r_0}^r \frac{ds}{w(s)}, \quad r > 0,\ r_0 > 0, \quad (3.5.10)$$

Ω^{-1} is the inverse of Ω and $\bar{t} \in R_+$ is chosen so that

$$\Omega(p(t)) + \int_0^t f(s)\,ds \in \text{Dom}(\Omega^{-1}),$$

for all $t \in R_+$ lying in the interval $0 \leq t \leq \bar{t}$.

$\square$

Proof: Let $c > 0$ and define a function $z(t)$ by the right side of (3.5.8). Then $z(0) = c^2$, $u(t) \leq \sqrt{z(t)}$ and

$$z'(t) = 2[f(t)u(t)w(u(t)) + g(t)u(t)]$$
$$\leq 2\sqrt{z(t)}[f(t)w(\sqrt{z(t)}) + g(t)]. \qquad (3.5.11)$$

From (3.5.11) we have

$$\frac{z'(t)}{\sqrt{z(t)}} \leq 2[f(t)w(\sqrt{z(t)}) + g(t)]. \qquad (3.5.12)$$

By taking $t = s$ in (3.5.12) and then integrating it from 0 to t we have

$$\sqrt{z(t)} \leq p(t) + \int_0^t f(s)w(\sqrt{z(s)})\,ds. \qquad (3.5.13)$$

Let $T \in R_+$ be any arbitrary number. Now from (3.5.13) we have for all $0 \leq t \leq T$,

$$\sqrt{z(t)} \leq p(T) + \int_0^t f(s)w(\sqrt{z(s)})\,ds.$$

Now an application of Bihari's inequality given in Theorem 2.3.1 gives

$$\sqrt{z(t)} \leq \Omega^{-1}\left[\Omega(p(T)) + \int_0^t f(s)\,ds\right], \quad 0 \leq t \leq T \leq \bar{t}.$$

By taking $t = T$ in the above inequality and using the fact that $u(t) \leq \sqrt{z(t)}$ is true for $t = T$, we obtain

$$u(T) \leq \Omega^{-1}\left[\Omega(p(T)) + \int_0^T f(s)\,ds\right]. \qquad (3.5.14)$$

Since T is arbitrary, the conclusion (3.5.9) is clear from (3.5.14). The proof of the case when $c \geq 0$ can be completed as mentioned in the proof of Theorem 3.4.1.

∎

Another interesting and useful inequality given by Pachpatte (in press h) is embodied in the following theorem.

Theorem 3.5.3 *Let u, f, g and c be as in Theorem 3.5.1. Let $L: R_+^2 \to R_+$ be a continuous function which satisfies the condition*

$$0 \leq L(t, x) - L(t, y) \leq k(t, y)(x - y), \qquad (L_0)$$

for $t \in R_+$ and $x \geq y \geq 0$, where $k: R_+^2 \to R_+$ is a continuous function. If

$$u^2(t) \leq c^2 + 2 \int_0^t [f(s)u(s)L(s, u(s)) + g(s)u(s)]\,ds, \qquad (3.5.15)$$

for $t \in R_+$, then

$$u(t) \leq p(t) + \int_0^t f(s)L(s, p(s)) \exp\left(\int_s^t f(\sigma)k(\sigma, p(\sigma))\,d\sigma\right) ds, \qquad (3.5.16)$$

for $t \in R_+$, where $p(t)$ is defined by (3.5.3).

□

Proof: Assume that $c > 0$ and define a function $z(t)$ by the right-hand side of (3.5.15). Then $z(0) = c^2$, $u(t) \leq \sqrt{z(t)}$ and

$$z'(t) = 2[f(t)u(t)L(t, u(t)) + g(t)u(t)]$$
$$\leq 2\sqrt{z(t)}[f(t)L(t, \sqrt{z(t)}) + g(t)]. \qquad (3.5.17)$$

From (3.5.17) we have

$$\frac{z'(t)}{\sqrt{z(t)}} \leq 2[f(t)L(t, \sqrt{z(t)}) + g(t)]. \qquad (3.5.18)$$

By taking $t = s$ in (3.5.18) and integrating it from 0 to t we get

$$\sqrt{z(t)} \leq p(t) + \int_0^t f(s)L(s, \sqrt{z(s)})\,ds. \qquad (3.5.19)$$

Define a function $m(t)$ by

$$m(t) = \int_0^t f(s)L(s, \sqrt{z(s)})\,ds. \qquad (3.5.20)$$

From (3.5.20) and using the fact that $\sqrt{z(t)} \leq p(t) + m(t)$, and the hypothesis (L_0) we observe that

$$\begin{aligned}
m'(t) &= f(t)L(t, \sqrt{z(t)}) \\
&\leq f(t)L(t, p(t) + m(t)) \\
&= f(t)[L(t, p(t) + m(t)) - L(t, p(t))] + f(t)L(t, p(t)) \\
&\leq f(t)k(t, p(t))m(t) + f(t)L(t, p(t)). \qquad (3.5.21)
\end{aligned}$$

The inequality (3.5.21) implies the estimate

$$m(t) \leq \int_0^t f(s)L(s, p(s)) \exp\left(\int_s^t f(\sigma)k(\sigma, p(\sigma))\,d\sigma\right)ds. \qquad (3.5.22)$$

The desired inequality in (3.5.16) now follows by using (3.5.22) in (3.5.19) and then in $u(t) \leq \sqrt{z(t)}$. The proof of the case when $c \geq 0$ can be completed as mentioned in Theorem 3.4.1.

∎

Pachpatte (1994c, 1995a) established the following inequalities, which are more convenient in certain applications.

Theorem 3.5.4 *Let u, v, g_i, h_i ($i = 1, 2, 3, 4$) be nonnegative continuous functions defined on R_+, and c_1, c_2, μ be nonnegative constants. If*

$$u^2(t) \leq c_1 + \int_0^t [g_1(s)u^2(s) + h_1(s)u(s)]\,ds$$

$$+ \int_0^t [g_2(s)\bar{v}^2(s) + h_2(s)\bar{v}(s)]\,ds, \qquad (3.5.23)$$

$$v^2(t) \leq c_2 + \int_0^t [g_3(s)\bar{u}^2(s) + h_3(s)\bar{u}(s)]\,ds$$

$$+ \int_0^t [g_4(s)v^2(s) + h_4(s)v(s)]\,ds, \qquad (3.5.24)$$

for $t \in R_+$, where $\bar{u}(t) = e^{-\mu t}u(t)$, $\bar{v}(t) = e^{\mu t}v(t)$ for $t \in R_+$, then

$$u(t) \leq e^{\mu t} p^*(t) \exp\left(\int_0^t G(s)\,ds\right), \quad (3.5.25)$$

$$v(t) \leq p^*(t) \exp\left(\int_0^t G(s)\,ds\right), \quad (3.5.26)$$

for $t \in R_+$, where

$$p^*(t) = \sqrt{2(c_1 + c_2)} + \int_0^t H(s)\,ds, \quad (3.5.27)$$

in which

$$G(t) = \max\{[g_1(t) + g_3(t)],\ [g_2(t) + g_4(t)]\}, \quad (3.5.28)$$

$$H(t) = \max\{[h_1(t) + h_3(t)],\ [h_2(t) + h_4(t)]\}, \quad (3.5.29)$$

for $t \in R_+$.

□

Proof: Multiplying both sides of (3.5.23) by $\exp(-2\mu t)$, it is easy to observe that

$$\bar{u}^2(t) \leq c_1 + \int_0^t [g_1(s)\bar{u}^2(s) + h_1(s)\bar{u}(s)]\,ds$$

$$+ \int_0^t [g_2(s)v^2(s) + h_2(s)v(s)]\,ds. \quad (3.5.30)$$

Define

$$F(t) = \bar{u}(t) + v(t). \quad (3.5.31)$$

Now by squaring both sides of (3.5.31) and using the elementary inequalities $(a+b)^2 \leq 2(a^2 + b^2)$, $a^2 + b^2 \leq (a+b)^2$, $a \geq 0$, $b \geq 0$ reals, and using (3.5.30), (3.5.24), (3.5.28), (3.5.29), it is easy to observe that

$$F^2(t) \leq 2(c_1 + c_2) + 2\int_0^t [G(s)F^2(s) + H(s)F(s)]\,ds, \quad (3.5.32)$$

The bounds in (3.5.25), (3.5.26) follow from an application of the inequality given in Theorem 3.5.1 to (3.5.32) and splitting. ∎

Theorem 3.5.5 *Let u, f, g and c be as in Theorem 3.5.1 and $p > 1$ be a constant. If*

$$u^p(t) \leq c^p + p \int_0^t [f(s)u^p(s) + g(s)u(s)]\,ds, \qquad (3.5.33)$$

for $t \in R_+$, then

$$u(t) \leq \exp\left(\int_0^t f(s)\,ds\right)\left[c^{p-1} + (p-1)\int_0^t g(s)\right.$$

$$\left.\times \exp\left(-(p-1)\int_0^s f(\sigma)\,d\sigma\right)ds\right]^{1/(p-1)}, \qquad (3.5.34)$$

for $t \in R_+$. ∎

Proof: Let $c > 0$ and define a function $z(t)$ by the right-hand side of (3.5.33). Then $z(0) = c^p$, $u(t) \leq (z(t))^{1/p}$ and

$$z'(t) = p[f(t)u^p(t) + g(t)u(t)]$$

$$\leq p[f(t)z(t) + g(t)(z(t))^{1/p}]. \qquad (3.5.35)$$

The inequality (3.5.35) implies the estimate

$$z(t) \leq \exp\left(p\int_0^t f(s)\,ds\right)\left[c^{p-1} + (p-1)\int_0^t g(s)\right.$$

$$\left.\times \exp\left(-(p-1)\int_0^s f(\sigma)\,d\sigma\right)ds\right]^{p/(p-1)}. \qquad (3.5.36)$$

Using (3.5.36) in $u(t) \leq (z(t))^{1/p}$, we get the required inequality in (3.5.34). The proof of the case when $c \geq 0$ can be completed as indicated in the proof of Theorem 3.4.1. ∎

3.6 Pachpatte's Inequalities III

Pachpatte (1995a, in press h,m) has established some general inequalities which can be used as handy tools in the analysis of certain integral and integro-differential equations for which the inequalities given earlier do not apply. This section presents some inequalities given in the above cited papers.

The inequalities in the following two theorems are established by Pachpatte (1995a).

Theorem 3.6.1 *Let u, f, g and h be nonnegative continuous functions defined on R_+ and c be a nonnegative constant.*

(a_1) *If*

$$u^2(t) \leq c^2 + 2\int_0^t \left[f(s)u(s) \left(u(s) + \int_0^s g(\sigma)u(\sigma)\,d\sigma \right) + h(s)u(s) \right] ds, \tag{3.6.1}$$

for $t \in R_+$, then

$$u(t) \leq p(t) \left[1 + \int_0^t f(s)\exp\left(\int_0^s [f(\sigma) + g(\sigma)]\,d\sigma \right) ds \right], \tag{3.6.2}$$

for $t \in R_+$, where

$$p(t) = c + \int_0^t h(s)\,ds, \tag{3.6.3}$$

for $t \in R_+$.

(a_2) *If*

$$u^2(t) \leq c^2 + 2\int_0^t \left[f(s)u(s) \left(\int_0^s g(\sigma)u(\sigma)\,d\sigma \right) + h(s)u(s) \right] ds, \tag{3.6.4}$$

for $t \in R_+$, then

$$u(t) \leq p(t)\exp\left(\int_0^t f(s) \left(\int_0^s g(\sigma)\,d\sigma \right) ds \right), \tag{3.6.5}$$

for $t \in R_+$, where $p(t)$ is defined by (3.6.3).

$\square$

Theorem 3.6.2 *Let u, g, h and c be as in Theorem 3.6.1. Let w(u) be a continuous nondecreasing function defined on R_+ and $w(u) > 0$ for $u > 0$.*

(a_3) *If*

$$u^2(t) \leq c^2 + 2 \int_0^t \left[g(s)u(s) \left(u(s) + \int_0^s g(\sigma)w(u(\sigma))\,d\sigma \right) + h(s)u(s) \right] ds, \quad (3.6.6)$$

for $t \in R_+$, then for $0 \leq t \leq \bar{t}$,

$$u(t) \leq p(t) + \int_0^t g(s) G^{-1} \left[G(p(t)) + \int_0^s g(\sigma)\,d\sigma \right], \quad (3.6.7)$$

where p(t) is defined by (3.6.3) and

$$G(r) = \int_{r_0}^r \frac{ds}{s + w(s)}, \quad r > 0, r_0 > 0, \quad (3.6.8)$$

G^{-1} is the inverse of G, and $\bar{t} \in R_+$ is chosen so that

$$G(p(t)) + \int_0^t g(\sigma)\,d\sigma \in \text{Dom}(G^{-1}),$$

for all $t \in R_+$ lying in the interval $0 \leq t \leq \bar{t}$.

(a_4) *If*

$$u^2(t) \leq c^2 + 2 \int_0^t \left[f(s)u(s) \left(\int_0^s g(\sigma)w(u(\sigma))\,d\sigma \right) + h(s)u(s) \right] ds, \quad (3.6.9)$$

for $t \in R_+$, then for $0 \leq t \leq t_1$,

$$u(t) \leq \Omega^{-1} \left[\Omega(p(t)) + \int_0^t f(s) \left(\int_0^s g(\sigma)\,d\sigma \right) ds \right], \quad (3.6.10)$$

where p(t) is defined by (3.6.3) and

$$\Omega(r) = \int_{r_0}^r \frac{ds}{w(s)}, \quad r > 0, r_0 > 0, \quad (3.6.11)$$

Ω^{-1} is the inverse of Ω and $t_1 \in R_+$ is chosen so that

$$\Omega(p(t)) + \int_0^t f(s) \left(\int_0^s g(\sigma) \, d\sigma \right) ds \in \text{Dom}(\Omega^{-1}),$$

for all $t \in R_+$ lying in the interval $0 \le t \le t_1$. $\square$

Pachpatte (in press h) obtained the inequalities given in the following theorem, which are effective in certain applications.

Theorem 3.6.3 Let u, f, g, h and c be as in Theorem 3.6.1. Let $L: R_+^2 \to R_+$ be a continuous function which satisfies the condition

$$0 \le L(t, x) - L(t, y) \le k(t, y)(x - y), \quad \text{(L)}$$

for $t \in R_+$ and $x \ge y \ge 0$, where $k: R_+^2 \to R_+$ is a continuous function.

(a_5) If

$$u^2(t) \le c^2 + 2 \int_0^t \left[f(s)u(s) \left(u(s) + \int_0^s g(\sigma)L(\sigma, u(\sigma)) \, d\sigma \right) + h(s)u(s) \right] ds,$$
(3.6.12)

for $t \in R_+$, then

$$u(t) \le p(t) + \int_0^t [f(s)p(s) + g(s)L(s, p(s))]$$

$$\times \exp\left(\int_s^t [f(\sigma) + g(\sigma)k(\sigma, p(\sigma))] \, d\sigma \right) ds, \quad (3.6.13)$$

for $t \in R_+$, where $p(t)$ is defined by (3.6.3).

(a_6) If

$$u^2(t) \le c^2 + 2 \int_0^t \left[f(s)u(s) \left(\int_0^s g(\sigma)L(\sigma, u(\sigma)) \, d\sigma \right) + h(s)u(s) \right] ds,$$
(3.6.14)

for $t \in R_+$, then

$$u(t) \leq p(t) + \int_0^t f(s) \left(\int_0^s g(\sigma) L(\sigma, p(\sigma)) \, d\sigma \right)$$

$$\times \exp \left(\int_s^t f(\sigma) \left(\int_0^\sigma g(\tau) k(\tau, p(\tau)) \, d\tau \right) d\sigma \right) ds, \quad (3.6.15)$$

for $t \in R_+$, where $p(t)$ is defined by (3.6.3).

$\square$

Proofs of Theorems 3.6.1–3.6.3: We give the details of the proofs for (a_1), (a_3) and (a_5) only; the proofs of (a_2), (a_4) and (a_6) can be completed by following the proofs of the above mentioned inequalities. In each case, it is sufficient to assume that c is positive, since the standard limiting arguments can be used to treat the remaining cases.

(a_1) Let $c > 0$ and define a function $z(t)$ by the right-hand side of (3.6.1). Then $z(0) = c^2$, $u(t) \leq \sqrt{z(t)}$ and

$$z'(t) \leq 2\sqrt{z(t)} \left[f(t) \left(\sqrt{z(t)} + \int_0^t g(\sigma) \sqrt{z(\sigma)} \, d\sigma \right) + h(t) \right]. \quad (3.6.16)$$

Differentiating $\sqrt{z(t)}$ and using (3.6.16) we have

$$\frac{d}{dt}(\sqrt{z(t)}) = \frac{z'(t)}{2\sqrt{z(t)}}$$

$$\leq \left[f(t) \left(\sqrt{z(t)} + \int_0^t g(\sigma) \sqrt{z(\sigma)} \, d\sigma \right) + h(t) \right]. \quad (3.6.17)$$

By taking $t = s$ in (3.6.17) and integrating it from 0 to t we have

$$\sqrt{z(t)} \leq p(t) + \int_0^t f(s) \left(\sqrt{z(s)} + \int_0^s g(\sigma) \sqrt{z(\sigma)} \, d\sigma \right) ds, \quad (3.6.18)$$

where $p(t)$ is defined by (3.6.3). Since $p(t)$ is positive and monotonic nondecreasing in $t \in R_+$, by using Theorem 1.7.4, the inequality (3.6.18) implies the estimate

$$\sqrt{z(t)} \leq p(t) \left[1 + \int_0^t f(s) \exp\left(\int_0^s [f(\sigma) + g(\sigma)] \, d\sigma \right) ds \right]. \quad (3.6.19)$$

Now by using (3.6.19) in $u(t) \leq \sqrt{z(t)}$ we get the required inequality in (3.6.2).

(a_3) Let $c > 0$ and define a function $z(t)$ by the right-hand side of (3.6.6). Then $z(0) = c^2$, $u(t) \leq \sqrt{z(t)}$ and

$$z'(t) \leq 2\sqrt{z(t)} \left[g(t) \left(\sqrt{z(t)} + \int_0^t g(\sigma) w(\sqrt{z(\sigma)}) \, d\sigma \right) + h(t) \right]. \quad (3.6.20)$$

Now differentiating $\sqrt{z(t)}$ and using (3.6.20) we get

$$\frac{d}{dt}(\sqrt{z(t)}) = \frac{z'(t)}{2\sqrt{z(t)}} \leq \left[g(t) \left(\sqrt{z(t)} + \int_0^t g(\sigma) w(\sqrt{z(\sigma)}) \, d\sigma \right) + h(t) \right].$$

$$(3.6.21)$$

By taking $t = s$ in (3.6.21) and integrating from 0 to t we have

$$\sqrt{z(t)} \leq p(t) + \int_0^t g(s) \left(\sqrt{z(s)} + \int_0^s g(\sigma) w(\sqrt{z(\sigma)}) \, d\sigma \right) ds. \quad (3.6.22)$$

For an arbitrary $T \in R_+$, it follows from (3.6.22) that

$$\sqrt{z(t)} \leq p(T) + \int_0^t g(s) \left(\sqrt{z(s)} + \int_0^s g(\sigma) w(\sqrt{z(\sigma)}) \, d\sigma \right) ds, \quad 0 \leq t \leq T,$$

$$(3.6.23)$$

Now an application of Theorem 2.8.1 we have

$$\sqrt{z(t)} \leq p(T) + \int_0^t g(s) G^{-1} \left[G(p(T)) + \int_0^s g(\sigma) \, d\sigma \right] ds, \quad 0 \leq t \leq T \leq \bar{t}.$$

$$(3.6.24)$$

Now by taking $t = T$ in (3.6.24) and using the fact that $u(t) \leq \sqrt{z(t)}$ for $t = T$ we obtain

$$u(T) \leq p(T) + \int_0^T g(s)G^{-1}\left[G(p(T)) + \int_0^s g(\sigma)\,d\sigma\right]ds. \quad (3.6.25)$$

Since T is arbitrary, the conclusion (3.6.7) is clear from (3.6.25).

(a_5) Let $c > 0$ and define a function $z(t)$ by the right-hand side of (3.6.12). Then $z(0) = c^2$, $u(t) \leq \sqrt{z(t)}$ and

$$z'(t) \leq 2\sqrt{z(t)}\left[f(t)\left(\sqrt{z(t)} + \int_0^t g(\sigma)L(\sigma, \sqrt{z(\sigma)})\,d\sigma\right) + h(t)\right]. \quad (3.6.26)$$

Now differentiating $\sqrt{z(t)}$ and using (3.6.26) we have

$$\frac{d}{dt}(\sqrt{z(t)}) = \frac{z'(t)}{2\sqrt{z(t)}}$$

$$\leq f(t)\left(\sqrt{z(t)} + \int_0^t g(\sigma)L(\sigma, \sqrt{z(\sigma)})\,d\sigma\right) + h(t). \quad (3.6.27)$$

Define a function $m(t)$ by

$$m(t) = \sqrt{z(t)} + \int_0^t g(\sigma)L(\sigma, \sqrt{z(\sigma)})\,d\sigma. \quad (3.6.28)$$

Then $m(0) = c$, $(d/dt)(\sqrt{z(t)}) \leq f(t)m(t) + h(t)$ from (3.6.27), $\sqrt{z(t)} \leq m(t)$ from (3.6.28) and

$$m'(t) \leq f(t)m(t) + h(t) + g(t)L(t, m(t)). \quad (3.6.29)$$

By taking $t = s$ in (3.6.29) and then integrating from 0 to t we have

$$m(t) \leq p(t) + \int_0^t [f(s)m(s) + g(s)L(s, m(s))]\,ds, \quad (3.6.30)$$

where $p(t)$ is defined by (3.6.3). Define a function $v(t)$ by

$$v(t) = \int_0^t [f(s)m(s) + g(s)L(s, m(s))]\,ds. \quad (3.6.31)$$

From (3.6.31) and using the fact that $m(t) \leq p(t) + v(t)$, and the hypothesis (L) we observe that

$$\begin{aligned}v'(t) &= f(t)m(t) + g(t)L(t, m(t)) \\ &\leq f(t)[p(t) + v(t)] + g(t)L(t, p(t) + v(t)) \\ &= f(t)[p(t) + v(t)] + g(t)[L(t, p(t) + v(t)) - L(t, p(t))] \\ &\quad + g(t)L(t, p(t)) \\ &\leq f(t)[p(t) + v(t)] + g(t)k(t, p(t))v(t) + g(t)L(t, p(t)) \\ &= [f(t)p(t) + g(t)L(t, p(t))] + [f(t) + g(t)k(t, p(t))]v(t). \end{aligned} \quad (3.6.32)$$

The inequality (3.6.32) implies the estimate

$$v(t) \leq \int_0^t [f(s)p(s) + g(s)L(s, p(s))] \exp\left(\int_s^t [f(\sigma) + g(\sigma)k(\sigma, p(\sigma))]\, d\sigma\right) ds. \quad (3.6.33)$$

The desired inequality in (3.6.13) now follows by using the facts that $u(t) \leq \sqrt{z(t)} \leq m(t) \leq p(t) + v(t)$ and (3.6.33).

∎

Pachpatte (in press m) has established the following theorem, which deals with the inequalities analogous to the results given in Theorem 3.6.1.

Theorem 3.6.4 Let u, f, g and h be nonnegative continuous functions defined on R_+ and c be a positive constant.

(b_1) If

$$u^2(t) \leq c^2 + 2\int_0^t \left[f(s)u^2(s) + g(s)u(s)\left(u(s) + \int_0^s h(\sigma)u(\sigma)\, d\sigma\right)\right] ds, \quad (3.6.34)$$

for $t \in R_+$, then

$$u(t) \leq c \left[\exp\left(\int_0^t f(s)\, ds\right) + \int_0^t g(s) \exp\left(\int_s^t f(\sigma)\, d\sigma\right) \right.$$
$$\left. \times \exp\left(\int_0^s [f(\sigma) + g(\sigma) + h(\sigma)]\, d\sigma\right) ds \right], \quad (3.6.35)$$

for $t \in R_+$.

(b_2) If

$$u^2(t) \leq c^2 + 2 \int_0^t \left[f(s)u^2(s) + g(s)u^2(s) \left(u(s) + \int_0^s h(\sigma)u(\sigma)\,d\sigma \right) \right] ds, \qquad (3.6.36)$$

for $t \in R_+$ and

$$A(t) = 1 - c \int_0^t g(\tau) \exp\left(\int_0^\tau [f(\sigma) + h(\sigma)]\,d\sigma \right) d\tau > 0, \qquad (3.6.37)$$

for $t \in R_+$, then

$$u(t) \leq c \exp\left(\int_0^t [f(s) + g(s)Q(s)]\,ds \right), \qquad (3.6.38)$$

for $t \in R_+$, where

$$Q(t) = (1/A(t))c \exp\left(\int_0^t [f(\sigma) + h(\sigma)]\,d\sigma \right), \qquad (3.6.39)$$

for $t \in R_+$.

$\square$

Proof: We will give the details of the proof of (b_2) only; the proof of (b_1) can be completed similarly. Let $c > 0$ and define a function $z(t)$ by the right-hand side of (3.6.36). Then $z(0) = c^2$, $u(t) \leq \sqrt{z(t)}$ and

$$z'(t) \leq 2\sqrt{z(t)} \left[f(t)\sqrt{z(t)} + g(t)\sqrt{z(t)} \left(\sqrt{z(t)} + \int_0^t h(\sigma)\sqrt{z(\sigma)}\,d\sigma \right) \right]. \qquad (3.6.40)$$

Differentiating $\sqrt{z(t)}$ and using (3.6.40) we have

$$\frac{d}{dt}(\sqrt{z(t)}) = \frac{z'(t)}{2\sqrt{z(t)}}$$

$$\leq f(t)\sqrt{z(t)} + g(t)\sqrt{z(t)} \left(\sqrt{z(t)} + \int_0^t h(\sigma)\sqrt{z(\sigma)}\,d\sigma \right). \qquad (3.6.41)$$

Define a function $v(t)$ by

$$v(t) = \sqrt{z(t)} + \int_0^t h(\sigma)\sqrt{z(\sigma)}\,d\sigma. \qquad (3.6.42)$$

Differentiating (3.6.42) and then using (3.6.41) and the fact that $\sqrt{z(t)} \leq v(t)$ we have

$$v'(t) \leq [f(t) + h(t)]v(t) + g(t)v^2(t). \qquad (3.6.43)$$

The inequality (3.6.43) implies the estimate (Pachpatte, 1976b)

$$v(t) \leq Q(t), \qquad (3.6.44)$$

where $Q(t)$ is defined by (3.6.39). Using (3.6.44) in (3.6.41) we have

$$\frac{d}{dt}(\sqrt{z(t)}) \leq [f(t) + g(t)Q(t)]\sqrt{z(t)}. \qquad (3.6.45)$$

The inequality (3.6.45) implies the estimate

$$\sqrt{z(t)} \leq c \exp\left(\int_0^t [f(s) + g(s)Q(s)]\,ds\right). \qquad (3.6.46)$$

Now by using (3.6.46) in $u(t) \leq \sqrt{z(t)}$ we get the required inequality in (3.6.38). The proof of the case when $c \geq 0$ can be completed as mentioned in the proof of Theorem 3.4.1.

∎

3.7 Pachpatte's Inequalities IV

In the study of certain differential, integral and integro-differential equations some specific nonlinear integral inequalities are needed in various situations. In this section we shall give some inequalities established by Pachpatte (1995a,d, 1996a,b, in press a) which can be used as basic tools in certain applications.

Our first theorem deals with the inequalities established by Pachpatte (1996b) which basically involves the comparison principle.

Theorem 3.7.1 Let u, f and g be nonnegative continuous functions defined on R_+ and c be a nonnegative constant. Let $w(t,r)$ be a nonnegative continuous function defined for $t \in R_+$, $0 \le r < \infty$, and monotonic nondecreasing with respect to r for any fixed $t \in R_+$.

(a_1) If
$$u^2(t) \le c^2 + 2\int_0^t u(s)w(s,u(s))\,ds, \qquad (3.7.1)$$

for $t \in R_+$, then
$$u(t) \le r(t), \quad t \in R_+, \qquad (3.7.2)$$

where $r(t)$ is the maximal solution of
$$r'(t) = w(t, r(t)), \quad r(0) = c, \qquad (3.7.3)$$

for $t \in R_+$.

(a_2) If
$$u^2(t) \le c^2 + 2\int_0^t u(s)[f(s)u(s) + w(s,u(s))]\,ds, \qquad (3.7.4)$$

for $t \in R_+$, then
$$u(t) \le A(t)r(t), \quad t \in R_+, \qquad (3.7.5)$$

where
$$A(t) = \exp\left(\int_0^t f(s)\,ds\right), \quad t \in R_+, \qquad (3.7.6)$$

and $r(t)$ is the maximal solution of
$$r'(t) = w(t, A(t)r(t)), \quad r(0) = c, \qquad (3.7.7)$$

for $t \in R_+$.

(a_3) If
$$u^2(t) \le c^2 + 2\int_0^t u(s)\left[f(s)\left(u(s) + \int_0^s g(\sigma)u(\sigma)\,d\sigma\right) + w(s,u(s))\right]ds, \qquad (3.7.8)$$

for $t \in R_+$, then
$$u(t) \leq B(t)r(t), \quad t \in R_+, \qquad (3.7.9)$$

where
$$B(t) = 1 + \int_0^t f(s) \exp\left(\int_0^s [f(\sigma) + g(\sigma)] \, d\sigma\right) ds, \quad t \in R_+, \qquad (3.7.10)$$

and $r(t)$ is the maximal solution of
$$r'(t) = w(t, B(t)r(t)), \quad r(0) = c, \qquad (3.7.11)$$

for $t \in R_+$.

(a_4) If
$$u^2(t) \leq c^2 + 2\int_0^t u(s) \left[f(s) \left(\int_0^s g(\sigma)u(\sigma) \, d\sigma\right) + w(s, u(s)) \right] ds, \qquad (3.7.12)$$

for $t \in R_+$, then
$$u(t) \leq E(t)r(t), \quad t \in R_+, \qquad (3.7.13)$$

where
$$E(t) = \exp\left(\int_0^t f(s) \left(\int_0^s g(\sigma) \, d\sigma\right) ds\right), \quad t \in R_+, \qquad (3.7.14)$$

and $r(t)$ is the maximal solution of
$$r'(t) = w(t, E(t)r(t)), \quad r(0) = c, \qquad (3.7.15)$$

for $t \in R_+$. $\square$

Proof: Since the proofs resemble one another, we give the details for (a_1) and (a_2); the proofs of (a_3) and (a_4) can be completed by following the proofs of the above mentioned results.

(a_1) Let $c > 0$ and define a function $z(t)$ by the right-hand side of (3.7.1). Then $z(0) = c^2$, $u(t) \leq \sqrt{z(t)}$ and
$$z'(t) \leq 2\sqrt{z(t)} w(t, \sqrt{z(t)}). \qquad (3.7.16)$$

Differentiating $\sqrt{z(t)}$ and using (3.7.16) we have

$$\frac{d}{dt}(\sqrt{z(t)}) = \frac{z'(t)}{2\sqrt{z(t)}} \leq w(t, \sqrt{z(t)}). \qquad (3.7.17)$$

Now a suitable application of Theorem 2.2.2 to (3.7.17) and (3.7.3) yields

$$\sqrt{z(t)} \leq r(t), \quad t \in R_+, \qquad (3.7.18)$$

where $r(t)$ is the maximal solution of (3.7.3). Now by using (3.7.18) in $u(t) \leq \sqrt{z(t)}$ we get the desired inequality in (3.7.2). The proof of the case when $c \geq 0$ can be completed as mentioned in the proof of Theorem 3.4.1.

(a_2) By assuming that $c > 0$, defining a function $z(t)$ by the right-hand side of (3.7.4) and following the same steps as in the proof of (a_1) we have

$$\frac{d}{dt}(\sqrt{z(t)}) \leq f(t)\sqrt{z(t)} + w(t, \sqrt{z(t)}). \qquad (3.7.19)$$

From (3.7.19) it is easy to observe that

$$\sqrt{z(t)} \leq c + \int_0^t f(s)\sqrt{z(s)}\,ds + \int_0^t w(s, \sqrt{z(s)})\,ds. \qquad (3.7.20)$$

Define a function $m(t)$ by

$$m(t) = c + \int_0^t w(s, \sqrt{z(s)})\,ds, \qquad (3.7.21)$$

then (3.7.20) can be restated as

$$\sqrt{z(t)} \leq m(t) + \int_0^t f(s)\sqrt{z(s)}\,ds. \qquad (3.7.22)$$

Since $m(t)$ is positive and monotonic nondecreasing for $t \in R_+$, by applying Theorem 1.3.1 yields

$$\sqrt{z(t)} \leq A(t)m(t), \quad t \in R_+, \qquad (3.7.23)$$

where $A(t)$ is defined by (3.7.6). From (3.7.21) and using (3.7.23) we observe that

$$m'(t) \leq w(t, A(t)m(t)), \qquad (3.7.24)$$

for $t \in R_+$. Now a suitable application of Theorem 2.2.2 to (3.7.24) and (3.7.7) yields

$$m(t) \leq r(t), \quad t \in R_+, \tag{3.7.25}$$

where $r(t)$ is the maximal solution of (3.7.7). From (3.7.23) and (3.7.25) we have

$$\sqrt{z(t)} \leq A(t) r(t), \quad t \in R_+. \tag{3.7.26}$$

Now using the fact that $u(t) \leq \sqrt{z(t)}$ and (3.7.26) we get the required inequality in (3.7.5). The proof of the case when $c \geq 0$ can be completed by following the limiting arguments.

∎

Pachpatte (1995a) has proved the following inequality.

Theorem 3.7.2 *Let u, f and h be nonnegative continuous functions defined on R_+, and c_1 and c_2 be nonnegative constants. If*

$$u^2(t) \leq \left(c_1^2 + 2 \int_0^t f(s) u(s) \, ds\right) \left(c_2^2 + 2 \int_0^t h(s) u(s) \, ds\right), \tag{3.7.27}$$

for $t \in R_+$, then

$$u(t) \leq p(t) \exp\left(2 \int_0^t \left[h(s) \left(\int_0^s f(\sigma) \, d\sigma\right) + f(s) \left(\int_0^s h(\sigma) \, d\sigma\right)\right] ds\right), \tag{3.7.28}$$

for $t \in R_+$, where

$$p(t) = c_1 c_2 + \int_0^t [c_1^2 h(s) + c_2^2 f(s)] \, ds, \tag{3.7.29}$$

for $t \in R_+$.

□

Proof: Let $c_1 > 0$, $c_2 > 0$ and define a function $z(t)$ by

$$z^2(t) = \left(c_1^2 + 2 \int_0^t f(s) u(s) \, ds\right) \left(c_2^2 + 2 \int_0^t h(s) u(s) \, ds\right). \tag{3.7.30}$$

Differentiating (3.7.30) and using the fact that $u(t) \leq z(t)$ we observe that

$$z'(t) \leq [c_1^2 h(t) + c_2^2 f(t)] + 2 \left[h(t) \left(\int_0^t f(\sigma) z(\sigma) \, d\sigma \right) \right.$$

$$\left. + f(t) \left(\int_0^t h(\sigma) z(\sigma) \, d\sigma \right) \right]. \qquad (3.7.31)$$

By taking $t = s$ in (3.7.31) and integrating it from 0 to t we have

$$z(t) \leq p(t) + 2 \int_0^t \left[h(s) \left(\int_0^s f(\sigma) z(\sigma) \, d\sigma \right) + f(s) \left(\int_0^s h(\sigma) z(\sigma) \, d\sigma \right) \right] ds, \qquad (3.7.32)$$

where $p(t)$ is defined by (3.7.29). Since $p(t)$ is positive and monotone nondecreasing for $t \in R_+$, from (3.7.32) we observe that

$$\frac{z(t)}{p(t)} \leq 1 + 2 \int_0^t \left[h(s) \left(\int_0^s f(\sigma) \frac{z(\sigma)}{p(\sigma)} \, d\sigma \right) + f(s) \left(\int_0^s h(\sigma) \frac{z(\sigma)}{p(\sigma)} \, d\sigma \right) \right] ds. \qquad (3.7.33)$$

Define a function $v(t)$ by the right-hand side of (3.7.33). Then using the facts that $z(t)/p(t) \leq v(t)$ and $v(t)$ is monotonic nondecreasing for $t \in R_+$, we observe that

$$v'(t) \leq 2v(t) \left[h(t) \left(\int_0^t f(\sigma) \, d\sigma \right) + f(t) \left(\int_0^t h(\sigma) \, d\sigma \right) \right]. \qquad (3.7.34)$$

The inequality (3.7.34) implies the estimate

$$v(t) \leq \exp \left(2 \int_0^t \left[h(s) \left(\int_0^s f(\sigma) \, d\sigma \right) + f(s) \left(\int_0^s h(\sigma) \, d\sigma \right) \right] ds \right). \qquad (3.7.35)$$

Now by using (3.7.35) in (3.7.33) and the fact that $u(t) \leq z(t)$ we get the desired inequality in (3.7.28).

If c_1, c_2 are nonnegative, we carry out the above procedure with $c_1 + \epsilon$ and $c_2 + \epsilon$ instead of c_1 and c_2, where $\epsilon > 0$ is an arbitrary small constant and subsequently pass to the limit as $\epsilon \to 0$ to obtain (3.7.28).

∎

Pachpatte (1995d) proved the following useful inequality.

Theorem 3.7.3 *Let u, f and g be nonnegative continuous functions defined on R_+ and c_1 and c_2 be positive constants. If*

$$u(t) \le \left(c_1 + \int_0^t f(s)u(s)\,ds\right)\left(c_2 + \int_0^t g(s)u(s)\,ds\right), \qquad (3.7.36)$$

for $t \in R_+$ and $H(t) = 1 - c_1 c_2 \int_0^t R(s)Q(s)\,ds > 0$, for all $t \in R_+$, where

$$R(t) = g(t)\int_0^t f(\sigma)\,d\sigma + f(t)\int_0^t g(\sigma)\,d\sigma, \qquad (3.7.37)$$

$$Q(t) = \exp\left(\int_0^t [c_1 g(\sigma) + c_2 f(\sigma)]\,d\sigma\right), \qquad (3.7.38)$$

for $t \in R_+$, then

$$u(t) \le (1/H(t))\, c_1 c_2 Q(t), \qquad (3.7.39)$$

for $t \in R_+$.

□

Proof: Define a function $z(t)$ by

$$z(t) = \left(c_1 + \int_0^t f(s)u(s)\,ds\right)\left(c_2 + \int_0^t g(s)u(s)\,ds\right). \qquad (3.7.40)$$

Differentiating (3.7.40) and using the facts that $u(t) \le z(t)$ and $z(t)$ is monotonic nondecreasing for $t \in R_+$, we observe that

$$z'(t) \le [c_1 g(t) + c_2 f(t)]z(t) + R(t)z^2(t). \qquad (3.7.41)$$

The inequality (3.7.41) implies the estimate for $z(t)$ such that

$$z(t) \le (1/H(t))\, c_1 c_2 Q(t). \qquad (3.7.42)$$

The desired inequality in (3.7.39) now follows by using (3.7.42) in $u(t) \leq z(t)$.

∎

In order to establish our next result, we use the following notation to simplify the details of the presentation. For some suitable functions a, b, y and $t \in R_+$, we set

$$E[t, a, b, y] = \left(\int_0^t a(\sigma) y(\sigma) \, d\sigma\right) \left(\int_0^t b(\sigma) y(\sigma) \, d\sigma\right).$$

The inequalities in the following theorem are established by Pachpatte (in press a).

Theorem 3.7.4 *Let u, a, b and h be nonnegative continuous functions defined for $t \in R_+$ and c be a nonnegative constant.*

(b_1) Let $f : R_+^2 \to R_+$ be a continuous function which satisfies the condition

$$0 \leq f(t, x) - f(t, y) \leq k(t, y)(x - y), \qquad (3.7.43)$$

for $t \in R_+$ and $x \geq y \geq 0$, where $k : R_+^2 \to R_+$ is a continuous function. If

$$u^2(t) \leq c^2 + 2E[t, a, b, u] + 2\int_0^t u(s) f(s, u(s)) \, ds, \qquad (3.7.44)$$

for $t \in R_+$, then

$$u(t) \leq Q(t) \left[c + \int_0^t f(s, cQ(s)) \exp\left(\int_s^t k(\sigma, cQ(\sigma)) Q(\sigma) \, d\sigma\right) ds\right],$$
$$(3.7.45)$$

for $t \in R_+$, where

$$Q(t) = \exp\left(\int_0^t \left[a(s) \left(\int_0^s b(\sigma) \, d\sigma\right) + b(s) \left(\int_0^s a(\sigma) \, d\sigma\right)\right] ds\right),$$
$$(3.7.46)$$

for $t \in R_+$.

(b_2) Let $w(u)$ be a continuous, nondecreasing and submultiplicative function defined for $u \in R_+$ and $w(u) > 0$ for $u > 0$. If

$$u^2(t) \leq c^2 + 2E[t, a, b, u] + 2\int_0^t h(s)u(s)w(u(s))\,ds, \qquad (3.7.47)$$

for $t \in R_+$, then for $0 \leq t \leq \bar{t}$,

$$u(t) \leq Q(t)\Omega^{-1}\left[\Omega(c) + \int_0^t h(s)w(Q(s))\,ds\right], \qquad (3.7.48)$$

where $Q(t)$ is defined by (3.7.46) and

$$\Omega(r) = \int_{r_0}^r \frac{ds}{w(s)}, \quad r > 0, r_0 > 0, \qquad (3.7.49)$$

Ω^{-1} is the inverse of Ω and $t \in R_+$ is chosen so that

$$\Omega(c) + \int_0^t h(s)w(Q(s))\,ds \in \mathrm{Dom}(\Omega^{-1}),$$

for all $t \in R_+$ lying in the interval $0 \leq t \leq \bar{t}$.

(b_3) Let $w(t, r)$ be a nonnegative continuous function defined for $t \in R_+, 0 \leq r < \infty$, and monotonic nondecreasing with respect to r for any fixed $t \in R_+$. If

$$u^2(t) \leq c^2 + 2E[t, a, b, u] + 2\int_0^t u(s)w(s, u(s))\,ds, \qquad (3.7.50)$$

for $t \in R_+$, then

$$u(t) \leq Q(t)r(t), \quad t \in R_+, \qquad (3.7.51)$$

where $Q(t)$ is defined by (3.7.46) and $r(t)$ is a maximal solution of

$$r'(t) = w(t, Q(t)r(t)), \quad r(0) = c, \qquad (3.7.52)$$

for $t \in R_+$.

□

Proof: It is sufficient to assume that c is positive, since a standard limiting argument can be used to treat the remaining case.

(b_1) Let $c > 0$ and define a function $z(t)$ by the right-hand side of (3.7.44). Differentiating $z(t)$ and using the fact that $u(t) \leq \sqrt{z(t)}$ we observe that

$$z'(t) \leq 2\sqrt{z(t)} \left[a(t) \left(\int_0^t b(\sigma)\sqrt{z(\sigma)}\,d\sigma \right) + b(t) \left(\int_0^t a(\sigma)\sqrt{z(\sigma)}\,d\sigma \right) \right.$$
$$\left. + f(t, \sqrt{z(t)}) \right]. \tag{3.7.53}$$

Differentiating $\sqrt{z(t)}$ and using (3.7.53) we have

$$\frac{d}{dt}(\sqrt{z(t)}) = \frac{z'(t)}{2\sqrt{z(t)}}$$

$$\leq \left[a(t) \left(\int_0^t b(\sigma)\sqrt{z(\sigma)}\,d\sigma \right) + b(t) \left(\int_0^t a(\sigma)\sqrt{z(\sigma)}\,d\sigma \right) \right.$$
$$\left. + f(t, \sqrt{z(t)}) \right]. \tag{3.7.54}$$

By setting $t = s$ in (3.7.54) and integrating from 0 to t we have

$$\sqrt{z(t)} \leq m(t) + \int_0^t \left[a(s) \left(\int_0^s b(\sigma)\sqrt{z(\sigma)}\,d\sigma \right) \right.$$
$$\left. + b(s) \left(\int_0^s a(\sigma)\sqrt{z(\sigma)}\,d\sigma \right) \right] ds, \tag{3.7.55}$$

where

$$m(t) = c + \int_0^t f(s, \sqrt{z(s)})\,ds. \tag{3.7.56}$$

Since $m(t)$ is positive and monotonic nondecreasing for $t \in R_+$, from (3.7.55) we observe that

$$\frac{\sqrt{z(t)}}{m(t)} \le 1 + \int_0^t \left[a(s) \left(\int_0^s b(\sigma) \frac{\sqrt{z(\sigma)}}{m(\sigma)} \, d\sigma \right) + b(s) \left(\int_0^s a(\sigma) \frac{\sqrt{z(\sigma)}}{m(\sigma)} \, d\sigma \right) \right] ds. \tag{3.7.57}$$

Define a function $v(t)$ by the right-hand side of (3.7.57). Differentiating $v(t)$ and then using (3.7.57) and the fact that $v(t)$ is monotonic nondecreasing for $t \in R_+$, we observe that

$$v'(t) \le \left[a(t) \left(\int_0^t b(\sigma) \, d\sigma \right) + b(t) \left(\int_0^t a(\sigma) \, d\sigma \right) \right] v(t). \tag{3.7.58}$$

The inequality (3.7.58) implies the estimate

$$v(t) \le Q(t), \quad t \in R_+, \tag{3.7.59}$$

where $Q(t)$ is defined by (3.7.46). By (3.7.57) and (3.7.59) we have

$$\sqrt{z(t)} \le Q(t)m(t), \quad t \in R_+. \tag{3.7.60}$$

Define a function $n(t)$ by

$$n(t) = \int_0^t f(s, \sqrt{z(s)}) \, ds. \tag{3.7.61}$$

From (3.7.61), (3.7.60) and using the fact that $m(t) = c + n(t)$ from (3.7.56) and the condition (3.7.43) we observe that

$$n'(t) \le f(t, Q(t)(c + n(t)))$$
$$= f(t, cQ(t) + Q(t)n(t)) - f(t, cQ(t)) + f(t, cQ(t))$$
$$\le k(t, cQ(t))Q(t)n(t) + f(t, cQ(t)). \tag{3.7.62}$$

The inequality (3.7.62) implies the estimate

$$n(t) \le \int_0^t f(s, cQ(s)) \exp\left(\int_s^t k(\sigma, cQ(\sigma))Q(\sigma) \, d\sigma \right) ds. \tag{3.7.63}$$

Using (3.7.63) in $m(t) = c + n(t)$ we have

$$m(t) \le c + \int_0^t f(s, cQ(s)) \exp\left(\int_s^t k(\sigma, cQ(\sigma))Q(\sigma) \, d\sigma \right) ds. \tag{3.7.64}$$

The required inequality in (3.7.45) follows by using (3.7.64) in (3.7.60) and then using the fact that $u(t) \leq \sqrt{z(t)}$.

(b_2) Assuming that $c > 0$ and defining a function $z(t)$ by the right-hand side of (3.7.47) and the function $m(t)$ by

$$m(t) = c + \int_0^t h(s)w(\sqrt{z(s)})\,ds, \qquad (3.7.65)$$

and following the same steps as in the proof of (b_1) we have

$$\sqrt{z(t)} \leq Q(t)m(t), \quad t \in R_+, \qquad (3.7.66)$$

where $Q(t)$ is defined by (3.7.46). From (3.7.65), (3.7.66) and using the conditions on w we observe that

$$m'(t) \leq h(t)w(Q(t))w(m(t)). \qquad (3.7.67)$$

From (3.7.49) and (3.7.67) we observe that

$$\frac{d}{dt}\Omega(m(t)) \leq h(t)w(Q(t)). \qquad (3.7.68)$$

Integrating both sides of (3.7.68) from 0 to t we have

$$\Omega(m(t)) \leq \Omega(c) + \int_0^t h(s)w(Q(s))\,ds. \qquad (3.7.69)$$

From (3.7.69) we have

$$m(t) \leq \Omega^{-1}\left[\Omega(c) + \int_0^t h(s)w(Q(s))\,ds\right]. \qquad (3.7.70)$$

Using (3.7.70) in (3.7.66) and then the fact that $u(t) \leq \sqrt{z(t)}$ we get the required inequality in (3.7.48).

(b_3) Assuming that $c > 0$ and defining a function $z(t)$ by the right-hand side of (3.7.50) and the function $m(t)$ by

$$m(t) = c + \int_0^t w(s, \sqrt{z(s)})\,ds, \qquad (3.7.71)$$

and following the same steps as in the proof of (b_1) we have

$$\sqrt{z(t)} \leq Q(t)m(t), \quad t \in R_+, \quad (3.7.72)$$

where $Q(t)$ is defined by (3.7.46). From (3.7.71) and (3.7.72) we observe that

$$m'(t) \leq w(t, Q(t)m(t)). \quad (3.7.73)$$

Now a suitable application of the comparison Theorem 2.2.2 to (3.7.73) and (3.7.52) yields

$$m(t) \leq r(t), \quad t \in R_+, \quad (3.7.74)$$

where $r(t)$ is the maximal solution of (3.7.52). Using (3.7.74) in (3.7.72) and then the fact $u(t) \leq \sqrt{z(t)}$, we get the desired inequality in (3.7.51).

∎

The following theorem deals with the integro-differential inequalities established by Pachpatte (1996a).

Theorem 3.7.5 *Let u, u' and a be nonnegative continuous functions defined on $I = [t_0, \infty)$, $t_0 \geq 0$ with $u(t_0) = k$, a constant and c a nonnegative constant.*

(c_1) *If*

$$(u'(t))^2 \leq c^2 + 2 \int_{t_0}^{t} a(s)u'(s)(u(s) + u'(s))\,ds, \quad (3.7.75)$$

for $t \in I$, then

$$u(t) \leq k + \int_{t_0}^{t} \left[c + (k+c) \int_{t_0}^{s} a(\tau) \exp\left(\int_{t_0}^{\tau} [1 + a(\sigma)]\,d\sigma \right) d\tau \right] ds, \quad (3.7.76)$$

for $t \in I$.

(c_2) *Let $L: I \times R_+ \times R_+ \to R_+$ be a continuous function which satisfies the condition*

$$0 \leq L(t, u, v) - L(t, \bar{u}, \bar{v}) \leq M(t, \bar{u}, \bar{v})[(u - \bar{u}) + (v - \bar{v})], \quad (3.7.77)$$

for $t \in I$ and $u \geq \bar{u} \geq 0$, $v \geq \bar{v} \geq 0$, where $M: I \times R_+ \times R_+ \to R_+$ is a continuous function. If

$$(u'(t))^2 \leq c^2 + 2 \int_{t_0}^{t} a(s) u'(s) L(s, u(s), u'(s)) \, ds, \qquad (3.7.78)$$

for $t \in I$, then

$$u(t) \leq k + c \int_{t_0}^{t} \exp\left(\int_{t_0}^{s} [1 + a(\tau) M(\tau, k, c)] \, d\tau\right) ds$$

$$+ \int_{t_0}^{t} \int_{t_0}^{s} a(\tau) L(\tau, k, c) \exp\left(\int_{\tau}^{s} [1 + a(\sigma) M(\sigma, k, c)] \, d\sigma\right) d\tau \, ds,$$

$$(3.7.79)$$

for $t \in I$.

□

Proof: As noted in the proof of Theorem 3.7.4, it is enough to prove the results for $c > 0$ only.

(c_1) Let $c > 0$ and define a function $z(t)$ by the right-hand side of (3.7.75). Then $u'(t) \leq \sqrt{z(t)}$, and hence integrating this inequality from t_0 to t we have $u(t) \leq k + \int_{t_0}^{t} \sqrt{z(s)} \, ds$. Differentiating the function $z(t)$ and using the above facts we have

$$z'(t) \leq 2a(t)\sqrt{z(t)} \left(k + \int_{t_0}^{t} \sqrt{z(s)} \, ds + \sqrt{z(t)}\right). \qquad (3.7.80)$$

Dividing both sides of (3.7.80) by $\sqrt{z(t)}$ and then integrating the resulting inequality from t_0 to t we have

$$\sqrt{z(t)} \leq c + \int_{t_0}^{t} a(s) \left(k + \int_{t_0}^{s} \sqrt{z(\tau)} \, d\tau + \sqrt{z(s)}\right) ds. \qquad (3.7.81)$$

Define a function $v(t)$ by the right-hand side of (3.7.81). Then $\sqrt{z(t)} \leq v(t)$ and

$$v'(t) \leq a(t)\left(k + \int_{t_0}^{t} v(\tau)\,d\tau + v(t)\right), \qquad (3.7.82)$$

Define a function $m(t)$ by

$$m(t) = k + \int_{t_0}^{t} v(\tau)\,d\tau + v(t). \qquad (3.7.83)$$

Differentiating (3.7.83) and using (3.7.82) and the fact that $v(t) \leq m(t)$ we have

$$m'(t) \leq [1 + a(t)]m(t). \qquad (3.7.84)$$

The inequality (3.7.84) implies the estimate

$$m(t) \leq (k+c)\exp\left(\int_{t_0}^{t}[1+a(\sigma)]\,d\sigma\right). \qquad (3.7.85)$$

Using (3.7.85) in (3.7.82) and integrating it from t_0 to t we have

$$v(t) \leq c + (k+c)\int_{t_0}^{t} a(s)\exp\left(\int_{t_0}^{s}[1+a(\sigma)]\,d\sigma\right)ds. \qquad (3.7.86)$$

Using (3.7.86) in (3.7.81) and the fact that $u'(t) \leq \sqrt{z(t)}$ we get

$$u'(t) \leq c + (k+c)\int_{t_0}^{t} a(s)\exp\left(\int_{t_0}^{s}[1+a(\sigma)]\,d\sigma\right)ds. \qquad (3.7.87)$$

Integrating (3.7.87) from t_0 to t we get the desired inequality in (3.7.76).

(c_2) Let $c > 0$ and define a function $z(t)$ by the right-hand side of (3.7.78). Differentiating $z(t)$ and then using the facts that $u'(t) \leq \sqrt{z(t)}$ and $u(t) \leq k + \int_{t_0}^{t}\sqrt{z(\tau)}\,d\tau$ we have

$$z'(t) \leq 2a(t)\sqrt{z(t)}\,L\left(t, k + \int_{t_0}^{t}\sqrt{z(\tau)}\,d\tau, \sqrt{z(t)}\right). \qquad (3.7.88)$$

Dividing both sides of (3.7.88) by $\sqrt{z(t)}$, setting $t = s$ and integrating the resulting inequality from t_0 to t we have

$$\sqrt{z(t)} \leq c + \int_{t_0}^{t} a(s) L\left(s, k + \int_{t_0}^{s} \sqrt{z(\tau)}\, d\tau, \sqrt{z(s)}\right) ds. \qquad (3.7.89)$$

Define a function $v(t)$ by the right-hand side of (3.7.89). Differentiating $v(t)$ and using the fact that $\sqrt{z(t)} \leq v(t)$ and the condition (3.7.77) we observe that

$$v'(t) \leq a(t) L\left(t, k + \int_{t_0}^{t} v(\tau)\, d\tau, c + v(t)\right) - a(t) L(t, k, c) + a(t) L(t, k, c)$$

$$\leq a(t) M(t, k, c) \left[v(t) + \int_{t_0}^{t} v(\tau)\, d\tau\right] + a(t) L(t, k, c). \qquad (3.7.90)$$

Define a function $m(t)$ by

$$m(t) = v(t) + \int_{t_0}^{t} v(\tau)\, d\tau. \qquad (3.7.91)$$

Differentiating (3.7.91) and using (3.7.90) and the fact that $v(t) \leq m(t)$ we have

$$m'(t) \leq [1 + a(t) M(t, k, c)] m(t) + a(t) L(t, k, c). \qquad (3.7.92)$$

The inequality (3.7.92) implies the estimate

$$m(t) \leq c \exp\left(\int_{t_0}^{t} [1 + a(\tau) M(\tau, k, c)]\, d\tau\right) + \int_{t_0}^{t} a(\tau) L(\tau, k, c)$$

$$\times \exp\left(\int_{\tau}^{t} [1 + a(\sigma) M(\sigma, k, c)]\, d\sigma\right) d\tau. \qquad (3.7.93)$$

Using the facts that $u'(t) \leq \sqrt{z(t)} \leq v(t) \leq m(t)$ we have

$$u'(t) \leq c \exp\left(\int_{t_0}^{t} [1 + a(\tau) M(\tau, k, c)]\, d\tau\right) + \int_{t_0}^{t} a(\tau) L(\tau, k, c)$$

$$\times \exp\left(\int_{\tau}^{t} [1 + a(\sigma) M(\sigma, k, c)]\, d\sigma\right) d\tau. \qquad (3.7.94)$$

By taking $t = s$ in (3.7.94) and integrating it from t_0 to t we get the required inequality in (3.7.79).

∎

3.8 The Inequalities of Haraux and Engler

The mathematical formulation of certain physical phenomena often leads to some specific type of nonlinear differential or integral equations. In the study of the qualitative behaviour of solutions of such equations some special types of integral inequalities are needed. This section deals with some inequalities which depart somewhat from the structure of the inequalities given in earlier sections.

Haraux (1981) used the following inequality to study the wave equation with logarithmic nonlinearity.

Theorem 3.8.1 *Let $u: I \to R_+$ be a continuous function and $a \geq 1$, $u_0 \geq 0$ be constants, where $I = [0, T]$. If*

$$u(t) \leq u_0 + a \int_0^t u(s) \log(a + u(s)) \, ds, \tag{3.8.1}$$

for $t \in I$, then

$$u(t) \leq (a + u_0)^{\exp(at)}, \tag{3.8.2}$$

for $t \in I$.

□

Proof: Define a function $z(t)$ by the right-hand side of (3.8.1). Then $z(0) = u_0$, $u(t) \leq z(t)$ and

$$z'(t) = au(t) \log(a + u(t))$$
$$\leq az(t) \log(a + z(t)). \tag{3.8.3}$$

From (3.8.3) it is easy to observe that

$$\frac{z'(t)}{a + z(t)} \leq \frac{z'(t)}{z(t)} \leq a \log(a + z(t)). \tag{3.8.4}$$

By taking $t = s$ in (3.8.4) and integrating from 0 to t we get

$$\log(a + z(t)) \leq \log(a + u_0) + a \int_0^t \log(a + z(s))\, ds. \tag{3.8.5}$$

Now an application of Bellman's inequality given in Theorem 1.2.2 to (3.8.5) yields

$$\log(a + z(t)) \leq (\log(a + u_0)) \exp(at)$$
$$= \log(a + u_0)^{\exp(at)}. \tag{3.8.6}$$

From (3.8.6) we observe that

$$a + z(t) \leq (a + u_0)^{\exp(at)}. \tag{3.8.7}$$

From (3.8.7) and using the fact that $u(t) \leq z(t)$, we get the required inequality in (3.8.2).

∎

Remark 3.8.1 We note that Haraux (1981) used the hypothesis $u \in L^\infty(0, T)$ and the inequality (3.8.1) holds a.e. on I and concluded that (3.8.2) also holds a.e. on I. From (3.8.7) it is easy to observe that one can obtain a better bound than the one given in (3.8.2) in the following form

$$u(t) \leq (a + u_0)^{\exp(at)} - a, \tag{3.8.8}$$

for all $t \in I$.

△

Engler (1989) used the following slight variant of the inequality given in Theorem 3.8.1 while studying the global regular solutions for the dynamic antiplane shear problem in nonlinear viscoelasticity.

Theorem 3.8.2 *Let* $a: I \to R_+$, $u: I \to R_1$ *be continuous functions and* $c \geq 1$ *be a constant, where* $I = [0, T]$, $R_1 = [1, \infty)$. *If*

$$u(t) \leq c \left[1 + \int_0^t a(s) u(s) \log u(s)\, ds \right], \tag{3.8.9}$$

for $t \in I$, *then*

$$u(t) \leq c^{\exp\left(c \int_0^t a(s)\, ds\right)}, \tag{3.8.10}$$

for $t \in I$.

□

Proof: Define a function $z(t)$ by the right-hand side of (3.8.9). Then $z(0) = c$, $u(t) \leq z(t)$ and

$$z'(t) = ca(t)u(t)\log u(t)$$
$$\leq ca(t)z(t)\log z(t). \tag{3.8.11}$$

From (3.8.11) we have

$$\frac{z'(t)}{z(t)} \leq ca(t)\log z(t). \tag{3.8.12}$$

By taking $t = s$ in (3.8.12) and integrating from 0 to t we get

$$\log z(t) \leq \log c + c\int_0^t a(s)\log z(s)\,ds. \tag{3.8.13}$$

Now by applying Bellman's inequality given in Theorem 1.2.2 to (3.8.13) we get

$$\log z(t) \leq (\log c)\exp\left(c\int_0^t a(s)\,ds\right)$$
$$= \log c^{\exp\left(c\int_0^t a(s)\,ds\right)}. \tag{3.8.14}$$

From (3.8.14) we observe that

$$z(t) \leq c^{\exp\left(c\int_0^t a(s)\,ds\right)}. \tag{3.8.15}$$

Now by using (3.8.15) in $u(t) \leq z(t)$ we get the required inequality in (3.8.10).

∎

Remark 3.8.2 Note that an essential difference between the above inequalities and the well-known Gronwall–Bellman inequality is that these inequalities contain an extra logarithmic factor in the integrand on the right-hand side of (3.8.1) and (3.8.9) and the bounds obtained in (3.8.2) and (3.8.10) are different from those given by the Gronwall–Bellman inequality. Obviously, the above inequalities are useful in the situations for which the other available inequalities do not apply directly.

△

3.9 Pachpatte's Inequalities V

In view of the interesting applications of the inequalities given by Haraux (1981) and Engler (1989), and other possible applications, Pachpatte (1991, 1994d) investigated a number of new inequalities of the type given in Theorems 3.8.1 and 3.8.2. This section gives the inequalities established in the above-cited papers, which can be used as tools in studying certain classes of nonlinear differential and integral equations.

Pachpatte (1994d) established the following generalization of the inequality given by Engler (1989) in Theorem 3.8.2.

Theorem 3.9.1 Let $a\colon I \to R_+$, $u\colon I \to R_1$ are continuous functions and $u_0 \geq 1$ be a constant, where $I = [0, T]$ and $R_1 = [1, \infty)$. Let $f(u)$ be a continuous nondecreasing function defined on R_+ and $f(u) > 0$ for $u > 0$. If

$$u(t) \leq u_0 + \int_0^t a(s)u(s)f(\log u(s))\,ds, \qquad (3.9.1)$$

for $t \in I$, then for $0 \leq t \leq t_1$,

$$u(t) \leq \exp\left(F^{-1}\left[F(\log u_0) + \int_0^t a(s)\,ds\right]\right), \qquad (3.9.2)$$

where

$$F(r) = \int_{r_0}^r \frac{ds}{f(s)}, \qquad r > 0,\ r_0 > 0, \qquad (3.9.3)$$

F^{-1} is the inverse of F and $t_1 \in I$ is chosen so that

$$F(\log u_0) + \int_0^t a(s)\,ds \in \mathrm{Dom}(F^{-1}),$$

for all $t \in I$ lying in the interval $0 \leq t \leq t_1$.

□

Proof: By taking $z(t)$ equal to the right-hand side of (3.9.1) and following the proof of Theorem 3.8.2 up to the inequality (3.8.13) with suitable changes we obtain

$$\log z(t) \leq \log u_0 + \int_0^t a(s) f(\log z(s))\, ds. \tag{3.9.4}$$

Now by applying Bihari's inequality given in Theorem 2.3.1 we have

$$\log z(t) \leq F^{-1}\left[F(\log u_0) + \int_0^t a(s)\, ds\right]. \tag{3.9.5}$$

From (3.9.5) we observe that

$$z(t) \leq \exp\left(F^{-1}\left[F(\log u_0) + \int_0^t a(s)\, ds\right]\right). \tag{3.9.6}$$

The desired inequality in (3.9.2) now follows by using (3.9.6) in $u(t) \leq z(t)$. The subinterval for $t \in I$ is obvious.

∎

Remark 3.9.1 We note that in Theorem 3.9.1, if we take $f(r) = r$, then (3.9.2) reduces to (3.8.10) with $c = u_0$ and if we take $f(r) = r^\alpha$, $0 < \alpha < 1$ is a constant, then (3.9.2) reduces to

$$u(t) \leq \exp\left(\left[(\log u_0)^{1-\alpha} + (1-\alpha)\int_0^t a(s)\, ds\right]^{1/(1-\alpha)}\right), \tag{3.9.7}$$

for $t \in I$.

△

A slightly different version of Theorem 3.9.1 is given in the following theorem.

Theorem 3.9.2 Let $u, b: I \to R_+$ be continuous functions, $a \geq 1$, $u_0 \geq 0$ be constants, f be as defined in Theorem 3.9.1, where $I = [0, T]$. If

$$u(t) \leq u_0 + \int_0^t b(s) u(s) f(\log(a + u(s)))\, ds, \tag{3.9.8}$$

for $t \in I$, then for $0 \le t \le t_2$,

$$u(t) \le \left[\exp\left(F^{-1}\left[F(\log(a+u_0)) + \int_0^t b(s)\,ds \right] \right) - a \right], \qquad (3.9.9)$$

where F, F^{-1} are as defined in Theorem 3.9.1 and $t_2 \in I$ is chosen so that

$$F(\log(a+u_0)) + \int_0^t b(s)\,ds \in \text{Dom}(F^{-1}),$$

for all $t \in I$ lying in the interval $0 \le t \le t_2$. $\square$

The proof of this theorem can be completed by following the proofs of Theorems 3.8.1 and 3.9.1 with suitable modifications. The details are omitted here.

A fairly general version of Theorem 3.8.1 established by Pachpatte (1994d) is given in the following theorem.

Theorem 3.9.3 Let $a, b: I \to R_+$, $u: I \to R_1$ be continuous functions, $u_0 \ge 1$ be a constant, where $I = [0, T]$ and $R_1 = [1, \infty)$. If

$$u(t) \le u_0 + \int_0^t a(s) u(s) \left(\log u(s) + \int_0^s b(\sigma) \log u(\sigma)\,d\sigma \right) ds, \qquad (3.9.10)$$

for $t \in I$, then

$$u(t) \le u_0^{\left[1 + \int_0^t a(s) \exp\left(\int_0^s [a(\sigma) + b(\sigma)]\,d\sigma \right) ds \right]}, \qquad (3.9.11)$$

for $t \in I$. $\square$

Proof: Define a function $z(t)$ by the right-hand side of (3.9.10). Then $z(0) = u_0$, $u(t) \le z(t)$ and

$$z'(t) \le a(t) z(t) \left(\log z(t) + \int_0^t b(\sigma) \log z(\sigma)\,d\sigma \right). \qquad (3.9.12)$$

From (3.9.12) we have

$$\frac{z'(t)}{z(t)} \le a(t)\left(\log z(t) + \int_0^t b(\sigma)\log z(\sigma)\,d\sigma\right). \quad (3.9.13)$$

By taking $t = s$ in (3.9.13) and integrating from 0 to t we get

$$\log z(t) \le \log u_0 + \int_0^t a(s)\left(\log z(s) + \int_0^s b(\sigma)\log z(\sigma)\,d\sigma\right)ds. \quad (3.9.14)$$

Now an application of Pachpatte's inequality given in Theorem 1.7.1 yields

$$\log z(t) \le \left[1 + \int_0^t a(s)\exp\left(\int_0^s [a(\sigma)+b(\sigma)]\,d\sigma\right)ds\right]\log u_0$$

$$= \log u_0^{\left[1+\int_0^t a(s)\exp\left(\int_0^s [a(\sigma)+b(\sigma)]\,d\sigma\right)ds\right]}. \quad (3.9.15)$$

From (3.9.15) we observe that

$$z(t) \le u_0^{\left[1+\int_0^t a(s)\exp\left(\int_0^s [a(\sigma)+b(\sigma)]\,d\sigma\right)ds\right]}, \quad (3.9.16)$$

for $t \in R_+$. Using (3.9.16) in $u(t) \le z(t)$ we get the required inequality in (3.9.11).

∎

Another interesting and useful inequality obtained by Pachpatte (1991) is embodied in the following theorem.

Theorem 3.9.4 Let $a, b: I \to R_+$, $u: I \to R_1$ be continuous functions and $u_0 \ge 1$ be a constant, where $I = [0, T]$ and $R_1 = [1, \infty)$. If

$$u(t) \le u_0 + \int_0^t a(s)u(s)\left(\int_0^s b(\sigma)\log u(\sigma)\,d\sigma\right)ds, \quad (3.9.17)$$

for $t \in I$, then

$$u(t) \le u_0^{\exp\left(\int_0^t a(s)\left(\int_0^s b(\sigma)\,d\sigma\right)ds\right)}, \quad (3.9.18)$$

for $t \in I$.

□

The proof of this theorem can be completed by following the proof of Theorem 3.9.3 with suitable modifications and hence we omit it here.

The following theorems deal with the inequalities established by Pachpatte (1994d) and (1991) respectively, which can be used in certain applications.

Theorem 3.9.5 *Let a, u, u_0 and f be as defined in Theorem 3.9.1. If*

$$u(t) \leq u_0 + \int_0^t a(s)u(s) \left(\log u(s) + \int_0^s a(\sigma) f(\log u(\sigma)) \, d\sigma \right) ds, \quad (3.9.19)$$

for $t \in I$, then for $0 \leq t \leq t_1$,

$$u(t) \leq u_0 \exp \left(\int_0^t a(s) \Omega^{-1} \left[\Omega(\log u_0) + \int_0^s a(\sigma) \, d\sigma \right] ds \right), \quad (3.9.20)$$

where

$$\Omega(r) = \int_{r_0}^r \frac{ds}{s + f(s)}, \quad r > 0, \, r_0 > 0, \quad (3.9.21)$$

Ω^{-1} *is the inverse of Ω and $t_1 \in I$ is chosen so that*

$$\Omega(\log u_0) + \int_0^t a(\sigma) \, d\sigma \in \text{Dom}(\Omega^{-1}),$$

for all $t \in I$ lying in the interval $0 \leq t \leq t_1$. □

The proof of this theorem follows by the same arguments as in the proof of Theorem 3.9.1 and applying Pachpatte's inequality given in Theorem 2.8.1 with suitable changes. The details are omitted here.

Theorem 3.9.6 *Let a, $b: I \to R_+$, $u: I \to R_1$ be continuous functions, $u_0 \geq 1$ be a constant and f be as in Theorem 3.9.1, where $I = [0, T]$ and $R_1 = [1, \infty)$. If*

$$u(t) \leq u_0 + \int_0^t a(s)u(s) \left(\int_0^s b(\sigma) f(\log u(\sigma)) \, d\sigma \right) ds, \quad (3.9.22)$$

for $t \in I$, then for $0 \leq t \leq t_2$,

$$u(t) \leq \exp\left(F^{-1}\left[F(\log u_0) + \int_0^t a(s)\left(\int_0^s b(\sigma)\,d\sigma\right)ds\right]\right), \quad (3.9.23)$$

where F, F^{-1} are as given in Theorem 3.9.1 and $t_2 \in I$ is chosen so that

$$F(\log u_0) + \int_0^t a(s)\left(\int_0^s b(\sigma)\,d\sigma\right)ds \in \mathrm{Dom}(F^{-1}),$$

for all $t \in I$ lying in the interval $0 \leq t \leq t_2$. $\square$

Proof: Define a function $z(t)$ by the right-hand side of (3.9.22). Then $z(0) = u_0$, $u(t) \leq z(t)$ and

$$z'(t) \leq a(t)z(t)\left(\int_0^t b(\sigma)f(\log z(\sigma))\,d\sigma\right). \quad (3.9.24)$$

From (3.9.24) we have

$$\frac{z'(t)}{z(t)} \leq a(t)\left(\int_0^t b(\sigma)f(\log z(\sigma))\,d\sigma\right). \quad (3.9.25)$$

By taking $t = s$ in (3.9.25) and integrating from 0 to t we get

$$\log z(t) \leq \log u_0 + \int_0^t a(s)\left(\int_0^s b(\sigma)f(\log z(\sigma))\,d\sigma\right)ds. \quad (3.9.26)$$

Define a function $v(t)$ by the right-hand side of (3.9.26). Differentiating the function $v(t)$ and using the facts that $\log z(t) \leq v(t)$ and $v(t)$ is a monotonic increasing function for $t \in I$, it is easy to observe that

$$v'(t) = a(t)\left(\int_0^t b(\sigma)f(\log z(\sigma))\,d\sigma\right)$$

$$\leq a(t)\left(\int_0^t b(\sigma)f(v(\sigma))\,d\sigma\right)$$

$$\le a(t) f(v(t)) \left(\int_0^t b(\sigma) \, d\sigma \right). \tag{3.9.27}$$

From (3.9.3) and (3.9.27) we have

$$\frac{d}{dt} F(v(t)) = \frac{v'(t)}{f(v(t))} \le a(t) \left(\int_0^t b(\sigma) \, d\sigma \right). \tag{3.9.28}$$

Now by setting $t = s$ in (3.9.28) and integrating it from 0 to t we get

$$F(v(t)) \le F(\log u_0) + \int_0^t a(s) \left(\int_0^s b(\sigma) \, d\sigma \right) ds. \tag{3.9.29}$$

Using the bound on $v(t)$ from (3.9.29) in (3.9.26) we have

$$\log z(t) \le F^{-1} \left[F(\log u_0) + \int_0^t a(s) \left(\int_0^s b(\sigma) \, d\sigma \right) ds \right]. \tag{3.9.30}$$

From (3.9.30) we observe that

$$z(t) \le \exp\left(F^{-1} \left[F(\log u_0) + \int_0^t a(s) \left(\int_0^s b(\sigma) \, d\sigma \right) ds \right] \right). \tag{3.9.31}$$

Now using (3.9.31) in $u(t) \le z(t)$ we get the required inequality in (3.9.23). The subinterval for $t \in I$ is obvious.

∎

3.10 Inequalities Involving Iterated Integrals

In the development of the theory of various classes of higher order differential and integro-differential equations one must often deal with certain new integral inequalities. This section presents some integral inequalities investigated by Pachpatte (1993, 1994a,b, 1995c, 1996d) which can be used as basic tools in certain new applications.

In what follows the definitions and notations as given in Section 1.9 are used to denote the iterated integrals. Furthermore, the following notation is used to simplify the details of the presentation. For $t \in R_+$ and for some functions $r_i(t)$, $i = 1, 2, \ldots, n$, we set:

$$F[t, r_1, r_2, \ldots, r_{n-1}, r_n] = r_1(t) \int_0^t r_2(s_2) \int_0^{s_2} r_3(s_3) \ldots \int_0^{s_{n-2}} r_{n-1}(s_{n-1}) \int_0^{s_{n-1}} r_n(s_n)$$
$$\times \, ds_n \, ds_{n-1} \ldots ds_3 \, ds_2$$
$$+ r_2(t) \int_0^t r_3(s_3) \int_0^{s_3} r_4(s_4) \ldots \int_0^{s_{n-2}} r_{n-1}(s_{n-1}) \int_0^{s_{n-1}} r_n(s_n)$$
$$\times \, ds_n \, ds_{n-1} \ldots ds_4 \, ds_3$$
$$+ \cdots$$
$$\vdots$$
$$+ r_{n-1}(t) \int_0^t r_n(s_n) \, ds_n$$
$$+ r_n(t).$$

Our first result contains the inequalities established by Pachpatte (1993, 1996).

Theorem 3.10.1 *Let* $u \geq 0$, $a \geq 0$, $b \geq 0$, $h \geq 0$, $r_j > 0$, $j = 1, 2, \ldots, n-1$ *be continuous functions defined on* R_+. *Let* $f : R_+^2 \to R_+$ *be a continuous function which satisfies the condition*

$$0 \leq f(t, x) - f(t, y) \leq k(t, y)(x - y), \quad (3.10.1)$$

for $t \in R_+$ *and* $x \geq y \geq 0$, *where* $k : R_+^2 \to R_+$ *is a continuous function.*

(a_1) *If*

$$u(t) \leq a(t) + b(t) M[t, r, f(t_n, u(t_n))], \quad (3.10.2)$$

for $t \in R_+$, *then*

$$u(t) \leq a(t) + b(t) M[t, r, f(t_n, a(t_n))]$$
$$\times \exp(M[t, r, k(t_n, a(t_n))b(t_n)]), \quad (3.10.3)$$

for $t \in R_+$.

(a_2) *Let* $g(u)$ *be a continuous, strictly increasing, convex submultiplicative function for* $u > 0$, $\lim_{u \to \infty} g(u) = \infty$, g^{-1} *denote the inverse function of*

g, $\alpha(t)$, $\beta(t)$ be continuous and positive functions on R_+ and $\alpha(t) + \beta(t) = 1$. If

$$u(t) \leq a(t) + b(t)g^{-1}(M[t, r, f(t_n, g(u(t_n)))]), \quad (3.10.4)$$

for $t \in R_+$, then

$$u(t) \leq a(t) + b(t)g^{-1}(M[t, r, f(t_n, \alpha(t_n)g(a(t_n)\alpha^{-1}(t_n)))]$$
$$\times \exp(M[t, r, k(t_n, \alpha(t_n)g(a(t_n)\alpha^{-1}(t_n)))\beta(t_n)g(b(t_n)\beta^{-1}(t_n))]), \quad (3.10.5)$$

for $t \in R_+$.

(a_3) Let $w(u)$ be a continuous, nondecreasing, subadditive and submultiplicative function defined on R_+ and $w(u) > 0$ for $u > 0$. If

$$u(t) \leq a(t) + b(t)M[t, r, f(t_n, w(u(t_n)))], \quad (3.10.6)$$

for $t \in R_+$, then for $0 \leq t \leq t_1$,

$$u(t) \leq a(t) + b(t)G^{-1}(G(M[t, r, f(t_n, w(a(t_n)))])$$
$$+ M[t, r, k(t_n, w(a(t_n)))w(b(t_n))]), \quad (3.10.7)$$

where

$$G(v) = \int_{v_0}^{v} \frac{ds}{w(s)}, \quad v > 0, v_0 > 0, \quad (3.10.8)$$

G^{-1} is the inverse of G and $t_1 \in R_+$ is chosen so that

$$G(M[t, r, f(t_n, w(a(t_n)))]) + M[t, r, k(t_n, w(a(t_n)))w(b(t_n))] \in \text{Dom}(G^{-1}),$$

for all $t \in R_+$ lying in the interval $0 \leq t \leq t_1$. $\square$

Proof: (a_1) Define a function $z(t)$ by

$$z(t) = M[t, r, f(t_n, u(t_n))]. \quad (3.10.9)$$

From (3.10.9) and using the fact that $u(t) \leq a(t) + b(t)z(t)$, and the condition (3.10.1), we observe that

$$z(t) \leq M[t, r, f(t_n, a(t_n) + b(t_n)z(t_n))]$$

$$= M[t, r, f(t_n, a(t_n))] + M[t, r, f(t_n, a(t_n)) + b(t_n)z(t_n))]$$
$$- M[t, r, f(t_n, a(t_n))]$$
$$\leq M[t, r, f(t_n, a(t_n))] + M[t, r, k(t_n, a(t_n))b(t_n)z(t_n)]$$
$$\leq c(t) + M[t, r, k(t_n, a(t_n))b(t_n)z(t_n)], \qquad (3.10.10)$$

where $c(t) = \epsilon + M[t, r, f(t_n, a(t_n))]$, in which $\epsilon > 0$ is an arbitrary small constant. Since $c(t)$ is positive and monotonic nondecreasing in $t \in R_+$, from (3.10.10) we observe that

$$\frac{z(t)}{c(t)} \leq 1 + M\left[t, r, k(t_n, a(t_n))b(t_n)\frac{z(t_n)}{c(t_n)}\right]. \qquad (3.10.11)$$

Define a function $m(t)$ by

$$m(t) = 1 + M\left[t, r, k(t_n, a(t_n))b(t_n)\frac{z(t_n)}{c(t_n)}\right]. \qquad (3.10.12)$$

From (3.10.12) it is easy to observe that

$$L_n m(t) = k(t, a(t))b(t)\frac{z(t)}{c(t)}, \qquad (3.10.13)$$

where L_i, $i = 0, 1, 2, \ldots, n$ are the operators as defined in Section 1.9. Using the fact that $z(t)/c(t) \leq m(t)$ in (3.10.13) we have

$$L_n m(t) \leq k(t, a(t))b(t)m(t). \qquad (3.10.14)$$

From (3.10.14) and using the same steps as in the proof of Theorem 1.9.1 part (b_1) below the inequality (1.9.27) up to (1.9.32) we have

$$m(t) \leq \exp(M[t, r, k(t_n, a(t_n))b(t_n)]). \qquad (3.10.15)$$

Using (3.10.15) in (3.10.11) we get

$$z(t) \leq c(t) \exp(M[t, r, k(t_n, a(t_n))b(t_n)]). \qquad (3.10.16)$$

Now, by letting $\epsilon \to 0$ in (3.10.16) we get

$$z(t) \leq M[t, r, f(t_n, a(t_n))] \exp(M[t, r, k(t_n, a(t_n))b(t_n)]). \qquad (3.10.17)$$

Now by using (3.10.17) in $u(t) \leq a(t) + b(t)z(t)$ we get the desired inequality in (3.10.3).

(a_2) Rewrite (3.10.4) as

$$u(t) \leq \alpha(t)a(t)\alpha^{-1}(t) + \beta(t)(b(t)\beta^{-1}(t))g^{-1}(M[t, r, f(t_n, g(u(t_n)))]).$$
(3.10.18)

Since g is convex, submultiplicative and monotonic, from (3.10.18) we have

$$g(u(t)) \leq \alpha(t)g(a(t)\alpha^{-1}(t)) + \beta(t)g(b(t)\beta^{-1}(t))$$
$$\times M[t, r, f(t_n, g(u(t_n)))]. \quad (3.10.19)$$

The estimate in (3.10.5) follows by first applying the inequality given in part (a_1) to (3.10.19) and then applying g^{-1} to both sides of the resulting inequality.

(a_3) Define a function $z(t)$ by

$$z(t) = M[t, r, f(t_n, w(u(t_n)))]. \quad (3.10.20)$$

From (3.10.20) and using the fact that $u(t) \leq a(t) + b(t)z(t)$ and the conditions on the functions w and f we observe that

$$z(t) \leq M[t, r, f(t_n, w(a(t_n) + b(t_n)z(t_n)))]$$
$$\leq M[t, r, f(t_n, w(a(t_n)) + w(b(t_n))w(z(t_n)))]$$
$$= M[t, r, f(t_n, w(a(t_n)))] + M[t, r, f(t_n, w(a(t_n))$$
$$+ w(b(t_n))w(z(t_n)))] - M[t, r, f(t_n, w(a(t_n)))]$$
$$\leq M[t, r, f(t_n, w(a(t_n)))]$$
$$+ M[t, r, k(t_n, w(a(t_n)))w(b(t_n))w(z(t_n))]. \quad (3.10.21)$$

For an arbitrary $T \in R_+$, it follows from (3.10.21) that

$$z(t) \leq M[T, r, f(t_n, w(a(t_n)))]$$
$$+ M[t, r, k(t_n, w(a(t_n)))w(b(t_n))w(z(t_n))], \quad 0 \leq t \leq T.$$
(3.10.22)

It is sufficient to assume that $M[T, r, f(t, w(a(t_n)))]$ is positive, since the standard limiting argument can be used to treat the case when $M[T, r, f(t, w(a(t_n)))]$ is nonnegative. Define a function $m(t)$ by

$$m(t) = M[T, r, f(t_n, w(a(t_n)))]$$
$$+ M[t, r, k(t_n, w(a(t_n)))w(b(t_n))w(z(t_n))], \quad 0 \le t \le T.$$
(3.10.23)

From (3.10.23) and using the fact that $z(t) \le m(t)$ for $0 \le t \le T$ and the monotonic character of w, we observe that

$$m'(t) \le r_1(t) \int_0^t r_2(t_2) \int_0^{t_2} r_3(t_3) \ldots \int_0^{t_{n-2}} r_{n-1}(t_{n-1})$$
$$\times \int_0^{t_{n-1}} k(t_n, w(a(t_n)))w(b(t_n))w(m(t_n))\, dt_n\, dt_{n-1} \ldots dt_3\, dt_2$$
$$\le \left[r_1(t) \int_0^t r_2(t_2) \int_0^{t_2} r_3(t_3) \ldots \int_0^{t_{n-2}} r_{n-1}(t_{n-1}) \right.$$
$$\left. \times \int_0^{t_{n-1}} k(t_n, w(a(t_n)))w(b(t_n))\, dt_n\, dt_{n-1} \ldots dt_3\, dt_2 \right] w(m(t)),$$
$$0 \le t \le T. \quad (3.10.24)$$

From (3.10.8) and (3.10.24) we observe that

$$\frac{d}{dt} G(m(t)) = \frac{m'(t)}{w(m(t))} \le r_1(t) \int_0^t r_2(t_2) \int_0^{t_2} r_3(t_3) \ldots \int_0^{t_{n-2}} r_{n-1}(t_{n-1})$$
$$\times \int_0^{t_{n-1}} k(t_n, w(a(t_n)))w(b(t_n))\, dt_n\, dt_{n-1}\, dt_3\, dt_2,$$
$$0 \le t \le T \quad (3.10.25)$$

By taking $t = t_1$ in (3.10.25) and integrating from 0 to T we get

$$G(m(T)) \le G(m(0)) + M[T, r, k(t_n, w(a(t_n)))w(b(t_n))]. \quad (3.10.26)$$

Since $z(t) \leq m(t)$ is true for $t = T$, from (3.10.26) we have

$$G(z(T)) \leq G(M[T, r, f(t_n, w(a(t_n)))])$$
$$+ M[T, r, k(t_n, w(a(t_n)))w(b(t_n))]. \quad (3.10.27)$$

Since T is arbitrary, by replacing T by t in (3.10.27) we observe that

$$z(t) \leq G^{-1}[G(M[t, r, f(t_n, w(a(t_n)))])$$
$$+ M[t, r, k(t_n, w(a(t_n)))w(b(t_n))]]. \quad (3.10.28)$$

Using (3.10.28) in $u(t) \leq a(t) + b(t)z(t)$ we get the desired inequality in (3.10.7). The subinterval $0 \leq t \leq t_1$ for $t \in R_+$ is obvious.

■

Pachpatte (1995c) gave the following general version of Theorem 3.10.1, which is more convenient in some applications.

Theorem 3.10.2 *Let* u, a, b, p_i, $i = 1, 2, \ldots, n$, *be nonnegative continuous functions defined on* R_+. *Let* f *and* k *be as in Theorem 3.10.1.*

(b_1) *If*

$$u(t) \leq a(t) + b(t) \int_0^t F[s, p_1, p_2, \ldots, p_{n-1}, p_n(fu)]\,ds, \quad (3.10.29)$$

for $t \in R_+$, *where* $(fu)(t) = f(t, u(t))$, *then*

$$u(t) \leq a(t) + b(t) \int_0^t F[s, p_1, p_2, \ldots, p_{n-1}, p_n(fa)]$$
$$\times \exp\left(\int_s^t F[\tau, p_1, p_2, \ldots, p_{n-1}, p_n(ka)b]\,d\tau\right) ds,$$
$$(3.10.30)$$

for $t \in R_+$, *where* $(ka)(t) = k(t, a(t))$.

(b_2) *Let* g, g^{-1}, α, β *be as in* (a_2) *in Theorem 3.10.1. If*

$$u(t) \leq a(t) + b(t)g^{-1}\left(\int_0^t F[s, p_1, p_2, \ldots, p_{n-1}, p_n(f(g(u)))]\,ds\right),$$
$$(3.10.31)$$

for $t \in R_+$, *where* $(f(g(u)))(t) = f(t, g(u(t)))$, *then*

$$u(t) \leq a(t) + b(t)g^{-1}\left(\int_0^t F[s, p_1, p_2, \ldots, p_{n-1}, p_n(f(\alpha g(a\alpha^{-1})))]\right.$$

$$\times \exp\left(\int_s^t F[\tau, p_1, p_2, \ldots, p_{n-1}, p_n(k(\alpha g(a\alpha^{-1})))\right.$$

$$\left.\left.\times (\beta g(b\beta^{-1}))]\,d\tau\right)ds\right), \qquad (3.10.32)$$

for $t \in R_+$, where $(f(\alpha g(a\alpha^{-1})))(t) = f(t, \alpha(t)g(a(t)\alpha^{-1}(t)))$, $(kg(a\alpha^{-1})))(t) = k(t, \alpha(t)g(a(t)\alpha^{-1}(t)))$, $(\beta g(b\beta^{-1}))(t) = \beta(t)g(b(t)\beta^{-1}(t))$.

(b_3) Let w be as in (a_3) in Theorem 3.10.1. If

$$u(t) \leq a(t) + b(t)\int_0^t F[s, p_1, p_2, \ldots, p_{n-1}, p_n(f(w(u)))]\,ds, \qquad (3.10.33)$$

for $t \in R_+$, where $(f(w(u)))(t) = f(t, w(u(t)))$, then for $0 \leq t \leq t_2$,

$$u(t) \leq a(t) + b(t)G^{-1}\left[G(\bar{c}(t)) + \int_0^t F[s, p_1, p_2, \ldots, p_{n-1},\right.$$

$$\left.p_n(k(w(a)))w(b)]\,ds\right], \qquad (3.10.34)$$

where G, G^{-1} are as in (a_3) in Theorem 3.10.1, $(k(w(a)))(t) = k(t, w(a(t)))$, $w(b)(t) = w(b(t))$,

$$\bar{c}(t) = \int_0^t F[s, p_1, p_2, \ldots, p_{n-1}, p_n(f(w(a)))]\,ds, \qquad (3.10.35)$$

and $t_2 \in R_+$ is chosen so that

$$G(\bar{c}(t)) + \int_0^t F[s, p_1, p_2, \ldots, p_{n-1}, p_n(k(w(a)))w(b)]\,ds \in \text{Dom}(G^{-1}),$$

for all $t \in R_+$ lying in the interval $0 \leq t \leq t_2$. $\qquad \square$

Proof: (b_1) Define a function $z(t)$ by

$$z(t) = \int_0^t F[s, p_1, p_2, \ldots, p_{n-1}, p_n(fu)] \, ds. \tag{3.10.36}$$

From (3.10.36) and using the condition (3.10.1) and the facts that $u(t) \leq a(t) + b(t)z(t)$ and $z(t)$ is monotonic nondecreasing for $t \in R_+$, we observe that

$$\begin{aligned} z'(t) &\leq F[t, p_1, p_2, \ldots, p_{n-1}, p_n(f(a+bz))] \\ &= F[t, p_1, p_2, \ldots, p_{n-1}, p_n\{(f(a+bz)) - (f(a)) + (f(a))\}] \\ &\leq F[t, p_1, p_2, \ldots, p_{n-1}, p_n(fa)] \\ &\quad + F[t, p_1, p_2, \ldots, p_{n-1}, p_n(ka)b]z(t). \end{aligned} \tag{3.10.37}$$

The inequality (3.10.37) implies the estimate

$$z(t) \leq \int_0^t F[s, p_1, p_2, \ldots, p_{n-1}, p_n(fa)] \\ \times \exp\left(\int_s^t F[\tau, p_1, p_2, \ldots, p_{n-1}, p_n(ka)b] \, d\tau\right) ds. \tag{3.10.38}$$

Using (3.10.38) in $u(t) \leq a(t) + b(t)z(t)$ we get the desired inequality in (3.10.30).

The proofs of (b_2) and (b_3) can be completed by following the proofs of (a_2) and (a_3) of Theorem 3.10.1 and closely looking at the proof of (b_1) given above. The details are omitted here. ∎

The following theorem is established by Pachpatte (1994b).

Theorem 3.10.3 *Let $u \geq 0$, $f \geq 0$, $g \geq 0$, $r_i > 0$, for $i = 1, 2, \ldots, n-1$, be continuous functions defined on R_+ and c be a nonnegative constant.*

(c_1) *If*

$$u^2(t) \leq c^2 + 2M[t, r, f(t_n)u^2(t_n) + g(t_n)u(t_n)], \tag{3.10.39}$$

for $t \in R_+$, then
$$u(t) \leq p(t)\exp(M[t, r, f(t_n)]). \tag{3.10.40}$$

for $t \in R_+$, where
$$p(t) = c + M[t, r, g(t_n)], \tag{3.10.41}$$

for $t \in R_+$.

(c_2) Let w be as in (a_3) in Theorem 3.10.1. If
$$u^2(t) \leq c^2 + 2M[t, r, f(t_n)u(t_n)w(u(t_n)) + g(t_n)u(t_n)], \tag{3.10.42}$$

for $t \in R_+$, then for $0 \leq t \leq t_3$,
$$u(t) \leq G^{-1}[G(p(t)) + M[t, r, f(t_n)]], \tag{3.10.43}$$

where $p(t)$ is defined by (3.10.41), G, G^{-1} are as defined in (a_3) in Theorem 3.10.1 and $t_3 \in R_+$ is chosen so that
$$G(p(t)) + M[t, r, f(t_n)] \in \text{Dom}(G^{-1}),$$

for all $t \in R_+$ lying in the interval $0 \leq t \leq t_3$.

(c_3) Let $L: R_+^2 \to R_+$ be a continuous function satisfying the condition
$$0 \leq L(t, x) - L(t, y) \leq k(t, y)(x - y), \tag{3.10.44}$$

for $t \in R_+$ and $x \geq y \geq 0$, where $k: R_+^2 \to R_+$ is a continuous function. If
$$u^2(t) \leq c^2 + 2M[t, r, f(t_n)u(t_n)L(t_n, u(t_n)) + g(t_n)u(t_n)], \tag{3.10.45}$$

for $t \in R_+$, then
$$u(t) \leq p(t) + q(t)\exp(M[t, r, f(t_n)k(t_n, p(t_n))]), \tag{3.10.46}$$

for $t \in R_+$, where $p(t)$ is defined by (3.10.41) and
$$q(t) = M[t, r, f(t_n)L(t_n, p(t_n))], \tag{3.10.47}$$

for $t \in R_+$.

$\square$

Proof: It is sufficient to assume that c is positive, since the limiting argument can be used to treat the remaining case.

(c_1) Let $c > 0$ and define a function $z(t)$ by the right-hand side of (3.10.39). Then $z(0) = c^2$, $u(t) \leq \sqrt{z(t)}$ and

$$L_n z(t) \leq 2\sqrt{z(t)}[f(t)\sqrt{z(t)} + g(t)], \quad (3.10.48)$$

where L_i, $i = 0, 1, 2, \ldots, n$, are the operators defined in Section 1.9. From (3.10.48) and using the facts that $\sqrt{z(t)} > 0$, $(d/dt)z(t) \geq 0$, $L_{n-1}z(t) \geq 0$ for $t \in R_+$, we observe that

$$\frac{L_n z(t)}{\sqrt{z(t)}} \leq 2[f(t)\sqrt{z(t)} + g(t)] + \frac{1}{2}\frac{((d/dt)z(t))L_{n-1}z(t)}{(z(t))^{3/2}},$$

i.e.

$$\frac{d}{dt}\left(\frac{L_{n-1}z(t)}{\sqrt{z(t)}}\right) \leq 2[f(t)\sqrt{z(t)} + g(t)]. \quad (3.10.49)$$

By setting $t = t_n$ in (3.10.49) and integrating it from 0 to t and using the fact that $L_{n-1}z(0) = 0$ for $n = 2, 3, \ldots$, we obtain

$$\frac{(d/dt)(L_{n-2}z(t))}{\sqrt{z(t)}} \leq 2r_{n-1}(t)\int_0^t [f(t_n)\sqrt{z(t_n)} + g(t_n)]\,dt_n. \quad (3.10.50)$$

Again, as above from (3.10.50) we observe that

$$\frac{d}{dt}\left(\frac{L_{n-2}z(t)}{\sqrt{z(t)}}\right) \leq 2r_{n-1}(t)\int_0^t [f(t_n)\sqrt{z(t_n)} + g(t_n)]\,dt_n. \quad (3.10.51)$$

By setting $t = t_{n-1}$ in (3.10.51) and integrating from 0 to t and using the fact that $L_{n-2}z(0) = 0$, for $n = 3, 4, \ldots$, we get

$$\frac{L_{n-2}z(t)}{\sqrt{z(t)}} \leq 2\int_0^t r_{n-1}(t_{n-1})\int_0^{t_{n-1}} [f(t_n)\sqrt{z(t_n)} + g(t_n)]\,dt_n\,dt_{n-1}.$$

Continuing in this way, we obtain

$$\frac{(d/dt)(z(t))}{\sqrt{z(t)}} \leq 2r_1(t)\int_0^t r_2(t_2)\int_0^{t_2} r_3(t_3)\ldots\int_0^{t_{n-2}} r_{n-1}(t_{n-1})$$

$$\times \int_0^{t_{n-1}} [f(t_n)\sqrt{z(t_n)} + g(t_n)]\,dt_n\,dt_{n-1}\ldots dt_3\,dt_2.$$

$$(3.10.52)$$

By setting $t = t_1$ in (3.10.52) and integrating from 0 to t and using the fact that $\sqrt{z(0)} = c$, we have

$$\sqrt{z(t)} \leq p(t) + M[t, r, f(t_n)\sqrt{z(t_n)}]. \tag{3.10.53}$$

Since $p(t)$ is positive and monotonic nondecreasing in t, from (3.10.53) we observe that

$$\frac{\sqrt{z(t)}}{p(t)} \leq 1 + M\left[t, r, f(t_n)\frac{\sqrt{z(t_n)}}{p(t_n)}\right]. \tag{3.10.54}$$

Now an application of Theorem 1.9.1 part (a_1) yields

$$\frac{\sqrt{z(t)}}{p(t)} \leq \exp(M[t, r, f(t_n)]). \tag{3.10.55}$$

The required inequality in (3.10.40) follows by using (3.10.55) in $u(t) \leq \sqrt{z(t)}$.

(c_2) Let $c > 0$ and define a function $z(t)$ by the right-hand side of (3.10.42). Then $z(0) = c^2$, $u(t) \leq \sqrt{z(t)}$ and

$$L_n z(t) \leq 2\sqrt{z(t)}[f(t)w(\sqrt{z(t)}) + g(t)]. \tag{3.10.56}$$

Now by following the same steps as in the proof of part (c_1) below (3.10.48) up to (3.10.53) with suitable changes, we have

$$\sqrt{z(t)} \leq p(t) + M[t, r, f(t_n)w(\sqrt{z(t_n)})]. \tag{3.10.57}$$

The rest of the proof follows by the same arguments as in the proof of (a_3) in Theorem 3.10.1, below the inequality (3.10.21), with suitable modifications. We omit the further details.

(c_3) Let $c > 0$ and define a function $z(t)$ by the right-hand side of (3.10.45). Then $z(0) = c^2$, $u(t) \leq \sqrt{z(t)}$ and

$$L_n z(t) \leq 2\sqrt{z(t)}[f(t)L(t, \sqrt{z(t)}) + g(t)]. \tag{3.10.58}$$

Now by following the same steps as in the proof of part (c_1) below (3.10.48) up to (3.10.53) we have

$$\sqrt{z(t)} \leq p(t) + M[t, r, f(t_n)L(t_n, \sqrt{z(t_n)})]. \tag{3.10.59}$$

Define a function $v(t)$ by

$$v(t) = M[t, r, f(t_n)L(t_n, \sqrt{z(t_n)})]. \tag{3.10.60}$$

From (3.10.60) and using the fact that $\sqrt{z(t)} \leq p(t) + v(t)$, and the condition (3.10.44) we have

$$\begin{aligned} L_n v(t) &= f(t) L(t, \sqrt{z(t)}) \\ &\leq f(t) L(t, p(t) + v(t)) \\ &= f(t)[L(t, p(t) + v(t)) - L(t, p(t))] + f(t) L(t, p(t)) \\ &\leq f(t) k(t, p(t)) v(t) + f(t) L(t, p(t)). \end{aligned} \quad (3.10.61)$$

From (3.10.61) we can very easily obtain

$$v(t) \leq q(t) + M[t, r, f(t_n) k(t_n, p(t_n)) v(t_n)], \quad (3.10.62)$$

where $q(t)$ is defined by (3.10.47). It is sufficient to assume that $q(t)$ is positive, since the standard limiting argument can be used to treat the case when $q(t)$ is nonnegative. Since $q(t)$ is monotonic nondecreasing for $t \in R_+$ and positive, from (3.10.62) we observe that

$$\frac{v(t)}{q(t)} \leq 1 + M \left[t, r, f(t_n) k(t_n, p(t_n)) \frac{v(t_n)}{q(t_n)} \right]. \quad (3.10.63)$$

Now an application of Theorem 1.9.1 part (a_1) yields

$$v(t) \leq q(t) \exp(M[t, r, f(t_n) k(t_n, p(t_n))]). \quad (3.10.64)$$

The desired inequality in (3.10.46) now follows by using (3.10.64) in (3.10.59) and the fact that $u(t) \leq \sqrt{z(t)}$.

∎

The following two theorems were established by Pachpatte (1994a).

Theorem 3.10.4 *Let $u \geq 0$, $g \geq 0$, $r_i > 0$, for $i = 1, 2, \ldots, n-1$, be continuous functions defined on R_+ and let $p > 1$ be a constant. If*

$$u^p(t) \leq c + M[t, r, g(t_n) u(t_n)], \quad (3.10.65)$$

for $t \in R_+$, where $c \geq 0$ is a constant, then

$$u(t) \leq [(c)^{(p-1)/p} + ((p-1)/p) M[t, r, g(t_n)]]^{1/(p-1)}, \quad (3.10.66)$$

for $t \in R_+$.

□

Proof: We first assume that $c > 0$ and define a function $z(t)$ by the right-hand side of (3.10.65). Then $z(0) = c$, $u(t) \leq \sqrt[p]{z(t)}$ and

$$L_n z(t) = g(t) u(t) \leq g(t) \sqrt[p]{z(t)}. \tag{3.10.67}$$

From (3.10.67) and using the facts that $z(t) > 0$, $(d/dt)[\sqrt[p]{z(t)}] \geq 0$ and $L_{n-1} z(t) \geq 0$ for $t \in R_+$, we observe that

$$\frac{L_n z(t)}{\sqrt[p]{z(t)}} \leq g(t) + \frac{[(d/dt)(\sqrt[p]{z(t)})] L_{n-1} z(t)}{[\sqrt[p]{z(t)}]^2},$$

i.e.

$$\frac{d}{dt}\left[\frac{L_{n-1} z(t)}{\sqrt[p]{z(t)}}\right] \leq g(t). \tag{3.10.68}$$

By setting $t = t_n$ in (3.10.68), integrating from 0 to t and using the fact that $L_{n-1} z(0) = 0$ for $n = 2, 3, \ldots$, we obtain

$$\frac{L_{n-1} z(t)}{\sqrt[p]{z(t)}} \leq \int_0^t g(t_n) \, dt_n,$$

which implies

$$\frac{(d/dt) L_{n-2} z(t)}{\sqrt[p]{z(t)}} \leq r_{n-1}(t) \int_0^t g(t_n) \, dt_n. \tag{3.10.69}$$

Again, as above from (3.10.69), we observe that

$$\frac{d}{dt}\left[\frac{L_{n-2} z(t)}{\sqrt[p]{z(t)}}\right] \leq r_{n-1}(t) \int_0^t g(t_n) \, dt_n. \tag{3.10.70}$$

By setting $t = t_{n-1}$ in (3.10.70) and integrating it from 0 to t and using the fact that $L_{n-2} z(0) = 0$ for $n = 3, 4, \ldots$, we get

$$\frac{L_{n-2} z(t)}{\sqrt[p]{z(t)}} \leq \int_0^t r_{n-1}(t_{n-1}) \int_0^{t_{n-1}} g(t_n) \, dt_n.$$

Continuing in this way we obtain

$$\frac{(d/dt) z(t)}{\sqrt[p]{z(t)}} \leq r_1(t) \int_0^t r_2(t_2) \int_0^{t_2} r_3(t_3) \ldots \int_0^{t_{n-2}} r_{n-1}(t_{n-1})$$

$$\times \int_0^{t_{n-1}} g(t_n) \, dt_n \, dt_{n-1} \ldots dt_3 \, dt_2. \tag{3.10.71}$$

By setting $t = t_1$ in (3.10.71) and integrating from 0 to t we have

$$[\sqrt[p]{z(t)}]^{p-1} - [\sqrt[p]{z(0)}]^{p-1} \leq ((p-1)/p)M[t, r, g(t_n)]. \quad (3.10.72)$$

From (3.10.72) and using the fact that $u(t) \leq \sqrt[p]{z(t)}$, we obtain the required inequality in (3.10.66). The proof of the case when $c \geq 0$ can be completed by following the limiting arguments.

∎

Theorem 3.10.5 *Let $u \geq 0$, $v \geq 0$, $r_i > 0$ for $i = 1, 2, \ldots, n-1$, $h_j \geq 0$ for $j = 1, 2, 3, 4$, be continuous functions defined on R_+ and let $p > 1$ be a constant. If c_1, c_2 and μ are nonnegative constants and*

$$u^p(t) \leq c_1 + M[t, r, h_1(t_n)u(t_n)] + M[t, r, h_2(t_n)\bar{v}(t_n)], \quad (3.10.73)$$

$$v^p(t) \leq c_2 + M[t, r, h_3(t_n)\bar{u}(t_n)] + M[t, r, h_4(t_n)v(t_n)], \quad (3.10.74)$$

for $t \in R_+$, where $\bar{u}(t) = \exp(-p\mu t)u(t)$ and $\bar{v}(t) = \exp(p\mu t)v(t)$ for $t \in R_+$, then

$$u(t) \leq \exp(\mu t)[(2^{p-1}(c_1 + c_2))^{(p-1)/p}$$
$$+ 2^{p-1}((p-1)/p)M[t, r, h(t_n)]]^{1/(p-1)}, \quad (3.10.75)$$

$$v(t) \leq [(2^{p-1}(c_1 + c_2))^{(p-1)/p}$$
$$+ 2^{p-1}((p-1)/p)M[t, r, h(t_n)]]^{1/(p-1)}, \quad (3.10.76)$$

for $t \in R_+$, where

$$h(t) = \max\{[h_1(t) + h_3(t)], [h_2(t) + h_4(t)]\}, \quad (3.10.77)$$

for $t \in R_+$.

□

Proof: We first multiply (3.10.73) by $\exp(-p\mu t)$ and observe that

$$\exp(-p\mu t)u^p(t) \leq c_1 + M[t, r, h_1(t_n)\bar{u}(t_n)] + M[t, r, h_2(t_n)v(t_n)]. \quad (3.10.78)$$

Define

$$F(t) = \exp(-\mu t)u(t) + v(t). \quad (3.10.79)$$

By taking the pth power on both sides of (3.10.79) and using the elementary inequality $(d_1 + d_2)^q \leq 2^{q-1}(d_1^q + d_2^q)$, where d_1, d_2 are nonnegative reals and $q > 1$, and (3.10.78) and (3.10.74), we observe that

$$F^p(t) \leq 2^{p-1}[\exp(-p\mu t)u^p(t) + v^p(t)]$$
$$\leq 2^{p-1}[(c_1 + c_2) + M[t, r, [h_1(t_n) + h_3(t_n)]\bar{u}(t_n)]$$
$$+ M[t, r, [h_2(t_n) + h_4(t_n)]v(t_n)]]. \qquad (3.10.80)$$

Now by using the fact that $\exp(-p\mu t) \leq \exp(-\mu t)$ and (3.10.77) in (3.10.80) we observe that

$$F^p(t) \leq 2^{p-1}(c_1 + c_2) + M[t, r, 2^{p-1}h(t_n)F(t_n)]. \qquad (3.10.81)$$

The bounds in (3.10.75) and (3.10.76) follow from an application of Theorem 3.10.4 to (3.10.81) and splitting.

∎

In concluding, we note that the results in this section are formulated corresponding to the iterated integral defined by $M[t, r, h]$ only. Similar results can be formulated corresponding to the other iterated integrals $A[t, g, h]$, $B[t, g, h]$ and $E[t, g, h]$ as defined in Section 1.9. The formulations of such results are left for the reader to fill in where needed.

3.11 Applications

This section presents applications of some of the inequalities given earlier to study the qualitative behaviour of the solutions of certain differential, integral and integro-differential equations. Many of the inequalities given here are recently developed, and it is hoped that these inequalities will have a significant impact on the development of the theory of certain new classes of differential and integral equations.

3.11.1 Volterra Integral Equations

This section considers the following Volterra integral equation

$$x(t) = g(t) + \int_\alpha^t F(t, s, x(s))\, ds, \qquad \text{(V)}$$

where $g: I \to R^n$, $F: I^2 \times R^n \to R^n$ are continuous functions, $I = [\alpha, \beta)$ and R^n is the n-dimensional Euclidean space with convenient norm $|\cdot|$. Further, we shall assume that the integral equation (V) has a solution in $C(I, R^n)$. Dragomir (1987a) used his inequality given in Theorem 3.2.1 to study the stability, boundedness and asymptotic behaviour of the solutions of equation (V). In the following two theorems we will present the results given by Dragomir (1987a), which provide explicit bounds on the solutions of equation (V).

Theorem 3.11.1 *Suppose that F satisfies the condition*

$$|F(t, s, x)| \leq b(t) L(s, |x|), \qquad (3.11.1)$$

for $t, s \in I$, $x \in R^n$, where $b(t)$ is a nonnegative continuous function defined on I, $L: I \times R_+ \to R_+$ be continuous and

$$0 \leq L(t, u) - L(t, v) \leq M(t, v)(u - v), \qquad (L)$$

for $t \in I$, $u \geq v \geq 0$, in which $M: I \times R_+ \to R_+$ is a continuous function. If $x \in C(I, R^n)$ is a solution of the equation (V), then

$$|x(t)| \leq |g(t)| + b(t) \int_\alpha^t L(s, |g(s)|) \exp\left(\int_s^t M(\sigma, |g(\sigma)|) b(\sigma) \, d\sigma\right) ds, \qquad (3.11.2)$$

for $t \in I$.

$\square$

Proof: Let $x \in C(I, R^n)$ be a solution of (V). Then we have

$$|x(t)| \leq |g(t)| + \int_\alpha^t |F(t, s, x(s))| \, ds, \qquad (3.11.3)$$

Using (3.11.1) in (3.11.3) we get

$$|x(t)| \leq |g(t)| + b(t) \int_\alpha^t L(s, |x(s)|) \, ds, \qquad (3.11.4)$$

for $t \in I$. Now an application of Theorem 3.2.1 gives the desired estimation in (3.11.2).

∎

The following theorem is established next, which gives the estimate for the solution of equation (V) by assuming that the function F in (V) satisfies the Lipschitz-type condition.

Theorem 3.11.2 *Suppose that the function F satisfies the condition*

$$|F(t, s, x) - F(t, s, y)| \leq b(t)L(s, |x - y|), \qquad (3.11.5)$$

for $t \in I$, $x, y \in R^n$, where b and L are as in Theorem 3.11.1 with L satisfying the condition (L) and M being the same function as given therein. If $x \in C(I, R^n)$ is any solution of (V), then

$$|x(t) - g(t)| \leq k(t) + b(t) \int_\alpha^t L(s, k(s)) \exp\left(\int_s^t M(\sigma, k(\sigma))b(\sigma)\,d\sigma\right) ds, \qquad (3.11.6)$$

for $t \in I$, where

$$k(t) = \int_\alpha^t |F(t, s, g(s))|\,ds, \qquad (3.11.7)$$

for $t \in I$.

□

Proof: Let $x \in C(I, R^n)$ be a solution of (V). Then we have

$$|x(t) - g(t)| \leq \int_\alpha^t |F(t, s, x(s))|\,ds$$

$$= \int_\alpha^t |F(t, s, x(s)) - F(t, s, g(s)) + F(t, s, g(s))|\,ds$$

$$\leq \int_\alpha^t [|F(t, s, x(s)) - F(t, s, g(s))| + |F(t, s, g(s))|]\,ds$$

$$\leq k(t) + b(t) \int_\alpha^t L(s, |x(s) - g(s)|)\,ds,$$

for $t \in I$. Now an application of Theorem 3.2.1 gives the required estimation in (3.11.6).

■

As an application of Theorem 3.3.4 given by Pachpatte (in press k), we obtain a bound on the solution of the Volterra equation of the form

$$x(t) = g(t) + \int_\alpha^t F(t, s, x(s), h(s, x(s)))\,ds, \qquad (3.11.8)$$

where $x, g: I \to R^n$, $h: I \times R^n \to R^n$, $F: I^2 \times R^n \times R^n \to R^n$ are continuous functions. We assume that

$$|g(t)| \leq a(t), \qquad (3.11.9)$$

$$|F(t, s, x, z)| \leq b(t)k(t, s)[c(s)|x| + |z|], \qquad (3.11.10)$$

$$|h(t, x)| \leq L(t, |x|), \qquad (3.11.11)$$

where a, b, c, k, L and M are as in Theorem 3.3.4. From (3.11.8)–(3.11.11) we observe that

$$|x(t)| \leq a(t) + b(t)\int_\alpha^t k(t, s)[c(s)|x(s)| + L(s, |x(s)|)]\,ds. \qquad (3.11.12)$$

Now an application of Theorem 3.3.4 to (3.11.12) yields

$$|x(t)| \leq a(t) + b(t)\int_\alpha^t A(\sigma)\exp\left(\int_\sigma^t B(\tau)\,d\tau\right)d\sigma, \qquad (3.11.13)$$

where $A(t)$ and $B(t)$ are defined by (3.3.16) and (3.3.17) in Theorem 3.3.4. The inequality (3.11.13) gives the bound on the solution $x(t)$ of equation (3.11.8) in terms of the known functions.

3.11.2 On Some Epidemic Models

This section considers the following integral equation

$$x(t) = k\left(p(t) - \int_0^t A(t-s)x(s)\,ds\right)\left(f(t) + \int_0^t a(t-s)x(s)\,ds\right). \qquad (D)$$

Gripenberg (1981) studied the qualitative behaviour of solutions of equation (D). This equation arises in the study of the spread of an infectious disease that does not induce permanent immunity. For detailed meanings of the various functions arising in (D), see Gripenberg (1980) and also Bailey (1975) and Diekmann (1979). Gripenberg (1981) studied the existence of a unique bounded continuous and nonnegative solution of (D) for $t \in R_+$, under appropriate assumptions on A and a and also obtained sufficient conditions for the convergence of the solution of a limit when $t \to \infty$. Aside from the various physical meanings of the functions arising in equation (D), we believe that equations like (D) are of great interest in their own right.

Pachpatte (1995d) used the inequality established in Theorem 3.7.3 to study the certain properties of the solutions of equation (D). In the following theorems we present the results related to the equation (D) given in Pachpatte (1995d). In what follows, we assume that the functions x, p, f, A and a in (D) are real-valued, continuous and defined on R_+, and k is a positive real constant and restrict our consideration to the solutions of (D) which exist on R_+.

Theorem 3.11.3 *Assume that*

$$|p(t)| \leq c_1, \quad |f(t)| \leq c_2, \qquad (3.11.14)$$

$$|A(t-s)| \leq M_1 q(s), \quad |a(t-s)| \leq M_2 h(s), \qquad (3.11.15)$$

for $0 \leq s \leq t$, $s, t \in R_+$ where c_1, c_2, M_1, M_2 are positive real constants, and q, h are real-valued nonnegative continuous functions defined on R_+. If $H_1(t) = 1 - c_1 c_2 k \int_0^t R_1(s) Q_1(s) \, ds > 0$ for all $t \in R_+$, where

$$R_1(t) = M_2 h(t) \int_0^t M_1 k q(\tau) \, d\tau + M_1 k q(t) \int_0^t M_2 h(\tau) \, d\tau, \quad (3.11.16)$$

$$Q_1(t) = \exp\left(\int_0^t [c_1 k M_2 h(\tau) + c_2 M_1 k q(\tau)] \, d\tau\right), \qquad (3.11.17)$$

for all $t \in R_+$ and

$$E_1(t) = (1/H_1(t)) c_1 c_2 k Q_1(t) < \infty, \qquad (3.11.18)$$

for all $t \in R_+$, then every solution $x(t)$ of (D) existing on R_+ is bounded.

□

Proof: From (D) and using the hypotheses (3.11.14), (3.11.15) we have

$$|x(t)| \leq \left(c_1 k + \int_0^t M_1 k q(s)|x(s)|\, ds\right)\left(c_2 + \int_0^t M_2 h(s)|x(s)|\, ds\right). \tag{3.11.19}$$

Now an application of Theorem 3.7.3 to (3.11.19) yields

$$|x(t)| \leq E_1(t), \tag{3.11.20}$$

for all $t \in R_+$. In view of the hypothesis (3.11.18), the estimation in (3.11.20) implies the boundedness of the solution $x(t)$ of (D) on R_+.

∎

Theorem 3.11.4 *Assume that*

$$|p(t)| \leq c_1 \exp(-\mu t), \quad |f(t)| \leq c_2 \exp(-\mu t), \tag{3.11.21}$$

$$|A(t-s)| \leq M_1 q(s) \exp(-\mu(t-2s)), \quad |a(t-s)| \leq M_2 h(s) \exp(-\mu(t-2s)), \tag{3.11.22}$$

for $0 \leq s \leq t$, $t, s \in R_+$, where c_1, c_2, M_1, M_2, q, h are as defined in Theorem 3.11.3 and $\mu \geq 0$ is a real constant. If (3.11.18) holds, then all solutions of (D) approach zero as $t \to \infty$.

□

Proof: From (D) and using the hypotheses (3.11.21) and (3.11.22) we observe that

$$|x(t)| \leq \exp(-2\mu t)\left(c_1 k + \int_0^t M_1 k q(s)|x(s)|\exp(2\mu s)\, ds\right)$$

$$\times \left(c_2 + \int_0^t M_2 h(s)|x(s)|\exp(2\mu s)\, ds\right). \tag{3.11.23}$$

Multiplying both sides of (3.11.23) by $\exp(2\mu t)$, applying Theorem 3.7.3 with $u(t) = |x(t)|\exp(2\mu t)$, and then multiplying the resulting inequality by $\exp(-2\mu t)$, we obtain

$$|x(t)| \leq E_1(t)\exp(-2\mu t), \tag{3.11.24}$$

for all $t \in R_+$. In view of the hypothesis (3.11.18), the inequality (3.11.24) yields the desired result and the proof is complete.

∎

Theorem 3.11.5 *Assume that*

$$|p(t)| \leq c_1 \exp(\mu t), \quad |f(t)| \leq c_2 \exp(\mu t), \quad (3.11.25)$$

$$|A(t-s)| \leq M_1 q(s) \exp(\mu(t-2s)),$$

$$|a(t-s)| \leq M_2 h(s) \exp(\mu(t-2s)), \quad (3.11.26)$$

for $0 \leq s \leq t$, s, $t \in R_+$, where c_1, c_2, M_1, M_2, μ, q and h are as defined in Theorem 3.11.4. If (3.11.18) holds for all $t \in R_+$, then all solutions of (D) *are slowly growing.*

□

The proof of this theorem can be completed by following the arguments used in the proof of Theorem 3.11.4 with suitable modifications. Hence the details are omitted.

3.11.3 Certain Integral and Differential Equations

This section first considers the nonlinear integral equation of the form

$$u^p(t) = f(t) + \int_0^t k(t,s) g(s, u(s)) \, ds, \quad (E)$$

where $f: R_+ \to R$, $k: R_+ \times R_+ \to R$, $g: R_+ \times R \to R$ are continuous functions and $p > 1$ is a constant. Okrasinski (1980) studied the problem of existence and uniqueness of the solutions of the variant of equation (E) written in the form

$$u^p = k * u + L, \quad p > 1, \quad (3.11.27)$$

where k, L are known smooth functions depending on physical parameters and the convolution on the right-hand side is well defined. For an interesting discussion concerning the occurrence of equation (3.11.27) in the theory of water percolation phenomena and its physical meaning, see Okrasinski (1980) and some of the references cited therein. Pachpatte (1994a) studied the boundedness and asymptotic behaviour of the solutions of equation (E) by using the inequality investigated by himself in Theorem 3.10.4. Here the results given in Pachpatte (1994a) are presented. In what follows it is assumed that every solution $u(t)$ of (E) under discussion exists on R_+.

We assume that the functions f, k and g in (E) satisfy the conditions

$$|f(t)| \leq c_1, \quad |k(t,s)| \leq c_2, \quad |g(t,u)| \leq r(t)|u|, \quad (3.11.28)$$

where c_1, c_2 are nonnegative constants and $r: R_+ \to R_+$ is a continuous function. From (E) and using (3.11.28) it is easy to observe that

$$|u(t)|^p \leq c_1 + \int_0^t c_2 r(s)|u(s)| \, ds. \quad (3.11.29)$$

Now an application of Theorem 3.10.4 with $n = 1$ yields

$$|u(t)| \leq \left[(c_1)^{(p-1)/p} + ((p-1)/p) \int_0^t c_2 r(s) \, ds \right]^{1/(p-1)}. \quad (3.11.30)$$

The right-hand side of (3.11.30) gives the bound on the solution $u(t)$ of (E) in terms of the known quantities.

We now consider the equation (E) under the following conditions on the functions f, k and g in (E):

$$|f(t)| \leq c_1 e^{-pt}, \quad |k(t,s)| \leq h(t) e^{-pt}, \quad |g(t,u)| \leq r(t)|u|, \quad (3.11.31)$$

where c_1 and $r(t)$ are as defined above, $h: R_+ \to R_+$ is a continuous function and

$$\int_0^\infty h(s) r(s) e^{-s} \, ds < \infty. \quad (3.11.32)$$

From (E) and using (3.11.31) it is easy to observe that

$$(e^t |u(t)|)^p \leq c_1 + \int_0^t h(s) r(s) e^{-s} (e^s |u(s)|) \, ds. \quad (3.11.33)$$

Now an application of Theorem 3.10.4 with $n = 1$ yields

$$e^t |u(t)| \leq \left[(c_1)^{(p-1)/p} + ((p-1)/p) \int_0^t h(s) r(s) e^{-s} \, ds \right]^{1/(p-1)},$$

$$(3.11.34)$$

From (3.11.34) and (3.11.32) we observe that

$$|u(t)| \leq c^* e^{-t}, \tag{3.11.35}$$

where $c^* \geq 0$ is a constant. From (3.11.35), we see that the solution $u(t)$ of equation (E) approaches zero as $t \to \infty$.

Next, consider the following second-order differential equation of the form

$$(a(t)\psi(u(t))u'(t))' + r(t)u(t) = 0, \tag{F}$$

with the given initial conditions

$$u(0) = c_1, \quad u'(0) = c_2, \tag{F_0}$$

where $a(t)$, $r(t)$ are real-valued continuous functions defined on R_+, $a(t) > 0$ for $t \in R_+$, $\psi(u) > 0$ for $u \neq 0$ (defined by $\psi(u) = u^{p-1}$ for $p > 1$ a fixed real number) and $c_1 > 0$, $c_2 \geq 0$ are real constants. In the past few years, many results have appeared in the literature which deal with the various properties of the solutions of equation (F) and its more general versions; see Bobisud (1992), Graef and Spikes (1987) and references given therein. The problem of existence of a solution for the more general version of equation (F) with given boundary conditions is studied by Bobisud (1992). Pachpatte (1994a) obtained an explicit bound on the solution of equation (F) by using the inequality given in Theorem 3.10.4. The results obtained by Pachpatte (1994a) related to the equation (F) are presented here. In what follows it is assumed that every solution $u(t)$ of (F) with (F_0) exists on R_+ and is nontrivial.

Integrating (F) twice from 0 to t and using the initial conditions in (F_0) we see that the problem (F)–(F_0) is equivalent to the following integral equation

$$u^p(t) = c_1^p + pa(0)c_1^{p-1}c_2 \int_0^t \frac{1}{a(s)}\,ds - p \int_0^t \frac{1}{a(s)} \left(\int_0^s r(\tau)u(\tau)\,d\tau \right) ds. \tag{3.11.36}$$

We suppose that the function $a(t)$ in (F) satisfies the condition

$$\int_0^\infty \frac{1}{a(s)}\,ds < \infty. \tag{3.11.37}$$

From (3.11.36) and (3.11.37) we observe that

$$|u(t)|^p \le c^* + \int_0^t \frac{p}{a(s)} \left(\int_0^s |r(\tau)||u(\tau)|\, d\tau \right) ds, \qquad (3.11.38)$$

where $c^* > 0$ is a constant. Now a suitable application of Theorem 3.10.4 with $n = 2$ yields

$$|u(t)| \le \left[(c^*)^{(p-1)/p} + ((p-1)/p) \int_0^t \frac{p}{a(s)} \left(\int_0^s |r(\tau)|\, d\tau \right) ds \right]^{1/(p-1)}. \qquad (3.11.39)$$

The right-hand side in (3.11.39) gives the estimate on the solution $u(t)$ of (F)–(F$_0$) in terms of the known quantities.

Note that the inequality established in Theorem 3.10.5 can be used to obtain the estimate on the solution of the following system of integral equations

$$u^p(t) = f_1(t) + \int_0^t k_1(t,s) g_1(s, u(s), v(s))\, ds, \qquad (3.11.40)$$

$$v^p(t) = f_2(t) + \int_0^t k_2(t,s) g_2(s, u(s), v(s))\, ds, \qquad (3.11.41)$$

under some suitable conditions on the functions involved in (3.11.40)–(3.11.41) where $p > 1$ is a constant. Furthermore, we also note that Theorem 3.10.5 can be used to obtain the estimate on the solution of the system of differential equations

$$(a_1(t)\psi(u(t))u'(t))' + q_1(t)u(t) + q_2(t)v(t) = 0, \qquad (3.11.42)$$

$$(a_2(t)\psi(v(t))v'(t))' + q_3(t)u(t) + q_4(t)v(t) = 0, \qquad (3.11.43)$$

with the given initial conditions

$$u(0) = c_1, \quad u'(0) = c_2, \qquad (3.11.44)$$

$$v(0) = c_3, \quad v'(0) = c_4, \qquad (3.11.45)$$

by using suitable conditions on the functions involved in (3.11.42)–(3.11.43), where $c_1 > 0$, $c_2 \ge 0$, $c_3 > 0$, $c_4 \ge 0$ are constants and ψ is as defined above.

3.11.4 Certain Integro-differential and Differential Equations

In this section we present the applications of the inequalities established by Pachpatte (1995a, 1996a) in Theorems 3.6.1 and 3.7.5 to obtain bounds on the solutions of certain integro-differential and differential equations given in Pachpatte (1995a, 1996a).

As a first application we obtain a bound on the solution of the integro-differential equation

$$x'(t) - F\left(t, x(t), \int_0^t k(t, s, x(s))\,ds\right) = h(t), \quad x(0) = x_0, \qquad (A)$$

where $h: R_+ \to R$, $k: R_+^2 \times R \to R$, $F: R_+ \times R^2 \to R$ are continuous functions. Here we assume that the solution $x(t)$ of (A) exists on R_+. Multiplying both sides of equation (A) by $x(t)$, substituting $t = s$ and then integrating from 0 to t we have

$$x^2(t) = x_0^2 + 2\int_0^t \left[x(s)F\left(s, x(s), \int_0^s k(s, \tau, x(\tau))\,d\tau\right) + h(s)x(s)\right]ds. \tag{3.11.46}$$

We assume that

$$|k(t, s, x(s))| \leq f(t)g(s)|x(s)|, \tag{3.11.47}$$

$$|F(t, x(t), v)| \leq f(t)|x(t)| + |v|, \tag{3.11.48}$$

where f and g are real-valued nonnegative continuous functions defined on R_+. From (3.11.46)–(3.11.48) we observe that

$$|x(t)|^2 \leq |x_0|^2 + 2\int_0^t \left[f(s)|x(s)|\left(|x(s)| + \int_0^s g(\tau)|x(\tau)|\,d\tau\right)\right.$$

$$\left. + |h(s)||x(s)|\right]ds. \tag{3.11.49}$$

Now an application of the inequality given in Theorem 3.6.1 part (a_1) yields

$$|x(t)| \leq p_1(t)\left[1 + \int_0^t f(s)\exp\left(\int_0^s [f(\tau) + g(\tau)]\,d\tau\right)ds\right], \tag{3.11.50}$$

where

$$p_1(t) = |x_0| + \int_0^t |h(s)|\, ds,$$

for $t \in R_+$. The inequality (3.11.50) gives the bound on the solution $x(t)$ of equation (A) in terms of the known functions.

As a second application we obtain bounds on the solutions of differential equations of the form

$$x'' = f(t, x, x'), \qquad \text{(P)}$$

with the given initial conditions

$$x(t_0) = k, \quad x'(t_0) = c, \qquad (\text{P}_0)$$

where $f: I \times R \times R \to R$ is a continuous function and k, c are real constants, $I = [t_0, \infty)$, $t_0 \geq 0$. Here we assume that the solution $x(t)$ of (P)–(P$_0$) exists on R_+. Multiplying both sides of (P) by $x'(t)$, substituting $t = s$ and then integrating it from t_0 to t and using (P$_0$) we have

$$(x'(t))^2 = c^2 + 2 \int_{t_0}^t x'(s) f(s, x(s), x'(s))\, ds. \qquad (3.11.51)$$

We assume that the function f in (P) satisfies the condition

$$|f(t, x, x')| \leq a(t)(|x| + |x'|), \qquad (3.11.52)$$

or

$$|f(t, x, x')| \leq a(t) L(t, |x|, |x'|), \qquad (3.11.53)$$

where $a(t)$ is a real-valued nonnegative continuous function defined on I and $L: I \times R_+ \times R_+ \to R_+$ is a continuous function satisfying the condition

$$0 \leq L(t, u, v) - L(t, \bar{u}, \bar{v}) \leq M(t, \bar{u}, \bar{v})[(u - \bar{u}) + (v - \bar{v})], \qquad \text{(L)}$$

for $t \in I$ and $u \geq \bar{u} \geq 0$, $v \geq \bar{v} \geq 0$, where $M: I \times R_+ \times R_+ \to R_+$ is a continuous function. From (3.11.51) and (3.11.52) we observe that

$$|x'(t)|^2 \leq |c|^2 + 2 \int_{t_0}^t a(s) |x'(s)| (|x(s)| + |x'(s)|)\, ds. \qquad (3.11.54)$$

Now an application of Theorem 3.7.5 part (c_1) yields

$$|x(t)| \leq |k| + \int_{t_0}^{t}\left[|c| + (|k| + |c|)\int_{t_0}^{s} a(\tau)\exp\left(\int_{t_0}^{\tau}[1 + a(\xi)]\,d\xi\right)d\tau\right]ds, \quad (3.11.55)$$

for $t \in I$. Furthermore, from (3.11.51) and (3.11.53) we observe that

$$|x'(t)|^2 \leq |c|^2 + 2\int_{t_0}^{t} a(s)|x'(s)|L(s, |x(s)|, |x'(s)|)\,ds. \quad (3.11.56)$$

Now an application of the inequality given in Theorem 3.7.5 part (c_2) yields

$$|x(t)| \leq |k| + |c|\int_{t_0}^{t} \exp\left(\int_{t_0}^{s}[1 + a(\tau)M(\tau, |k|, |c|)]\,d\tau\right)ds$$

$$+ \int_{t_0}^{t}\int_{t_0}^{s} a(\tau)L(\tau, |k|, |c|)$$

$$\times \exp\left(\int_{\tau}^{s}\left[1 + a(\xi)M(\xi, |k|, |c|)\right]d\xi\right)d\tau\,ds, \quad (3.11.57)$$

for $t \in I$. The inequalities (3.11.55) and (3.11.57) gives us the bounds on the solutions $x(t)$ of equations (P)–(P_0) in terms of the known functions.

3.12 Miscellaneous Inequalities

3.12.1 Dragomir (1987a)

Let $u, a, b: I \to R_+$ be continuous functions, where $I = [\alpha, \beta)$. Let $D: I \times R_+ \to R_+$ be a continuous function such that D is differentiable on $(\alpha, \beta) \times (0, \infty)$, $(\partial/\partial x)D(t, x)$ is nonnegative on $(\alpha, \beta) \times (0, \infty)$ and there exists a continuous function $P: I \times R_+ \to R_+$ such that $(\partial/\partial x)D(t, u) \leq P(t, v)$ for any $t \in (\alpha, \beta)$ and $u \geq v \geq 0$. If

$$u(t) \leq a(t) + b(t)\int_{\alpha}^{t} D(s, u(s))\,ds,$$

for all $t \in I$, then

$$u(t) \leq a(t) + b(t) \int_\alpha^t D(s, a(s)) \exp\left(\int_s^t P(\sigma, a(\sigma))b(\sigma)\, d\sigma\right) ds,$$

for all $t \in I$.

3.12.2 Pachpatte (in press k)

Let $u, a, b, c, k(t, s), (\partial/\partial t)k(t, s), L$ and M be as in Theorem 3.3.4. Let $g(u)$ be a nonnegative continuous strictly increasing convex and submultiplicative function for $u > 0$, $\lim_{u \to \infty} g(u) = \infty$, g^{-1} denote the inverse function of g, $\alpha(t), \beta(t)$ be continuous and positive functions on R_+ and $\alpha(t) + \beta(t) = 1$. If

$$u(t) \leq a(t) + b(t)g^{-1}\left(\int_0^t k(t, s)[c(s)g(u(s)) + L(s, g(u(s)))]\, ds\right),$$

for all $t \in R_+$, then

$$u(t) \leq a(t) + b(t)g^{-1}\left(\int_0^t \overline{A}(\sigma) \exp\left(\int_\sigma^t \overline{B}(\tau)\, d\tau\right) d\sigma\right),$$

for all $t \in R_+$, where $\overline{A}(t)$ and $\overline{B}(t)$ are defined by the right sides of (3.3.16) and (3.3.17) in Theorem 3.3.4 by replacing $a(t)$ by $\alpha(t)g(a(t)\alpha^{-1}(t))$ and $b(t)$ by $\beta(t)g(b(t)\beta^{-1}(t))$ for $t \in R_+$.

3.12.3 Pachpatte (1996d)

Let u, f and g be nonnegative continuous functions defined on R_+ and $c \geq 0$ be a constant. Let $L: R_+^2 \to R_+$ be a continuous function which satisfies the condition

$$0 \leq L(t, x) - L(t, y) \leq k(t, y)(x - y), \tag{L}$$

for $t \in R_+$ and $x \geq y \geq 0$, where $k: R_+^2 \to R_+$ is a continuous function.

(i) If

$$u^2(t) \leq c^2 + 2\int_0^t u(s)[f(s)u(s) + g(s)L(s, u(s))]\,ds,$$

for $t \in R_+$, then

$$u(t) \leq c\exp\left(\int_0^t [f(s) + g(s)k(s, c)]\,ds\right)$$

$$+ \int_0^t g(s)L(s, c)\exp\left(\int_s^t [f(\sigma) + g(\sigma)k(\sigma, c)]\,d\sigma\right)ds,$$

for $t \in R_+$.

(ii) If

$$u^2(t) \leq c^2 + 2\int_0^t u(s)[f(s)u(s) + g(s)u^p(s)]\,ds,$$

for $t \in R_+$, where $0 \leq p < 1$ is a constant, then

$$u(t) \leq \exp\left(\int_0^t f(s)\,ds\right)\left[c^{1-p} + (1-p)\int_0^t g(s)\right.$$

$$\left.\times \exp\left(-(1-p)\int_0^s f(\sigma)\,d\sigma\right)ds\right]^{1/(1-p)}$$

for $t \in R_+$.

3.12.4 Pachpatte (1994c, in press i)

Let u, f and g be nonnegative continuous functions defined on R_+.

(i) Let $c \geq 0$, $p > 1$ be constants. If

$$u^p(t) \leq c^p + p\int_t^\infty [f(s)u^p(s) + g(s)u(s)]\,ds,$$

for $t \in R_+$ and $\int_0^\infty f(s)\,ds < \infty$, $\int_0^\infty g(s)\,ds < \infty$, then

$$u(t) \leq \exp\left(\int_t^\infty f(s)\,ds\right) \left[c^{p-1} + (p-1)\int_t^\infty g(s) \right.$$

$$\left. \times \exp\left(-(p-1)\int_s^\infty f(\sigma)\,d\sigma\right) ds \right]^{1/(p-1)},$$

for $t \in R_+$.

(ii) Let $c_1 \geq 0$, $c_2 \geq 0$ be constants. If

$$u^2(t) \leq \left(c_1^2 + 2\int_t^\infty f(s)u(s)\,ds\right)\left(c_2^2 + 2\int_t^\infty g(s)u(s)\,ds\right),$$

for $t \in R_+$ and $\int_0^\infty f(s)\,ds < \infty$, $\int_0^\infty g(s)\,ds < \infty$, then

$$u(t) \leq p(t)\exp\left(2\int_t^\infty \left[g(s)\int_s^\infty f(\sigma)\,d\sigma + f(s)\int_s^\infty g(\sigma)\,d\sigma\right] ds\right),$$

for $t \in R_+$, where

$$p(t) = c_1 c_2 + \int_t^\infty [c_1^2 g(s) + c_2^2 f(s)]\,ds,$$

for $t \in R_+$.

3.12.5 Pachpatte (in press d)

Let $u, a, b, c: I = [\alpha, \beta) \to R_+$ be continuous functions. Let L and M be the functions as in Theorem 3.2.1. If

$$u(t) \leq a(t) + b(t)\int_\alpha^t c(s)\left(u(s) + b(s)\int_\alpha^s L(\sigma, u(\sigma))\,d\sigma\right) ds,$$

for $t \in I$, then

$$u(t) \leq a(t) + b(t) \int_\alpha^t c(s) \left[a(s) + b(s) \int_\alpha^s [a(\tau)c(\tau) + L(\tau, a(\tau))] \right.$$
$$\left. \times \exp\left(\int_\tau^s b(\sigma)[c(\sigma) + M(\sigma, a(\sigma))] \, d\sigma \right) d\tau \right] ds,$$

for $t \in I$.

3.12.6 Pachpatte (1990)

Let $u, a, b, p: I = [\alpha, \beta) \to R_+$ be continuous functions. Let $k: I \times R_+ \to R_+$, $h: I \times R_+ \times R_+ \to R_+$ be continuous functions such that

$$0 \leq k(t, x) - k(t, y) \leq N(t, y)(x - y),$$
$$0 \leq h(t, x, Bx) - h(t, y, By) \leq M(t, y)[(x - y) + (Bx - By)],$$

for $t \in I$ and $x \geq y \geq 0$, $Bx \geq By \geq 0$, where $N, M: I \times R_+ \to R_+$ are continuous functions and the operator Bz for $z \in R_+$ is defined by

$$Bz = \int_\alpha^t k(s, z(s)) \, ds, \quad t \in I.$$

If

$$u(t) \leq a(t) + b(t) \int_\alpha^t h(s, u(s), p(s)Bu(s)) \, ds,$$

for $t \in I$, then

$$u(t) \leq a(t) + b(t) \int_\alpha^t [h(s, a(s), p(s)Ba(s)) + Q(s)] \, ds,$$

for $t \in I$, where

$$Q(t) = \mu(t) \left[\int_\alpha^t h(\tau, a(\tau)p(\tau)Ba(\tau)) \exp\left(\int_\tau^t \lambda(\sigma) \, d\sigma \right) d\tau \right],$$

in which

$$\mu(t) = M(t, a(t))[p^2(t) + b^2(t)]^{1/2},$$
$$\lambda(t) = \mu(t) + N(t, a(t))b(t),$$

for $t \in I$.

3.12.7 Pachpatte (1996e)

Let u, f, g and h be nonnegative continuous functions defined on R_+ and $c \geq 0$ be a constant.

(i) If

$$u^2(t) \leq c^2 + 2 \int_t^\infty \left[f(s)u(s) \left(u(s) + \int_s^\infty g(\sigma)u(\sigma)\,d\sigma \right) + h(s)u(s) \right] ds,$$

for $t \in R_+$ and $\int_0^\infty h(s)\,ds < \infty$, $\int_0^\infty [f(s) + g(s)]\,ds < \infty$, then

$$u(t) \leq c \exp\left(\int_t^\infty [f(s) + g(s)]\,ds \right) + \int_t^\infty h(s) \exp\left(\int_t^s [f(\sigma) + g(\sigma)]\,d\sigma \right) ds,$$

for $t \in R_+$.

(ii) If

$$u^2(t) \leq c^2 + 2 \int_t^\infty \left[f(s)u(s) \left(\int_s^\infty g(\sigma)u(\sigma)\,d\sigma \right) + h(s)u(s) \right] ds,$$

for $t \in R_+$ and $\int_0^\infty f(s)\,ds < \infty$, $\int_0^\infty g(s)\,ds < \infty$, $\int_0^\infty h(s)\,ds < \infty$, then

$$u(t) \leq c \exp\left(\int_t^\infty f(s) \left(\int_s^\infty g(\sigma)\,d\sigma \right) ds \right.$$
$$\left. + \int_t^\infty h(s) \exp\left(\int_t^s f(\tau) \left(\int_\tau^\infty g(\sigma)\,d\sigma \right) d\tau \right) ds \right),$$

for $t \in R_+$.

3.12.8 Pachpatte (in press f)

Let u, f, g, h and k be nonnegative continuous functions defined on R_+ and $c \geq 0$ be a constant. If

$$u^2(t) \leq c^2 + 2 \int_0^t u(s) \left[f(s)u(s) + g(s) + h(s) \left(u(s) + \int_0^s k(\sigma)u(\sigma)\,d\sigma \right) \right] ds,$$

for $t \in R_+$, then

$$u(t) \leq c \exp\left(\int_0^t f(s)\,ds\right) + \int_0^t [g(s) + h(s)A(s)] \exp\left(\int_s^t f(\sigma)\,d\sigma\right) ds,$$

for $t \in R_+$, where

$$A(t) = c \exp\left(\int_0^t [f(\sigma) + h(\sigma) + k(\sigma)]\,d\sigma\right)$$

$$+ \int_0^t g(\tau) \exp\left(\int_\tau^t [f(\sigma) + h(\sigma) + k(\sigma)]\,d\sigma\right) d\tau,$$

for $t \in R_+$.

3.12.9 Pachpatte (in press m)

Let u, f, g and h be nonnegative continuous functions defined on R_+ and $c > 0$ be a constant.

(i) If

$$u^2(t) \leq c^2 + 2 \int_0^t \left[f(s)u^2(s) + g(s)u^{\alpha+1}(s) \left(u(s) + \int_0^s h(\sigma)u(\sigma)\,d\sigma \right) \right] ds,$$

for $t \in R_+$ and $0 < \alpha < 1$ a constant and

$$B(t) = 1 - c^\alpha \alpha \int_0^t g(\tau) \exp\left(\alpha \int_0^\tau [f(\sigma) + h(\sigma)]\,d\sigma\right) d\tau > 0,$$

for all $t \in R_+$, then

$$u(t) \leq \exp\left(\int_0^t f(s)\,ds\right)\left[c^{1-\alpha} + (1-\alpha)\int_0^t g(s)L(s)\right.$$
$$\left.\times \exp\left(-(1-\alpha)\int_0^s f(\sigma)\,d\sigma\right)ds\right]^{1/(1-\alpha)},$$

for $t \in R_+$, where

$$L(t) = [1/(B(t))^{1/\alpha}]c\exp\left(\int_0^t [f(\sigma) + h(\sigma)]\,d\sigma\right),$$

for $t \in R_+$.

(ii) If

$$u^2(t) \leq c^2 + 2\int_0^t \left[f(s)u^2(s) + g(s)u^{\alpha+1}(s)\left(u(s) + \int_0^s h(\sigma)u(\sigma)\,d\sigma\right)\right]ds,$$

for $t \in R_+$, $\alpha > 1$ a constant and $B(t)$, $L(t)$ defined in part (i), and

$$D(t) = c^{1-\alpha} - (\alpha - 1)\int_0^t g(s)L(s)\exp\left((\alpha - 1)\int_0^s f(\sigma)\,d\sigma\right)ds > 0,$$

for $t \in R_+$, then

$$u(t) \leq [1/(D(t))^{1/(\alpha-1)}]\exp\left(\int_0^t f(s)\,ds\right),$$

for $t \in R_+$.

3.12.10 Pachpatte (1996a)

Let u, u', a and b be nonnegative continuous functions defined on $I = [t_0, \infty)$, $t_0 \geq 0$, with $u(t_0) = k$ a constant and $c \geq 0$ a constant.

(i) If

$$(u'(t))^2 \leq c^2 + 2\int_{t_0}^t a(s)u'(s)\left(u(s) + u'(s) + \int_{t_0}^s b(\tau)u'(\tau)\,d\tau\right)ds,$$

for $t \in I$, then

$$u(t) \leq k + \int_{t_0}^{t} \left[c + (k+c) \int_{t_0}^{s} a(\tau) \exp\left(\int_{t_0}^{\tau} [1 + a(\sigma) + b(\sigma)] \, d\sigma \right) d\tau \right] ds,$$

for $t \in I$.

(ii) If

$$(u'(t))^2 \leq c^2 + 2 \int_{t_0}^{t} a(s) u'(s) \left(u(s) + u'(s) + \int_{t_0}^{s} b(\sigma)(u'(\sigma))^\alpha \, d\sigma \right) ds,$$

for $t \in I$, where $0 \leq \alpha < 1$ is a constant, then

$$u(t) \leq k + \int_{t_0}^{t} \left[c + \int_{t_0}^{s} a(\tau) \exp\left(\int_{t_0}^{\tau} [1 + a(\sigma)] \, d\sigma \right) \left[(k+c)^{1-\alpha} \right. \right.$$

$$\left. \left. + (1-\alpha) \int_{t_0}^{\tau} b(\eta) \exp\left(-(1-\alpha) \int_{t_0}^{\eta} [1 + a(\sigma)] \, d\sigma \right) d\eta \right]^{1/(1-\alpha)} d\tau \right] ds,$$

for $t \in I$.

(iii) If

$$(u'(t))^2 \leq c^2 + 2 \int_{t_0}^{t} a(s) u'(s) \left(u(s) + u'(s) + \int_{t_0}^{s} b(\sigma)(u(\sigma) + u'(\sigma)) \, d\sigma \right) ds,$$

for $t \in I$, then

$$u(t) \leq k + \int_{t_0}^{t} \left[c + \int_{t_0}^{s} a(\tau) \left[(k+c) \exp\left(\int_{t_0}^{\tau} [1 + a(\eta)] \, d\eta \right) \right. \right.$$

$$+ (2k+c) \int_{t_0}^{\tau} b(\eta) \exp\left(\int_{\eta}^{\tau} [1 + a(\xi)] \, d\xi \right)$$

$$\left. \left. \times \exp\left(\int_{t_0}^{\eta} [2 + a(\xi) + b(\xi)] \, d\xi \right) d\eta \right] d\tau \right] ds,$$

for $t \in I$.

3.12.11 Pachpatte (1995b)

Suppose that

(H$_1$) $\sigma \in C(R_+, R)$ with $\sigma(t) \leq t$, $t \in R_+$;
(H$_2$) $f, g \in C(R_+, R_+)$;
(H$_3$) $f, g, h \in C(R_+, R_+)$;
(H$_4$) $u \in C([a, \infty), R_+)$, $u(t) \geq u_0 \geq 0$, $c \geq 0$, where $a = \min_{t \in R_+} \sigma(t) < 0$ and u_0, c are constants;
(H$_5$) $u \in C([a, \infty), R_1)$, $u(t) \geq u_0 \geq 1$, $c \geq 1$, where a, u_0, c are as in (H$_4$), $R_1 = [1, \infty)$;
(H$_6$) $w \in C([u_0, \infty), R_+)$, nondecreasing, $w(u) > 0$ on (u_0, ∞) and $w(u_0) = 0$.

Consider the following delay integral inequalities

$$u^2(t) \leq c^2 + 2 \int_0^t [f(s)u(\sigma(s))w(u(\sigma(s))) + h(s)u(\sigma(s))]\,ds, \quad \text{(A)}$$

$$u^2(t) \leq c^2 + 2 \int_0^t \left[f(s)u(\sigma(s)) \left(\int_0^s g(\tau)w(u(\sigma(\tau)))\,d\tau \right) \right.$$
$$\left. + h(s)u(\sigma(s)) \right] ds, \quad \text{(B)}$$

$$u^2(t) \leq c^2 + 2 \int_0^t \left[f(s)u^2(\sigma(s)) \left(\int_0^s g(\tau)w(\log u(\sigma(\tau)))\,d\tau \right) \right.$$
$$\left. + h(s)u(\sigma(s)) \right] ds, \quad \text{(L)}$$

for $t \in [0, \infty)$ with the condition

$$u(t) = \psi(t), \quad t \in [0, a], \quad \psi(\sigma(t)) \leq c \quad \text{for every} \quad t \geq 0 \quad \text{and} \quad \sigma(t) \leq 0. \tag{Q}$$

(i) The inequality (A), with (Q) and assumptions (H$_1$), (H$_2$), (H$_4$) and (H$_6$), implies

$$u(t) \le G^{-1}\left[G\left(c + \int_0^t h(s)\,ds\right) + \int_0^t f(s)\,ds\right], \quad 0 \le t \le \beta_1, \quad (\overline{A})$$

(ii) The inequality (B), with (Q) and assumptions (H_1), (H_3), (H_4) and (H_6) implies

$$u(t) \le G^{-1}\left[G\left(c + \int_0^t h(s)\,ds\right) + \int_0^t f(s)\left(\int_0^s g(\tau)\,d\tau\right)ds\right], \quad 0 \le t \le \beta_2,$$
$$(\overline{B})$$

(iii) The inequality (L), with (Q) and assumptions (H_1), (H_3), (H_5) and (H_6) implies

$$u(t) \le \exp\left(G^{-1}\left[G\left(\log c + \int_0^t h(s)\,ds\right) + \int_0^t f(s)\left(\int_0^s g(\tau)\,d\tau\right)ds\right]\right),$$
$$0 \le t \le \beta_3, \quad (\overline{L})$$

where

$$G(r) = \int_{r_0}^r \frac{ds}{w(s)}, \quad r \ge r_0 \quad \text{with} \quad r_0 > u_0,$$

G^{-1} is the inverse of G and the numbers $\beta_1, \beta_2, \beta_3$ are chosen so that the quantity in square brackets in $(\overline{A})$, $(\overline{B})$, $(\overline{L})$ is in the range of G.

3.12.12 Pachpatte (in press d)

Let $f, g: I \to R_+$, $u: I \to R_1$ be continuous functions, where $I = [0, T]$, $R_1 = [1, \infty)$ and $u_0 \ge 1$ is a constant. If

$$u(t) \le u_0 + \int_0^t f(s)u(s)[\log u(s) + g(s)]\,ds,$$

for $t \in I$, then

$$u(t) \le \exp\left(p(t)\exp\left(\int_0^t f(s)\,ds\right)\right),$$

for $t \in I$, where

$$p(t) = \log u_0 + \int_0^t f(s)g(s)\,ds,$$

for $t \in I$.

3.12.13 Pachpatte (in press d)

Let $u, f, g: I \to R_+$ be continuous functions and $a \geq 1$, $u_0 \geq 0$ are constants, where $I = [0, T]$. If

$$u(t) \leq u_0 + \int_0^t f(s)u(s)[\log(a + u(s)) + g(s)]\,ds,$$

for $t \in I$, then

$$u(t) \leq \left[\exp\left(q(t)\exp\left(\int_0^t f(s)\,ds\right)\right) - a\right],$$

for $t \in I$, where

$$q(t) = \log(a + u_0) + \int_0^t f(s)g(s)\,ds,$$

for $t \in I$.

3.12.14 Pachpatte (in press d)

Let $u, b, c: I \to R_+$ be continuous functions and $a \geq 1$, $u_0 \geq 0$ are constants, where $I = [0, T]$. If

$$u(t) \leq u_0 + \int_0^t b(s)u(s)\left(\log(a + u(s)) + \int_0^s c(\tau)\log(a + u(\tau))\,d\tau\right)ds,$$

for $t \in I$, then

$$u(t) \leq \left[(a + u_0)^{\left[1 + \int_0^t b(s)\exp\left(\int_0^s [b(\tau) + c(\tau)]\,d\tau\right)ds\right]} - a\right],$$

for $t \in I$.

3.12.15 Pachpatte (in press d)

Let $u, b, c \colon I \to R_+$ be continuous functions and $a \geq 1$, $u_0 \geq 0$ be constants, where $I = [0, T]$. If

$$u(t) \leq u_0 + \int_0^t b(s) u(s) \left(\int_0^s c(\tau) \log(a + u(\tau)) \, d\tau \right) ds,$$

for $t \in I$, then

$$u(t) \leq \left[(a + u_0)^{\exp\left(\int_0^t b(s) \left(\int_0^s c(\tau) d\tau \right) ds \right)} - a \right],$$

for $t \in I$.

3.12.16 Pachpatte (in press d)

Let $u, b, c \colon I \to R_+$ be continuous functions and $a \geq 1$, $u_0 \geq 0$ be constants, where $I = [0, T]$. Let $f(u)$ be the same function as in Theorem 3.9.1. If

$$u(t) \leq u_0 + \int_0^t b(s) u(s) \left(\int_0^s c(\tau) f(\log(a + u(\tau))) \, d\tau \right) ds,$$

for $t \in I$, then for $0 \leq t \leq \bar{t}$,

$$u(t) \leq \left[\exp\left[F^{-1} \left(F(\log u_0) + \int_0^t b(s) \left(\int_0^s c(\tau) d\tau \right) ds \right) \right] - a \right],$$

where F, F^{-1} are as defined in Theorem 3.9.1 and $\bar{t} \in I$ is chosen so that

$$F(\log u_0) + \int_0^t b(s) \left(\int_0^s c(\tau) d\tau \right) ds \in \mathrm{Dom}(F^{-1}),$$

for all $t \in I$ lying in the interval $0 \leq t \leq \bar{t}$.

3.12.17 Pachpatte (1995c)

Let u, a, b, p_i, $i = 1, 2, \ldots, n$, be nonnegative continuous functions defined on R_+. Let $F[t, p_1, p_2, \ldots, p_{n-1}, p_n]$ be as defined in Section 3.10.

(i) If

$$u(t) \leq a(t) + b(t) \int_0^t F[s, p_1, p_2, \ldots, p_{n-1}, p_n u] \, ds,$$

for $t \in R_+$, then

$$u(t) \leq a(t) + b(t) \int_0^t F[s, p_1, p_2, \ldots, p_{n-1}, p_n a]$$

$$\times \exp\left(\int_s^t F[\tau, p_1, p_2, \ldots, p_{n-1}, p_n b] \, d\tau\right) ds,$$

for $t \in R_+$.

(ii) Let g, g^{-1}, α, β be as in (a_2) in Theorem 3.10.1. If

$$u(t) \leq a(t) + b(t) g^{-1}\left(\int_0^t F[s, p_1, p_2, \ldots, p_{n-1}, p_n g(u)] \, ds\right),$$

for $t \in R_+$, then

$$u(t) \leq a(t) + b(t) g^{-1}\left(\int_0^t F[s, p_1, p_2, \ldots, p_{n-1}, p_n \alpha g(a\alpha^{-1})]\right.$$

$$\left.\times \exp\left(\int_s^t F[\tau, p_1, p_2, \ldots, p_{n-1}, p_n \beta g(b\beta^{-1})] \, d\tau\right) ds\right),$$

for $t \in R_+$.

(iii) Let w, G, G^{-1} be as in (a_3) in Theorem 3.10.1. If

$$u(t) \leq a(t) + b(t) \int_0^t F[s, p_1, p_2, \ldots, p_{n-1}, p_n w(u)] \, ds,$$

for $t \in R_+$, then for $0 \leq t \leq t_1$,

$$u(t) \leq a(t) + b(t) G^{-1}\left[G(c(t)) + \int_0^t F[s, p_1, p_2, \ldots, p_{n-1}, p_n w(b)] \, ds\right],$$

where
$$c(t) = \int_0^t F[s, p_1, p_2, \ldots, p_{n-1}, p_n w(a)] \, ds,$$
and $t_1 \in R_+$ is chosen so that
$$G(c(t)) + \int_0^t F[s, p_1, p_2, \ldots, p_n w(b)] \, ds \in \text{Dom}(G^{-1}),$$
for all $t \in R_+$ lying in the interval $0 \leq t \leq t_1$.

3.12.18 Pachpatte (in press e)

Let $u \geq 0$, $f_i \geq 0$, $p_i \geq 0$ for $i = 1, 2, \ldots, n$ be continuous functions defined on R_+ and $c \geq 0$ is a constant.

(i) If
$$u^2(t) \leq c^2 + 2 \sum_{k=1}^{n} \left(\int_0^t p_1(s_1) \int_0^{s_1} p_2(s_2) \ldots \int_0^{s_{k-2}} p_{k-1}(s_{k-1}) \right.$$
$$\left. \times \int_0^{s_{k-1}} p_k(s_k) f_k(s_k) u(s_k) \, ds_k \, ds_{k-1} \ldots ds_2 \, ds_1 \right),$$
for $t \in R_+$ with $s_0 = t$, then
$$u(t) \leq c + \sum_{k=1}^{n} \left(\int_0^t p_1(s_1) \int_0^{s_1} p_2(s_2) \ldots \int_0^{s_{k-2}} p_{k-1}(s_{k-1}) \right.$$
$$\left. \times \int_0^{s_{k-1}} p_k(s_k) f_k(s_k) \, ds_k \, ds_{k-1} \ldots ds_2 \, ds_1 \right),$$
for $t \in R_+$.

(ii) Let $w(u)$ be a continuous nondecreasing function defined for $u \geq 0$ and $w(u) > 0$ for $u > 0$ and $w'(u) \geq 0$ for $u \geq 0$. If
$$u^2(t) \leq c^2 + 2 \sum_{k=1}^{n} \left(\int_0^t p_1(s_1) \int_0^{s_1} p_2(s_2) \ldots \int_0^{s_{k-2}} p_{k-1}(s_{k-1}) \right.$$
$$\left. \times \int_0^{s_{k-1}} p_k(s_k) f_k(s_k) u(s_k) w(u(s_k)) \, ds_k \, ds_{k-1} \ldots ds_2 \, ds_1 \right),$$

for $t \in R_+$, with $s_0 = t$, then for $0 \leq t \leq \bar{t}$,

$$u(t) \leq G^{-1}\left[G(c) + \sum_{k=1}^{n}\left(\int_0^t p_1(s_1)\int_0^{s_1} p_2(s_2)\ldots \int_0^{s_{k-2}} p_{k-1}(s_{k-1})\right.\right.$$
$$\left.\left. \times \int_0^{s_{k-1}} p_k(s_k)f_k(s_k)\,ds_k\,ds_{k-1}\ldots ds_2\,ds_1\right)\right],$$

where

$$G(r) = \int_{r_0}^{r} \frac{ds}{w(s)}, \quad r > 0, r_0 > 0,$$

G^{-1} is the inverse of G and $\bar{t} \in R_+$ is chosen so that

$$G(c) + \sum_{k=1}^{n}\left(\int_0^t p_1(s_1)\int_0^{s_1} p_2(s_2)\ldots \int_0^{s_{k-2}} p_{k-1}(s_{k-1})\right.$$
$$\left. \times \int_0^{s_{k-1}} p_k(s_k)f_k(s_k)\,ds_k\,ds_{k-1}\ldots ds_2\,ds_1\right) \in \mathrm{Dom}(G^{-1}),$$

for all $t \in R_+$ lying in the interval $0 \leq t \leq \bar{t}$.

3.12.19 Pachpatte (in press j)

Let $u \geq 0$, $p \geq 0$, $q \geq 0$, $r_i > 0$ for $i = 1, 2, \ldots, n-1$ be continuous functions defined on R_+, $c \geq 0$ be a constant and $n \geq 2$ be a natural number. Let $M[t, r, h]$ be as defined in Section 1.9.

(i) If

$$u^2(t) \leq c^2 + 2M[t, r, p(t_n)u(t_n)(u(t_n) + M[t_n, r, q(t_n)u(t_n)])],$$

for $t \in R_+$, then

$$u(t) \leq c[1 + M[t, r, p(t_n)\exp(M[t_n, r, (p(t_n) + q(t_n))])]],$$

for $t \in R_+$.

(ii) Let $w(u)$ be a continuous nondecreasing function defined on R_+ and $w(u) > 0$ for $u > 0$ and $w'(u) \geq 0$ for $u \geq 0$. If

$$u^2(t) \leq c^2 + 2M[t, r, p(t_n)u(t_n)(u(t_n) + M[t_n, r, q(s_n)w(u(s_n))])],$$

for $t \in R_+$, then for $0 \leq t \leq t_1$,

$$u(t) \leq c + M[t, r, p(t_n)G^{-1}[G(c) + M[t_n, r, (p(s_n) + q(s_n))]]],$$

where

$$G(v) = \int_{v_0}^{v} \frac{ds}{s + w(s)}, \quad v > 0, v_0 > 0,$$

G^{-1} is the inverse of G and $t_1 \in R_+$ is chosen so that

$$G(c) + M[t, r, (p(s_n) + q(s_n))] \in \text{Dom}(G^{-1}),$$

for all $t \in R_+$ lying in the interval $0 \leq t \leq t_1$.

3.12.20 Pachpatte (1992)

Let $u: R_+ \to R_1 = [1, \infty)$, $b: R_+ \to R_+$, $r_i: R_+ \to (0, \infty)$, $i = 1, 2, \ldots, n-1$, be continuous functions and $u_0 \geq 1$ is a constant. Let $w(u)$ be a continuously differentiable function defined on R_+ and $w(u) > 0$ for $u > 0$ and $w'(u) \geq 0$ for $u \geq 0$. Let $M[t, r, h]$ be as defined in Section 1.9.

(i) If

$$u(t) \leq u_0 + M[t, r, p(t_n)u(t_n) \log u(t_n)],$$

for $t \in R_+$, then

$$u(t) \leq u_0^{\exp(M[t,r,p(t_n)])},$$

for $t \in R_+$.

(ii) If

$$u(t) \leq u_0 + M[t, r, p(t_n)u(t_n)w(\log u(t_n))],$$

for $t \in R_+$, then for $0 \leq t \leq t_2$,

$$u(t) \leq \exp(G^{-1}[G(\log u_0) + M[t, r, p(t_n)]]),$$

where

$$G(v) = \int_{v_0}^{v} \frac{ds}{w(s)}, \quad v > 0, v_0 > 0,$$

G^{-1} is the inverse of G and $t_2 \in R_+$ is chosen so that

$$G(\log u_0) + M[t, r, p(t_n)] \in \text{Dom}(G^{-1}),$$

for all $t \in R_+$ lying in the interval $0 \le t \le t_2$.

3.13 Notes

The integral inequalities which are dealt with in Section 3.2 are due to Dragomir (1987a,b). Theorem 3.2.1 and Corollaries 3.2.1 and 3.2.2 present some useful generalizations of the Gronwall–Bellman inequality, which are very effective in the qualitative theory of differential and integral equations. The inequalities given in Theorems 3.2.2 and 3.2.3 are also given by Dragomir (1992) which are the further generalizations of Theorem 3.2.1.

The inequalities given in Section 3.3 have recently been investigated by Pachpatte (in press k, l) which are the further generalizations of Dragomir's inequality given in Theorem 3.2.1. Theorems 3.3.1 and 3.3.3 are taken from Pachpatte (in press l), which in turn are the generalizations of the inequalities due to Gollwitzer (1969) and Bihari (1956). Theorem 3.3.2 is motivated by the results given by Pachpatte (1975c,h) and is new. Theorem 3.3.4 is taken from Pachpatte (in press k), which deals with another useful generalization of Dragomir's inequality in Theorem 3.2.1.

Section 3.4 deals with some fundamental inequalities investigated by Ou-Iang (1957) and Dafermos (1979). The inequality in Theorem 3.4.1 was first given by Ou-Iang (1957) while studying the boundedness of solutions of certain second-order differential equations. The inequalities in Theorems 3.4.2 and 3.4.3 are due to Dafermos (1979) and were developed in order to establish a different connection between stability and the second law of thermodynamics.

The inequalities given in Sections 3.5–3.7 were discovered by Pachpatte (1994c, 1995a,d, 1996a,b, in press a,h,m) in order to apply them as basic tools in various new applications. Theorems 3.5.1, 3.5.2, 3.5.4 are taken from Pachpatte (1994c, 1995a) while Theorems 3.5.3 and 3.5.5 are taken respectively from Pachpatte (in press h) and Pachpatte (1994c). Theorems 3.6.1 and 3.6.2 are taken from Pachpatte (1995a), Theorem 3.6.3 is taken from Pachpatte (in press h), and Theorem 3.6.4 is taken from Pachpatte (in press m). Theorem 3.7.1 is taken from Pachpatte (1996b) which basically involves the comparison principle. Theorems 3.7.2 and 3.7.3 are taken from Pachpatte (1995a) and (1995d) respectively, while Theorems 3.7.4 and 3.7.5 are taken respectively from Pachpatte (in press a) and (1996a).

The results given in Section 3.8 are due to Haraux (1981) and Engler (1989), which are motivated by certain applications. The inequality in Theorem 3.8.1 is given by Haraux (1981) while studying the existence of solutions of wave equation with logarithmic nonlinearities. The inequality given in Theorem 3.8.2 is a slight variant of the inequality given by Haraux in Theorem 3.8.1 and was given by Engler (1989) while studying the global regular solutions for the dynamic antiplane shear problem in nonlinear viscoelasticity.

Section 3.9 is devoted to the various inequalities investigated by Pachpatte (1991, 1994d), which in fact are motivated by the inequalities given in Theorems 3.8.1 and 3.8.2. These inequalities can be considered as further generalizations of the inequalities given in Section 3.8 and can be used as effective tools in certain new applications. Theorems 3.9.1, 3.9.3 and 3.9.5 are taken from Pachpatte (1994d), while Theorems 3.9.4 and 3.9.6 are taken from Pachpatte (1991) and Theorem 3.9.2 is a slight variant of Theorem 3.9.1 and is new.

The inequalities given in Section 3.10 are established by Pachpatte (1993, 1994a,b, 1995c, 1996d) in order to study the qualitative behaviour of solutions of certain higher order differential and integro-differential equations. The inequalities in Theorem 3.10.1 are taken from Pachpatte (1993, 1996d), while Theorems 3.10.2 and 3.10.3 are taken respectively from Pachpatte (1995c) and (1994b). The results in Theorems 3.10.4 and 3.10.5 are taken from Pachpatte (1994a).

Section 3.11 contains applications of some of the inequalities given in earlier sections to study the qualitative behaviour of the solutions of certain differential, integral and integro-differential equations. Section 3.12 deals with miscellaneous inequalities which may be useful in certain new applications.

Chapter Four

Multidimensional Linear Integral Inequalities

4.1 Introduction

In analysing the dynamics of physical systems governed by various nonlinear partial differential equations one often needs some new ideas and methods. It is well known that the method of differential and integral inequalities plays an important role in the qualitative theory of partial differential, integral and integro-differential equations. During the past few years, many papers have appeared in the literature which deal with integral inequalities in more than one independent variable which are motivated by certain applications in the theory of hyperbolic partial differential and integral equations. This chapter presents a number of multidimensional linear integral inequalities developed in the literature which can be used as ready and powerful tools in the analysis of various classes of hyperbolic partial differential, integral and integro-differential equations. Applications of some of the inequalities are also presented and some miscellaneous inequalities which can be used readily in certain applications are given.

4.2 Wendroff's Inequality

Beckenbach and Bellman (1961) stated without proof a two independent variable generalization of the well-known Gronwall–Bellman inequality due to Wendroff which has its origin in the field of partial differential equations. In this section we present the basic inequality due to Wendroff given in Beckenbach and Bellman (1961) and some of its variants which can be used in certain applications.

The main result due to Wendroff, given in Beckenbach and Bellman (1961 p. 154) is embodied in the following theorem.

Theorem 4.2.1 Let $u(x, y), c(x, y)$ be nonnegative continuous functions defined for $x, y \in R_+$. If

$$u(x, y) \leq a(x) + b(y) + \int_0^x \int_0^y c(s, t) u(s, t)\, ds\, dt, \qquad (4.2.1)$$

for $x, y \in R_+$, where $a(x), b(y)$ are positive continuous functions for $x, y \in R_+$, having derivatives such that $a'(x) \geq 0, b'(y) \geq 0$ for $x, y \in R_+$, then

$$u(x, y) \leq E(x, y) \exp\left(\int_0^x \int_0^y c(s, t)\, ds\, dt\right), \qquad (4.2.2)$$

for $x, y \in R_+$, where

$$E(x, y) = [a(x) + b(0)][a(0) + b(y)]/[a(0) + b(0)], \qquad (4.2.3)$$

for $x, y \in R_+$.

□

Proof: Define a function $z(x, y)$ by the right-hand side of (4.2.1). Then $z(0, y) = a(0) + b(y)$, $z(x, 0) = a(x) + b(0)$, $u(x, y) \leq z(x, y)$ and

$$z_{xy}(x, y) = c(x, y) u(x, y) \leq c(x, y) z(x, y). \qquad (4.2.4)$$

From (4.2.4) and using the facts that $z(x, y) > 0, z_x(x, y) > 0, z_y(x, y) \geq 0$, we observe that

$$\frac{z_{xy}(x, y)}{z(x, y)} \leq c(x, y) + \frac{z_x(x, y) z_y(x, y)}{z^2(x, y)},$$

i.e.

$$\frac{\partial}{\partial y}\left(\frac{(\partial/\partial x) z(x, y)}{z(x, y)}\right) \leq c(x, y), \qquad (4.2.5)$$

for $x, y \in R_+$. Now keeping x fixed in (4.2.5), set $y = t$ and integrate with respect to t from 0 to y to obtain the estimate

$$\frac{(\partial/\partial x)z(x, y)}{z(x, y)} - \frac{(\partial/\partial x)z(x, 0)}{z(x, 0)} \leq \int_0^y c(x, t)\, dt. \tag{4.2.6}$$

Keeping y fixed in (4.2.6), set $x = s$ and integrate with respect to s from 0 to x to obtain the estimate

$$\log z(x, y) - \log z(0, y) - \log z(x, 0) + \log z(0, 0) \leq \int_0^x \int_0^y c(s, t)\, ds\, dt. \tag{4.2.7}$$

From (4.2.7) it is easy to observe that

$$z(x, y) \leq E(x, y) \exp\left(\int_0^x \int_0^y c(s, t)\, ds\, dt\right). \tag{4.2.8}$$

Using (4.2.8) in $u(x, y) \leq z(x, y)$ we obtain the desired bound in (4.2.2).

∎

Remark 4.2.1 If we take in (4.2.1), $a(x) + b(y) = k$, where k is a positive constant, then the bound obtained in (4.2.2) reduces to

$$u(x, y) \leq k \exp\left(\int_0^x \int_0^y c(s, t)\, ds\, dt\right),$$

for $x, y \in R_+$. The requirement $k > 0$ can be weakened to $k \geq 0$. If $k = 0$, then the result holds with $k = \epsilon > 0$, and the conclusion follows by letting $\epsilon \to 0$.

△

An interesting and useful generalization of Wendroff's inequality is given in the following theorem.

Theorem 4.2.2 *Let $u(x, y)$, $n(x, y)$ and $c(x, y)$ be nonnegative continuous functions defined for $x, y \in R_+$, and let $n(x, y)$ be nondecreasing in each variable $x, y \in R_+$. If*

$$u(x, y) \leq n(x, y) + \int_0^x \int_0^y c(s, t) u(s, t)\, ds\, dt, \tag{4.2.9}$$

for $x, y \in R_+$, then

$$u(x, y) \leq n(x, y) \exp\left(\int_0^x \int_0^y c(s, t) \, ds \, dt\right), \qquad (4.2.10)$$

for $x, y \in R_+$.

Proof: First we assume that $n(x, y) > 0$ for $x, y \in R_+$. From (4.2.9) we observe that

$$\frac{u(x, y)}{n(x, y)} \leq 1 + \int_0^x \int_0^y c(s, t) \frac{u(s, t)}{n(s, t)} \, ds \, dt. \qquad (4.2.11)$$

Now an application of a special version of Theorem 4.2.1 yields the required inequality in (4.2.10). If $n(x, y) = 0$, then from (4.2.9) we observe that

$$u(x, y) \leq \epsilon + \int_0^x \int_0^y c(s, t) u(s, t) \, ds \, dt,$$

where $\epsilon > 0$ is an arbitrary small constant. A suitable application of Theorem 4.2.1 yields

$$u(x, y) \leq \epsilon \exp\left(\int_0^x \int_0^y c(s, t) \, ds \, dt\right). \qquad (4.2.12)$$

Now by letting $\epsilon \to 0$ in (4.2.12) we have $u(x, y) = 0$ and hence (4.2.10) holds. The proof is complete.

■

As an immediate consequence of Theorem 4.2.2 we have the following.

Corollary 4.2.1 If, in (4.2.9), $n(x, y) = a(x) + b(y)$, where $a(x)$ and $b(y)$ are nonnegative continuous and nondecreasing functions for $x, y \in R_+$, then

$$u(x, y) \leq [a(x) + b(y)] \exp\left(\int_0^x \int_0^y c(s, t) \, ds \, dt\right), \qquad (4.2.13)$$

for $x, y \in R_+$.

Δ

Remark 4.2.2 Note that under the conditions of Theorem 4.2.2, (4.2.13) is better than (4.2.2), since $a(x) + b(y) \leq E(x, y)$ for $x, y \in R_+$. For a slight variant of Theorem 4.2.2, see Bondge et al. (1980).

△

4.3 Pachpatte's Inequalities I

Wendroff's inequality given in Theorem 4.2.1 is very effective in the study of various properties of the solutions of certain hyperbolic partial differential equations. In the past few years many authors have obtained various interesting and useful generalizations and extensions of this inequality. In this section we offer some basic inequalities established by Pachpatte (1980b, 1995e), which claim their origin in Wendroff's inequality.

Pachpatte (1995e) established the following inequality, which can be widely used in various applications.

Theorem 4.3.1 Let $u(x, y)$, $p(x, y)$ and $q(x, y)$ be nonnegative continuous functions defined for $x, y \in R_+$. Let $k(x, y, s, t)$ and its partial derivatives $k_x(x, y, s, t)$, $k_y(x, y, s, t)$, $k_{xy}(x, y, s, t)$ be nonnegative continuous functions for $0 \leq s \leq x < \infty, 0 \leq t \leq y < \infty$. If

$$u(x, y) \leq p(x, y) + q(x, y) \int_0^x \int_0^y k(x, y, s, t) u(s, t) \, ds \, dt, \qquad (4.3.1)$$

for $x, y \in R_+$, then

$$u(x, y) \leq p(x, y) + q(x, y) \left(\int_0^x \int_0^y A(\sigma, \tau) \, d\sigma \, d\tau \right)$$

$$\times \exp \left(\int_0^x \int_0^y B(\sigma, \tau) \, d\sigma \, d\tau \right), \qquad (4.3.2)$$

for $x, y \in R_+$, where

$$A(x, y) = k(x, y, x, y) p(x, y) + \int_0^x k_x(x, y, s, y) p(s, y) \, ds$$

$$+ \int_0^y k_y(x, y, x, t) p(x, t) \, dt + \int_0^x \int_0^y k_{xy}(x, y, s, t) p(s, t) \, ds \, dt, \qquad (4.3.3)$$

$$B(x, y) = k(x, y, x, y)q(x, y) + \int_0^x k_x(x, y, s, y)q(s, y)\,ds$$

$$+ \int_0^y k_y(x, y, x, t)q(x, t)\,dt + \int_0^x\int_0^y k_{xy}(x, y, s, t)q(s, t)\,ds\,dt, \quad (4.3.4)$$

for $x, y \in R_+$.

Proof: Define a function $z(x, y)$ by

$$z(x, y) = \int_0^x\int_0^y k(x, y, s, t)u(s, t)\,ds\,dt. \quad (4.3.5)$$

From (4.3.5) and using $u(x, y) \leq p(x, y) + q(x, y)z(x, y)$, and the fact that $z(x, y)$ is monotonic nondecreasing in both the variables x and y and (4.3.3) and (4.3.4), we observe that

$$z_{xy}(x, y) \leq A(x, y) + B(x, y)z(x, y). \quad (4.3.6)$$

From (4.3.6) it is easy to observe that

$$z(x, y) \leq c(x, y) + \int_0^x\int_0^y B(s, t)z(s, t)\,ds\,dt, \quad (4.3.7)$$

where

$$c(x, y) = \int_0^x\int_0^y A(\sigma, \tau)\,d\sigma\,d\tau. \quad (4.3.8)$$

It is easy to observe that $c(x, y)$ is nonnegative and nondecreasing in each variable $x, y \in R_+$. An application of Theorem 4.2.2 to (4.3.7) yields

$$z(x, y) \leq c(x, y)\exp\left(\int_0^x\int_0^y B(\sigma, \tau)\,d\sigma\,d\tau\right). \quad (4.3.9)$$

Now using (4.3.8), (4.3.9) in $u(x, y) \leq p(x, y) + q(x, y)z(x, y)$ we get the required inequality in (4.3.2).

∎

As an immediate consequence of Theorem 4.3.1 we have the following.

Corollary 4.3.1 Let $u(x, y)$, $p(x, y)$, $q(x, y)$ and $k(x, y)$ be nonnegative continuous functions defined for $x, y \in R_+$. If

$$u(x, y) \leq p(x, y) + q(x, y) \int_0^x \int_0^y k(s, t) u(s, t) \, ds \, dt, \qquad (4.3.10)$$

for $x, y \in R_+$, then

$$u(x, y) \leq p(x, y) + q(x, y) \left(\int_0^x \int_0^y k(s, t) p(s, t) \, ds \, dt \right)$$

$$\times \exp\left(\int_0^x \int_0^y k(s, t) q(s, t) \, ds \, dt \right), \qquad (4.3.11)$$

for $x, y \in R_+$.

△

The following theorem deals with a useful two independent variable generalization of the inequality given by Greene (1977).

Theorem 4.3.2 Let $u(x, y)$, $v(x, y)$, $h_i(x, y)$, $i = 1, 2, 3, 4$, be nonnegative continuous functions defined for $x, y \in R_+$ and c_1, c_2 and μ be nonnegative constants. If

$$u(x, y) \leq c_1 + \int_0^x \int_0^y h_1(s, t) u(s, t) \, ds \, dt + \int_0^x \int_0^y h_2(s, t) \bar{v}(s, t) \, ds \, dt,$$

$$(4.3.12)$$

$$v(x, y) \leq c_2 + \int_0^x \int_0^y h_3(s, t) \bar{u}(s, t) \, ds \, dt + \int_0^x \int_0^y h_4(s, t) v(s, t) \, ds \, dt,$$

$$(4.3.13)$$

for $x, y \in R_+$, where $\bar{u}(x, y) = \exp(-\mu(x + y)) u(x, y)$ and $\bar{v}(x, y) = \exp(\mu(x + y)) v(x, y)$ for $x, y \in R_+$, then

$$u(x, y) \leq (c_1 + c_2) \exp\left(\mu(x + y) + \int_0^x \int_0^y h(s, t) \, ds \, dt \right), \qquad (4.3.14)$$

$$v(x, y) \leq (c_1 + c_2) \exp\left(\int_0^x \int_0^y h(s, t) \, ds \, dt\right), \qquad (4.3.15)$$

for $x, y \in R_+$, where

$$h(x, y) = \max\{[h_1(x, y) + h_3(x, y)], [h_2(x, y) + h_4(x, y)]\},$$

for $x, y \in R_+$.

□

Proof: From (4.3.12) we observe that

$$\exp(-\mu(x + y))u(x, y) \leq c_1 + \int_0^x \int_0^y h_1(s, t)\bar{u}(s, t) \, ds \, dt$$

$$+ \int_0^x \int_0^y h_2(s, t)v(s, t) \, ds \, dt. \qquad (4.3.16)$$

Define
$$F(x, y) = \exp(-\mu(x + y))u(x, y) + v(x, y). \qquad (4.3.17)$$

Using (4.3.16) and (4.3.13) in (4.3.17) we observe that

$$F(x, y) \leq (c_1 + c_2) + \int_0^x \int_0^y h_1(s, t)\bar{u}(s, t) \, ds \, dt + \int_0^x \int_0^y h_2(s, t)v(s, t) \, ds \, dt$$

$$+ \int_0^x \int_0^y h_3(s, t)\bar{u}(s, t) \, ds \, dt + \int_0^x \int_0^y h_4(s, t)v(s, t) \, ds \, dt$$

$$\leq (c_1 + c_2) + \int_0^x \int_0^y h(s, t)F(s, t) \, ds \, dt. \qquad (4.3.18)$$

Now an application of the special version of Theorem 4.2.1 to (4.3.18) yields

$$F(x, y) \leq (c_1 + c_2) \exp\left(\int_0^x \int_0^y h(s, t) \, ds \, dt\right). \qquad (4.3.19)$$

The bounds in (4.3.14) and (4.3.15) follow by using (4.3.19) in (4.3.17) and splitting.

■

A useful two independent variable integro-differential inequality established by Pachpatte (1980b) is embodied in the following theorem.

Theorem 4.3.3 *Let $u(x, y)$, $u_{xy}(x, y)$ and $c(x, y)$ be nonnegative continuous functions defined for $x, y \in R_+$, and $u(x, 0) = u(0, y) = 0$. If*

$$u_{xy}(x, y) \leq a(x) + b(y) + \int_0^x \int_0^y c(s, t)(u(s, t) + u_{st}(s, t)) \, ds \, dt, \quad (4.3.20)$$

for $x, y \in R_+$, where $a(x) > 0, b(y) > 0$ are continuous functions for $x, y \in R_+$, having derivatives such that $a'(x) \geq 0, b'(y) \geq 0$ for $x, y \in R_+$, then

$$u_{xy}(x, y) \leq a(x) + b(y) + \int_0^x \int_0^y c(s, t) E(s, t)$$

$$\times \exp\left(\int_0^s \int_0^t [1 + c(\sigma, \eta)] \, d\sigma \, d\eta\right) ds \, dt, \quad (4.3.21)$$

for $x, y \in R_+$, where $E(x, y)$ is defined by (4.2.3) in Theorem 4.2.1. □

Proof: Define a function $z(x, y)$ by the right-hand side of (4.3.20). Then $z(0, y) = a(0) + b(y)$, $z(x, 0) = a(x) + b(0)$,

$$u_{xy}(x, y) \leq z(x, y), \quad (4.3.22)$$

and

$$z_{xy}(x, y) = c(x, y)(u(x, y) + u_{xy}(x, y)), \quad (4.3.23)$$

From (4.3.22) it is easy to observe that

$$u(x, y) \leq \int_0^x \int_0^y z(s, t) \, ds \, dt. \quad (4.3.24)$$

Using (4.3.22) and (4.3.24) in (4.3.23) we obtain

$$z_{xy}(x, y) \leq c(x, y)\left(z(x, y) + \int_0^x \int_0^y z(s, t) \, ds \, dt\right). \quad (4.3.25)$$

Define a function $v(x, y)$ by

$$v(x, y) = z(x, y) + \int_0^x \int_0^y z(s, t)\, ds\, dt. \tag{4.3.26}$$

Then $v(0, y) = z(0, y)$, $v(x, 0) = z(x, 0)$ and

$$v_{xy}(x, y) = z_{xy}(x, y) + z(x, y). \tag{4.3.27}$$

Using the facts that $z_{xy}(x, y) \leq c(x, y)v(x, y)$ from (4.3.25) and $z(x, y) \leq v(x, y)$ from (4.3.26) in (4.3.27), we get

$$v_{xy}(x, y) \leq [1 + c(x, y)]v(x, y). \tag{4.3.28}$$

By following an argument used in the proof of Theorem 4.2.1, the inequality (4.3.28) yields

$$v(x, y) \leq E(x, y) \exp\left(\int_0^x \int_0^y [1 + c(\sigma, \eta)]\, d\sigma\, d\eta\right). \tag{4.3.29}$$

Using (4.3.29) in (4.3.25) we get

$$z_{xy}(x, y) \leq c(x, y) E(x, y) \exp\left(\int_0^x \int_0^y [1 + c(\sigma, \eta)]\, d\sigma\, d\eta\right). \tag{4.3.30}$$

From (4.3.30) it is easy to obtain

$$z(x, y) \leq a(x) + b(y) + \int_0^x \int_0^y c(s, t) E(s, t)$$

$$\times \exp\left(\int_0^s \int_0^t [1 + c(\sigma, \eta)]\, d\sigma\, d\eta\right) ds\, dt. \tag{4.3.31}$$

Now using (4.3.31) in (4.3.22) we get the required inequality in (4.3.21).

∎

Remark 4.3.1 In the special case when $a(x) + b(y) = k$, in (4.3.20), for $x, y \in R_+$, where $k > 0$ is a constant, the bound obtained in (4.3.21) reduces to

$$u_{xy}(x,y) \le k\left[1 + \int_0^x\int_0^y c(s,t)\exp\left(\int_0^s\int_0^t [1+c(\sigma,\eta)]\,d\sigma\,d\eta\right)ds\,dt\right].$$
(4.3.32)

The inequality (4.3.21) or (4.3.32) gives us a completely known bound on the right-hand side, which majorizes $u_{xy}(x,y)$ and consequently $u(x,y)$ after integration. The requirement $k > 0$ can be weakened to $k \ge 0$ as noted in Remark 4.2.1.

△

Another interesting and useful integro-differential inequality given by Pachpatte (1980b) reads as follows.

Theorem 4.3.4 Let $u(x,y)$, $u_x(x,y)$, $u_y(x,y)$, $u_{xy}(x,y)$ and $c(x,y)$ be nonnegative continuous functions defined for $x, y \in R_+$, and $u(x, 0) = u(0, y) = 0$. If

$$u_{xy}(x,y) \le a(x) + b(y) + M\left[u(x,y) + \int_0^x\int_0^y c(s,t)(u(s,t) + u_{st}(s,t))\,ds\,dt\right]$$
(4.3.33)

for $x, y \in R_+$, where $a(x) > 0, b(y) > 0$ are continuous functions for $x, y \in R_+$, having derivatives such that $a'(x) \ge 0, b'(y) \ge 0$ for $x, y \in R_+$, and $M \ge 0$ is a constant, then

$$u_{xy}(x,y) \le E(x,y)\exp\left(\int_0^x\int_0^y [M + (1+M)c(s,t)]\,ds\,dt\right),$$
(4.3.34)

for $x, y \in R_+$, where $E(x,y)$ is defined by (4.2.3) in Theorem 4.2.1.

□

Proof: Define a function $z(x,y)$ by the right-hand side of (4.3.33). Then $z(0,y) = a(0) + b(y)$, $z(x,0) = a(x) + b(0)$, and

$$z_{xy}(x,y) = M[u_{xy}(x,y) + c(x,y)(u(x,y) + u_{xy}(x,y))]. \quad (4.3.35)$$

Using the facts that $u_{xy}(x,y) \le z(x,y)$ from (4.3.33) and $Mu(x,y) \le z(x,y)$ from the definition of $z(x,y)$, in (4.3.35) we see that the inequality

$$z_{xy}(x,y) \le [M + (1+M)c(x,y)]z(x,y),$$

is satisfied, which implies the estimate for $z(x, y)$ such that

$$z(x, y) \leq E(x, y) \exp\left(\int_0^x \int_0^y [M + (1+M)c(s, t)] \, ds \, dt\right). \qquad (4.3.36)$$

Now by using (4.3.36) in $u_{xy}(x, y) \leq z(x, y)$, we get the required inequality in (4.3.34).

∎

Remark 4.3.2 We note that in the special case when $a(x) + b(y) = k$, for $x, y \in R_+$, in (4.3.33), where $k > 0$ is a constant, the bound obtained in (4.3.34) reduces to

$$u_{xy}(x, y) \leq k \exp\left(\int_0^x \int_0^y [M + (1+M)c(s, t)] \, ds \, dt\right),$$

for $x, y \in R_+$. The requirement $k > 0$ can be weakened to $k \geq 0$ as noted in Remark 4.2.1.

△

4.4 Pachpatte's Inequalities II

During the past few years some new inequalities of the Wendroff type have been developed which provide a natural and effective means for the further development of the theory of partial integro-differential and integral equations. This section presents some inequalities of the Wendroff type given by Pachpatte (1980b) and analogous inequalities which can be used in the study of qualitative properties of the solutions of certain integro-differential and integral equations.

Pachpatte (1980b) established the following inequality.

Theorem 4.4.1 *Let $u(x, y)$, $f(x, y)$ and $g(x, y)$ be nonnegative continuous functions defined for $x, y \in R_+$. If*

$$u(x, y) \leq a(x) + b(y) + \int_0^x \int_0^y f(s, t) \left(u(s, t) \right.$$

$$\left. + \int_0^s \int_0^t g(\sigma, \eta) u(\sigma, \eta) \, d\sigma \, d\eta \right) ds \, dt, \qquad (4.4.1)$$

for $x, y \in R_+$, where $a(x) > 0$, $b(y) > 0$ are continuous functions for $x, y \in R_+$, having derivatives such that $a'(x) \geq 0$, $b'(y) \geq 0$ for $x, y \in R_+$, then

$$u(x, y) \leq a(x) + b(y) + \int_0^x \int_0^y f(s, t) E(s, t)$$

$$\times \exp\left(\int_0^s \int_0^t [f(\sigma, \eta) + g(\sigma, \eta)] \, d\sigma \, d\eta\right) ds \, dt, \quad (4.4.2)$$

for $x, y \in R_+$, where $E(x, y)$ is defined by (4.2.3) in Theorem 4.2.1. $\square$

Proof: Define a function $z(x, y)$ by the right-hand side of (4.4.1). Then $z(0, y) = a(0) + b(y)$, $z(x, 0) = a(x) + b(0)$, $u(x, y) \leq z(x, y)$ and

$$z_{xy}(x, y) = f(x, y)\left(u(x, y) + \int_0^x \int_0^y g(\sigma, \eta) u(\sigma, \eta) \, d\sigma \, d\eta\right)$$

$$\leq f(x, y)\left(z(x, y) + \int_0^x \int_0^y g(\sigma, \eta) z(\sigma, \eta) \, d\sigma \, d\eta\right). \quad (4.4.3)$$

Define a function $v(x, y)$ by

$$v(x, y) = z(x, y) + \int_0^x \int_0^y g(\sigma, \eta) z(\sigma, \eta) \, d\sigma \, d\eta. \quad (4.4.4)$$

Then $v(0, y) = a(0) + b(y)$, $v(x, 0) = a(x) + b(0)$, $z_{xy}(x, y) \leq f(x, y) v(x, y)$, $z(x, y) \leq v(x, y)$ and

$$v_{xy}(x, y) = z_{xy}(x, y) + g(x, y) z(x, y)$$

$$\leq [f(x, y) + g(x, y)] z(x, y). \quad (4.4.5)$$

Now by following the proof of Theorem 4.2.1, the inequality (4.4.5) implies the estimate

$$v(x, y) \leq E(x, y) \exp\left(\int_0^x \int_0^y [f(\sigma, \eta) + g(\sigma, \eta)] \, d\sigma \, d\eta\right). \quad (4.4.6)$$

Using (4.4.6) in (4.4.3) we have

$$z_{xy}(x, y) \leq f(x, y)E(x, y)\exp\left(\int_0^x\int_0^y [f(\sigma, \eta) + g(\sigma, \eta)]\,d\sigma\,d\eta\right). \quad (4.4.7)$$

From (4.4.7) it follows that

$$z(x, y) \leq a(x) + b(y) + \int_0^x\int_0^y f(s, t)E(s, t)$$

$$\times \exp\left(\int_0^s\int_0^t [f(\sigma, \eta) + g(\sigma, \eta)]\,d\sigma\,d\eta\right)\,ds\,dt, \quad (4.4.8)$$

Using (4.4.8) in $u(x, y) \leq z(x, y)$ we get the required inequality in (4.4.2).

∎

Remark 4.4.1 In the special case when $a(x) + b(y) = k$, for $x, y \in R_+$, in (4.4.1), where $k > 0$ is a constant, then the bound obtained in (4.4.2) reduces to

$$u(x, y) \leq k\left[1 + \int_0^x\int_0^y f(s, t)\exp\left(\int_0^s\int_0^t [f(\sigma, \eta) + g(\sigma, \eta)]\,d\sigma\,d\eta\right)\,ds\,dt\right],$$

for $x, y \in R_+$. The requirement $k > 0$ can be weakened to $k \geq 0$ as noted in Remark 4.2.1. We note that the inequality given in Theorem 4.4.1 is a two independent variable generalization of the inequality established by Pachpatte (1973a).

△

A useful generalization of Theorem 4.4.1 is given in the following theorem.

Theorem 4.4.2 Let $u(x, y)$, $f(x, y)$, $g(x, y)$ and $c(x, y)$ be nonnegative continuous functions defined for $x, y \in R_+$, and let $c(x, y)$ be nondecreasing in each variable $x, y \in R_+$. If

$$u(x, y) \leq c(x, y) + \int_0^x\int_0^y f(s, t)\left(u(s, t) + \int_0^s\int_0^t g(\sigma, \eta)u(\sigma, \eta)\,d\sigma\,d\eta\right)ds\,dt, \quad (4.4.9)$$

for $x, y \in R_+$, then

$$u(x, y) \leq c(x, y)H(x, y), \quad (4.4.10)$$

for $x, y \in R_+$, where

$$H(x, y) = 1 + \int_0^x \int_0^y f(s, t) \exp\left(\int_0^s \int_0^t [f(\sigma, \eta) + g(\sigma, \eta)] \, d\sigma \, d\eta\right) ds \, dt,$$
(4.4.11)

for $x, y \in R_+$. $\square$

The proof of this theorem follows by the same argument as in the proof of Theorem 4.2.2 and using a suitable version of Theorem 4.4.1. The details are omitted here.

A two independent variable analogue of Theorem 1.7.2 is given in the following theorem.

Theorem 4.4.3 Let $u(x, y)$, $f(x, y)$, $g(x, y)$, $h(x, y)$ and $p(x, y)$ be nonnegative continuous functions defined for $x, y \in R_+$ and u_0 be a nonnegative constant.

(a_1) If

$$u(x, y) \leq u_0 + \int_0^x \int_0^y [f(s, t)u(s, t) + p(s, t)] \, ds \, dt$$

$$+ \int_0^x \int_0^y f(s, t) \left(\int_0^s \int_0^t g(\sigma, \eta) u(\sigma, \eta) \, d\sigma \, d\eta\right) ds \, dt, \quad (4.4.12)$$

for $x, y \in R_+$, then

$$u(x, y) \leq \left(u_0 + \int_0^x \int_0^y p(s, t) \, ds \, dt\right) H(x, y), \quad (4.4.13)$$

for $x, y \in R_+$, where $H(x, y)$ is defined by (4.4.11) in Theorem 4.4.2.

(a_2) If

$$u(x, y) \leq u_0 + \int_0^x \int_0^y f(s, t) u(s, t) \, ds \, dt + \int_0^x \int_0^y f(s, t)$$

$$\times \left(\int_0^s \int_0^t [g(\sigma, \eta) u(\sigma, \eta) + p(\sigma, \eta)] \, d\sigma \, d\eta\right) ds \, dt, \quad (4.4.14)$$

for $x, y \in R_+$, then

$$u(x, y) \leq \left(u_0 + \int_0^x \int_0^y f(s, t) \left(\int_0^s \int_0^t p(\sigma, \eta) \, d\sigma \, d\eta \right) ds \, dt \right) H(x, y), \quad (4.4.15)$$

for $x, y \in R_+$, where $H(x, y)$ is defined by (4.4.11) in Theorem 4.4.2.

(a_3) If

$$u(x, y) \leq u_0 + \int_0^x \int_0^y h(s, t) u(s, t) \, ds \, dt + \int_0^x \int_0^y f(s, t)$$

$$\times \left(u(s, t) + \int_0^s \int_0^t g(\sigma, \eta) u(\sigma, \eta) \, d\sigma \, d\eta \right) ds \, dt, \quad (4.4.16)$$

for $x, y \in R_+$, then

$$u(x, y) \leq u_0 \exp\left(\int_0^x \int_0^y h(s, t) \, ds \, dt \right) H(x, y), \quad (4.4.17)$$

for $x, y \in R_+$, where $H(x, y)$ is defined by (4.4.11) in Theorem 4.4.2.

(a_4) If

$$u(x, y) \leq h(x, y) + p(x, y) \int_0^x \int_0^y f(s, t) \left(u(s, t) + p(s, t) \right.$$

$$\left. \times \int_0^s \int_0^t g(\sigma, \eta) u(\sigma, \eta) \, d\sigma \, d\eta \right) ds \, dt, \quad (4.4.18)$$

for $x, y \in R_+$, then

$$u(x, y) \leq h(x, y) + p(x, y) M(x, y) \left[1 + \int_0^x \int_0^y f(s, t) p(s, t) \right.$$

$$\left. \times \exp\left(\int_0^s \int_0^t [f(\sigma, \eta) + g(\sigma, \eta)] p(\sigma, \eta) \, d\sigma \, d\eta \right) ds \, dt \right], \quad (4.4.19)$$

for $x, y \in R_+$, where

$$M(x, y) = \int_0^x \int_0^y f(s, t) \left(h(s, t) + p(s, t) \int_0^s \int_0^t g(\sigma, \eta) h(\sigma, \eta) \, d\sigma \, d\eta \right) ds \, dt \tag{4.4.20}$$

for $x, y \in R_+$.

□

Proof: We give the details for the proofs of (a_1) and (a_4) only; the proofs of (a_2) and (a_3) can be completed by following the proofs of (a_1) and (a_4) and the proof of Theorem 4.4.1.

(a_1) From (4.4.12) we have

$$u(x, y) \leq a(x, y) + \int_0^x \int_0^y f(s, t) \left(u(s, t) + \int_0^s \int_0^t g(\sigma, \eta) u(\sigma, \eta) \, d\sigma \, d\eta \right) ds \, dt, \tag{4.4.21}$$

where

$$a(x, y) = u_0 + \int_0^x \int_0^y p(s, t) \, ds \, dt.$$

Clearly $a(x, y)$ is nonnegative and nondecreasing in each variable $x, y \in R_+$. Now an application of Theorem 4.4.2 to (4.4.21) yields the required inequality in (4.4.13).

(a_4) Define a function $z(x, y)$ by

$$z(x, y) = \int_0^x \int_0^y f(s, t) \left(u(s, t) + p(s, t) \int_0^s \int_0^t g(\sigma, \eta) u(\sigma, \eta) \, d\sigma \, d\eta \right) ds \, dt. \tag{4.4.22}$$

Using the fact that $u(x, y) \leq h(x, y) + p(x, y) z(x, y)$ in (4.4.22) we have

$$z(x, y) \leq \int_0^x \int_0^y f(s, t) \left(h(s, t) + p(s, t) z(s, t) + p(s, t) \right.$$
$$\left. \times \int_0^s \int_0^t g(\sigma, \eta) (h(\sigma, \eta) + p(\sigma, \eta) z(\sigma, \eta)) \, d\sigma \, d\eta \right) ds \, dt$$

$$= M(x, y) + \int_0^x \int_0^y f(s, t) p(s, t)$$

$$\times \left(z(s, t) + \int_0^s \int_0^t g(\sigma, \eta) p(\sigma, \eta) z(\sigma, \eta) \, d\sigma \, d\eta \right) ds \, dt. \quad (4.4.23)$$

Clearly $M(x, y)$ is nonnegative and nondecreasing in each variable $x, y \in R_+$. Now an application of Theorem 4.4.2 to (4.4.23) yields

$$z(x, y) \le M(x, y) \left[1 + \int_0^x \int_0^y f(s, t) p(s, t) \right.$$

$$\left. \times \exp \left(\int_0^s \int_0^t [f(\sigma, \eta) + g(\sigma, \eta)] p(\sigma, \eta) \, d\sigma \, d\eta \right) ds \, dt \right]. \quad (4.4.24)$$

Using (4.4.24) in $u(x, y) \le h(x, y) + p(x, y) z(x, y)$ we get the required inequality in (4.4.19).

∎

A generalization of Theorem 1.7.5 to the case of functions of two independent variables is contained in the following theorem.

Theorem 4.4.4 Let $u(x, y)$, $f(x, y)$, $g(x, y)$, $h(x, y)$ and $p(x, y)$ be nonnegative continuous functions defined for $x, y \in R_+$, $f(x, y)$, $g(x, y)$ be positive and sufficiently smooth functions for $x, y \in R_+$ and $u_0 \ge 0$ be a constant.

(b_1) If

$$u(x, y) \le u_0 + \int_0^x \int_0^y f(s, t) \left(h(s, t) + \int_0^s \int_0^t p(\sigma, \eta) u(\sigma, \eta) \, d\sigma \, d\eta \right) ds \, dt,$$

$$(4.4.25)$$

for $x, y \in R_+$, then

$$u(x, y) \le Q(x, y) \exp \left(\int_0^x \int_0^y f(s, t) \left(\int_0^s \int_0^t p(\sigma, \eta) \, d\sigma \, d\eta \right) ds \, dt \right),$$

$$(4.4.26)$$

for $x, y \in R_+$, where

$$Q(x, y) = u_0 + \int_0^x \int_0^y f(s, t) h(s, t) \, ds \, dt, \quad (4.4.27)$$

for $x, y \in R_+$.

(b_2) If

$$u(x, y) \leq u_0 + \int_0^x \int_0^y f(s, t) \left(h(s, t) + \int_0^s \int_0^t g(\sigma, \eta) \right.$$

$$\left. \times \left(\int_0^\sigma \int_0^\eta p(s_1, t_1) u(s_1, t_1) \, ds_1 \, dt_1 \right) d\sigma \, d\eta \right) ds \, dt, \quad (4.4.28)$$

for $x, y \in R_+$, then

$$u(x, y) \leq Q(x, y) \exp \left(\int_0^x \int_0^y f(s, t) \left(\int_0^s \int_0^t g(\sigma, \eta) \right. \right.$$

$$\left. \left. \times \left(\int_0^\sigma \int_0^\eta p(s_1, t_1) \, ds_1 \, dt_1 \right) d\sigma \, d\eta \right) ds \, dt \right), \quad (4.4.29)$$

for $x, y \in R_+$, where $Q(x, y)$ is defined by (4.4.27). $\square$

Proof: We shall give details of the proof of (b_1) only; the proof of (b_2) can be completed similarly.

(b_1) First we assume that u_0 is positive. From (4.4.25) we have

$$u(x, y) \leq Q(x, y) + \int_0^x \int_0^y f(s, t) \left(\int_0^s \int_0^t p(\sigma, \eta) u(\sigma, \eta) \, d\sigma \, d\eta \right) ds \, dt. \quad (4.4.30)$$

Clearly $Q(x, y)$ is a positive and nondecreasing function in both the variables $x, y \in R_+$ and hence from (4.4.30) we observe that

$$\frac{u(x, y)}{Q(x, y)} \leq 1 + \int_0^x \int_0^y f(s, t) \left(\int_0^s \int_0^t p(\sigma, \eta) \frac{u(\sigma, \eta)}{Q(\sigma, \eta)} d\sigma \, d\eta \right) ds \, dt. \quad (4.4.31)$$

Define a function $z(x, y)$ by the right-hand side of (4.4.31). Then $z(0, y) = z(x, 0) = 1$ and

$$z_{xy}(x, y) = f(x, y) \int_0^x \int_0^y p(\sigma, \eta) \frac{u(\sigma, \eta)}{Q(\sigma, \eta)} \, d\sigma \, d\eta. \qquad (4.4.32)$$

From (4.4.32) it is easy to observe that

$$\frac{\partial^2}{\partial x \partial y} \left(\frac{z_{xy}(x, y)}{f(x, y)} \right) = p(x, y) \frac{u(x, y)}{Q(x, y)}. \qquad (4.4.33)$$

Now by using the fact that $u(x, y)/Q(x, y) \leq z(x, y)$ in (4.4.33) we have

$$\frac{\frac{\partial}{\partial y} \left(\frac{\partial}{\partial x} \left(\frac{z_{xy}(x, y)}{f(x, y)} \right) \right)}{z(x, y)} \leq p(x, y). \qquad (4.4.34)$$

Since

$$\frac{\partial}{\partial x} \left(\frac{z_{xy}(x, y)}{f(x, y)} \right) \geq 0, \quad \frac{\partial}{\partial y} z(x, y) \geq 0 \quad \text{and} \quad z(x, y) > 0,$$

from (4.4.34) we observe that

$$\frac{\frac{\partial}{\partial y} \left(\frac{\partial}{\partial x} \left(\frac{z_{xy}(x, y)}{f(x, y)} \right) \right)}{z(x, y)} \leq p(x, y) + \frac{\frac{\partial}{\partial x} \left(\frac{z_{xy}(x, y)}{f(x, y)} \right) \frac{\partial}{\partial y} z(x, y)}{z^2(x, y)},$$

i.e.

$$\frac{\partial}{\partial y} \left(\frac{\frac{\partial}{\partial x} \left(\frac{z_{xy}(x, y)}{f(x, y)} \right)}{z(x, y)} \right) \leq p(x, y). \qquad (4.4.35)$$

Now, keeping x fixed in (4.4.35), set $y = \eta$ and integrate with respect to η from 0 to y to obtain the estimate

$$\frac{\frac{\partial}{\partial x} \left(\frac{z_{xy}(x, y)}{f(x, y)} \right)}{z(x, y)} \leq \int_0^y p(x, \eta) \, d\eta. \qquad (4.4.36)$$

Since

$$\frac{z_{xy}(x, y)}{f(x, y)} \geq 0, \quad z_x(x, y) \geq 0 \quad \text{and} \quad z(x, y) > 0,$$

again as above from (4.4.36) we observe that

$$\frac{\partial}{\partial x}\left(\frac{\left(\frac{z_{xy}(x,y)}{f(x,y)}\right)}{z(x,y)}\right) \leq \int_0^y p(x,\eta)\,d\eta. \tag{4.4.37}$$

Keeping y fixed in (4.4.37), set $x = \sigma$ and integrate with respect to σ from 0 to x to obtain the estimate

$$\frac{z_{xy}(x,y)}{z(x,y)} \leq f(x,y) \int_0^x \int_0^y p(\sigma,\eta)\,d\sigma\,d\eta. \tag{4.4.38}$$

Since $z_x(x,y) \geq 0$, $z_y(x,y) \geq 0$ and $z(x,y) > 0$, from (4.4.38) we observe that

$$\frac{\partial}{\partial y}\left(\frac{\frac{\partial}{\partial x}z(x,y)}{z(x,y)}\right) \leq f(x,y)\left(\int_0^x \int_0^y p(\sigma,\eta)\,d\sigma\,d\eta\right). \tag{4.4.39}$$

Now, keeping x fixed in (4.4.39), set $y = t$ and integrate with respect to t from 0 to y to obtain the estimate

$$\frac{\frac{\partial}{\partial x}z(x,y)}{z(x,y)} \leq \int_0^y f(x,t)\left(\int_0^x \int_0^t p(\sigma,\eta)\,d\sigma\,d\eta\right)dt. \tag{4.4.40}$$

Keeping y fixed in (4.4.40), set $x = s$ and integrate with respect to s from 0 to x to obtain the estimate

$$z(x,y) \leq \exp\left(\int_0^x \int_0^y f(s,t)\left(\int_0^s \int_0^t p(\sigma,\eta)\,d\sigma\,d\eta\right)ds\,dt\right). \tag{4.4.41}$$

Now by using (4.4.41) in (4.4.31) we get the required inequality in (4.4.26). The proof of the case when u_0 is nonnegative can be carried out as above with $u_0 + \epsilon$ instead of u_0, where $\epsilon > 0$ is an arbitrary small constant and subsequently passing to the limit as $\epsilon \to 0$ to obtain (4.4.26).

∎

4.5 Pachpatte's Inequalities III

In the qualitative analysis of some classes of partial differential and integro-differential equations, the bounds provided by the inequalities in earlier sections are inadequate and it is necessary to seek some new Wendroff-like inequalities in order to achieve a diversity of desired goals. In this section we shall give some Wendroff-like integral inequalities established by Pachpatte (1988b, c) which can be used as ready tools in the study of certain fourth-order partial differential and integro-differential equations.

First, we specify some notations and definitions which will be used in this section. The ordinary first and second derivatives of a function $r(x)$ defined for $x \geq 0$ are denoted by $r'(x)$ and $r''(x)$ respectively. The partial derivatives of a function $m(x, y)$ with respect to the variables x and y are denoted by $m_x(x, y)$ and $m_y(x, y)$ or $(\partial/\partial x)m(x, y)$ and $(\partial/\partial y)m(x, y)$ respectively. The second-order partial derivatives of the function $m(x, y)$ with respect to the variables x and y are denoted by $m_{xx}(x, y)$ and $m_{yy}(x, y)$ or $(\partial^2/\partial x^2)m(x, y)$ and $(\partial^2/\partial y^2)m(x, y)$ respectively. The other partial derivatives of $m(x, y)$ can be denoted in the usual way. For $x, y \in R_+$, and some function $h(x, y)$ defined for $x, y \in R_+$, we set

$$A[x, y, h(s_1, t_1)] = \int_0^x \int_0^s \int_0^y \int_0^t h(s_1, t_1)\,dt_1\,dt\,ds_1\,ds.$$

Let $m \geq 1$, $n \geq 1$ be integers. We denote the differential operators D_1^n and D_2^m by

$$D_1^n z(x, y) = \frac{\partial^n z(x, y)}{\partial x^n}, \quad D_2^m z(x, y) = \frac{\partial^m z(x, y)}{\partial y^m},$$

where $z(x, y)$ is some function defined for $x, y \in R_+$. For $x, y \in R_+$, we set

$$B[x, y, h(s, t)] = \int_0^x \int_0^{s_{n-1}} \cdots \int_0^{s_1} \int_0^y \int_0^{t_{m-1}} \cdots \int_0^{t_1} h(s, t)$$
$$\times dt\,dt_1 \ldots dt_{m-1}\,ds\,ds_1 \ldots ds_{n-1},$$

where $s_0 = x$ and $t_0 = y$. It is easy to observe that $A[x, y, h] = A[y, x, h]$ and $B[x, y, h] = B[y, x, h]$.

A two independent variable Wendroff-like integral inequality established by Pachpatte (1988c) is given in the following theorem.

Theorem 4.5.1 *Let $a(x)$, $b(y)$, $p(x)$ and $q(y)$ be positive and twice continuously differentiable functions defined for $x, y \in R_+$, $a'(x)$, $b'(y)$,*

$p'(x)$ and $q'(y)$ be nonnegative for $x, y \in R_+$ and define $c(x, y) = a(x) + b(y) + yp(x) + xq(y)$ for $x, y \in R_+$. Let $u(x, y)$ and $h(x, y)$ be nonnegative continuous functions defined for $x, y \in R_+$ and

$$u(x, y) \leq c(x, y) + A[x, y, h(s_1, t_1)u(s_1, t_1)], \tag{4.5.1}$$

for $x, y \in R_+$.

(i) If $a''(x)$, $p''(x)$ are nonnegative for $x \in R_+$, then

$$u(x, y) \leq c(0, y) \exp\left[x\left(\frac{c_x(0, y)}{c(0, y)}\right) + \int_0^x \int_0^s \frac{a''(s_1) + yp''(s_1)}{c(s_1, 0)} ds_1 \, ds\right.$$

$$\left. + A[x, y, h(s_1, t_1)]\right], \tag{4.5.2}$$

for $x, y \in R_+$.

(ii) If $b''(y)$, $q''(y)$ are nonnegative for $y \in R_+$, then

$$u(x, y) \leq c(x, 0) \exp\left[y\left(\frac{c_y(x, 0)}{c(x, 0)}\right) + \int_0^y \int_0^t \frac{b''(t_1) + xq''(t_1)}{c(t_1, 0)} dt_1 \, dt\right.$$

$$\left. + A[y, x, h(s_1, t_1)]\right], \tag{4.5.3}$$

for $x, y \in R_+$.

$\square$

Proof: Define a function $z(x, y)$ by

$$z(x, y) = c(x, y) + A[x, y, h(s_1, t_1)u(s_1, t_1)]. \tag{4.5.4}$$

From (4.5.4) it is easy to observe that

$$z(0, y) = c(0, y) = a(0) + b(y) + yp(0), \tag{4.5.5}$$

$$z(x, 0) = c(x, 0) = a(x) + b(0) + xq(0), \tag{4.5.6}$$

$$z_x(0, y) = c_x(0, y) = a'(0) + yp'(0) + q(y), \tag{4.5.7}$$

$$z_{xx}(x, 0) = c_{xx}(x, 0) = a''(x), \tag{4.5.8}$$

$$z_{xxy}(x, 0) = c_{xxy}(x, 0) = p''(x), \tag{4.5.9}$$

and
$$z_{xxyy}(x, y) = h(x, y)u(x, y). \qquad (4.5.10)$$

Using the fact that $u(x, y) \leq z(x, y)$ in (4.5.10) we have
$$z_{xxyy}(x, y) \leq h(x, y)z(x, y). \qquad (4.5.11)$$

From (4.5.11) and the facts that $z_{xxy}(x, y) \geq 0$, $z_y(x, y) \geq 0$, $z(x, y) > 0$, we observe that
$$\frac{z_{xxyy}(x, y)}{z(x, y)} \leq h(x, y) + \frac{z_{xxy}(x, y)z_y(x, y)}{z^2(x, y)},$$

i.e.
$$\frac{\partial}{\partial y}\left(\frac{z_{xxy}(x, y)}{z(x, y)}\right) \leq h(x, y). \qquad (4.5.12)$$

By keeping x fixed in the inequality (4.5.12), we set $y = t_1$ and then integrating with respect to t_1 from 0 to y and using (4.5.6) and (4.5.9) we have
$$\frac{z_{xxy}(x, y)}{z(x, y)} \leq \frac{p''(x)}{c(x, 0)} + \int_0^y h(x, t_1)\, dt_1. \qquad (4.5.13)$$

Again as above, from (4.5.13) and the facts that $z_{xx}(x, y) \geq 0$, $z_y(x, y) \geq 0$, $z(x, y) > 0$ for $x, y \in R_+$, we observe that
$$\frac{\partial}{\partial y}\left(\frac{z_{xx}(x, y)}{z(x, y)}\right) \leq \frac{p''(x)}{c(x, 0)} + \int_0^y h(x, t_1)\, dt_1. \qquad (4.5.14)$$

By keeping x fixed in (4.5.14), set $y = t$ and then integrating with respect to t from 0 to y and using (4.5.6) and (4.5.8) we have
$$\frac{z_{xx}(x, y)}{z(x, y)} \leq \frac{a''(x) + yp''(x)}{c(x, 0)} + \int_0^y \int_0^t h(x, t_1)\, dt_1\, dt. \qquad (4.5.15)$$

As above, from (4.5.15) and the fact that $z_x(x, y) \geq 0$, $z(x, y) > 0$ for $x, y \in R_+$, we observe that
$$\frac{\partial}{\partial x}\left(\frac{z_x(x, y)}{z(x, y)}\right) \leq \frac{a''(x) + yp''(x)}{c(x, 0)} + \int_0^y \int_0^t h(x, t_1)\, dt_1\, dt. \qquad (4.5.16)$$

Now, keeping y fixed in (4.5.16), we set $x = s_1$, and then integrating with respect to s_1 from 0 to x and using (4.5.5) and (4.5.7) we have

$$\frac{z_x(x,y)}{z(x,y)} \le \frac{c_x(0,y)}{c(0,y)} + \int_0^x \frac{a''(s_1) + yp''(s_1)}{c(s_1,0)} ds_1$$

$$+ \int_0^x \int_0^y \int_0^t h(s_1, t_1) dt_1 \, dt \, ds_1. \quad (4.5.17)$$

By keeping y fixed in (4.5.17), set $x = s$ and then integrating with respect to s from 0 to x and using (4.5.5) we obtain

$$z(x,y) \le c(0,y) \exp\left[x\left(\frac{c_x(0,y)}{c(0,y)}\right) + \int_0^x \int_0^s \frac{a''(s_1) + yp''(s_1)}{c(s_1,0)} ds_1 \, ds \right.$$

$$\left. + A[x, y, h(s_1, t_1)] \right]. \quad (4.5.18)$$

Using (4.5.18) in $u(x,y) \le z(x,y)$ we obtain the inequality (4.5.2).

The proof of the inequality (4.5.3) follows by rewriting the definition of $z(x,y)$ in (4.5.4) in the form

$$z(x,y) = c(x,y) + A[y, x, h(s_1, t_1)u(s_1, t_1)],$$

and closely looking at the proof of the inequality (4.5.2) given above with suitable modifications. We omit the details. ∎

The following theorem deals with a more general Wendroff-like inequality established by Pachpatte (1988c).

Theorem 4.5.2 *Let $a(x)$, $b(y)$, $p(x)$, $q(y)$, $a'(x)$, $b'(y)$, $p'(x)$, $q'(y)$, $c(x,y)$, $u(x,y)$ and $h(x,y)$ be as in Theorem 4.5.1. Let $k(x,y)$ be a nonnegative continuous function defined for $x, y \in R_+$ and*

$$u(x,y) \le c(x,y) + A[x, y, h(s_1, t_1)(u(s_1, t_1)$$
$$+ A[s_1, t_1, k(\sigma_1, \eta_1)u(\sigma_1, \eta_1)])], \quad (4.5.19)$$

for $x, y \in R_+$.

(i) If $a''(x)$ and $p''(x)$ are nonnegative for $t \in R_+$, then

$$u(x, y) \leq c(0, y) + xc_x(0, y) + \int_0^x \int_0^s [a''(s_1) + yp''(s_1)]\, ds_1\, ds$$
$$+ A[x, y, h(s_1, t_1) F(s_1, t_1)], \qquad (4.5.20)$$

for $x, y \in R_+$, where

$$F(x, y) = c(0, y) + \exp\left[x\left(\frac{c_x(0, y)}{c(0, y)}\right) + \int_0^x \int_0^{s_2} \frac{a''(s_3) + yp''(s_3)}{c(s_3, 0)}\, ds_3\, ds_2\right.$$
$$\left. + A[x, y, h(\sigma_1, \eta_1) + k(\sigma_1, \eta_1)]\right], \qquad (4.5.21)$$

for $x, y \in R_+$.

(ii) If $b''(y)$ and $q''(y)$ are nonnegative for $y \in R_+$, then

$$u(x, y) \leq c(x, 0) + yc_y(x, 0) + \int_0^y \int_0^t [b''(t_1) + xq''(t_1)]\, dt_1\, dt$$
$$+ A[y, x, h(s_1, t_1) G(s_1, t_1)], \qquad (4.5.22)$$

for $x, y \in R_+$, where

$$G(x, y) = c(x, 0) + \exp\left[y\left(\frac{c_y(x, 0)}{c(x, 0)}\right) + \int_0^y \int_0^{t_2} \frac{b''(t_3) + xq''(t_3)}{c(0, t_3)}\, dt_3\, dt_2\right.$$
$$\left. + A[y, x, h(\sigma_1, \eta_1) + k(\sigma_1, \eta_1)]\right], \qquad (4.5.23)$$

for $x, y \in R_+$. $\square$

Proof: Define a function $z(x, y)$ by

$$z(x, y) = c(x, y) + A[x, y, h(s_1, t_1)(u(s_1, t_1)$$
$$+ A[s_1, t_1, k(\sigma_1, \eta_1) u(\sigma_1, \eta_1)])]. \qquad (4.5.24)$$

From (4.5.24) we see that the relations (4.5.5)–(4.5.9) hold and

$$z_{xxyy}(x, y) = h(x, y)(u(x, y) + A[x, y, k(\sigma_1, \eta_1)u(\sigma_1, \eta_1)]). \quad (4.5.25)$$

Using the fact that $u(x, y) \leq z(x, y)$ in (4.5.25) we have

$$z_{xxyy}(x, y) \leq h(x, y)(z(x, y) + A[x, y, k(\sigma_1, \eta_1)z(\sigma_1, \eta_1)]). \quad (4.5.26)$$

Define a function $v(x, y)$ by

$$v(x, y) = z(x, y) + A[x, y, k(\sigma_1, \eta_1)z(\sigma_1, \eta_1)]. \quad (4.5.27)$$

From (4.5.27) it is easy to observe that $v(0, y) = z(0, y)$, $v(x, 0) = z(x, 0)$, $v_x(0, y) = z_x(0, y)$, $v_{xx}(x, 0) = z_{xx}(x, 0)$, $v_{xxy}(x, 0) = z_{xxy}(x, 0)$ and hence the relations in (4.5.5)–(4.5.9) are true and

$$v_{xxyy}(x, y) = z_{xxyy}(x, y) + k(x, y)z(x, y). \quad (4.5.28)$$

Using the facts that $z_{xxyy}(x, y) \leq h(x, y)v(x, y)$ from (4.5.26) and $z(x, y) \leq v(x, y)$ from (4.5.27) in (4.5.28) we have

$$v_{xxyy}(x, y) \leq [h(x, y) + k(x, y)]v(x, y). \quad (4.5.29)$$

Now by following the same argument as in the proof of inequality (4.5.2) in Theorem 4.5.1 we obtain the estimate

$$v(x, y) \leq F(x, y), \quad (4.5.30)$$

where $F(x, y)$ is defined by (4.5.21). Using (4.5.30) in (4.5.26) and following the last argument as in the proof of (4.5.2) of Theorem 4.5.1 with suitable modifications we obtain the estimate

$$z(x, y) \leq c(0, y) + xc_x(0, y) + \int_0^x \int_0^s [a''(s_1) + yp''(s_1)] \, ds_1 \, ds$$
$$+ A[x, y, h(s_1, t_1)F(s_1, t_1)]. \quad (4.5.31)$$

Using (4.5.31) in $u(x, y) \leq z(x, y)$ we get the required inequality in (4.5.20). The proof of the inequality (4.5.22) follows by the argument as mentioned at the end of the proof of Theorem 4.5.1.

∎

Pachpatte (1988c) gave the following inequality, which can be used in certain applications.

Theorem 4.5.3 Let $a_i(x)$, $b_i(y)$, $p_i(x)$, $q_i(y)$ for $i = 1, 2$, be positive and twice continuously differentiable functions for $x, y \in R_+$, $a_i'(x)$, $b_i'(y)$, $p_i'(x)$, $q_i'(y)$ for $i = 1, 2$, are nonnegative for $x, y \in R_+$ and define $c_i(x, y) = a_i(x) + b_i(y) + yp_i(x) + xq_i(y)$ for $i = 1, 2$ and $c(x, y) = c_1(x, y) + c_2(x, y)$ for $x, y \in R_+$. Let $u(x, y)$, $v(x, y)$, $h_j(x, y)$, $j = 1, 2, 3, 4$, are nonnegative continuous functions defined for $x, y \in R_+$ and define $h(x, y) = \max\{[h_1(x, y) + h_3(x, y)], [h_2(x, y) + h_4(x, y)]\}$, for $x, y \in R_+$ and suppose

$$u(x, y) \leq c_1(x, y) + A[x, y, h_1(s_1, t_1)u(s_1, t_1)]$$
$$+ A[x, y, h_2(s_1, t_1)\bar{v}(s_1, t_1)], \qquad (4.5.32)$$

$$v(x, y) \leq c_2(x, y) + A[x, y, h_3(s_1, t_1)\bar{u}(s_1, t_1)]$$
$$+ A[x, y, h_4(s_1, t_1)v(s_1, t_1)], \qquad (4.5.33)$$

holds for $x, y \in R_+$, where $\bar{u}(x, y) = \exp(-\mu(x + y))u(x, y)$, $\bar{v}(x, y) = \exp(\mu(x + y))v(x, y)$ for $x, y \in R_+$ and μ is a nonnegative constant.

(i) If $a_i''(x)$, $p_i''(x)$ for $i = 1, 2$, are nonnegative for $x \in R_+$ and $a(x) = a_1(x) + a_2(x)$, $p(x) = p_1(x) + p_2(x)$ for $x \in R_+$, then

$$u(x, y) \leq c(0, y) \exp\left[\mu(x + y) + x\left(\frac{c_x(0, y)}{c(0, y)}\right)\right.$$
$$\left. + \int_0^x \int_0^s \frac{a''(s_1) + yp''(s_1)}{c(s_1, 0)} ds_1\, ds + A[x, y, h(s_1, t_1)]\right], \quad (4.5.34)$$

$$v(x, y) \leq c(0, y) \exp\left[x\left(\frac{c_x(0, y)}{c(0, y)}\right) + \int_0^x \int_0^s \frac{a''(s_1) + yp''(s_1)}{c(s_1, 0)} ds_1\, ds\right.$$
$$\left. + A[x, y, h(s_1, t_1)]\right], \qquad (4.5.35)$$

for $x, y \in R_+$.

(ii) If $b_i''(y)$, $q_i''(y)$ for $i = 1, 2$, are nonnegative for $y \in R_+$ and $b(y) = b_1(y) + b_2(y)$, $q(y) = q_1(y) + q_2(y)$ for $y \in R_+$, then

$$u(x, y) \le c(x, 0) \exp\left[\mu(x+y) + y\left(\frac{c_y(x,0)}{c(x,0)}\right)\right.$$

$$\left. + \int_0^y \int_0^t \frac{b''(t_1) + xq''(t_1)}{c(0, t_1)} dt_1\, dt + A[y, x, h(s_1, t_1)]\right], \quad (4.5.36)$$

$$v(x, y) \le c(x, 0) \exp\left[y\left(\frac{c_y(x,0)}{c(x,0)}\right) + \int_0^y \int_0^t \frac{b''(t_1) + xq''(t_1)}{c(0, t_1)} dt_1\, dt\right.$$

$$\left. + A[y, x, h(s_1, t_1)]\right], \quad (4.5.37)$$

for $x, y \in R_+$.

□

Proof: From (4.5.32) it is easy to observe that

$$\bar{u}(x, y) \le c_1(x, y) + A[x, y, h_1(s_1, t_1)\bar{u}(s_1, t_1)]$$
$$+ A[x, y, h_2(s_1, t_1)v(s_1, t_1)]. \quad (4.5.38)$$

Define $z(x, y) = \bar{u}(x, y) + v(x, y)$; then (4.5.38) and (4.5.33) lead to

$$z(x, y) \le c(x, y) + A[x, y, h(s_1, t_1)z(s_1, t_1)]. \quad (4.5.39)$$

The bounds in (4.5.34), (4.5.35) and (4.5.36), (4.5.37) follow from an application of Theorem 4.5.1 and splitting. The proof is complete.

■

The inequalities in the following theorem are established by Pachpatte (1988b), and can be used in certain new applications.

Theorem 4.5.4 *Let $u(x, y)$, $a(x, y)$, $h(x, y)$ and $k(x, y)$ be nonnegative continuous functions defined for $x, y \in R_+$ and let $a(x, y)$ be nondecreasing in each variable $x, y \in R_+$.*

(i) If

$$u(x, y) \le a(x, y) + B[x, y, h(s, t)u(s, t)], \quad (4.5.40)$$

for $x, y \in R_+$, then

$$u(x, y) \le a(x, y) \exp(B[x, y, h(s, t)]), \quad (4.5.41)$$

for $x, y \in R_+$.

(ii) If

$$u(x, y) \leq a(x, y) + B[x, y, h(s, t)(u(s, t) + B[s, t, k(\sigma, \eta)u(\sigma, \eta)])], \quad (4.5.42)$$

for $x, y \in R_+$, then

$$u(x, y) \leq a(x, y)[1 + B[x, y, h(s, t)\exp(B[s, t, (h(\sigma, \eta) + k(\sigma, \eta))])]], \quad (4.5.43)$$

for $x, y \in R_+$.

$\square$

Proof: (i) First we assume that $a(x, y) > 0$ for $x, y \in R_+$. Then from (4.5.40) we observe that

$$\frac{u(x, y)}{a(x, y)} \leq 1 + B\left[x, y, h(s, t)\frac{u(s, t)}{a(s, t)}\right]. \quad (4.5.44)$$

Define a function $v(x, y)$ by

$$v(x, y) = 1 + B\left[x, y, h(s, t)\frac{u(s, t)}{a(s, t)}\right]. \quad (4.5.45)$$

From (4.5.45) it is easy to observe that

$$D_2^m D_1^n v(x, y) = h(x, y)\frac{u(x, y)}{a(x, y)}. \quad (4.5.46)$$

Using the fact that $u(x, y)/a(x, y) \leq v(x, y)$ in (4.5.46) we have

$$D_2^m D_1^n v(x, y) \leq h(x, y)v(x, y). \quad (4.5.47)$$

From (4.5.47) and using the facts that $v(x, y)$ is positive, $D_2 v(x, y)$ and $D_2^{m-1} D_1^n v(x, y)$ are nonnegative for $x, y \in R_+$, we observe that

$$\frac{D_2^m D_1^n v(x, y)}{v(x, y)} \leq h(x, y) + \frac{[D_2 v(x, y)][D_2^{m-1} D_1^n v(x, y)]}{v^2(x, y)},$$

i.e.

$$D_2\left(\frac{D_2^{m-1} D_1^n v(x, y)}{v(x, y)}\right) \leq h(x, y). \quad (4.5.48)$$

By keeping x fixed in (4.5.48), we set $y = t$; then integrating with respect to t from 0 to y and using the fact that $D_2^{m-1} D_1^n v(x, 0) = 0$, for $m = 2, 3, \ldots,$ we have

$$\frac{D_2^{m-1}D_1^n v(x,y)}{v(x,y)} \leq \int_0^y h(x,t)\,dt. \tag{4.5.49}$$

Again, as above, from (4.5.49) and the fact that $v(x,y)$ is positive, $D_2 v(x,y)$ and $D_2^{m-2}D_1^n v(x,y)$ are nonnegative for $x, y \in R_+$, we observe that

$$D_2\left(\frac{D_2^{m-2}D_1^n v(x,y)}{v(x,y)}\right) \leq \int_0^y h(x,t)\,dt. \tag{4.5.50}$$

By keeping x fixed in (4.5.50), set $y = t_1$; then integrating with respect to t_1 from 0 to y and using the fact that $D_2^{m-2}D_1^n v(x,0) = 0$, $m = 3, 4, \ldots$, we have

$$\frac{D_2^{m-2}D_1^n v(x,y)}{v(x,y)} \leq \int_0^y \int_0^{t_1} h(x,t)\,dt\,dt_1.$$

Continuing in this way we obtain

$$\frac{D_1^n v(x,y)}{v(x,y)} \leq \int_0^y \int_0^{t_{m-1}} \cdots \int_0^{t_1} h(x,t)\,dt\,dt_1\ldots dt_{m-1}. \tag{4.5.51}$$

From (4.5.51) and the facts that $v(x,y)$ is positive, $D_1 v(x,y)$ and $D_1^{n-1} v(x,y)$ are nonnegative for $x, y \in R_+$, we observe that

$$\frac{D_1^n v(x,y)}{v(x,y)} \leq \int_0^y \int_0^{t_{m-1}} \cdots \int_0^{t_1} h(x,t)\,dt\,dt_1\ldots dt_{m-1}$$
$$+ \frac{[D_1 v(x,y)][D_1^{n-1} v(x,y)]}{v^2(x,y)},$$

i.e.

$$D_1\left(\frac{D_1^{n-1} v(x,y)}{v(x,y)}\right) \leq \int_0^y \int_0^{t_{m-1}} \cdots \int_0^{t_1} h(x,t)\,dt\,dt_1\ldots dt_{m-1}. \tag{4.5.52}$$

Now keeping y fixed in (4.5.52), set $x = s$; then integrating with respect to s from 0 to x and using the fact that $D_1^{n-1} v(0, y) = 0$, $n = 2, 3, \ldots$, we have

$$\frac{D_1^{n-1} v(x, y)}{v(x, y)} \leq \int_0^x \int_0^y \int_0^{t_{m-1}} \cdots \int_0^{t_1} h(s, t) \, dt \, dt_1 \ldots dt_{m-1} \, ds.$$

Continuing in this way, we obtain

$$\frac{D_1 v(x, y)}{v(x, y)} \leq \int_0^x \int_0^{s_{n-2}} \cdots \int_0^{s_1} \int_0^y \int_0^{t_{m-1}} \cdots \int_0^{t_1} h(s, t) \, dt \, dt_1 \ldots dt_{m-1}$$
$$\times ds \, ds_1 \ldots ds_{n-2}. \tag{4.5.53}$$

Now keeping y fixed in (4.5.53), set $x = s_{n-1}$; then integrating with respect to s_{n-1} from 0 to x and using the fact that $v(0, y) = 1$, we have

$$\log v(x, y) \leq B[x, y, h(s, t)],$$

which implies

$$v(x, y) \leq \exp(B[x, y, h(s, t)]). \tag{4.5.54}$$

Using (4.5.54) in (4.5.44), we get the required inequality in (4.5.41). If $a(x, y) = 0$; then, by following the argument as in the proof of Theorem 4.2.2, we conclude that (4.5.41) holds.

(ii) The proof can be completed by following the arguments used in the proof of (i) and in view of the proof of Theorem 4.5.2 given above. Here the details are left to the reader.

∎

4.6 Snow's Inequalities

Snow (1971, 1972) gave one of the first Gronwall-type integral inequalities involving two independent variables for scalar and vector functions by using the notion of a Riemann function. The inequalities given in Snow (1971, 1972) have significant applications in the study of various properties of the solutions of partial differential equations. This section deals with the inequalities obtained by Snow (1971, 1972). The proof of the main theorem in these references involves a second-order partial differential inequality which is integrated by using Riemann's method (see Sneddon, 1957, p. 120]).

Snow (1971) established a useful generalization of Theorem 4.2.1 in the following form.

Theorem 4.6.1 *Suppose $u(x, y)$, $a(x, y)$ and $b(x, y)$ are continuous on a domain D with $b \geq 0$ there. Let $P_0(x_0, y_0)$ and $P(x, y)$ be two points in D such that $(x - x_0)(y - y_0) \geq 0$ and let R be the rectangular region whose opposite corners are the points P_0 and P. Let $v(s, t; x, y)$ be the solution of the characteristic initial value problem*

$$L[v] = v_{st} - b(s, t)v = 0, \quad v(s, y) = v(x, t) = 1,$$

and let D^+ be a connected subdomain of D which contains P and on which $v \geq 0$ (Figure 4.1). If $R \subset D^+$ and $u(x, y)$ satisfies

$$u(x, y) \leq a(x, y) + \int_{x_0}^{x}\int_{y_0}^{y} b(s, t)u(s, t)\,ds\,dt, \qquad (4.6.1)$$

then $u(x, y)$ also satisfies

$$u(x, y) \leq a(x, y) + \int_{x_0}^{x}\int_{y_0}^{y} a(s, t)b(s, t)v(s, t; x, y)\,ds\,dt. \qquad (4.6.2)$$

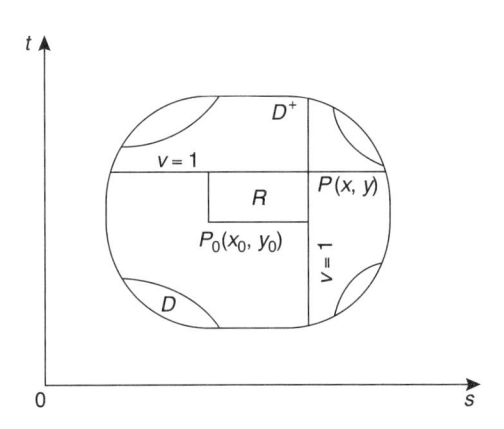

□

Figure 4.1

We note that the function $v(s, t; x, y)$ in (4.6.2) is a Riemann function relative to the point $P(x, y)$ for the self-adjoint operator L. There is a subdomain D^+ containing P on which $v \geq 0$ since $v = 1$ on the vertical and horizontal lines through P and since v is continuous.

Proof: Define a function $z(x, y)$ by

$$z(x, y) = \int_{x_0}^{x} \int_{y_0}^{y} b(s, t) u(s, t) \, ds \, dt. \tag{4.6.3}$$

Then $z(x_0, y) = z(x, y_0) = 0$, $u(x, y) \leq a(x, y) + z(x, y)$ and

$$z_{xy} = bu$$
$$\leq b(a + z),$$

i.e.

$$L[z] = z_{xy} - bz \leq ab. \tag{4.6.4}$$

The operator L is self-adjoint and hyperbolic (Sneddon, 1957). For any twice continuously differentiable z and v the operator L satisfies the identity

$$vL[z] - zL[v] = -(zv_y)_x + (vz_x)_y.$$

Let P_0 and P be any points as in the theorem and label the directed sides and corners of the rectangle as shown in Figure 4.2.

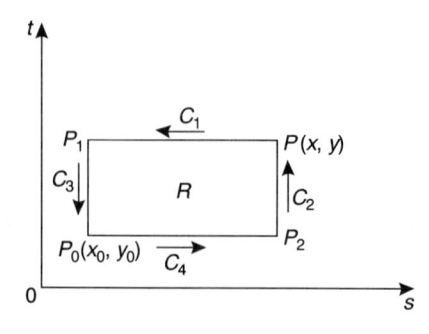

Figure 4.2

Using s and t as the independent variables, we integrate the identity over R and use Green's theorem to obtain

$$\iint_R (vL[z] - zL[v]) \, ds \, dt = - \int_{C_1+C_2+C_3+C_4} vz_s \, ds + zv_t \, dt$$

$$= - \int_{C_1+C_4} vz_s \, ds - \int_{C_2+C_3} zv_t \, dt.$$

This holds for any function in C^2.

For the particular function z defined in (4.6.3) we have $z = 0$ on C_3 and $z = z_s = 0$ on C_4, so the right-hand side reduces to

$$- \int_{C_1} vz_s \, ds - \int_{C_2} zv_t \, dt. \tag{4.6.5}$$

Now suppose v satisfies

$$L[v] = v_{st} - bv = 0, \tag{4.6.6}$$

$$v = 1 \text{ on } C_1, \tag{4.6.7}$$

$$v_t = 0 \text{ on } C_2. \tag{4.6.8}$$

Then (4.6.7) and (4.6.8) imply that

$$v = 1 \text{ on } C_2. \tag{4.6.9}$$

Since $v \geq 0$ on R and $z(P_0) = 0$, by using (4.6.4) identity (4.6.5) becomes

$$z(P) \leq \iint_R vab \, ds \, dt.$$

This gives us an upper bound on the integral term in (4.6.1) so that (4.6.2) results.

The proof will be complete when we show the existence of a function v satisfying conditions (4.6.6), (4.6.7) and (4.6.9). But (4.6.6), (4.6.7) and (4.6.9) specify the Riemann function for the operator L relative to the point P and the existence and continuity of the Riemann function is well known and may be demonstrated by the method of successive approximations (Courant and Hilbert, 1962). Continuity and the fact that $v = 1$ on C_1 and C_2 guarantee

358 MULTIDIMENSIONAL LINEAR INTEGRAL INEQUALITIES

that there is a domain D^+ on which $v \geq 0$. Properties and specific examples of v are discussed by Copson (1958).

■

We note that by the method of proof, the right-hand side of the final inequality (4.6.2) is the solution to the Volterra integral equation obtained by using equality in place of inequality in (4.6.1). Since this function is an upper bound for any solution of inequality (4.6.1), it is the maximal solution of the inequality. To use this method to solve a Volterra integral equation, the restriction $v \geq 0$, and hence the subdomain D^+, are unnecessary.

We give now three corollaries that follow easily from the above theorem.

Corollary 4.6.1 If

$$u \leq \int_{x_0}^{x} \int_{y_0}^{y} bu \, ds \, dt, \quad b \geq 0,$$

then $u \leq 0$.

△

Proof: Set $a = 0$ in Theorem 4.6.1.

■

Corollary 4.6.2 If a is a constant, then

$$u(x, y) \leq a v(x_0, y_0; x, y),$$

and (since the Riemann function is symmetric in its variables)

$$u(x, y) \leq a v(x, y; x_0, y_0).$$

△

Proof: For a a constant, Theorem 4.6.1 gives

$$u(x, y) \leq a \left(1 + \int_{x_0}^{x} \int_{y_0}^{y} b(s, t) v(s, t; x, y) \, ds \, dt \right).$$

Since v satisfies equation (4.6.6), the integrand here is v_{st}. Integrating and using the initial conditions for v, we obtain $u(x, y) \leq a v(x_0, y_0; x, y)$.

■

Corollary 4.6.3 If inequality (4.6.1) is reversed, then so is inequality (4.6.2).

△

Proof: Since $b \geq 0$, all the remaining inequalities in the proof are reversed.

∎

Snow (1972) gave a further generalization of the result given in Theorem 4.6.1 to a system of integral inequalities. The main result in Snow (1972) is obtained by reducing the vector inequality to a vector differential inequality and then integrating it by generalizing Riemann's method to apply to vector hyperbolic partial differential equations.

The main result established by Snow (1972) reads as follows.

Theorem 4.6.2 *Suppose $u(x, y)$ and $a(x, y)$ are continuous n-vector functions on a domain D and $B(x, y)$ is a continuous symmetric, nonnegative (i.e. $b_{ij} \geq 0$, all i and j), $n \times n$ matrix function on D. Let $P_0(x_0, y_0)$ and $P(x, y)$ be two points in D such that $(x - x_0)(y - y_0) \geq 0$ and let R be the rectangular region whose diagonal is the line joining the points P_0 and P. Let $V(s, t; x, y)$ be the $n \times n$ matrix function satisfying the matrix characteristic initial value problem*

$$L[V] = V_{st} - B(s, t)V = 0, \quad V(s, y) = V(x, t) = I,$$

where I is the identity matrix. Let D^+ be the connected subdomain of D containing P and on which $V(s, t; x, y)$ is nonnegative (Figure 4.3). If $R \subset D^+$ and u satisfies

$$u(x, y) \leq a(x, y) + \int_{x_0}^{x} \int_{y_0}^{y} B(s, t)u(s, t)\, ds\, dt, \quad (4.6.10)$$

where the inequality holds componentwise, then u also satisfies

$$u(x, y) \leq a(x, y) + \int_{x_0}^{x} \int_{y_0}^{y} V^T(s, t; x, y)B(s, t)a(s, t)\, ds\, dt. \quad (4.6.11)$$

□

The function $V(s, t; x, y)$ is a matrix generalization of a Riemann function relative to the point $P(x, y)$ for the self-adjoint operator L. There

![Figure 4.3](figure-4.3)

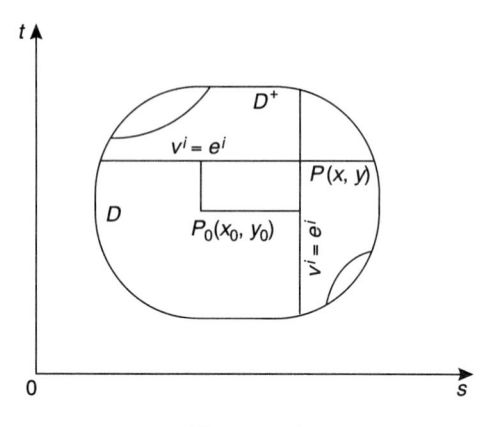

Figure 4.3

are such a function and region D^+ on which $V \geq 0$, as proven in the following Lemma.

Lemma 4.6.1 *Let $B(s, t)$ be a continuous matrix function. Then the matrix characteristic initial value problem*

$$L[V] = V_{st} - B(s, t)V = 0, \qquad (4.6.12)$$

$$V(s, y) = V(x, t) = I, \qquad (4.6.13)$$

where I is the identity matrix, has a unique solution $V(s, t; x, y)$ for s and t near to $P(x, y)$ and satisfying $(s - x)(t - y) \geq 0$. This solution is continuous and if $B(s, t)$ is nonnegative, so is $V(s, t; x, y)$.

△

Proof: The proof is by a successive approximations argument. It is easily seen that the integral equation

$$V(s, t) = I + \int_x^s \int_y^t B(\sigma, \eta) V(\sigma, \eta) \, d\sigma \, d\eta, \qquad (4.6.14)$$

is equivalent to equation (4.6.12) with conditions (4.6.13).

Let T represent the transformation

$$TV = \int_x^s \int_y^t B(\sigma, \eta) V(\sigma, \eta) \, d\sigma \, d\eta,$$

so equation (4.6.14) is $V = I + TV$.

Let $V_0(s,t) \equiv I$ and define $V_{n+1} = I + TV_n$. Since TV is continuous if V is, it follows immediately by induction that V_n is defined and continuous for all n. Let $\|\cdot\|$ be any matrix norm. Then since $(s-x)(t-y) \geq 0$,

$$\|TV\| = \left\| \int_x^s \int_y^t B(\sigma, \eta) V(\sigma, \eta) \, d\sigma \, d\eta \right\|$$

$$\leq \int_x^s \int_y^t \|B(\sigma, \eta)\| \|V(\sigma, \eta)\| \, d\sigma \, d\eta$$

$$\leq \tfrac{1}{2} \max \|V(\sigma, \eta)\|, \qquad (4.6.15)$$

if we restrict s and t to be close enough to (x, y).

Then

$$\|V_{k+1} - V_k\| = \|T(V_k - V_{k-1})\|$$

$$\leq \tfrac{1}{2} \max \|V_k - V_{k-1}\| \leq \cdots \leq 2^{-k} \max \|V_1 - V_0\|.$$

Since $V_{n+1} = V_0 + \sum_{k=0}^{n}(V_{k+1} - V_k)$, V_{n+1} is the nth partial sum of a matrix series dominated in norm by a convergent geometric series, namely $\max \|V_1 - V_0\| \sum_{k=0}^{\infty} 2^{-k}$.

Therefore the matrix sequence $\{V_n\}$ converges uniformly on the domain where (4.6.15) holds. Since each V_n is continuous, the limit function $V(s, t; x, y)$ is also. To see that V is a solution to (4.6.14), note that $I + TV = I + T(\lim V_n) = I + \lim TV_n = \lim(I + TV_n) = \lim V_{n+1} = V$. To see that V is unique, suppose W also a solution. Then $V - W = T(V - W)$, so

$$\|V - W\| = \|T(V - W)\| \leq \tfrac{1}{2} \max \|V - W\|,$$

which is possible only if $\|V - W\| = 0$, i.e. $V = W$.

Now suppose $B(s,t) \geq 0$. Then, $V \geq 0$, $TV \geq 0$, since $(s-x)(t-y) \geq 0$. Since $V_0 \equiv I \geq 0$, it follows by induction that $V_n \geq 0$ for all n. But then the limit function satisfies $V(s,t;x,y) \geq 0$ also.

∎

Proof of Theorem 4.6.2: Let

$$z(x, y) = \int_{x_0}^{x} \int_{y_0}^{y} B(s, t) u(s, t) \, ds \, dt. \qquad (4.6.16)$$

Then $z_{xy} = B(x, y)u(x, y)$ and since $B \geq 0$ and (4.6.10) holds,

$$z_{xy} = Bu \leq B(a + z),$$

or

$$L[z] = z_{xy} - Bz \leq Ba. \qquad (4.6.17)$$

This is a hyperbolic vector partial differential inequality for z. The initial conditions for z are

$$z(x_0, y) = z(x, y_0) = 0. \qquad (4.6.18)$$

The operator L is self-adjoint. We note that for any $z, v \in C^2$,

$$v^T L[z] - z^T L[v] = v^T z_{xy} - v^T Bz - z^T v_{xy} + z^T Bv.$$

The terms here are all scalars, and since B is symmetric, the second and fourth terms on the right cancel. The right-hand side is

$$= -(z^T v_y)_x + z_x^T v_y + (v^T z_x)_y - v_y^T z_x$$
$$= -(z^T v_y)_x + (v^T z_x)_y. \qquad (4.6.19)$$

For P_0 and P as required in the hypothesis we label the directed sides and corners of the rectangle R as shown in Figure 4.4.

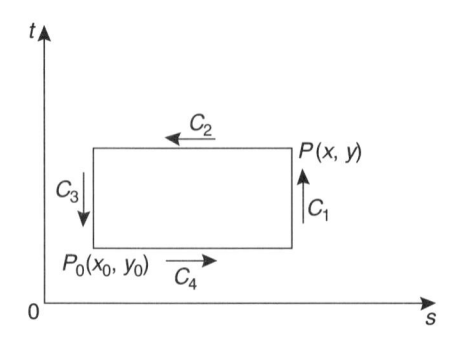

Figure 4.4

Using s and t as the independent variables in identity (4.6.19), integrating over R and using Green's theorem, we get

$$\iint_R (v^T L[z] - z^T L[v])\, ds\, dt = \iint_R (-(z^T v_t)_s + (v^T z_s)_t)\, ds\, dt$$

$$= -\int_C v^T z_s\, ds + z^T v_t\, dt$$

$$= -\int_{C_2+C_4} v^T z_s\, ds - \int_{C_1+C_3} z^T v_t\, dt.$$

(4.6.20)

This holds for any functions in C^2.

For any $z \in C^2$ which also satisfies the initial conditions (4.6.18), $z = 0$ on C_3 and $z = z_s = 0$ on C_4. Thus (4.6.20) reduces to

$$\iint_R (v^T L[z] - z^T L[v])\, ds\, dt = -\int_{C_2} v^T z_s\, ds - \int_{C_1} z^T v_t\, dt. \quad (4.6.21)$$

Now suppose the vectors $v^i(s, t; x, y)$ are the columns of the matrix $V(s, t; x, y)$ of Lemma 4.6.1. Then $L[v^i] = 0$ and $v^i(s, y) = v^i(x, t) = e^i$, the ith column of the identity matrix. Thus $v_t^i = 0$ on C_1, so (4.6.21) reduces to

$$\iint_R v^{iT} L[z]\, ds\, dt = -\int_{C_2} e^i z_s\, ds = z_i(P),$$

where the subscript refers to the component of the vector. Using the matrix V this becomes

$$z(P) = \iint_R V^T(s, t; x, y) L[z]\, ds\, dt.$$

For z defined by (4.6.16), since (4.6.17) holds and since $V \geq 0$ by Lemma 4.6.1, we get

$$z(P) \leq \iint_R V^T B a\, ds\, dt. \quad (4.6.22)$$

This gives an upper bound for the integral term in (4.6.10) so that (4.6.11) results. The proof is complete.

The matrix V is a generalization of a scalar Riemann function, and when u, a and B are scalars it reduces to the Riemann function relative to the point $P(x, y)$ for the operator L.

Note that, by the method of proof, if equality were to hold in (4.6.10) there would have been equality in (4.6.17) and (4.6.22) regardless of the nonnegativity of V, so the right-hand side of (4.6.11) is the solution to the Volterra integral equation corresponding to (4.6.10). Since the right-hand side of (4.6.11) is a solution of the inequality (4.6.10) and is an upper bound for all such solutions, it is the maximal solution of (4.6.10).

4.7 Generalizations of Snow's Inequalities

Ghoshal and Masood (1974a,b) obtained some generalizations of Snow's inequalities given in Snow (1971, 1972) which can be used as tools in the study of various properties of the solutions of some self-adjoint partial differential equations of the hyperbolic type. This section presents the results given by Ghoshal and Masood (1974a,b).

A generalization of Theorem 4.6.1 established by Ghoshal and Masood (1974a) is contained in the following.

Theorem 4.7.1 *Let $u(x, y)$, $g(x, y)$, $h(x, y)$, $p(x, y)$ and $q(x, y)$ be continuous functions with $h(x, y) \geq 0$ on a domain D and $P_0(x_0, y_0)$ and $P(x, y)$ be two points such that $(x - x_0)(y - y_0) \geq 0$ on D. Let R denote the rectangular region whose opposite corners are P_0 and P. Let $v(s, t; x, y)$ be the solution of the characteristic initial value problem*

$$M[v] = 0, \qquad (4.7.1)$$

where M is the adjoint operator of the operator L defined by

$$L[w] = w_{st} + aw_s + bw_t + cw, \qquad (4.7.2)$$

where $a = -h(x, y)q(x, y)$, $b = -h(x, y)p(x, y)$, $c = -h(x, y)$. The function $v(s, t; x, y)$ is called the well-known Riemann function for the partial differential operator L and satisfies all the properties of a Riemann function for an operator with continuous coefficients. Let D^+ be a connected subdomain of D which contains P and on which $v \geq 0$ (Figure 4.5). If $R \subset D^+$ and $u(x, y)$ satisfies

$$u(x, y) \le g(x, y) + p(x, y) \int_{x_0}^{x} h(s, y)u(s, y)\,ds$$

$$+ q(x, y) \int_{y_0}^{y} h(x, t)u(x, t)\,dt + \int_{x_0}^{x}\int_{y_0}^{y} h(s, t)u(s, t)\,ds\,dt, \quad (4.7.3)$$

then $u(x, y)$ also satisfies

$$u(x, y) \le g(x, y) + p(x, y) \int_{x_0}^{x} h(s, y)u(s, y)\,ds + q(x, y) \int_{y_0}^{y} h(x, t)u(x, t)\,dt$$

$$+ \int_{x_0}^{x}\int_{y_0}^{y} g(s, t)h(s, t)v(s, t; x, y)\,ds\,dt. \quad (4.7.4)$$

Further, if $q(x, y) = 0$ we obtain

$$u(x, y) \le g(x, y) + \int_{x_0}^{x}\int_{y_0}^{y} g(s, t)h(s, t)v(s, t; x, y)\,ds\,dt$$

$$+ p(x, y) \int_{x_0}^{x} h(s, y) \left[g(s, y) + \int_{x_0}^{s}\int_{y_0}^{y} g(\sigma, t)h(\sigma, t) \right.$$

$$\left. \times v(\sigma, t; s, y)\,d\sigma\,dt \right] \exp\left(\int_{s}^{x} h(\sigma, y)p(\sigma, y)\,d\sigma \right) ds. \quad (4.7.5)$$

Also, if $p(x, y) = 0$, then

$$u(x, y) \le g(x, y) + \int_{x_0}^{x}\int_{y_0}^{y} g(s, t)h(s, t)v(s, t; x, y)\,ds\,dt$$

$$+ q(x, y) \int_{y_0}^{y} h(x, t) \left[g(x, t) + \int_{x_0}^{x}\int_{y_0}^{t} g(s, \eta)h(s, \eta) \right.$$

$$\left. \times v(s, \eta; x, t)\,ds\,d\eta \right] \exp\left(\int_{t}^{y} h(x, \eta)q(x, \eta)\,d\eta \right) dt. \quad (4.7.6)$$

$\square$

Proof: Let

$$z(x, y) = \int_{x_0}^{x} \int_{y_0}^{y} h(s, t) u(s, t) \, ds \, dt. \qquad (4.7.7)$$

Then

$$z(x_0, y) = z(x, y_0) = 0, \qquad (4.7.8)$$

and

$$z_{xy}(x, y) = h(x, y) u(x, y)$$
$$\leq h(x, y)[g(x, y) + p(x, y) z_y(x, y) + q(x, y) z_x(x, y) + z(x, y)],$$

i.e.,

$$L[z] = z_{xy} + a z_x + b z_y + c z \leq hg, \qquad (4.7.9)$$

where $a = -hq$, $b = -hp$, $c = -h$.

Now for any two twice continuously differentiable functions z and v, the operators L and M satisfy the identity

$$vL[z] - zM[v] = \left(azv + \frac{v z_y}{2} - \frac{v_y z}{2} \right)_x + \left(bzv + \frac{v z_x}{2} - \frac{z v_x}{2} \right)_y, \qquad (4.7.10)$$

where M is the adjoint operator of L.

Let R be a rectangular region with corners $P_0(x_0, y_0)$, $P_1(x, y_0)$, $P(x, y)$ and $P_2(x_0, y)$, so that $P_0 P$ is the diagonal, as shown in Figure 4.5.

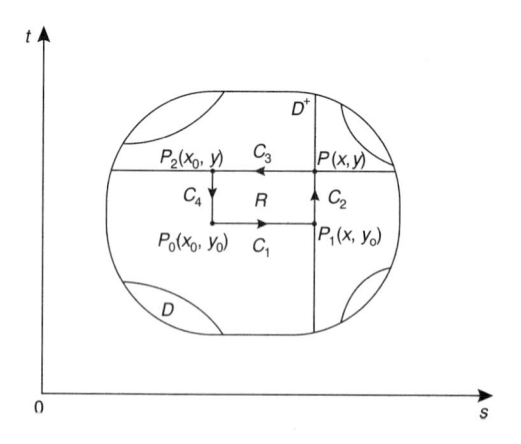

Figure 4.5

Since z and v are twice continuously differentiable functions, by using Green's theorem we obtain from (4.7.10)

$$\iint_R (vL[z] - zM[v])\,ds\,dt = \int_{C_1+C_2+C_3+C_4} \left[\left(azv + \frac{vz_t}{2} - \frac{zv_t}{2}\right)dt \right.$$
$$\left. - \left(bzv + \frac{vz_s}{2} - \frac{zv_s}{2}\right)ds\right].$$

Since z is zero on C_1 and C_4 and also ds does not vary on C_2 and dt does not vary on C_3, we get

$$\iint_R (vL[z] - zM[v])\,ds\,dt = \int_{C_2} \left(azv + \frac{vz_t}{2} - \frac{zv_t}{2}\right)dt$$
$$- \int_{C_3} \left(bzv + \frac{vz_s}{2} - \frac{zv_s}{2}\right)ds.$$

Integrating the right-hand side by parts along the characteristic segments C_2 and C_3 to eliminate partial derivatives of z, we obtain

$$\iint_R (vL[z] - zM[v])\,ds\,dt = \int_{C_2} (av - v_t)z\,dt - \int_{C_3} (bv - v_s)z\,ds$$
$$+ z(P)v(P) - \frac{z(P_2)v(P_2)}{2} - \frac{z(P_1)v(P_1)}{2}$$
$$= \int_{C_2} (av - v_t)z\,dt - \int_{C_3} (bv - v_s)z\,ds$$
$$+ z(P)v(P). \qquad (4.7.11)$$

Now, since $v(s, t; x, y)$ is the solution of the characteristic initial value problem $M[v] = 0$, it is by definition the Riemann function $v(s, t; x, y) = v(s, t)$ associated with the partial differential equation $L[z] = 0$ such that

$$v(x, y; x, y) = v(x, y) = v(P) = 1,$$

$$v_t = av \text{ on } C_2, \quad v(x, t) = \exp\left(\int_y^t a(x, \eta)\,d\eta\right),$$

and

$$v_s = bv \text{ on } C_3, \quad v(s, y) = \exp\left(\int_x^s b(\sigma, y) \, d\sigma\right).$$

So we get, from the identity (4.7.11)

$$z(x, y) = \int_{x_0}^{x} \int_{y_0}^{y} v(s, t; x, y) L[z] \, ds \, dt. \tag{4.7.12}$$

From (4.7.12) and (4.7.9) we get

$$z(x, y) \leq \int_{x_0}^{x} \int_{y_0}^{y} v(s, t; x, y) g(s, t) h(s, t) \, ds \, dt.$$

Thus

$$u(x, y) \leq g(x, y) + p(x, y) \int_{x_0}^{x} h(s, y) u(s, y) \, ds + q(x, y) \int_{y_0}^{y} h(x, t) u(x, t) \, dt$$

$$+ \int_{x_0}^{x} \int_{y_0}^{y} g(s, t) h(s, t) v(s, t; x, y) \, ds \, dt,$$

which is the required inequality in (4.7.4).

Now let $q(x, y) = 0$ and

$$m(x, y) = g(x, y) + \int_{x_0}^{x} \int_{y_0}^{y} g(s, t) h(s, t) v(s, t; x, y) \, ds \, dt. \tag{4.7.13}$$

Then the inequality (4.7.4) reduces to

$$u(x, y) \leq m(x, y) + p(x, y) \int_{x_0}^{x} h(s, y) u(s, y) \, ds. \tag{4.7.14}$$

MULTIDIMENSIONAL LINEAR INTEGRAL INEQUALITIES   369

The inequality (4.7.14) may be treated as a one-dimensional inequality for any fixed y, which by applying Theorem 1.3.2 yields

$$u(x, y) \leq m(x, y) + p(x, y) \int_{x_0}^{x} h(s, y) m(s, y) \exp\left(\int_{s}^{x} h(\sigma, y) p(\sigma, y) \,d\sigma\right) ds,$$
(4.7.15)

Using (4.7.13) in (4.7.15) we get the required inequality in (4.7.5).

Further, substituting $p(x, y) = 0$ in (4.7.4) and following a similar argument to the above we obtain the desired bound in (4.7.6).

∎

Remark 4.7.1 Note that in the special case when $p(x, y) = q(x, y) = 0$, the result established in Theorem 4.7.1 reduces to the inequality given by Snow (1971) in Theorem 4.6.1.

△

The following theorem established by Ghoshal and Masood (1974b) gives a generalization of a result given by Snow (1972).

Theorem 4.7.2 *Let $u(x, y)$ and $g(x, y)$ be continuous n-vector functions on a domain D, and $p(x, y)$, $h(x, y)$, $q(x, y)$ and $r(x, y)$ be continuous symmetric $n \times n$ matrix functions with $h(x, y)$ nonnegative on D ($r(x, y)$ may be a constant matrix K). Let $P_0(x_0, y_0)$ and $P(x, y)$ be two points in D such that $(x - x_0)(y - y_0) \geq 0$, R be the rectangular region whose diagonal is PP_0, and $V(s, t; x, y) = V(s, t)$ be the $n \times n$ matrix function satisfying the matrix characteristic initial value problem*

$$M[V] = 0,$$
(4.7.16)

where M is the adjoint of the operator L given by

$$L[w] = w_{st} + aw_s + bw_t + cw$$
(4.7.17)

where $a = -hq$, $b = -hp$, $c = -hr = -m$, with the boundary conditions

$$V(x, y; x, y) = V(x, y) = V(P) = I,$$
(4.7.18)

$$V(x, t) = \exp\left(\int_{y}^{t} a(x, \eta) \,d\eta\right),$$
(4.7.19)

$$V(s, y) = \exp\left(\int_x^s b(\sigma, y) \, d\sigma\right), \tag{4.7.20}$$

where I is the identity matrix and $V(s, t; x, y)$ is called Riemann's function. Let D^+ be a connected subdomain of D, which contains P and on which $V \geq 0$ (Figure 4.5). If $R \subset D^+$ and $u(x, y)$ satisfies

$$u(x, y) \leq g(x, y) + p(x, y) \int_{x_0}^x h(s, y) u(s, y) \, ds + q(x, y) \int_{y_0}^y h(x, t) u(x, t) \, dt$$

$$+ r(x, y) \int_{x_0}^x \int_{y_0}^y h(s, t) u(s, t) \, ds \, dt, \tag{4.7.21}$$

where the inequality holds componentwise, then $u(x, y)$ satisfies

$$u(x, y) \leq g(x, y) + p(x, y) \int_{x_0}^x h(s, y) u(s, y) \, ds + q(x, y) \int_{y_0}^y h(x, t) u(x, t) \, dt$$

$$+ r(x, y) \int_{x_0}^x \int_{y_0}^y V^T(s, t) h(s, t) g(s, t) \, ds \, dt. \tag{4.7.22}$$

Here the relation holds for every column of $V(s, t; x, y)$. Further if $q(x, y) = 0$, then

$$u(x, y) \leq g(x, y) + r(x, y) \int_{x_0}^x \int_{y_0}^y V^T(s, t) h(s, t) g(s, t) \, ds \, dt$$

$$+ p(x, y) \int_{x_0}^x h(s, y) \left[g(s, y) + r(s, y) \int_{x_0}^s \int_{y_0}^y V^T(\sigma, t) \right.$$

$$\left. \times h(\sigma, t) g(\sigma, t) \, d\sigma \, dt \right] \exp\left(\int_s^x h(\xi, y) p(\xi, y) \, d\xi\right) ds. \tag{4.7.23}$$

Also, if $p(x, y) = 0$, then

$$u(x, y) \leq g(x, y) + r(x, y) \int_{x_0}^{x} \int_{y_0}^{y} V^T(s, t) h(s, t) g(s, t) \, ds \, dt$$

$$+ q(x, y) \int_{y_0}^{y} h(x, t) \left[g(x, t) + r(x, t) \int_{x_0}^{x} \int_{y_0}^{t} V^T(s, t') \right.$$

$$\left. \times h(s, t') g(s, t') \, ds \, dt' \right] \exp\left(\int_{t}^{y} h(x, \eta) q(x, \eta) \, d\eta \right) dt. \quad (4.7.24)$$

$\square$

The function $V(s, t)$ is the matrix generalization of Riemann's function (Garabedian 1964) relative to the point $P(x, y)$ for the non-self-adjoint operator M. In the proof of the above theorem we need the following general version of Lemma 4.6.1 given by Ghoshal and Masood (1974b).

Lemma 4.7.1 *Let $m(s, t) = h(s, t) r(s, t)$, $a(s, t)$ and $b(s, t)$ be continuous matrix functions. Then the matrix characteristic initial value problem (4.7.16) i.e. $M[V] = 0$, which is the adjoint of the equation $L[w] = w_{st} + aw_s + bw_t - mw$ with the boundary conditions (4.7.18)–(4.7.20) has a unique solution $V(s, t; x, y)$ for (s, t) near to $P(x, y)$ and satisfying $(s - x)(t - y) \geq 0$. The solution is continuous, and if a, b, m are nonnegative, so is V.*

$\triangle$

Proof: Equation (4.7.16) together with the conditions (4.7.18)–(4.7.20) is equivalent to the Volterra integral equation

$$V(s, t) = I + \int_{x}^{s} b(\sigma, y) V(\sigma, y) \, d\sigma + \int_{y}^{t} a(x, \eta) V(x, \eta) \, d\eta$$

$$+ \int_{x}^{s} \int_{y}^{t} m(\sigma, \eta) V(\sigma, \eta) \, d\sigma \, d\eta, \quad (4.7.25)$$

since $M[V] = V_{xy} - (aV)_x - (bV)_y - mV = 0$.

Let T be the transformation

$$TV = \int_x^s b(\sigma, y)V(\sigma, y)\,d\sigma + \int_y^t a(x, \eta)V(x, \eta)\,d\eta$$

$$+ \int_x^s \int_y^t m(\sigma, \eta)V(\sigma, \eta)\,d\sigma\,d\eta, \qquad (4.7.26)$$

so that the integral equation (4.7.25) can be written as

$$V = I + TV. \qquad (4.7.27)$$

The rest of the proof can be completed by following the same steps as in the proof of Lemma 4.6.1, with suitable modifications.

■

Proof of Theorem 4.7.2: Let

$$z(x, y) = \int_{x_0}^x \int_{y_0}^y h(s, t)u(s, t)\,ds\,dt. \qquad (4.7.28)$$

Then

$$z(x_0, y) = z(x, y_0) = 0, \qquad (4.7.29)$$

and

$$z_{xy}(x, y) = h(x, y)u(x, y)$$
$$\leq h(x, y)[g(x, y) + p(x, y)z_y(x, y) + q(x, y)z_x(x, y)$$
$$+ r(x, y)z(x, y)],$$

or

$$L[z] = z_{xy} + az_x + bz_y + cz \leq hg, \qquad (4.7.30)$$

where $a = -hq$, $b = -hp$, $c = -hr$. Now for any two functions $z, V \in C^2$, we have

$$V^T L[z] - z^T M[V] = [V^T(z_{xy} + az_x + bz_y + cz)$$
$$- z^T(V_{xy} - aV_x - bV_y - (c - a_x - b_y)V].$$

MULTIDIMENSIONAL LINEAR INTEGRAL INEQUALITIES

Here the relations are scalar and hold true for each column of V. If a, b, c are symmetric $n \times n$ matrices then we can show that

$$V^T L[z] - z^T M[V] = \left(z^T a V + \frac{V^T z_y}{2} - \frac{z^T V_y}{2} \right)_x$$
$$+ \left(V^T bz + \frac{V^T z_x}{2} - \frac{z^T V_x}{2} \right)_y.$$

Taking the region R as in Figure 4.5 and applying Green's theorem, we get

$$\iint_R (V^T L[z] - z^T M[V]) \, ds \, dt = \int_{C_1+C_2+C_3+C_4} \left[\left(z^T a V + \frac{V^T z_t}{2} - \frac{z^T V_t}{2} \right) dt \right.$$
$$\left. - \left(V^T bz + \frac{V^T z_s}{2} - \frac{z^T V_s}{2} \right) ds \right].$$

Since z is zero on C_1 and C_4 and also ds does not vary on C_2 and dt does not vary on C_3 (i.e. $z_t = 0$ on C_4, $z_s = 0$ on C_1), we get

$$\iint_R (V^T L[z] - z^T M[V]) \, ds \, dt = \int_{C_2} (z^T (aV - V_t) + \tfrac{1}{2}(V^T z)_t) \, dt$$
$$- \int_{C_3} (z^T (bV - V_s) + \tfrac{1}{2}(V^T z)_s) \, ds.$$

Here the relations are scalar relations and hold for each column of V. If V is the Riemann function with the conditions (4.7.18)–(4.7.20), i.e. $aV - V_t = 0$ on C_2, $bV - V_s = 0$ on C_3 and $V(x, y) = I = V^T(P)$, and $M[V] = 0$, then we obtain

$$\iint_R V^T L[z] \, ds \, dt = V^T(x, y)z(x, y) - \tfrac{1}{2} z^T(x, y_0) V(x, y_0)$$
$$- \tfrac{1}{2} z(x_0, y) V^T(x_0, y)$$
$$= z(x, y),$$

so that

$$z(x, y) = \int_{x_0}^{x} \int_{y_0}^{y} V^T(s, t) L[z] \, ds \, dt$$

$$\leq \int_{x_0}^{x}\int_{y_0}^{y} V^T(s,t)h(s,t)g(s,t)\,ds\,dt.$$

Hence

$$u(x,y) \leq g(x,y) + p(x,y)\int_{x_0}^{x} h(s,y)u(s,y)\,ds$$

$$+ q(x,y)\int_{y_0}^{y} h(x,t)u(x,t)\,dt$$

$$+ r(x,y)\int_{x_0}^{x}\int_{y_0}^{y} V^T(s,t)h(s,t)g(s,t)\,ds\,dt,$$

which is the required inequality in (4.7.22).

The proofs of the inequalities in (4.7.23) and (4.7.24) can be completed by the last arguments used in the proof of Theorem 4.7.1 with suitable changes. The details are omitted here.

∎

Remark 4.7.2 In the special case when $p(x,y) = q(x,y) = 0$, the inequality given in Theorem 4.7.2 reduces to the inequality established by Snow (1972) in Theorem 4.6.2.

△

4.8 Pachpatte's Inequalities IV

Pachpatte (1979a, 1980c,d) investigated a number of new integral and integro-differential inequalities in two independent variables which can be used in the analysis of various problems in the theory of partial differential and integral equations as ready and powerful tools. This section deals with some basic inequalities given in Pachpatte (1979a, 1980c,d). An elementary technique of reducing the integral inequality to second-order partial differential inequality and then integrating it by Riemann's method is used to establish the results.

Pachpatte (1979a) established the following inequality.

Theorem 4.8.1 *Suppose $u(x, y)$, $a(x, y)$, $b(x, y)$, $c(x, y)$ and $\sigma(x, y)$ are nonnegative continuous functions on a domain D. Let $P_0(x_0, y_0)$ and $P(x, y)$ be two points in D such that $(x - x_0)(y - y_0) \geq 0$ and let R be the rectangular region whose opposite corners are the points P_0 and P. Let $v(s, t; x, y)$ be the solution of the characteristic initial value problem*

$$L[v] = v_{st} - [b(s, t) + c(s, t)]v = 0, \quad v(s, y) = v(x, t) = 1,$$

and let D^+ be a connected subdomain of D which contains P and on which $v \geq 0$ (Figure 4.1). If $R \subset D^+$ and $u(x, y)$ satisfies

$$u(x, y) \leq a(x, y) + \int_{x_0}^{x} \int_{y_0}^{y} b(s, t) u(s, t) \, ds \, dt$$

$$+ \int_{x_0}^{x} \int_{y_0}^{y} b(s, t) \left[\sigma(s, t) + \int_{x_0}^{s} \int_{y_0}^{t} c(\xi, \eta) u(\xi, \eta) \, d\xi \, d\eta \right] ds \, dt, \quad (4.8.1)$$

then $u(x, y)$ also satisfies

$$u(x, y) \leq a(x, y) + \int_{x_0}^{x} \int_{y_0}^{y} b(s, t) \left[a(s, t) + \sigma(s, t) \right.$$

$$+ \int_{x_0}^{s} \int_{y_0}^{t} [a(\xi, \eta) c(\xi, \eta) + b(\xi, \eta)(a(\xi, \eta) + \sigma(\xi, \eta))]$$

$$\left. \times v(\xi, \eta; s, t) \, d\xi \, d\eta \right] ds \, dt, \quad (4.8.2)$$

where the function $v(s, t; x, y)$ involved in (4.8.2) is the Riemann function relative to the point $P(x, y)$ for self-adjoint operator L. □

Proof: Define a function $\phi(x, y)$ such that

$$\phi(x, y) = \int_{x_0}^{x} \int_{y_0}^{y} b(s, t) u(s, t) \, ds \, dt$$

$$+ \int_{x_0}^{x} \int_{y_0}^{y} b(s, t) \left[\sigma(s, t) + \int_{x_0}^{s} \int_{y_0}^{t} c(\xi, \eta) u(\xi, \eta) \, d\xi \, d\eta \right] ds \, dt,$$

then $\phi(x_0, y) = \phi(x, y_0) = 0$, $u(x, y) \leq a(x, y) + \phi(x, y)$ from (4.8.1) and

$$\phi_{xy}(x, y) = b(x, y) \left[u(x, y) + \sigma(x, y) + \int_{x_0}^{x} \int_{y_0}^{y} c(\xi, \eta) u(\xi, \eta) \, d\xi \, d\eta \right]$$

$$\leq b(x, y) \left[a(x, y) + \phi(x, y) + \sigma(x, y) \right.$$

$$\left. + \int_{x_0}^{x} \int_{y_0}^{y} c(\xi, \eta) [a(\xi, \eta) + \phi(\xi, \eta)] \, d\xi \, d\eta \right]. \quad (4.8.3)$$

If we put

$$\psi(x, y) = \phi(x, y) + \int_{x_0}^{x} \int_{y_0}^{y} c(\xi, \eta) [a(\xi, \eta) + \phi(\xi, \eta)] \, d\xi \, d\eta, \quad (4.8.4)$$

then $\psi(x_0, y) = \psi(x, y_0) = 0$, and

$$\psi_{xy}(x, y) = \phi_{xy}(x, y) + c(x, y)[a(x, y) + \phi(x, y)]. \quad (4.8.5)$$

Using $\phi_{xy}(x, y) \leq b(x, y)[a(x, y) + \sigma(x, y) + \psi(x, y)]$ from (4.8.3) and $\phi(x, y) \leq \psi(x, y)$ from (4.8.4) in (4.8.5) we have

$$\psi_{xy}(x, y) \leq [b(x, y) + c(x, y)]\psi(x, y) + a(x, y)c(x, y)$$
$$+ b(x, y)[a(x, y) + \sigma(x, y)],$$

i.e.

$$L[\psi] = \psi_{xy}(x, y) - [b(x, y) + c(x, y)]\psi(x, y)$$
$$\leq a(x, y)c(x, y) + b(x, y)[a(x, y) + \sigma(x, y)]. \quad (4.8.6)$$

The operator L is self-adjoint and hyperbolic. For any twice continuously differentiable functions ψ and v the operator L satisfies the identity

$$vL[\psi] - \psi L[v] = -(\psi v_y)_x + (v\psi_x)_y.$$

Let P_0 and P be any points as in Theorem 4.8.1 and label the directed sides and corners of the rectangle R as shown in Figure 4.2.

Using s and t as the independent variables, we integrate the identity over R and use Green's theorem to obtain

$$\iint_R (vL[\psi] - \psi L[v])\, ds\, dt = -\int_{C_1+C_2+C_3+C_4} (v\psi_s\, ds + \psi v_t\, dt)$$

$$= -\int_{C_1+C_4} v\psi_s\, ds - \int_{C_2+C_3} \psi v_t\, dt.$$

This holds for any functions in C^2.

For the particular function ψ defined earlier we have $\psi = 0$ on C_3 and $\psi = \psi_s = 0$ on C_4, so the right-hand side reduces to

$$-\int_{C_1} v\psi_s\, ds - \int_{C_2} \psi v_t\, dt. \tag{4.8.7}$$

Now suppose v satisfies

$$L[v] = v_{st} - [b(s,t) + c(s,t)]v = 0, \tag{4.8.8}$$

$$v = 1 \text{ on } C_1, \tag{4.8.9}$$

$$v_t = 0 \text{ on } C_2. \tag{4.8.10}$$

Then (4.8.9) and (4.8.10) imply that

$$v = 1 \text{ on } C_2. \tag{4.8.11}$$

Since $v \geq 0$ on R and $\psi(P_0) = 0$, by using (4.8.6) identity (4.8.7) becomes

$$\psi(P) \leq \iint_R v[a(s,t)c(s,t) + b(s,t)(a(s,t) + \sigma(s,t))]\, ds\, dt,$$

i.e.

$$\psi(x,y) \leq \int_{x_0}^x \int_{y_0}^y [a(s,t)c(s,t) + b(s,t)(a(s,t) + \sigma(s,t))]v(s,t;x,y)\, ds\, dt.$$

Substituting this bound on $\psi(x,y)$ in (4.8.3) we obtain

$$\phi_{xy}(x,y) \leq b(x,y) \left[a(x,y) + \sigma(x,y) + \int_{x_0}^x \int_{y_0}^y [a(s,t)c(s,t) \right.$$

$$\left. + b(s,t)(a(s,t) + \sigma(s,t))]v(s,t;x,y)\, ds\, dt \right],$$

which implies the estimation for $\phi(x,y)$ such that

$$\phi(x, y) \leq \int_{x_0}^{x}\int_{y_0}^{y} b(s, t) \left[a(s, t) + \sigma(s, t) + \int_{x_0}^{s}\int_{y_0}^{t} [a(\xi, \eta)c(\xi, \eta) \right.$$

$$\left. + b(\xi, \eta)(a(\xi, \eta) + \sigma(\xi, \eta))]v(\xi, \eta; s, t)\,d\xi\,d\eta \right] ds\,dt.$$

Now, substituting this bound on $\phi(x, y)$ in $u(x, y) \leq a(x, y) + \phi(x, y)$, we get the desired bound in (4.8.2).

∎

Remark 4.8.1 Note that, as in Snow (1971), Theorem 4.8.1 in the special case when $a = 0$, $\sigma = 0$ or $a = $ constant can be used in some applications.

△

Pachpatte (1980c) has investigated the following useful inequality.

Theorem 4.8.2 *Suppose $u(x, y)$, $a(x, y)$, $b(x, y)$, $c(x, y)$, $p(x, y)$ and $q(x, y)$ are nonnegative continuous functions defined on a domain D. Let $P_0(x_0, y_0)$ and $P(x, y)$ be two points in D such that $(x - x_0)(y - y_0) \geq 0$ and R the rectangular region whose opposite corners are the points P_0 and P. Let $v(s, t; x, y)$ and $w(s, t; x, y)$ be the solutions of the characteristic initial value problems*

$$L[v] = v_{st} - [p(s, t) + b(s, t)(c(s, t) + q(s, t))]v = 0,$$

$$v(s, y) = v(x, t) = 1,$$

and

$$M[w] = w_{st} - [b(s, t)c(s, t) - p(s, t)]w = 0,$$

$$w(s, y) = w(x, t) = 1,$$

respectively and let D^+ be a connected subdomain of D which contains P and on which $v \geq 0$ and $w \geq 0$ (Figure 4.1). If $R \subset D^+$ and $u(x, y)$ satisfies

$$u(x, y) \leq a(x, y) + b(x, y) \left[\int_{x_0}^{x}\int_{y_0}^{y} c(s, t)u(s, t)\,ds\,dt \right.$$

$$\left. + \int_{x_0}^{x}\int_{y_0}^{y} p(s, t) \left(\int_{x_0}^{s}\int_{y_0}^{t} q(\xi, \eta)u(\xi, \eta)\,d\xi\,d\eta \right) ds\,dt \right], \quad (4.8.12)$$

then $u(x, y)$ also satisfies

$$u(x, y) \le a(x, y) + b(x, y) \left[\int_{x_0}^{x} \int_{y_0}^{y} w(s, t; x, y) \left\{ a(s, t)c(s, t) \right. \right.$$

$$\left. \left. + p(s, t) \int_{x_0}^{s} \int_{y_0}^{t} a(\xi, \eta)[c(\xi, \eta) + q(\xi, \eta)]v(\xi, \eta; s, t) \, d\xi \, d\eta \right\} ds \, dt \right].$$

(4.8.13)

$\square$

Proof: Define a function $\phi(x, y)$ such that

$$\phi(x, y) = \int_{x_0}^{x} \int_{y_0}^{y} c(s, t)u(s, t) \, ds \, dt$$

$$+ \int_{x_0}^{x} \int_{y_0}^{y} p(s, t) \left(\int_{x_0}^{s} \int_{y_0}^{t} q(\xi, \eta)u(\xi, \eta) d\xi \, d\eta \right) ds \, dt,$$

then $\phi(x_0, y) = \phi(x, y_0) = 0$, $u(x, y) \le a(x, y) + b(x, y)\phi(x, y)$ from (4.8.12) and

$$\phi_{xy}(x, y) = c(x, y)u(x, y) + p(x, y) \left(\int_{x_0}^{x} \int_{y_0}^{y} q(\xi, \eta)u(\xi, \eta) \, d\xi \, d\eta \right)$$

$$\le c(x, y)[a(x, y) + b(x, y)\phi(x, y)]$$

$$+ p(x, y) \left(\int_{x_0}^{x} \int_{y_0}^{y} q(\xi, \eta)[a(\xi, \eta) + b(\xi, \eta)\phi(\xi, \eta)] \, d\xi \, d\eta \right).$$

Adding $p(x, y)\phi(x, y)$ to both sides of the above inequality we have

$$\phi_{xy}(x, y) + p(x, y)\phi(x, y) \le c(x, y)[a(x, y) + b(x, y)\phi(x, y)]$$

$$+ p(x, y) \left[\phi(x, y) + \int_{x_0}^{x} \int_{y_0}^{y} q(\xi, \eta) \right.$$

$$\left. \times [a(\xi, \eta) + b(\xi, \eta)\phi(\xi, \eta)] \, d\xi \, d\eta \right], (4.8.14)$$

If we put

$$\psi(x, y) = \phi(x, y) + \int_{x_0}^{x} \int_{y_0}^{y} q(\xi, \eta)[a(\xi, \eta) + b(\xi, \eta)\phi(\xi, \eta)] \, d\xi \, d\eta, \quad (4.8.15)$$

then $\psi(x_0, y) = \psi(x, y_0) = 0$, and

$$\psi_{xy}(x, y) = \phi_{xy}(x, y) + q(x, y)[a(x, y) + b(x, y)\phi(x, y)]. \quad (4.8.16)$$

Using $\phi_{xy}(x, y) \leq c(x, y)[a(x, y) + b(x, y)\phi(x, y)] + p(x, y)\psi(x, y)$ from (4.8.14) and $\phi(x, y) \leq \psi(x, y)$ from (4.8.15) in (4.8.16) we have

$$\psi_{xy}(x, y) \leq a(x, y)[c(x, y) + q(x, y)] + [p(x, y) + b(x, y)$$
$$\times (c(x, y) + q(x, y)]\psi(x, y),$$

i.e.

$$L[\psi] = \psi_{xy}(x, y) - [p(x, y) + b(x, y)(c(x, y) + q(x, y))]\psi(x, y)$$
$$\leq a(x, y)[c(x, y) + q(x, y)]. \quad (4.8.17)$$

The operator L is self-adjoint and hyperbolic. For any twice continuously differentiable functions ψ and v the operator L satisfies the identity

$$vL[\psi] - \psi L[v] = -(\psi v_y)_x + (v\psi_x)_y. \quad (4.8.18)$$

Let P_0 and P be any points as in Theorem 4.8.2 and label the directed sides and corners of the rectangle R as shown in Figure 4.2.

Using s and t as the independent variables, we integrate the identity (4.8.18) over R and use Green's theorem to obtain

$$\iint_R (vL[\psi] - \psi L[v]) \, ds \, dt = -\int_{C_1+C_2+C_3+C_4} (v\psi_s \, ds + \psi v_t \, dt)$$
$$= -\int_{C_1+C_4} v\psi_s \, ds - \int_{C_2+C_3} \psi v_t \, dt.$$

This holds for any functions in C^2.

For the particular function ψ defined earlier we have $\psi = 0$ on C_3 and $\psi = \psi_s = 0$ on C_4, so the right-hand side in the above identity reduces to

$$-\int_{C_1} v\psi_s \, ds - \int_{C_2} \psi v_t \, dt. \qquad (4.8.19)$$

Now suppose v satisfies

$$L[v] = v_{st} - [p(s,t) + b(s,t)(c(s,t) + q(s,t))]v = 0, \qquad (4.8.20)$$

$$v = 1 \text{ on } C_1, \qquad (4.8.21)$$

$$v_t = 0 \text{ on } C_2. \qquad (4.8.22)$$

Then (4.8.21) and (4.8.22) imply that

$$v = 1 \text{ on } C_2. \qquad (4.8.23)$$

Since $v \geq 0$ on R and $\psi(P_0) = 0$, by using (4.8.17) identity (4.8.19) becomes

$$\psi(P) \leq \iint_R v[a(s,t)[c(s,t) + q(s,t)]] \, ds \, dt,$$

i.e.

$$\psi(x,y) \leq \int_{x_0}^{x} \int_{y_0}^{y} a(s,t)[c(s,t) + q(s,t)]v(s,t;x,y) \, ds \, dt.$$

Substituting this bound on $\psi(x,y)$ in (4.8.14) we obtain

$$M[\phi] = \phi_{xy}(x,y) - [b(x,y)c(x,y) - p(x,y)]\phi(x,y)$$

$$\leq \left[a(x,y)c(x,y) + p(x,y) \int_{x_0}^{x} \int_{y_0}^{y} a(s,t)[c(s,t) + q(s,t)] \right.$$

$$\left. \times v(s,t;x,y) \, ds \, dt \right].$$

Again, by following the argument as above we obtain the estimate for $\phi(x,y)$ such that

$$\phi(x, y) \le \int_{x_0}^{x}\int_{y_0}^{y} w(s, t; x, y)\left[a(s, t)c(s, t) + p(s, t)\int_{x_0}^{s}\int_{y_0}^{t} a(\xi, \eta)\right.$$

$$\left. \times [c(\xi, \eta) + q(\xi, \eta)]v(\xi, \eta; s, t)\,d\xi\,d\eta\right]ds\,dt.$$

Now substituting this bound on $\phi(x, y)$ in $u(x, y) \le a(x, y) + b(x, y)\phi(x, y)$ we obtain the desired bound in (4.8.13). ∎

Another interesting and useful inequality established by Pachpatte (1980c) is embodied in the following theorem.

Theorem 4.8.3 *Suppose $u(x, y)$, $a(x, y)$, $b(x, y)$, $c(x, y)$, $p(x, y)$ and $q(x, y)$ are nonnegative continuous functions defined on a domain D. Let $P_0(x_0, y_0)$ and $P(x, y)$ be two points in D such that $(x - x_0)(y - y_0) \ge 0$ and R be the rectangular region whose opposite corners are the points P_0 and P. Let $v(s, t; x, y)$ and $w(s, t; x, y)$ be the solutions of the characteristic initial value problems*

$$L[v] = v_{st} - b(s, t)[c(s, t) + p(s, t) + q(s, t)]v = 0,$$

$$v(s, y) = v(x, t) = 1,$$

and

$$M[w] = w_{st} - b(s, t)c(s, t)w = 0,$$

$$w(s, y) = w(x, t) = 1,$$

respectively and let D^+ be a connected subdomain of D which contains P and on which $v \ge 0$ and $w \ge 0$ (Figure 4.1). If $R \subset D^+$ and $u(x, y)$ satisfies

$$u(x, y) \le a(x, y) + b(x, y)\left[\int_{x_0}^{x}\int_{y_0}^{y} c(s, t)u(s, t)\,ds\,dt\right.$$

$$\left. + \int_{x_0}^{x}\int_{y_0}^{y} p(s, t)\left(u(s, t) + b(s, t)\int_{x_0}^{s}\int_{y_0}^{t} q(\xi, \eta)u(\xi, \eta)\,d\xi\,d\eta\right)ds\,dt\right],$$

(4.8.24)

then $u(x, y)$ also satisfies

$$u(x, y) \leq a(x, y) + b(x, y) \left[\int_{x_0}^{x} \int_{y_0}^{y} w(s, t; x, y) \left\{ a(s, t)(c(s, t) + p(s, t)) \right. \right.$$

$$+ b(s, t) p(s, t) \int_{x_0}^{s} \int_{y_0}^{t} a(\xi, \eta)[c(\xi, \eta) + p(\xi, \eta) + q(\xi, \eta)]$$

$$\left. \left. \times v(\xi, \eta; s, t) \, d\xi \, d\eta \right\} ds \, dt \right]. \tag{4.8.25}$$

$\square$

The proof of this theorem follows by an argument similar to that in the proof of Theorem 4.8.2, with suitable modifications. The details are omitted here.

The following inequalities established by Pachpatte (1979a) can be used in certain situations.

Theorem 4.8.4 *Suppose $u(x, y)$, $a(x, y)$, $b(x, y)$, $c(x, y)$ and $h(x, y)$ are nonnegative continuous functions on a domain D. Let $P_0(x_0, y_0)$ and $P(x, y)$ be two points in D such that $(x - x_0)(y - y_0) \geq 0$ and let R be the rectangular region whose opposite corners are the points P_0 and P. Let $v(s, t; x, y)$ and $w(s, t; x, y)$ be the solutions of the characteristic initial value problems*

$$L[v] = v_{st} - [b(s, t) + c(s, t) + h(s, t)]v = 0, \quad v(s, y) = v(x, t) = 1,$$

and

$$M[w] = w_{st} - b(s, t)w = 0, \quad w(s, y) = w(x, t) = 1,$$

respectively and let D^+ be a connected subdomain of D which contains P and on which $v \geq 0$ and $w \geq 0$ (Figure 4.1). If $R \subset D^+$ and $u(x, y)$ satisfies

$$u(x, y) \leq a(x, y) + \int_{x_0}^{x} \int_{y_0}^{y} b(s, t) u(s, t) \, ds \, dt$$

$$+ \int_{x_0}^{x} \int_{y_0}^{y} b(s, t) \left(\int_{x_0}^{s} \int_{y_0}^{t} c(\xi, \eta) u(\xi, \eta) \, d\xi \, d\eta \right) ds \, dt$$

$$+ \int_{x_0}^{x} \int_{y_0}^{y} b(s,t) \left(\int_{x_0}^{s} \int_{y_0}^{t} c(\xi, \eta) \right.$$

$$\left. \times \left(\int_{x_0}^{\xi} \int_{y_0}^{\eta} h(\xi_1, \eta_1) u(\xi_1, \eta_1) \, d\xi_1 \, d\eta_1 \right) d\xi \, d\eta \right) ds \, dt, \quad (4.8.26)$$

then $u(x, y)$ also satisfies

$$u(x, y) \leq a(x, y) + \int_{x_0}^{x} \int_{y_0}^{y} b(s,t) \left[a(s,t) + \int_{x_0}^{s} \int_{y_0}^{t} w(\xi, \eta; s, t) \right.$$

$$\times \left[a(\xi, \eta) b(\xi, \eta) + c(\xi, \eta) \right.$$

$$\times \left\{ a(\xi, \eta) + \int_{x_0}^{\xi} \int_{y_0}^{\eta} v(\xi_1, \eta_1; \xi, \eta) a(\xi_1, \eta_1) [b(\xi_1, \eta_1) \right.$$

$$\left. + c(\xi_1, \eta_1) + h(\xi_1, \eta_1)] \, d\xi_1 \, d\eta_1 \right\} \bigg] d\xi \, d\eta \bigg] ds \, dt. \quad (4.8.27)$$

$\square$

Theorem 4.8.5 *Suppose $u(x, y)$, $a(x, y)$, $b(x, y)$, $c(x, y)$, $h(x, y)$, $p(x, y)$ and $q(x, y)$ are nonnegative continuous functions on a domain D. Let $P_0(x_0, y_0)$ and $P(x, y)$ be two points in D such that $(x - x_0)(y - y_0) \geq 0$ and let R be the rectangular region whose opposite corners are the points P_0 and P. Let $v(s, t; x, y)$ and $w(s, t; x, y)$ and $e(s, t; x, y)$ be the solutions of the characteristic initial value problems*

$$L[v] = v_{st} - [b(s,t) + c(s,t) + h(s,t) + p(s,t) + q(s,t)]v = 0,$$

$$v(s, y) = v(x, t) = 1,$$

and

$$M[w] = w_{st} - [b(s,t) + c(s,t) + h(s,t) - p(s,t)]w = 0,$$

$$w(s, y) = w(x, t) = 1,$$

and

$$N[e] = e_{st} - [b(s,t) - c(s,t)]e = 0, \quad e(s,y) = e(x,t) = 1,$$

respectively and let D^+ be a connected subdomain of D which contains P and on which $v \geq 0$, $w \geq 0$ and $e \geq 0$ (Figure 4.1). If $R \subset D^+$ and $u(x,y)$ satisfies

$$u(x,y) \leq a(x,y) + \int_{x_0}^{x}\int_{y_0}^{y} b(s,t)u(s,t)\,ds\,dt$$

$$+ \int_{x_0}^{x}\int_{y_0}^{y} c(s,t) \left(\int_{x_0}^{s}\int_{y_0}^{t} h(\xi,\eta)u(\xi,\eta)\,d\xi\,d\eta \right) ds\,dt$$

$$+ \int_{x_0}^{x}\int_{y_0}^{y} c(s,t) \left(\int_{x_0}^{s}\int_{y_0}^{t} p(\xi,\eta) \right.$$

$$\left. \times \left(\int_{x_0}^{\xi}\int_{y_0}^{\eta} q(\xi_1,\eta_1)u(\xi_1,\eta_1)\,d\xi_1\,d\eta_1 \right) d\xi\,d\eta \right) ds\,dt, \quad (4.8.28)$$

then $u(x,y)$ also satisfies

$$u(x,y) \leq a(x,y) + \int_{x_0}^{x}\int_{y_0}^{y} e(s,t;x,y) \left[a(s,t)b(s,t) \right.$$

$$+ c(s,t) \int_{x_0}^{s}\int_{y_0}^{t} w(\xi,\eta;s,t) \left[a(\xi,\eta)(b(\xi,\eta) + h(\xi,\eta)) \right.$$

$$+ p(\xi,\eta) \int_{x_0}^{\xi}\int_{y_0}^{\eta} v(\xi_1,\eta_1;\xi,\eta)a(\xi_1,\eta_1)[b(\xi_1,\eta_1) + h(\xi_1,\eta_1)$$

$$\left. \left. + q(\xi_1,\eta_1)]\,d\xi_1\,d\eta_1 \right] d\xi\,d\eta \right] ds\,dt. \quad (4.8.29)$$

□

The details of the proofs of Theorems 4.8.4 and 4.8.5 follow by closely looking at the proofs of Theorems 4.8.1 and 4.8.2, with suitable modifications. The details are omitted here.

In the following theorems we present the hyperbolic integro-differential inequalities established by Pachpatte (1979a), which can be used in the analysis of certain hyperbolic integro-differential equations.

Theorem 4.8.6 *Suppose $u(x, y)$, $u_{xy}(x, y)$, $a(x, y)$, $b(x, y)$ are nonnegative continuous functions on a domain D and $h(x, y) = u(x, y_0) + u(x_0, y) - u(x_0, y_0)$ is a nonnegative continuous function on D. Let $P_0(x_0, y_0)$ and $P(x, y)$ be two points in D such that $(x - x_0)(y - y_0) \geq 0$ and let R be the rectangular region whose opposite corners are the points P_0 and P. Let $v(s,t;x,y)$ be the solution of the characteristic initial value problem*

$$L[v] = v_{st} - [1 + b(s,t)]v = 0, \quad v(s, y) = v(x, t) = 1,$$

and let D^+ be a connected subdomain of D which contains P and on which $v \geq 0$ (Figure 4.1). If $R \subset D^+$ and $u(x, y)$ satisfies

$$u_{xy}(x, y) \leq a(x, y) + \int_{x_0}^{x} \int_{y_0}^{y} b(s, t)[u(s, t) + u_{st}(s, t)] \, ds \, dt, \quad (4.8.30)$$

then $u(x, y)$ also satisfies

$$u(x, y) \leq h(x, y) + \int_{x_0}^{x} \int_{y_0}^{y} \left[a(s, t) + \int_{x_0}^{s} \int_{y_0}^{t} b(\xi, \eta) \left[a(\xi, \eta) + h(\xi, \eta) \right. \right.$$

$$+ \int_{x_0}^{\xi} \int_{y_0}^{\eta} v(\xi_1, \eta_1; \xi, \eta)\{b(\xi_1, \eta_1)[a(\xi_1, \eta_1) + h(\xi_1, \eta_1)]$$

$$\left. \left. + a(\xi_1, \eta_1)\} \, d\xi_1 \, d\eta_1 \right] d\xi \, d\eta \right] ds \, dt. \quad (4.8.31)$$

□

Proof: Define a function $\phi(x, y)$ such that

$$\phi(x, y) = \int_{x_0}^{x} \int_{y_0}^{y} b(s, t)[u(s, t) + u_{st}(s, t)] \, ds \, dt,$$

then $\phi(x_0, y) = \phi(x, y_0) = 0$, and

$$\phi_{xy}(x, y) = b(x, y)[u(x, y) + u_{xy}(x, y)]. \tag{4.8.32}$$

Using the definition of $\phi(x, y)$, (4.8.30) can be restated as

$$u_{xy}(x, y) \leq a(x, y) + \phi(x, y). \tag{4.8.33}$$

Integrating both sides of (4.8.33) on R we obtain

$$u(x, y) \leq h(x, y) + \int_{x_0}^{x}\int_{y_0}^{y} [a(s, t) + \phi(s, t)]\, ds\, dt, \tag{4.8.34}$$

where $h(x, y) = u(x, y_0) + u(x_0, y) - u(x_0, y_0)$. Using (4.8.33) and (4.8.34) in (4.8.32) we have

$$\phi_{xy}(x, y) \leq b(x, y)[a(x, y) + h(x, y) + \phi(x, y)$$

$$+ \int_{x_0}^{x}\int_{y_0}^{y} [a(s, t) + \phi(s, t)]\, ds\, dt]. \tag{4.8.35}$$

Define

$$\psi(x, y) = \phi(x, y) + \int_{x_0}^{x}\int_{y_0}^{y} [a(s, t) + \phi(s, t)]\, ds\, dt, \tag{4.8.36}$$

then $\psi(x_0, y) = \psi(x, y_0) = 0$, and

$$\psi_{xy}(x, y) = \phi_{xy}(x, y) + a(x, y) + \phi(x, y). \tag{4.8.37}$$

Using $\phi_{xy}(x, y) \leq b(x, y)[a(x, y) + h(x, y) + \psi(x, y)]$ from (4.8.35) and $\phi(x, y) \leq \psi(x, y)$ from (4.8.36) in (4.8.37) we have

$$\psi_{xy}(x, y) \leq b(x, y)[a(x, y) + h(x, y) + \psi(x, y)] + a(x, y) + \psi(x, y)$$

i.e.

$$L[\psi] = \psi_{xy}(x, y) - [1 + b(x, y)]\psi(x, y)$$

$$\leq b(x, y)[a(x, y) + h(x, y)] + a(x, y). \tag{4.8.38}$$

Now by following the same steps as in the proof of Theorem 4.8.1 we obtain the estimate for $\psi(x, y)$ such that

$$\psi(x, y) \leq \int_{x_0}^{x} \int_{y_0}^{y} v(s, t; x, y)[b(s, t)(a(s, t) + h(s, t)) + a(s, t)] \, ds \, dt.$$

Substituting this bound on $\psi(x, y)$ in (4.8.35) we have

$$\phi_{xy}(x, y) \leq b(x, y) \left[a(x, y) + h(x, y) + \int_{x_0}^{x} \int_{y_0}^{y} v(s, t; x, y) \right.$$

$$\left. \times [b(s, t)(a(s, t) + h(s, t)) + a(s, t)] \, ds \, dt \right],$$

which implies the estimation for $\phi(x, y)$ such that

$$\phi(x, y) \leq \int_{x_0}^{x} \int_{y_0}^{y} b(s, t) \left[a(s, t) + h(s, t) + \int_{x_0}^{s} \int_{y_0}^{t} v(\xi, \eta; s, t) \right.$$

$$\left. \times [b(\xi, \eta)(a(\xi, \eta) + h(\xi, \eta)) + a(\xi, \eta)] \, d\xi \, d\eta \right] ds \, dt.$$

Now substituting this bound on $\phi(x, y)$ in (4.8.33) and integrating both sides on R we obtain the desired bound in (4.8.31).

∎

Theorem 4.8.7 *Suppose* $u(x, y)$, $u_{xy}(x, y)$, $a(x, y)$, $c(x, y)$ *and* $p(x, y)$ *are nonnegative continuous functions on a domain D and* $h(x, y) = u(x, y_0) + u(x_0, y) - u(x_0, y_0)$ *is a nonnegative continuous function on D. Let* $P_0(x_0, y_0)$ *and* $P(x, y)$ *be two points in D such that* $(x - x_0)(y - y_0) \geq 0$ *and let R be the rectangular region whose opposite corners are the points* P_0 *and P. Let* $v(s, t; x, y)$ *and* $w(s, t; x, y)$ *be the solutions of the characteristic initial value problems*

$$L[v] = v_{st} - [1 + b(s, t) + c(s, t) + p(s, t)]v = 0,$$

$$v(s, y) = v(x, t) = 1,$$

and

$$M[w] = w_{st} - [1 + b(s,t) - c(s,t)]w = 0, \quad w(s,y) = w(x,t) = 1,$$

respectively and let D^+ be a connected subdomain of D which contains P and on which $v \geq 0$ and $w \geq 0$ (Figure 4.1). If $R \subset D^+$ and $u(x,y)$ satisfies

$$u_{xy}(x,y) \leq a(x,y) + \int_{x_0}^{x}\int_{y_0}^{y} b(s,t)[u(s,t) + u_{st}(s,t)]\,ds\,dt$$

$$+ \int_{x_0}^{x}\int_{y_0}^{y} c(s,t)\left[\int_{x_0}^{s}\int_{y_0}^{t} p(\xi,\eta)[u(\xi,\eta) + u_{\xi\eta}(\xi,\eta)]\,d\xi\,d\eta\right]\,ds\,dt, \qquad (4.8.39)$$

then $u(x,y)$ also satisfies

$$u(x,y) \leq h(x,y) + \int_{x_0}^{x}\int_{y_0}^{y}\left[a(s,t) + \int_{x_0}^{s}\int_{y_0}^{t}\left[b(\xi,\eta)\right.\right.$$

$$\times (a(\xi,\eta) + c(\xi,\eta) + Q(\xi,\eta))$$

$$+ c(\xi,n)\int_{x_0}^{\xi}\int_{y_0}^{\eta} p(\xi_1,\eta_1)(a(\xi_1,\eta_1) + h(\xi_1,\eta_1)$$

$$+ Q(\xi_1,\eta_1))\,d\xi_1\,d\eta_1\bigg]\,d\xi\,d\eta\bigg]\,ds\,dt, \qquad (4.8.40)$$

where

$$Q(x,y) = \int_{x_0}^{x}\int_{y_0}^{y} w(s,t;x,y)\left[a(s,t) + b(s,t)(a(s,t) + h(s,t))\right.$$

$$+ c(s,t)\int_{x_0}^{s}\int_{y_0}^{t} v(\xi,\eta;s,t)\left[(a(\xi,\eta)\right.$$

$$+ h(\xi,\eta))(b(\xi,\eta) + p(\xi,\eta)) + a(\xi,\eta)\bigg]\,d\xi\,d\eta\bigg]\,ds\,dt.$$

□

The proof of this theorem follows by a similar argument as in the proofs of Theorems 4.8.6 and 4.8.2, with suitable modifications, and the details are left to the reader.

Remark 4.8.2 We note that the usefulness of the inequalities established in Theorems 4.8.6 and 4.8.7 becomes apparent if we consider, a, b, c, p and $h(x, y)$ are known and $u(x, y)$ and $u_{xy}(x, y)$ are unknown functions, i.e. the inequalities established in Theorems 4.8.6 and 4.8.7 give us bounds in terms of the known functions which majorize $u(x, y)$.

△

A fairly general and useful integral inequality investigated by Pachpatte (1980d) is given in the following theorem.

Theorem 4.8.8 *Suppose $u(x, y)$, $a(x, y)$, $b(x, y)$, $c(x, y)$, $p(x, y)$, $q(x, y)$, $r(x, y)$, $h(x, y)$ and $g(x, y)$ are nonnegative continuous functions defined on a domain D. Let $P_0(x_0, y_0)$ and $P(x, y)$ be two points in D such that $(x - x_0)(y - y_0) \geq 0$ and let R be the rectangular region whose opposite corners are the points P_0 and P. Let $V(s, t; x, y)$ be the solution of the characteristic initial value problem*

$$M[V] = 0, \quad (4.8.41)$$

where M is the adjoint operator of the operator L defined by

$$L[\psi] = \psi_{st} + a_1 \psi_s + a_2 \psi_t + a_3 \psi, \quad (4.8.42)$$

in which $a_1 = -bcq$, $a_2 = -bcp$, $a_3 = -[g + bc(r + h)]$. Let $W(s, t; x, y)$ be the solution of the characteristic initial value problem

$$N[W] = 0, \quad (4.8.43)$$

where N is the adjoint operator of the operator T defined by

$$T[\phi] = \phi_{st} + b_1 \phi_s + b_2 \phi_t + b_3 \phi, \quad (4.8.44)$$

in which $b_1 = -bcq$, $b_2 = -bcp$, $b_3 = -bc(r - h)$. Let D^+ be a connected subdomain of D which contains P and on which $V \geq 0$ and $W \geq 0$ (Figure 4.5). If $R \subset D^+$ and $u(x, y)$ satisfies

$$u(x, y) \leq a(x, y) + b(x, y) \left[p(x, y) \int_{x_0}^{x} c(s, y) u(s, y) \, ds \right.$$

$$+ q(x, y) \int_{y_0}^{y} c(x, t) u(x, t) \, dt + r(x, y) \int_{x_0}^{x} \int_{y_0}^{y} c(s, t) u(s, t) \, ds \, dt$$

$$+ h(x, y) \int_{x_0}^{x} \int_{y_0}^{y} g(s, t) \left(\int_{x_0}^{s} \int_{y_0}^{t} c(\xi, \eta) u(\xi, \eta) \, d\xi \, d\eta \right) ds \, dt \Bigg],$$

(4.8.45)

then $u(x, y)$ also satisfies

$$u(x, y) \leq a(x, y) + b(x, y) \Bigg[p(x, y) \int_{x_0}^{x} c(s, y) u(s, y) \, ds$$

$$+ q(x, y) \int_{y_0}^{y} c(x, t) u(x, t) \, dt + r(x, y) Q(x, y)$$

$$+ h(x, y) \int_{x_0}^{x} \int_{y_0}^{y} g(s, t) Q(s, t) \, ds \, dt \Bigg], \quad (4.8.46)$$

where

$$Q(x, y) = \int_{x_0}^{x} \int_{y_0}^{y} W(s, t; x, y) c(s, t) \Bigg[a(s, t) + b(s, t) h(s, t)$$

$$\times \left(\int_{x_0}^{s} \int_{y_0}^{t} V(\xi, \eta; s, t) a(\xi, \eta) c(\xi, \eta) \, d\xi \, d\eta \right) \Bigg] ds \, dt. \quad (4.8.47)$$

Further, if $q(x, y) = 0$, then

$$u(x, y) \leq f(x, y) + b(x, y) p(x, y) \Bigg[\int_{x_0}^{x} c(s, y) f(s, y)$$

$$\times \exp \left(\int_{s}^{x} c(\xi, y) b(\xi, y) p(\xi, y) \, d\xi \right) ds \Bigg], \quad (4.8.48)$$

where

$$f(x, y) = a(x, y) + b(x, y) \left[r(x, y)Q(x, y) \right.$$

$$\left. + h(x, y) \int_{x_0}^{x} \int_{y_0}^{y} g(s, t)Q(s, t) \, ds \, dt \right]. \qquad (4.8.49)$$

Again, if $p(x, y) = 0$, then

$$u(x, y) \le f(x, y) + b(x, y)q(x, y) \left[\int_{y_0}^{y} c(x, t)f(x, t) \right.$$

$$\left. \times \exp\left(\int_{t}^{y} c(x, \eta)b(x, \eta)q(x, \eta) \, d\eta \right) dt \right], \qquad (4.8.50)$$

where $f(x, y)$ is defined by (4.8.49) in which the function $Q(x, y)$ is defined by (4.8.47).

$\square$

Proof: Define a function $\phi(x, y)$ such that

$$\phi(x, y) = \int_{x_0}^{x} \int_{y_0}^{y} c(\xi, \eta) u(\xi, \eta) \, d\xi \, d\eta,$$

then $\phi(x_0, y) = \phi(x, y_0) = 0$, and

$$\phi_{xy}(x, y) = c(x, y)u(x, y),$$

which in view of the definition of $\phi(x, y)$ and (4.8.45) implies

$$\phi_{xy}(x, y) \le c(x, y) \left[a(x, y) + b(x, y) \left\{ p(x, y)\phi_y(x, y) + q(x, y)\phi_x(x, y) \right. \right.$$

$$\left. \left. + r(x, y)\phi(x, y) + h(x, y) \int_{x_0}^{x} \int_{y_0}^{y} g(s, t)\phi(s, t) \, ds \, dt \right\} \right].$$

Adding $b(x, y)c(x, y)h(x, y)\phi(x, y)$ to both sides of the above inequality we have

$$\phi_{xy}(x, y) + b(x, y)c(x, y)h(x, y)\phi(x, y)$$
$$\leq a(x, y)c(x, y) + b(x, y)c(x, y)\left[p(x, y)\phi_y(x, y) + q(x, y)\phi_x(x, y)\right.$$
$$\left. + r(x, y)\phi(x, y) + h(x, y)\left\{\phi(x, y) + \int_{x_0}^{x}\int_{y_0}^{y} g(s, t)\phi(s, t)\,ds\,dt\right\}\right].$$
(4.8.51)

If we put

$$\psi(x, y) = \phi(x, y) + \int_{x_0}^{x}\int_{y_0}^{y} g(s, t)\phi(s, t)\,ds\,dt, \qquad (4.8.52)$$

then $\psi(x_0, y) = \psi(x, y_0) = 0$, and

$$\psi_{xy}(x, y) = \phi_{xy}(x, y) + g(x, y)\phi(x, y). \qquad (4.8.53)$$

Using

$$\phi_{xy}(x, y) \leq a(x, y)c(x, y) + b(x, y)c(x, y)[p(x, y)\phi_y(x, y)$$
$$+ q(x, y)\phi_x(x, y) + r(x, y)\phi(x, y) + h(x, y)\psi(x, y)],$$

from (4.8.51) and $\phi(x, y) \leq \psi(x, y)$, $\phi_x(x, y) \leq \psi_x(x, y)$, $\phi_y(x, y) \leq \psi_y(x, y)$ from (4.8.52) in (4.8.53) we have

$$L[\psi] = \psi_{xy}(x, y) + a_1\psi_x(x, y) + a_2\psi_y(x, y) + a_3\psi(x, y)$$
$$\leq a(x, y)c(x, y), \qquad (4.8.54)$$

where a_1, a_2 and a_3 are as defined in (4.8.42).

Now for any two twice continuously differentiable functions ψ and V, the operators L and M satisfy the identity

$$VL[\psi] - \psi M[V] = (a_1\psi V + \tfrac{1}{2}V\psi_y - \tfrac{1}{2}V_y\psi)_x$$
$$+ (a_2\psi V + \tfrac{1}{2}V\psi_x - \tfrac{1}{2}\psi V_x)_y, \qquad (4.8.55)$$

where M is the adjoint operator of L. Let R be a rectangular region with corners $P_0(x_0, y_0)$, $P_1(x, y_0)$, $P(x, y)$, $P_2(x_0, y)$ so that P_0P is the diagonal as shown in Figure 4.5.

Using s and t as the independent variables, we integrate identity (4.8.55) over R and use Green's theorem to obtain

$$\iint_R (VL[\psi] - \psi M[V])\,ds\,dt = \int_{C_1+C_2+C_3+C_4} (a_1\psi V + \tfrac{1}{2}V\psi_t - \tfrac{1}{2}\psi V_t)\,dt$$
$$- (a_2\psi V + \tfrac{1}{2}V\psi_s - \tfrac{1}{2}\psi V_s)\,ds.$$

Since ψ is zero on C_1 and C_4, and also ds does not vary on C_2, and dt does not vary on C_3, we get

$$\iint_R (VL[\psi] - \psi M[V])\,ds\,dt = \int_{C_2} (a_1\psi V + \tfrac{1}{2}V\psi_t - \tfrac{1}{2}\psi V_t)\,dt$$
$$- \int_{C_3} (a_2\psi V + \tfrac{1}{2}V\psi_s - \tfrac{1}{2}\psi V_s)\,ds.$$

Integrating the right-hand side by parts along the characteristic segments C_2 and C_3 to eliminate partial derivatives of ψ, we obtain

$$\iint_R (VL[\psi] - \psi M[V])\,ds\,dt$$
$$= \int_{C_2} (a_1 V - V_t)\psi\,dt - \int_{C_3} (a_2 V - V_s)\psi\,ds$$
$$+ \psi(P)V(P) - \tfrac{1}{2}\psi(P_1)V(P_1) - \tfrac{1}{2}\psi(P_2)V(P_2)$$
$$= \int_{C_2} (a_1 V - V_t)\psi\,dt - \int_{C_3} (a_2 V - V_s)\psi\,ds + \psi(P)V(P). \quad (4.8.56)$$

Now since $V(s, t; x, y)$ is the solution of the characteristic initial value problem $M[V] = 0$, it is by definition the Riemann function $V(s, t; x, y) = V(s, t)$ associated with the partial differential equation $L[\psi] = 0$ such that

$$V(x, y; x, y) = V(x, y) = V(P) = 1,$$

MULTIDIMENSIONAL LINEAR INTEGRAL INEQUALITIES

$$V_t = a_1 V \text{ on } C_2, \quad V(x,t) = \exp\left(\int_y^t a_1(x,\eta)\,d\eta\right),$$

and

$$V_s = a_2 V \text{ on } C_3, \quad V(s,y) = \exp\left(\int_x^s a_2(\xi,y)\,d\xi\right).$$

So we get from identity (4.8.56) and inequality (4.8.54)

$$\psi(x,y) \le \int_{x_0}^x \int_{y_0}^y V(s,t;x,y) a(s,t) c(s,t)\,ds\,dt.$$

Now substituting this bound on $\psi(x,y)$ in (4.8.51) we have

$$T[\phi] = \phi_{xy}(x,y) + b_1 \phi_x(x,y) + b_2 \phi_y(x,y) + b_3 \phi(x,y)$$

$$\le c(x,y)\left[a(x,y) + b(x,y)h(x,y)\right.$$

$$\left.\times \left(\int_{x_0}^x \int_{y_0}^y V(s,t;x,y) a(s,t) c(s,t)\,ds\,dt\right)\right].$$

Now by following the same steps as above we obtain the estimate

$$\phi(x,y) \le \int_{x_0}^x \int_{y_0}^y W(s,t;x,y) c(s,t) \left[a(s,t) + b(s,t)h(s,t)\right.$$

$$\left.\times \left(\int_{x_0}^s \int_{y_0}^t V(\xi,\eta;s,t) a(\xi,\eta) c(\xi,\eta)\,d\xi\,d\eta\right)\right] ds\,dt = Q(x,y).$$

Now substituting this bound on $\phi(x,y)$ in (4.8.45) we obtain the desired bound in (4.8.46).

Now let $q(x,y) = 0$ in (4.8.46) and define a function $f(x,y)$ by (4.8.49). Then inequality (4.8.46) reduces to

$$u(x,y) \le f(x,y) + b(x,y) p(x,y) \int_{x_0}^x c(s,y) u(s,y)\,ds. \tag{5.8.57}$$

Inequality (4.8.57) may be treated as a one-dimensional inequality for any fixed y, which by applying Theorem 1.3.2 yields

$$u(x, y) \leq f(x, y) + b(x, y)p(x, y)\left[\int_{x_0}^{x} c(s, y)f(s, y)\right.$$

$$\left.\times \left(\int_{s}^{x} c(\xi, y)b(\xi, y)p(\xi, y)\,d\xi\right) ds\right], \quad (4.8.58)$$

which is the desired inequality in (4.8.48). Further, substituting $p(x, y) = 0$ in (4.8.46) and following a similar argument to that above, we obtain the required inequality in (4.8.50).

∎

Remark 4.8.3 Note that the integral inequality established in Theorem 4.8.8 is of more general type and contains as special cases the results of Snow (1971) and Ghoshal and Masood (1974a). It can be used to study nonlinear non-self-adjoint hyperbolic partial integro-differential equations of the more general type.

△

4.9 Inequalities in Several Variables

In the past few years, Wendroff's inequality (Beckenbach and Bellman, 1961, p. 154) has received considerable attention and a number of papers have appeared in the literature dealing with the various extensions, generalizations and applications. This section presents some basic inequalities of Wendroff's type in n independent variables investigated by Pachpatte and his co-workers in Bondge and Pachpatte (1979a) and Singore and Pachpatte (1981).

In what follows, we adopt the following definitions and notational conventions. Let Ω be an open bounded set in R^n and let a point $(x_1, \ldots, x_n)$ in Ω be denoted by x. Let x^0 and x ($x^0 < x$) be any two points in Ω and $\int_{x^0}^{x} \ldots d\xi$ denote the n-fold integral $\int_{x_1^0}^{x_1} \ldots \int_{x_n^0}^{x_n} \ldots d\xi_n \ldots d\xi_1$, $D_i = \partial/\partial x_i$, $1 \leq i \leq n$. For any pair x, s of points of Ω with $x < s$ we denote $D(x, s) = \{x \in R^n : x \leq \xi \leq s\} \subset \Omega$ and $\int_{x}^{s} \ldots d\xi$ denotes the n-fold integral $\int_{x_1}^{s_1} \ldots \int_{x_n}^{s_n} \ldots d\xi_n \ldots d\xi_1$.

Bondge and Pachpatte (1979a) investigated the inequalities in the following theorem.

Theorem 4.9.1 Let $u(x)$, $p(x)$, $q(x)$ be nonnegative continuous functions defined on Ω and $a_i(x_i) > 0$, $a'_i(x_i) \geq 0$ for $1 \leq i \leq n$, are continuous functions defined for $x_i \geq x_i^0$.

(i) If
$$u(x) \leq \sum_{i=1}^{n} a_i(x_i) + \int_{x^0}^{x} p(y)u(y)\,dy, \qquad (4.9.1)$$

for $x \in \Omega$, then
$$u(x) \leq E(x)\exp\left(\int_{x^0}^{x} p(y)\,dy\right), \qquad (4.9.2)$$

for $x \in \Omega$, where

$$E(x) = \frac{\left[\sum_{i=1}^{n} a_i(x_i) + a_1(x_1^0) - a_1(x_1)\right]\left[\sum_{i=1}^{n} a_i(x_i) + a_2(x_2^0) - a_2(x_2)\right]}{\left[\sum_{i=3}^{n} a_i(x_i) + a_1(x_1^0) + a_2(x_2^0)\right]}$$

(4.9.3)

(ii) If
$$u(x) \leq \sum_{i=1}^{n} a_i(x_i) + \int_{x^0}^{x} p(y)u(y)\,dy + \int_{x^0}^{x} p(y)\left(\int_{x^0}^{y} q(s)u(s)\,ds\right)dy, \qquad (4.9.4)$$

for $x \in \Omega$, then
$$u(x) \leq \sum_{i=1}^{n} a_i(x_i) + \int_{x^0}^{x} p(y)E(y)\exp\left(\int_{x^0}^{y} [p(s) + q(s)]\,ds\right)dy, \qquad (4.9.5)$$

for $x \in \Omega$, where $E(x)$ is defined by (4.9.3). $\square$

Proof: (i) Define a function $z(x)$ by
$$z(x) = \sum_{i=1}^{n} a_i(x_i) + \int_{x^0}^{x} p(y)u(y)\,dy, \qquad (4.9.6)$$

then
$$z(x) = \sum_{i=1}^{n} a_i(x_i) + a_j(x_j^0) - a_j(x_j) \text{ on } x_j = x_j^0, \quad 1 \leq j \leq n,$$

and
$$D_1 \ldots D_n z(x) = p(x) u(x). \tag{4.9.7}$$

Using the fact that $u(x) \leq z(x)$ in (4.9.7) we have
$$D_1 \ldots D_n z(x) \leq p(x) z(x). \tag{4.9.8}$$

From (4.9.8) we observe that
$$\frac{z(x) D_1 \ldots D_n z(x)}{z^2(x)} \leq p(x). \tag{4.9.9}$$

Further, from (4.9.9) we observe that
$$\frac{z(x) D_1 \ldots D_n z(x)}{z^2(x)} \leq p(x) + \frac{(D_n z(x))(D_1 \ldots D_{n-1} z(x))}{z^2(x)},$$

i.e.
$$D_n \left(\frac{D_1 \ldots D_{n-1} z(x)}{z(x)} \right) \leq p(x). \tag{4.9.10}$$

Now by setting $x_n = y_n$ in (4.9.10) and integrating it with respect to the component y_n from x_n^0 to x_n we have
$$\frac{D_1 \ldots D_{n-1} z(x)}{z(x)} \leq \int_{x_n^0}^{x_n} p(x_1, \ldots, x_{n-1}, y_n) \, dy_n. \tag{4.9.11}$$

Again from (4.9.11) we observe that
$$D_{n-1} \left(\frac{D_1 \ldots D_{n-2} z(x)}{z(x)} \right) \leq \int_{x_n^0}^{x_n} p(x_1, \ldots, x_{n-1}, y_n) \, dy_n. \tag{4.9.12}$$

By taking $x_{n-1} = y_{n-1}$ in (4.9.12) and integrating with respect to the component y_{n-1} from x_{n-1}^0 to x_{n-1} we have
$$\frac{D_1 \ldots D_{n-2} z(x)}{z(x)} \leq \int_{x_{n-1}^0}^{x_{n-1}} \int_{x_n^0}^{x_n} p(x_1, \ldots, x_{n-2}, y_{n-1}, y_n) \, dy_n \, dy_{n-1}.$$

Continuing in this way we obtain
$$\frac{D_1 D_2 z(x)}{z(x)} \leq \int_{x_3^0}^{x_3} \ldots \int_{x_n^0}^{x_n} p(x_1, x_2, y_3, \ldots, y_n) \, dy_n \ldots dy_3. \tag{4.9.13}$$

From (4.9.13) as above we observe that

$$D_2\left(\frac{D_1 z(x)}{z(x)}\right) \leq \int_{x_3^0}^{x_3} \cdots \int_{x_n^0}^{x_n} p(x_1, x_2, y_3, \ldots, y_n)\, dy_n \ldots dy_3. \quad (4.9.14)$$

By taking $x_2 = y_2$ in (4.9.14) and integrating from x_2^0 to x_2 we have

$$\frac{D_1 z(x)}{z(x)} - \frac{D_1 z(x_1, x_2^0, x_3, \ldots, x_n)}{z(x_1, x_2^0, x_3, \ldots, x_n)} \leq \int_{x_2^0}^{x_2} \cdots \int_{x_n^0}^{x_n} p(x_1, y_2, \ldots, y_n)$$
$$\times dy_n \ldots dy_2. \quad (4.9.15)$$

Now by taking $x_1 = y_1$ in (4.9.15) and integrating it from x_1^0 to x_1 we have

$$\log\left[\frac{z(x)}{z(x_1^0, x_2, x_3, \ldots, x_n)} \cdot \frac{z(x_1^0, x_2^0, x_3, \ldots, x_n)}{z(x_1, x_2^0, x_3, \ldots, x_n)}\right] \leq \int_{x^0}^{x} p(y)\, dy,$$

which implies

$$z(x) \leq E(x) \exp\left(\int_{x^0}^{x} p(y)\, dy\right).$$

Substituting this bound on $z(x)$ in $u(x) \leq z(x)$ we obtain the desired inequality in (4.9.2).

(ii) Define a function $z(x)$ by the right-hand side of (4.9.4); then we have

$$z(x) = \sum_{i=1}^{n} a_i(x_i) + a_j(x_j^0) - a_j(x_j) \text{ on } x_j = x_j^0, \quad 1 \leq j \leq n,$$

and

$$D_1 \ldots D_n z(x) = p(x)\left[u(x) + \int_{x^0}^{x} q(s)u(s)\, ds\right],$$

which by using the fact that $u(x) \leq z(x)$ implies

$$D_1 \ldots D_n z(x) \leq p(x)\left[z(x) + \int_{x^0}^{x} q(s)z(s)\, ds\right]. \quad (4.9.16)$$

If we put

$$v(x) = z(x) + \int_{x^0}^{x} q(s)z(s)\,ds, \qquad (4.9.17)$$

then

$$v(x) = z(x) \text{ on } x_j = x_j^0, \quad 1 \le j \le n,$$

and

$$D_1 \ldots D_n v(x) = D_1 \ldots D_n z(x) + q(x)z(x). \qquad (4.9.18)$$

Using the facts that $D_1 \ldots D_n z(x) \le p(x)v(x)$ from (4.9.16) and $z(x) \le v(x)$ from (4.9.17) in (4.9.18) we obtain

$$D_1 \ldots D_n v(x) \le [p(x) + q(x)]v(x).$$

Now by following a similar argument as in the proof of (i) given above we obtain the estimate for $v(x)$ such that

$$v(x) \le E(x) \exp\left(\int_{x^0}^{x} [p(s) + q(s)]\,ds\right). \qquad (4.9.19)$$

Using (4.9.19) in (4.9.16) we have

$$D_1 \ldots D_n z(x) \le p(x)E(x) \exp\left(\int_{x^0}^{x} [p(s) + q(s)]\,ds\right). \qquad (4.9.20)$$

From (4.9.20) it is easy to obtain

$$z(x) \le \sum_{i=1}^{n} a_i(x_i) + \int_{x^0}^{x} p(y)E(y) \exp\left(\int_{x^0}^{y} [p(s) + q(s)]\,ds\right) dy. \qquad (4.9.21)$$

Using (4.9.21) in $u(x) \le z(x)$ we get the required inequality in (4.9.5).

∎

A slight variant of Theorem 4.9.1 which can be used in some applications is given in the following theorem.

Theorem 4.9.2 *Let $u(x)$, $p(x)$ and $q(x)$ be as defined in Theorem 4.9.1 and let $a(x)$, $x \in \Omega$, be a continuous function which is nonnegative and nondecreasing in each component x_i of x.*

(i) If
$$u(x) \leq a(x) + \int_{x^0}^{x} p(y)u(y)\,dy, \qquad (4.9.22)$$

for $x \in \Omega$, then
$$u(x) \leq a(x) \exp\left(\int_{x^0}^{x} p(y)\,dy\right), \qquad (4.9.23)$$

for $x \in \Omega$.

(ii) If
$$u(x) \leq a(x) + \int_{x^0}^{x} p(y)\left(u(y) + \int_{x^0}^{y} q(s)u(s)\,ds\right) dy, \qquad (4.9.24)$$

for $x \in \Omega$, then
$$u(x) \leq a(x)\left[1 + \int_{x^0}^{x} p(y)\exp\left(\int_{x^0}^{y} [p(s) + q(s)]\,ds\right) dy\right]. \qquad (4.9.25)$$

for $x \in \Omega$,

□

The proof of this theorem follows by the same arguments as in the proof of Theorem 4.2.2 and in view of the proof of Theorem 4.9.1 given above. The details are omitted here.

The following theorem deals with general inequalities which can be used in certain situations.

Theorem 4.9.3 *Let $u(x)$, $a(x)$, $b(x)$, $p(x)$ and $q(x)$ be nonnegative continuous functions defined on Ω.*

(i) If
$$u(x) \leq a(x) + b(x)\int_{x^0}^{x} p(y)u(y)\,dy, \qquad (4.9.26)$$

for $x \in \Omega$, then
$$u(x) \leq a(x) + b(x)m(x)\exp\left(\int_{x^0}^{x} b(y)p(y)\,dy\right), \qquad (4.9.27)$$

for $x \in \Omega$, where
$$m(x) = \int_{x^0}^{x} a(y)p(y)\,dy, \qquad (4.9.28)$$

for $x \in \Omega$.

(ii) If
$$u(x) \le a(x) + b(x)\int_{x^0}^{x} p(y)\left(u(y) + \int_{x^0}^{y} q(s)u(s)\,ds\right) dy, \qquad (4.9.29)$$

for $x \in \Omega$, then
$$u(x) \le a(x) + b(x)r(x)\left[1 + \int_{x^0}^{x} p(y)b(y)\right.$$
$$\left. \times \exp\left(\int_{x^0}^{y} b(s)[p(s) + q(s)]\,ds\right) dy\right], \qquad (4.9.30)$$

for $x \in \Omega$, where
$$r(x) = \int_{x^0}^{x} p(y)\left(a(y) + b(y)\int_{x^0}^{y} q(s)a(s)\,ds\right) dy, \qquad (4.9.31)$$

for $x \in \Omega$.

□

Proof: (i) Define a function $z(x)$ by
$$z(x) = \int_{x^0}^{x} p(y)u(y)\,dy. \qquad (4.9.32)$$

Using the fact that $u(x) \le a(x) + b(x)z(x)$ in (4.9.32) we have
$$z(x) \le \int_{x^0}^{x} p(y)[a(y) + b(y)z(y)]\,dy$$
$$= m(x) + \int_{x^0}^{x} p(y)b(y)z(y)\,dy. \qquad (4.9.33)$$

Clearly $m(x)$ is nonnegative for $x \in \Omega$ and nondecreasing in each component x_i of $x \in \Omega$. Now applying Theorem 4.9.2 part (i) we have

$$z(x) \leq m(x) \exp\left(\int_{x^0}^{x} p(y)b(y)\,dy \right). \tag{4.9.34}$$

Using (4.9.34) in $u(x) \leq a(x) + b(x)z(x)$ we get the required inequality in (4.9.27).

(ii) The proof follows by the same arguments as in the proof of part (i) given above and applying the inequality in part (ii) of Theorem 4.9.2. The details are omitted here.

∎

Inspired by the inequalities given by Gollwitzer (1969) and Pachpatte (1975c), Singare and Pachpatte (1981) established the inequalities in the following theorem.

Theorem 4.9.4 *Let $\phi(x)$, $a(x)$, $b(x)$ and $c(x)$ be nonnegative continuous functions defined on Ω and $u(x)$ be a positive continuous function defined on Ω.*

(i) *If*

$$u(s) \geq \phi(x) - a(s) \int_{x}^{s} b(\xi)\phi(\xi)\,d\xi, \tag{4.9.35}$$

for $x \leq s$, $x, s \in \Omega$, then

$$u(s) \geq \phi(x) \exp\left(-a(s) \int_{x}^{s} b(\xi)\,d\xi \right), \tag{4.9.36}$$

for $x \leq s$, $x, s \in \Omega$.

(ii) *If*

$$u(s) \geq \phi(x) - a(s) \left[\int_{x}^{s} b(\xi)\phi(\xi)\,d\xi + \int_{x}^{s} b(\xi) \left(\int_{\xi}^{s} c(\rho)\phi(\rho)\,d\rho \right) d\xi \right], \tag{4.9.37}$$

for $x \leq s$, $x, s \in \Omega$, then

$$u(s) \geq \phi(x) \left[1 + a(s) \left(\int_x^s b(\xi) \exp \left(\int_x^s [a(s)b(\rho) + c(\rho)] \, d\rho \right) d\xi \right) \right], \tag{4.9.38}$$

for $x \leq s$, $x, s \in \Omega$.

□

Proof: (i) Rewrite (4.9.35) as

$$\phi(x) \leq u(s) + a(s) \int_x^s b(\xi)\phi(\xi) \, d\xi. \tag{4.9.39}$$

For fixed $s \in \Omega$, we define for $x \leq s$,

$$r(x) = u(s) + a(s) \int_x^s b(\xi)\phi(\xi) \, d\xi, \tag{4.9.40}$$

so that

$$r(s_1, x_2, \ldots, x_n) = \cdots = r(x_1, \ldots, x_{n-1}, s_n) = u(s_1, \ldots, s_n).$$

Then from (4.9.40) we have

$$D_1 r(x) = -a(s) \int_{x_2}^{s_2} \cdots \int_{x_n}^{s_n} b(x_1, \xi_2, \ldots, \xi_n) \phi(x_1, \xi_2, \ldots, \xi_n) \, d\xi_n \ldots d\xi_2, \tag{4.9.41}$$

and from (4.9.41) we have

$$D_1 D_2 r(x) = a(s) \int_{x_3}^{s_3} \cdots \int_{x_n}^{s_n} b(x_1, x_2, \xi_3, \ldots, \xi_n)$$
$$\times \phi(x_1, x_2, \xi_3, \ldots, \xi_n) \, d\xi_n \ldots d\xi_3, \tag{4.9.42}$$

and in general we have

$$D_1 \ldots D_k r(x) = (-1)^k \int_{x_{k+1}}^{s_{k+1}} \cdots \int_{x_n}^{s_n} b(x_1, \ldots, x_k, \xi_{k+1}, \ldots, \xi_n)$$
$$\times \phi(x_1, \ldots, x_k, \xi_{k+1}, \ldots, \xi_n) \, d\xi_n \ldots d\xi_{k+1}, \tag{4.9.43}$$

and continuing in this way, we obtain

$$D_1 \ldots D_n r(x) = (-1)^n a(s) b(x) \phi(x). \qquad (4.9.44)$$

Now, we consider the following two cases.

Case I: If the order n of the derivatives in (4.9.44) is even, then from (4.9.44) we have

$$D_1 \ldots D_n r(x) = a(s) b(x) \phi(x),$$

which in view of (4.9.39) implies

$$D_1 \ldots D_n r(x) \leq a(s) b(x) r(x),$$

i.e.

$$\frac{D_1 \ldots D_n r(x)}{r(x)} \leq a(s) b(x). \qquad (4.9.45)$$

From (4.9.45) we observe that

$$\frac{r(x)[D_1 \ldots D_n r(x)]}{r^2(x)} \leq a(s) b(x) + \frac{[D_n r(x)][D_1 \ldots D_{n-1} r(x)]}{r^2(x)}. \qquad (4.9.46)$$

By (4.9.40) we see that $D_n r(x)$ and $D_1 \ldots D_{n-1} r(x)$ are both nonpositive which implies that $[D_n r(x)][D_1 \ldots D_{n-1} r(x)]$ is nonnegative and hence (4.9.46) is true. Now (4.9.46) is equivalent to

$$D_n \left(\frac{D_1 \ldots D_{n-1} r(x)}{r(x)} \right) \leq a(s) b(x).$$

Now keeping $x_1, \ldots, x_{n-1}$ fixed in the above inequality, set $x_n = \xi_n$; then integrating with respect to ξ_n from x_n to s_n, we have

$$\frac{D_1 \ldots D_{n-1} r(x)}{r(x)} \geq -a(s) \int_{x_n}^{s_n} b(x_1, \ldots, x_{n-1}, \xi_n) \, d\xi_n. \qquad (4.9.47)$$

Again from (4.9.47), we observe that

$$\frac{r(x)[D_1 \ldots D_{n-1} r(x)]}{r^2(x)} \geq -a(s) \int_{x_n}^{s_n} b(x_1, \ldots, x_{n-1}, \xi_n) \, d\xi_n$$

$$+ \frac{[D_{n-1} r(x)][D_1 \ldots D_{n-2} r(x)]}{r^2(x)}. \qquad (4.9.48)$$

As before, we see that $D_{n-1}r(x)$ and $D_1 \ldots D_{n-2}r(x)$ are nonpositive and nonnegative respectively, which implies $[D_{n-1}r(x)][D_1 \ldots D_{n-2}r(x)]$ is nonpositive and hence (4.9.48) is true. But (4.9.48) is equivalent to

$$D_{n-1}\left(\frac{D_1 \ldots D_{n-2}r(x)}{r(x)}\right) \geq -a(s)\int_{x_n}^{s_n} b(x_1, \ldots, x_{n-1}, \xi_n)\,d\xi_n.$$

Now keeping $x_1, \ldots, x_{n-2}, x_n$ fixed in the above inequality, set $x_{n-1} = \xi_{n-1}$; then integrating with respect to ξ_{n-1} from x_{n-1} to s_{n-1} we have

$$\frac{D_1 \ldots D_{n-2}r(x)}{r(x)} \leq a(s) \int_{x_{n-1}}^{s_{n-1}}\int_{x_n}^{s_n} b(x_1, \ldots, x_{n-2}, \xi_{n-1}, \xi_n)\,d\xi_n\,d\xi_{n-1}. \quad (4.9.49)$$

Proceeding in this way, we finally obtain

$$\frac{D_1 r(x)}{r(x)} \geq -a(s)\int_{x_2}^{s_2}\ldots\int_{x_n}^{s_n} b(x_1, \xi_2, \ldots, \xi_n)\,d\xi_n \ldots d\xi_2. \quad (4.9.50)$$

Now keeping $x_2, \ldots, x_n$ fixed in (4.9.50), set $x_1 = \xi_1$; then integrating with respect to ξ_1 from x_1 to s_1, we have

$$r(x) \leq u(s)\exp\left(a(s)\int_x^s b(\xi)\,d\xi\right). \quad (4.9.51)$$

Substituting this bound on $r(x)$ in (4.9.39), we obtain the desired bound in (4.9.36).

Case II: If the order n of the derivatives in (4.9.44) is odd, then from (4.9.44), we have

$$D_1 \ldots D_n r(x) = -a(s)b(x)\phi(x), \quad (4.9.52)$$

which in view of (4.9.39) implies

$$D_1 \ldots D_n r(x) \geq -a(s)b(x)r(x),$$

i.e.

$$\frac{D_1 \ldots D_n r(x)}{r(x)} \geq -a(s)b(x). \quad (4.9.53)$$

The rest of the proof for Case II is precisely the same as for Case I and the final inequality (4.9.51) remains unchanged, since n is now odd.

(ii) Rewrite (4.9.37) as

$$\phi(x) \leq u(s) + a(s) \left[\int_x^s b(\xi)\phi(\xi)\,d\xi + \int_x^s b(\xi) \left(\int_\xi^s c(\rho)\phi(\rho)\,d\rho \right) d\xi \right]. \quad (4.9.54)$$

For fixed $s \in \Omega$, we define for $x \leq s, x \in \Omega$,

$$r(x) = u(s) + a(s) \left[\int_x^s b(\xi)\phi(\xi)\,d\xi + \int_x^s b(\xi) \left(\int_\xi^s c(\rho)\phi(\rho)\,d\rho \right) d\xi \right], \quad (4.9.55)$$

so that

$$r(s_1, x_2, \ldots, x_n) = \cdots = r(x_1, \ldots, x_{n-1}, s_n) = u(s_1, \ldots, s_n).$$

Then, by following the same argument as in the proof of (i) given above, we obtain in general, from (4.9.55) that

$$D_1 \ldots D_k r(x) = (-1)^k \left[\int_{x_{k+1}}^{s_{k+1}} \cdots \int_{x_n}^{s_n} b(x_1, \ldots, x_k, \xi_{k+1}, \ldots, \xi_n) \right.$$

$$\times \phi(x_1, \ldots, x_k, \xi_{k+1}, \ldots, \xi_n)\,d\xi_n \ldots d\xi_{k+1}$$

$$+ \int_{x_{k+1}}^{s_{k+1}} \cdots \int_{x_n}^{s_n} b(x_1, \ldots, x_k, \xi_{k+1}, \ldots, \xi_n)$$

$$\times \left(\int_{\xi_{k+1}}^{s_{k+1}} \cdots \int_{\xi_n}^{s_n} c(\xi_1, \ldots, \xi_k, \rho_{k+1}, \ldots, \rho_n) \right.$$

$$\times \phi(\xi_1, \ldots, \xi_k, \rho_{k+1}, \ldots, \rho_n)\,d\rho_n \ldots d\rho_{k+1} \Big)$$

$$\times d\xi_n \ldots d\xi_{k+1} \Bigg], \quad (4.9.56)$$

and continuing in this way, we obtain

$$D_1 \ldots D_n r(x) = (-1)^n a(s) b(x) \left[\phi(x) + \int_x^s c(\xi) \phi(\xi) \, d\xi \right]. \quad (4.9.57)$$

We now consider the following two cases.

Case I: If the order n of the derivatives in (4.9.57) is even, then from (4.9.57) and (4.9.54), we have

$$D_1 \ldots D_n r(x) \leq a(s) b(x) \left[r(x) + \int_x^s c(\xi) r(\xi) \, d\xi \right]. \quad (4.9.58)$$

If we put

$$v(x) = r(x) + \int_x^s c(\xi) r(\xi) \, d\xi, \quad (4.9.59)$$

so that

$$v(s_1, x_2, \cdots, x_n) = \cdots = v(x_1, \ldots, x_{n-1}, s_n) = r(s_1, \ldots, s_n)$$
$$= u(s_1, \ldots, s_n).$$

Then we have

$$D_1 \ldots D_n v(x) = D_1 \ldots D_n r(x) + c(x) r(x), \quad (4.9.60)$$

Since the order n of the derivative is even. Using (4.9.58) and the fact that $r(x) \leq v(x)$ in (4.9.60) we have

$$D_1 \ldots D_n v(x) \leq [a(s) b(x) + c(x)] v(x).$$

Now repeating the argument used in the proof of (i) given above, we obtain the estimate

$$v(x) \leq u(s) \exp \left(\int_x^s [a(s) b(\rho) + c(\rho)] \, d\rho \right).$$

Now substitute this bound for $v(x)$ in (4.9.58) and carry out n successive integrations, using the fact that $D_1 \ldots D_k r(x_1, \ldots, \xi_{k+1}, \ldots, x_n) = 0$ for $\xi_{k+1} =$

s_{k+1}, by (4.9.56), to obtain

$$r(x) \le u(s) \left[1 + a(s) \left(\int_x^s b(\xi) \exp \left(\int_\xi^s [a(s)b(\rho) + c(\rho)] \,d\rho \right) d\xi \right) \right].$$
(4.9.61)

Substituting this bound on $r(x)$ in (4.9.54), we obtain the desired bound in (4.9.38).

Case II: When n is odd, (4.9.57) becomes

$$D_1 \ldots D_n r(x) = -a(s)b(x) \left[r(x) + \int_x^s c(\xi)r(\xi) \,d\xi \right],$$
(4.9.62)

and the proof proceeds exactly as in Case I, again leading to (4.9.61).

∎

Remark 4.9.1 Note that the method employed in the proofs of the above theorems can also be used to obtain n independent variable versions of various inequalities given in earlier sections. Since this translation is quite straightforward in view of the results given in this section, it is left to the reader to fill in the details where needed.

△

4.10 Young's Inequality and its Generalizations

Young (1973) extended Snow's technique to the case of n independent variables. Bondge and Pachpatte (1980a) gave more general integral inequalities in n independent variables by using Young's technique. The inequalities given in Bondge and Pachpatte (1980a) are the further generalizations of the inequalities given by Pachpatte (1979a). This section deals with the inequalities given by Young (1973) and Bondge and Pachpatte (1980a), which have many important applications in the theory of partial differential and integro-differential equations in n independent variables. In what follows the notations and definitions as given in Section 4.9 will be used.

The following inequality is established by Young (1973).

Theorem 4.10.1 *Suppose $\phi(x)$, $a(x)$, and $b(x) \ge 0$ are continuous functions in $\Omega \subset R^n$. Let $v(\xi; x)$ be a solution of the characteristic initial value*

problem
$$(-1)^n v_{\xi_1\ldots\xi_n}(\xi; x) - b(\xi)v(\xi; x) = 0 \text{ in } \Omega,$$
$$v(\xi; x) = 1 \text{ on } \xi_i = x_i, \quad i = 1, \ldots, n \tag{4.10.1}$$

and let D^+ be a connected subdomain of Ω containing x such that $v \geq 0$ for all $\xi \in D^+$. If $D \subset D^+$ and

$$\phi(x) \leq a(x) + \int_{x^0}^{x} b(\xi)\phi(\xi)\,d\xi, \tag{4.10.2}$$

then

$$\phi(x) \leq a(x) + \int_{x^0}^{x} a(\xi)b(\xi)v(\xi; x)\,d\xi. \tag{4.10.3}$$

$\square$

Proof: Define a function $u(x)$ by

$$u(x) = \int_{x^0}^{x} b(\xi)\phi(\xi)\,d\xi, \tag{4.10.4}$$

then we have

$$D_1 \ldots D_n u(x) = b(x)\phi(x). \tag{4.10.5}$$

Since $b(x) \geq 0$ in D, it follows from (4.10.2), (4.10.4) and (4.10.5) that

$$Lu = D_1 \ldots D_n u(x) - b(x)u(x) \leq a(x)b(x), \tag{4.10.6}$$
$$u(x) = 0 \text{ on } x_i = x_i^0, 1 \leq i \leq n. \tag{4.10.7}$$

Furthermore, all pure mixed derivatives of u with respect to $x_1, \ldots, x_{i-1}, x_{i+1}, \ldots, x_n$ up to order $n-1$ vanish on $x_i = x_i^0, 1 \leq i \leq n$. If w is a function which is n times continuously differentiable in D, then

$$wLu - uMw = \sum_{k=1}^{n}(-1)^{k-1}D_k[(D_0D_1\ldots D_{k-1}w)(D_{k+1}\ldots D_n D_{n+1}u)], \tag{4.10.8}$$

where $Mw = (-1)^n D_1 \ldots D_n w(x) - b(x)w(x)$ with $D_0 \equiv D_{n+1} = I$ the identity. By integrating (4.10.8) over D, using ξ as the variable of integration, and noting that u vanishes together with all its mixed derivatives up to order

$n-1$ on $\xi_k = x_k^0$, $1 \le k \le n$, we then obtain

$$\int_D (wLu - uMw) \, d\xi$$

$$= \sum_{k=1}^n (-1)^{k-1} \int_{\xi_k = x_k} (D_1 \ldots D_{k-1} w)(D_{k+1} \ldots D_n u) \, d\xi', \quad (4.10.9)$$

where $d\xi' = d\xi_1 \ldots d\xi_{k-1} \, d\xi_{k+1} \ldots d\xi_n$.

Now let w be chosen as the function v satisfying (4.10.1). Since $v = 1$ on $\xi_k = x_k$, $1 \le k \le n$, it follows that $D_1 \ldots D_{k-1} v(\xi; x) = 0$ on $\xi_k = x_k$ for $2 \le k \le n$. Thus (4.10.9) becomes

$$\int_D v(\xi; x) Lu(\xi) \, d\xi = \int_{\xi_1 = x_1} v(\xi; x) D_2 \ldots D_n u(\xi) \, d\xi' = u(x). \quad (4.10.10)$$

By the continuity of v and by the fact that $v = 1$ on $\xi = x$, there is a domain D^+ containing x on which $v \ge 0$. Hence by multiplying (4.10.6) throughout by v and using (4.10.4) and (4.10.10), we obtain (4.10.3).

∎

Remark 4.10.1 We observe that the problem (4.10.1) defines precisely the so-called Riemann function for the operator L. The existence and regularity property of v can be deduced from Courant and Hilbert (1962) (see also Copson, 1958; Garbedian, 1964). Indeed, (4.10.1) is equivalent to the integral equation

$$v(\xi; x) = 1 + \int_\xi^x b(\eta) v(\eta; x) \, d\eta.$$

△

The following two theorems given by Bondge and Pachpatte (1980a) extend to the case of n independent variables the quite general results established by Pachpatte (1979a).

Theorem 4.10.2 *Suppose $\phi(x)$, $a(x)$, $b(x)$, $c(x)$ and $\sigma(x)$ are nonnegative continuous functions defined on $\Omega \subset R^n$. Let $v(\xi; x)$ be a solution of the characteristic initial value problem*

$$(-1)^n v_{\xi_1 \ldots \xi_n}(\xi; x) - [b(\xi) + c(\xi)]v(\xi; x) = 0 \text{ in } \Omega,$$
$$v(\xi; x) = 1 \text{ on } \xi_i = x_i, \ 1 \leq i \leq n,$$
(4.10.11)

and let D^+ be a connected subdomain of Ω containing x such that $v \geq 0$ for all $\xi \in D^+$. If $D \subset D^+$ and

$$\phi(x) \leq a(x) + \int_{x^0}^{x} b(\rho)\phi(\rho)\,d\rho + \int_{x^0}^{x} b(\rho) \left[\sigma(\rho) + \int_{x^0}^{\rho} c(\xi)\phi(\xi)\,d\xi \right] d\rho,$$
(4.10.12)

then

$$\phi(x) \leq a(x) + \int_{x^0}^{x} b(\rho) \left[a(\rho) + \sigma(\rho) + \int_{x^0}^{\rho} \{a(\xi)c(\xi) + b(\xi) \right.$$

$$\left. \times [a(\xi) + \sigma(\xi)]\}v(\xi; \rho)\,d\xi \right] d\rho.$$
(4.10.13)

□

Proof: Define a function $u(x)$ such that

$$u(x) = \int_{x^0}^{x} b(\rho)\phi(\rho)\,d\rho + \int_{x^0}^{x} b(\rho) \left[\sigma(\rho) + \int_{x^0}^{\rho} c(\xi)\phi(\xi)\,d\xi \right] d\rho,$$
(4.10.14)

$$u(x) = 0 \text{ on } x_i = x_i^0, \quad 1 \leq i \leq n,$$

then we have

$$D_1 \ldots D_n u(x) = b(x) \left[\phi(x) + \sigma(x) + \int_{x^0}^{x} c(\xi)\phi(\xi)\,d\xi \right],$$
(4.10.15)

Using the fact that $\phi(x) \leq a(x) + u(x)$ in (4.10.15) we have

$$D_1 \ldots D_n u(x) \leq b(x) \left[a(x) + \sigma(x) + u(x) + \int_{x^0}^{x} c(\xi)[a(\xi) + u(\xi)]\,d\xi \right].$$
(4.10.16)

If we put

$$\psi(x) = u(x) + \int_{x^0}^{x} c(\xi)[a(\xi) + u(\xi)]\,d\xi, \qquad (4.10.17)$$

$$\psi(x) = u(x) = 0 \text{ on } x_i = x_i^0, \quad 1 \le i \le n,$$

then we obtain

$$D_1 \ldots D_n \psi(x) = D_1 \ldots D_n u(x) + c(x)[a(x) + u(x)]. \qquad (4.10.18)$$

Using $D_1 \ldots D_n u(x) \le b(x)[a(x) + \sigma(x) + \psi(x)]$ from (4.10.16) and $u(x) \le \psi(x)$ from (4.10.17) in (4.10.18) we have

$$D_1 \ldots D_n \psi(x) \le b(x)[a(x) + \sigma(x) + \psi(x)] + c(x)[a(x) + \psi(x)],$$

i.e.

$$L\psi(x) = D_1 \ldots D_n \psi(x) - [b(x) + c(x)]\psi(x)$$
$$\le a(x)c(x) + b(x)[a(x) + \sigma(x)], \qquad (4.10.19)$$

and

$$\psi(x) = 0 \text{ on } x_i = x_i^0, \quad 1 \le i \le n.$$

Furthermore, all pure mixed derivatives of ψ with respect to $x_1, \ldots, x_{i-1}, x_{i+1}, \ldots, x_n$ up to order $n-1$ vanish on $x_i = x_i^0$, $1 \le i \le n$. If w is a function which is n times continuously differentiable in D, then

$$wL\psi - \psi Mw = \sum_{k=1}^{n} (-1)^{k-1} D_k [(D_0 D_1 \ldots D_{k-1} w)(D_{k+1} \ldots D_n D_{n+1} \psi)], \qquad (4.10.20)$$

where $Mw = (-1)^n D_1 \ldots D_n w(x) - [b(x) + c(x)]w(x)$ with $D_0 = D_{n+1} = I$ the identity. By integrating (4.10.20) over D, using ξ as variable of integration, and noting that ψ vanishes together with all its mixed derivatives up to order $n-1$ on $\xi_k = x_k^0$, $1 \le k \le n$, we then obtain

$$\int_D (wL\psi - \psi Mw)\,d\xi = \sum_{k=1}^{n} (-1)^{k-1} \int_{\xi_k = x_k} (D_1 \ldots D_{k-1} w)$$
$$\times (D_{k+1} \ldots D_n \psi)\,d\xi', \qquad (4.10.21)$$

where $d\xi' = d\xi_1 \ldots d\xi_{k-1} d\xi_{k+1} \ldots d\xi_n$.

Now let w be chosen as the function v satisfying (4.10.11). Since $v = 1$ on $\xi_k = x_k$, $1 \leq k \leq n$, it follows that $D_1 \ldots D_{k-1}v(\xi; x) = 0$ on $\xi_\kappa = x_k$ for $2 \leq k \leq n$. Thus (4.10.21) becomes

$$\int_D v(\xi; x)L\psi(\xi)\,d\xi = \int_{\xi_1 = x_1} v(\xi; x)D_2 \ldots D_n \psi(\xi)\,d\xi' = \psi(x). \quad (4.10.22)$$

By the continuity of v and by the fact that $v = 1$ on $\xi = x$, there is a domain D^+ containing x on which $v \geq 0$. Now multiplying (4.10.19) throughout by v and using (4.10.14) and (4.10.22), we obtain

$$\psi(x) \leq \int_{x^0}^{x} \{a(\xi)c(\xi) + b(\xi)[a(\xi) + \sigma(\xi)]\}v(\xi; x)\,d\xi.$$

Now substituting this bound on $\psi(x)$ in (4.10.16) we obtain

$$D_1 \ldots D_n u(x) \leq b(x)\left[a(x) + \sigma(x) + \int_{x^0}^{x} \{a(\xi)c(\xi) + b(\xi)\right.$$
$$\left. \times [a(\xi) + \sigma(\xi)]\}v(\xi; x)\,d\xi\right], \quad (4.10.23)$$

which implies the estimation for $u(x)$ such that

$$u(x) \leq \int_{x^0}^{x} b(\rho)\left[a(\rho) + \sigma(\rho) + \int_{x^0}^{\rho} \{a(\xi)c(\xi) + b(\xi)[a(\xi) + \sigma(\xi)]\}\right.$$
$$\left. \times v(\xi; \rho)\,d\xi\right] d\rho.$$

Using this bound on $u(x)$ in $\phi(x) \leq a(x) + u(x)$, we obtain the desired inequality in (4.10.13).

∎

Theorem 4.10.3 *Suppose $\phi(x)$, $a(x)$, $b(x)$, $c(x)$, $d(x)$ are nonnegative continuous functions defined on $\Omega \subset R^n$. Let $v(\xi; x)$ and $w(\xi; x)$ be the solutions of the characteristic initial value problems*

$$(-1)^n v_{\xi_1\ldots\xi_n}(\xi; x) - [b(\xi) + c(\xi) + k(\xi)]v(\xi; x) = 0 \text{ in } \Omega,$$
$$v(\xi; x) = 1 \text{ on } \xi_i = x_i, \quad 1 \leq i \leq n, \quad (4.10.24)$$

and

$$(-1)^n w_{\xi_1\ldots\xi_n}(\xi; x) - [b(\xi) - c(\xi)]w(\xi; x) = 0 \text{ in } \Omega,$$
$$w(\xi; x) = 1 \text{ on } \xi_i = x_i, \quad 1 \leq i \leq n, \quad (4.10.25)$$

respectively and let D^+ be a connected subdomain of Ω containing x such that $v \geq 0$, $w \geq 0$ for all $\xi \in D^+$. If $D \subset D^+$ and

$$\phi(x) \leq a(x) + \int_{x^0}^{x} b(\rho)\phi(\rho)\,d\rho + \int_{x^0}^{x} c(\rho)\left(\int_{x^0}^{\rho} k(\xi)\phi(\xi)\,d\xi\right) d\rho, \quad (4.10.26)$$

then

$$\phi(x) \leq a(x) + \int_{x^0}^{x} w(\rho; x)\left[a(\rho)b(\rho) + c(\rho)\int_{x^0}^{\rho} a(\xi)[b(\xi) + k(\xi)]\right.$$
$$\left.\times v(\xi; \rho)\,d\xi\right] d\rho. \quad (4.10.27)$$

□

The proof of this theorem can be completed by following the idea of the proof of Theorem 4.8.2 and closely looking at the proof of Theorem 4.10.2. The details are omitted here.

The inequalities established in the following two theorems by Bondge and Pachpatte (1980a) can be used in certain applications.

Theorem 4.10.4 *Suppose $\phi(x)$, $a(x)$, $b(x)$, $c(x)$ and $d(x)$ are nonnegative continuous functions defined on $\Omega \subset R^n$. Let $v(\xi; x)$ and $w(\xi; x)$ be the solutions of the characteristic initial value problems*

$$(-1)^n v_{\xi_1\ldots\xi_n}(\xi; x) - [b(\xi) + c(\xi) + d(\xi)]v(\xi; x) = 0 \text{ in } \Omega,$$
$$v(\xi; x) = 1 \text{ on } \xi_i = x_i, \quad 1 \leq i \leq n, \quad (4.10.28)$$

and
$$(-1)^n w_{\xi_1\ldots\xi_n}(\xi; x) - b(\xi)w(\xi; x) = 0 \text{ in } \Omega, \quad (4.10.29)$$
$$w(\xi; x) = 1 \text{ on } \xi_i = x_i, \quad 1 \leq i \leq n,$$

respectively and let D^+ be a connected subdomain of Ω containing x such that $v \geq 0$, $w \geq 0$, for all $\xi \in D^+$. If $D \subset D^+$ and

$$\phi(x) \leq a(x) + \int_{x^0}^{x} b(\eta)\phi(\eta)\,d\eta + \int_{x^0}^{x} b(\eta)\left(\int_{x^0}^{\eta} c(\rho)\phi(\rho)\,d\rho\right) d\eta$$
$$+ \int_{x^0}^{x} b(\eta)\left(\int_{x^0}^{\eta} c(\rho)\left(\int_{x^0}^{\rho} d(\xi)\phi(\xi)\,d\xi\right) d\rho\right) d\eta, \quad (4.10.30)$$

then

$$\phi(x) \leq a(x) + \int_{x^0}^{x} b(\eta)\left[a(\eta) + \int_{x^0}^{\eta} w(\rho; \eta)\left\{a(\rho)[b(\rho) + c(\rho)]\right.\right.$$
$$\left.\left. + c(\rho)\int_{x^0}^{\rho} a(\xi)[b(\xi) + c(\xi) + d(\xi)]v(\xi; \rho)\,d\xi\right\} d\rho\right] d\eta. \quad (4.10.31)$$

$\square$

Theorem 4.10.5 *Suppose $\phi(x)$, $a(x)$, $b(x)$, $c(x)$, $d(x)$, $p(x)$ and $q(x)$ are nonnegative continuous functions defined on $\Omega \subset R^n$. Let $v(\xi; x)$, $w(\xi; x)$ and $e(\xi; x)$ be the solutions of the characteristic initial value problems*

$$(-1)^n v_{\xi_1\ldots\xi_n}(\xi; x) - [b(\xi) + c(\xi) + d(\xi) + p(\xi) + q(\xi)]v(\xi; x) = 0 \text{ in } \Omega, \quad (4.10.32)$$
$$v(\xi; x) = 1 \text{ on } \xi_i = x_i, \quad 1 \leq i \leq n,$$

and
$$(-1)^n w_{\xi_1\ldots\xi_n}(\xi; x) - [b(\xi) + c(\xi) + d(\xi) - p(\xi)]w(\xi; x) = 0 \text{ in } \Omega, \quad (4.10.33)$$
$$w(\xi; x) = 1 \text{ on } \xi_i = x_i, \quad 1 \leq i \leq n,$$

and
$$(-1)^n e_{\xi_1\ldots\xi_n}(\xi; x) - [b(\xi) - c(\xi)]e(\xi; x) = 0 \text{ in } \Omega, \quad (4.10.34)$$
$$e(\xi; x) = 1 \text{ on } \xi_i = x_i, \quad 1 \leq i \leq n,$$

respectively and let D^+ be a connected subdomain of Ω containing x such that $v \geq 0$, $w \geq 0$, $e \geq 0$ for all $\xi \in D^+$. If $D \subset D^+$ and

$$\phi(x) \leq a(x) + \int_{x^0}^{x} b(\eta)\phi(\eta)\,d\eta + \int_{x^0}^{x} c(\eta)\left(\int_{x^0}^{\eta} d(\rho)\phi(\rho)\,d\rho\right)d\eta$$

$$+ \int_{x^0}^{x} c(\eta)\left(\int_{x^0}^{\eta} p(\rho)\left(\int_{x^0}^{\rho} q(\xi)\phi(\xi)\,d\xi\right)d\rho\right)d\eta, \quad (4.10.35)$$

then

$$\phi(x) \leq a(x) + \int_{x^0}^{x} e(\eta; x)\left[a(\eta)b(\eta) + c(\eta)\int_{x^0}^{\eta} w(\rho; \eta)\left(a(\rho)[b(\rho) + d(\rho)]\right.\right.$$

$$\left.\left. + p(\rho)\int_{x^0}^{\rho} a(\xi)[b(\xi) + d(\xi) + q(\xi)]v(\xi; \rho)\,d\xi\right)d\rho\right]d\eta. \quad (4.10.36)$$

$\square$

The details of the proofs of Theorems 4.10.4 and 4.10.5 follow by closely looking at the proof of Theorem 4.10.2, with suitable modifications. The details are omitted here.

4.11 Applications

In the past few years, the study of various classes of partial differential and integral equations has led to the investigation of a number of new inequalities contained in Sections 4.2–4.10. This section presents applications of some of the inequalities given in earlier sections which have recently appeared in the literature. It appears that these inequalities will have as many applications for partial differential, integral and integro-differential equations as the inequalities given in Chapter 1 have had for ordinary differential, integral and integro-differential equations.

4.11.1 Hyperbolic Partial Integro-differential Equations

This section presents applications of certain inequalities given in earlier sections to study the boundedness, uniqueness and continuous dependence of the solutions of some nonlinear hyperbolic partial integro-differential equations. These applications are given by Pachpatte (1979a, 1980c).

As a first application, we obtain the bound on the solution of a nonlinear hyperbolic partial integro-differential equation

$$u_{xy}(x, y) = f(x, y, u(x, y)) + h\left(x, y, u(x, y), \int_{x_0}^{x}\int_{y_0}^{y} k(x, y, s, t, u(s, t))\, ds\, dt\right)$$

(4.11.1)

with the given boundary conditions

$$u(x, y_0) = a_1(x), \quad u(x_0, y) = a_2(y), \quad a_1(x_0) = a_2(y_0) = 0, \quad (4.11.2)$$

where $u \in C[D, R]$, $f \in C[D \times R, R]$, $k \in C[D^2 \times R, R]$, $h \in C[D \times R^2, R]$, $R = (-\infty, \infty)$, $D \subset R_+ \times R_+$ such that

$$|f(x, y, u)| \leq c(x, y)|u|, \quad (4.11.3)$$

$$|k(x, y, s, t, u)| \leq q(s, t)|u|, \quad (4.11.4)$$

$$|h(x, y, u, v)| \leq p(x, y)[|u| + |v|], \quad (4.11.5)$$

where $c(x, y)$, $p(x, y)$ and $q(x, y)$ are nonnegative continuous real-valued functions defined on a domain D. The equation (4.11.1) with (4.11.2) is equivalent to the Volterra integral equation

$$u(x, y) = a_1(x) + a_2(y) + \int_{x_0}^{x}\int_{y_0}^{y} f(s, t, u(s, t))\, ds\, dt$$

$$+ \int_{x_0}^{x}\int_{y_0}^{y} h\left(s, t, u(s, t), \int_{x_0}^{s}\int_{y_0}^{t} k(s, t, \xi, \eta, u(\xi, \eta))\, d\xi\, d\eta\right) ds\, dt.$$

(4.11.6)

Let $u(x, y)$ be any solution of (4.11.1) with (4.11.2). Using (4.11.3)–(4.11.5) in (4.11.6) and assuming that $|a_1(x)| + |a_2(y)| \leq a(x, y)$, where $a(x, y)$ is a nonnegative continuous function defined on a domain D, we have

$$|u(x, y)| \leq a(x, y) + \int_{x_0}^{x}\int_{y_0}^{y} c(s, t)|u(s, t)|\, ds\, dt$$

$$+ \int_{x_0}^{x}\int_{y_0}^{y} p(s, t)\left(|u(s, t)| + \int_{x_0}^{s}\int_{y_0}^{t} q(\xi, \eta)|u(\xi, \eta)|\, d\xi\, d\eta\right) ds\, dt.$$

Let P_0, P, D^+, $\overline{R}$ (R replaced by $\overline{R}$) be as in Theorem 4.8.3. Let $v(s, t; x, y)$ and $w(s, t; x, y)$ be the solutions of the characteristic initial value problems given in Theorem 4.8.3 with $b(x, y) = 1$. Now an application of Theorem 4.8.3 with $b(x, y) = 1$ yields

$$|u(x, y)| \leq a(x, y) + \int_{x_0}^{x}\int_{y_0}^{y} w(s, t; x, y) \bigg\{ a(s, t)[c(s, t) + p(s, t)]$$

$$+ p(s, t) \int_{x_0}^{s}\int_{y_0}^{t} a(\xi, \eta)[c(\xi, \eta) + p(\xi, \eta) + q(\xi, \eta)]$$

$$\times v(\xi, \eta; s, t) \, d\xi \, d\eta \bigg\} \, ds \, dt, \qquad (4.11.7)$$

Thus the right-hand side in (4.11.7) gives us the bound on the solution $u(x, y)$ of (4.11.1)–(4.11.2) in terms of known functions.

If $|a_1(x)| + |a_2(y)| \leq \epsilon$, where $\epsilon > 0$ is arbitrary, then the bound obtained in (4.11.7) reduces to

$$|u(x, y)| \leq \epsilon \bigg\{ 1 + \int_{x_0}^{x}\int_{y_0}^{y} w(s, t; x, y) \bigg[[c(s, t) + p(s, t)]$$

$$+ p(s, t) \int_{x_0}^{s}\int_{y_0}^{t} [c(\xi, \eta) + p(\xi, \eta)$$

$$+ q(\xi, \eta)]v(\xi, \eta; s, t) \, d\xi \, d\eta \bigg] ds \, dt \bigg\}. \qquad (4.11.8)$$

In this case we note that (4.11.8) implies not only the boundedness but the stability of the solution $u(x, y)$ of (4.11.1)–(4.11.2) if the bound obtained on the right-hand side in (4.11.8) is small enough.

As a second application, we discuss the uniqueness of the solution of the nonlinear hyperbolic partial integro-differential equation (4.11.1) with the given boundary conditions (4.11.2). We assume that the functions f, k

and h in (4.11.1) satisfy

$$|f(x, y, u) - f(x, y, \bar{u})| \leq c(x, y)|u - \bar{u}|, \qquad (4.11.9)$$

$$|k(x, y, s, t, u) - k(x, y, s, t, \bar{u})| \leq q(s, t)|u - \bar{u}|, \qquad (4.11.10)$$

$$|h(x, y, u, r) - h(x, y, \bar{u}, \bar{r})| \leq p(x, y)[|u - \bar{u}| + |r - \bar{r}|], \qquad (4.11.11)$$

where $c(x, y)$, $p(x, y)$ and $q(x, y)$ are nonnegative continuous real-valued functions defined on a domain $D \subset R_+ \times R_+$. The problem (4.11.1)–(4.11.2) is equivalent to the Volterra integral equation (4.11.6). Now if $u(x, y)$ and $\bar{u}(x, y)$ are two solutions of the problem (4.11.1)–(4.11.2), then we have

$$u - \bar{u} = \int_{x_0}^{x} \int_{y_0}^{y} \{f(s, t, u) - f(s, t, \bar{u})\} \, ds \, dt$$

$$+ \int_{x_0}^{x} \int_{y_0}^{y} \left\{ h\left(s, t, u, \int_{x_0}^{s} \int_{y_0}^{t} k(s, t, \xi, \eta, u) \, d\xi \, d\eta \right) \right.$$

$$\left. - h\left(s, t, \bar{u}, \int_{x_0}^{s} \int_{y_0}^{t} k(s, t, \xi, \eta, \bar{u}) \, d\xi \, d\eta \right) \right\} ds \, dt. \qquad (4.11.12)$$

Using (4.11.9)–(4.11.11) in (4.11.12) we have

$$|u - \bar{u}| \leq \int_{x_0}^{x} \int_{y_0}^{y} c(s, t)|u - \bar{u}| \, ds \, dt$$

$$+ \int_{x_0}^{x} \int_{y_0}^{y} p(s, t) \left(|u - \bar{u}| + \int_{x_0}^{s} \int_{y_0}^{t} q(\xi, \eta)|u - \bar{u}| \, d\xi \, d\eta \right) ds \, dt.$$

Let P_0, P, D^+, $\bar{R}$ (R replaced by $\bar{R}$) $v(s, t; x, y)$ and $w(s, t; x, y)$ be as in Theorem 4.8.3 with $b(x, y) = 1$. Now an application of Theorem 4.8.3 with $a(x, y) = 0$ and $b(x, y) = 1$ yields $|u - \bar{u}| \leq 0$. Therefore $u = \bar{u}$, i.e. there is at most one solution of the problem (4.11.1)–(4.11.2).

Our third application deals with a continuous dependence of the solutions on the equation and given boundary data. We consider the boundary value problems

$$u_{xy}(x, y) = f(x, y, u(x, y), \phi(x, y)), \qquad (4.11.13)$$

in which
$$\phi(x, y) = \sigma(x, y) + \int_{x_0}^{x} \int_{y_0}^{y} k(x, y, s, t, u(s, t)) \, ds \, dt, \quad (4.11.14)$$

with the given boundary conditions
$$u(x_0, y) = g(y), \quad u(x, y_0) = h(x), \quad g(y_0) = h(x_0), \quad (4.11.15)$$

and
$$U_{xy}(x, y) = F(x, y, U(x, y), \psi(x, y)), \quad (4.11.16)$$

in which
$$\psi(x, y) = \rho(x, y) + \int_{x_0}^{x} \int_{y_0}^{y} k(x, y, s, t, U(s, t)) \, ds \, dt, \quad (4.11.17)$$

with the given boundary conditions
$$U(x_0, y) = G(y), \quad U(x, y_0) = H(x), \quad G(y_0) = H(x_0), \quad (4.11.18)$$

where all functions are real-valued and continuous on their respective domains of their definitions and $|g - G| \leq \epsilon$, $|h - H| \leq \epsilon$, and

$$\left| f\left(s, t, U, \sigma + \int_{x_0}^{s} \int_{y_0}^{t} k(s, t, \xi, \eta, U) \, d\xi \, d\eta \right) \right.$$
$$\left. - F\left(s, t, U, \rho + \int_{x_0}^{s} \int_{y_0}^{t} k(s, t, \xi, \eta, U) \, d\xi \, d\eta \right) \right| \leq \epsilon, \quad (4.11.19)$$

$$|k(s, t, \xi, \eta, u) - k(s, t, \xi, \eta, \bar{u})| \leq B|u - \bar{u}|, \quad (4.11.20)$$

$$|f(s, t, u, \phi) - f(s, t, \bar{u}, \psi)| \leq A[|u - \bar{u}| + |\phi - \psi|], \quad (4.11.21)$$

where $(s, t) \in D \subset R_+ \times R_+$, and ϵ, A and B are positive constants. The equivalent integral equations of (4.11.13)–(4.11.15) and (4.11.16)–(4.11.18) are

$$u(x, y) = g(y) + h(x) - g(y_0) + \int_{x_0}^{x} \int_{y_0}^{y} f\left(s, t, u(s, t), \sigma(s, t) \right.$$
$$\left. + \int_{x_0}^{s} \int_{y_0}^{t} k(s, t, \xi, \eta, u(\xi, \eta)) \, d\xi \, d\eta \right) ds \, dt,$$

and

$$U(x,y) = G(y) + H(x) - G(y_0) + \int_{x_0}^{x}\int_{y_0}^{y} F\left(s,t,U(s,t),\rho(s,t)\right.$$
$$\left. + \int_{x_0}^{s}\int_{y_0}^{t} k(s,t,\xi,\eta,U(\xi,\eta))\,d\xi\,d\eta\right) ds\,dt.$$

Then

$$u - U = (g - G) + (h - H) - (g(y_0) - G(y_0))$$
$$+ \int_{x_0}^{x}\int_{y_0}^{y}\left\{ f\left(s,t,u,\sigma + \int_{x_0}^{s}\int_{y_0}^{t} k(s,t,\xi,\eta,u)\,d\xi\,d\eta\right)\right.$$
$$\left. - F\left(s,t,U,\rho + \int_{x_0}^{s}\int_{y_0}^{t} k(s,t,\xi,\eta,U)\,d\xi\,d\eta\right)\right\} ds\,dt.$$

By adding and subtracting $f(s,t,U,\sigma + \int_{x_0}^{s}\int_{y_0}^{t} k(s,t,\xi,\eta,U)\,d\xi\,d\eta)$ in the integrand and if $(x - x_0)(y - y_0) \geq 0$, we obtain

$$|u - U| \leq |g - G| + |h - H| + |g(y_0) - G(y_0)|$$
$$+ \int_{x_0}^{x}\int_{y_0}^{y}\left| f\left(s,t,u,\sigma + \int_{x_0}^{s}\int_{y_0}^{t} k(s,t,\xi,\eta,u)\,d\xi\,d\eta\right)\right.$$
$$\left. - f\left(s,t,U,\sigma + \int_{x_0}^{s}\int_{y_0}^{t} k(s,t,\xi,\eta,U)\,d\xi\,d\eta\right)\right| ds\,dt$$
$$+ \int_{x_0}^{x}\int_{y_0}^{y}\left| f\left(s,t,U,\sigma + \int_{x_0}^{s}\int_{y_0}^{t} k(s,t,\xi,\eta,U)\,d\xi\,d\eta\right)\right.$$
$$\left. - F\left(s,t,U,\rho + \int_{x_0}^{s}\int_{y_0}^{t} k(s,t,\xi,\eta,U)\,d\xi\,d\eta\right)\right| ds\,dt$$

$$\leq \epsilon[3+(x-x_0)(y-y_0)] + \int_{x_0}^{x}\int_{y_0}^{y} A|u-U|\,ds\,dt$$

$$+ \int_{x_0}^{x}\int_{y_0}^{y} A \left(\int_{x_0}^{s}\int_{y_0}^{t} B|u-U|\,d\xi\,d\eta \right) ds\,dt.$$

Here we have used the hypotheses (4.11.19)–(4.11.21) to get the last inequality. Let P_0, P, D^+, R and $v(s,t;x,y)$ be as in Theorem 4.8.1 with $b(s,t) = A$ and $c(s,t) = B$. Now a suitable application of Theorem 4.8.1 yields

$$|u - U| \leq \epsilon \bigg\{ [3+(x-x_0)(y-y_0)] + \int_{x_0}^{x}\int_{y_0}^{y} A \bigg[[3+(s-x_0)(t-y_0)]$$

$$+ \int_{x_0}^{s}\int_{y_0}^{t} [3+(\xi-x_0)(\eta-y_0)](A+B)v(\xi,\eta;s,t)\,d\xi\,d\eta \bigg] ds\,dt \bigg\}.$$

On the compact set, the quantity in braces is bounded by some constant M^*. Therefore, $|u - U| \leq M^*\epsilon$ on this set, so the solution to such a boundary value problem depends continuously on f and the boundary values. If $E\epsilon \to 0$, then $|u - U| \to 0$ on the set.

We note that Theorems 4.8.4, 4.8.6 and 4.8.7 can be used to establish similar results as given above for nonlinear hyperbolic partial integro-differential equations of the forms

$$u_{xy} = f(x,y,u) + g\left(x, y, \int_{x_0}^{x}\int_{y_0}^{y} k\left(s, t, u, \int_{x_0}^{s}\int_{y_0}^{t} h(\xi,\eta,u)\,d\xi\,d\eta \right) ds\,dt \right), \quad (4.11.22)$$

$$u_{xy} = f(x,y) + \int_{x_0}^{x}\int_{y_0}^{y} k(s,t,u,u_{st})\,ds\,dt, \quad (4.11.23)$$

$$u_{xy} = f(x,y) + \int_{x_0}^{x}\int_{y_0}^{y} k(s,t,u,u_{st})\,ds\,dt$$

$$+ \int_{x_0}^{x}\int_{y_0}^{y} g\left(s, t, \int_{x_0}^{s}\int_{y_0}^{t} h(\xi,\eta,u,u_{\xi\eta})\,d\xi\,d\eta \right) ds\,dt, \quad (4.11.24)$$

respectively, with given boundary conditions and under some suitable conditions on the functions involved in (4.11.22)–(4.11.24).

4.11.2 Non-Self-Adjoint Hyperbolic Differential and Integro-differential Equations

This section presents an application of Theorem 4.8.8 to study the behavioural relations between the solutions of the nonlinear non-self-adjoint hyperbolic partial differential and integro-differential equations given by Pachpatte (1980d).

Consider the nonlinear non-self-adjoint hyperbolic partial differential equation

$$u_{xy}(x, y) = (c_0(x, y)u(x, y))_y + c_0(x, y)F(x, y, u(x, y)), \quad (4.11.25)$$

with the given boundary conditions

$$u(x_0, y) = g_0(y), \quad u(x, y_0) = h_0(x), \quad g_0(y_0) = h_0(x_0), \quad (4.11.26)$$

and the nonlinear non-self-adjoint hyperbolic partial integro-differential equation

$$z_{xy}(x, y) = (c_0(x, y)z(x, y))_y + c_0(x, y)F(x, y, z(x, y))$$
$$+ H\left(x, y, \int_{x_0}^{x}\int_{y_0}^{y} k(x, y, s, t, z(s, t))\,ds\,dt\right), \quad (4.11.27)$$

with the given boundary conditions

$$z(x_0, y) = g_1(y), \quad z(x, y_0) = h_1(x), \quad g_1(y_0) = h_1(x_0), \quad (4.11.28)$$

where $(x, y) \in D \subset R_+ \times R_+$ and all the functions are real-valued, continuous and defined on the respective domains of their definitions and

$$|c_0(x, y)| \le c(x, y), \quad (4.11.29)$$

$$|F(x, y, s, t, z) - F(x, y, u)| \le M_0|z - u|, \quad (4.11.30)$$

$$|k(x, y, s, t, z)| \le M_0 c(s, t)|z|, \quad (4.11.31)$$

$$|H(x, y, \bar{z})| \le g(x, y)|\bar{z}|, \quad (4.11.32)$$

$$|\lambda(x, y)| \le \epsilon, \quad (4.11.33)$$

where
$$\lambda(x, y) = g_1(y) - g_0(y) + h_1(x) - h_0(x) - [g_1(y_0) - g_0(y_0)]$$
$$- \int_{x_0}^{x} c(s, y)[h_1(s) - h_0(s)]\,ds,$$

and the functions $c(x, y)$ and $g(x, y)$ are real-valued nonnegative continuous and defined on a domain D and M_0 and ϵ are positive constants. Let $u(x, y)$ and $z(x, y)$ be the solutions of (4.11.25)–(4.11.26) and (4.11.27)–(4.11.28) respectively. Equations (4.11.25)–(4.11.26) and (4.11.27)–(4.11.28) are equivalent to the integral equations

$$u(x, y) = g_0(y) + h_0(x) - g_0(y_0) + \int_{x_0}^{x} c_0(s, y)u(s, y)\,ds$$
$$- \int_{x_0}^{x} c_0(s, y_0)u(s, y_0)\,ds + \int_{x_0}^{x}\int_{y_0}^{y} c_0(s, t)F(s, t, u(s, t))\,ds\,dt, \quad (4.11.34)$$

and

$$z(x, y) = g_1(y) + h_1(x) - g_1(y_0) + \int_{x_0}^{x} c_0(s, y)z(s, y)\,ds$$
$$- \int_{x_0}^{x} c_0(s, y_0)z(s, y_0)\,ds + \int_{x_0}^{x}\int_{y_0}^{y} c_0(s, t)F(s, t, z(s, t))\,ds\,dt$$
$$+ \int_{x_0}^{x}\int_{y_0}^{y} H\left(s, t, \int_{x_0}^{s}\int_{y_0}^{t} k(s, t, \xi, \eta, z(\xi, \eta))\,d\xi\,d\eta\right) ds\,dt. \quad (4.11.35)$$

From (4.11.34) and (4.11.35) we have
$$z(x, y) - u(x, y) = \lambda(x, y) + \int_{x_0}^{x} c_0(s, y)[z(s, y) - u(s, y)]\,ds$$
$$+ \int_{x_0}^{x}\int_{y_0}^{y} c_0(s, t)[F(s, t, z(s, t)) - F(s, t, u(s, t))]\,ds\,dt$$
$$+ \int_{x_0}^{x}\int_{y_0}^{y} H\left(s, t, \int_{x_0}^{s}\int_{y_0}^{t} k(s, t, \xi, \eta, z(\xi, \eta))\,d\xi\,d\eta\right) ds\,dt,$$
$$(4.11.36)$$

Using (4.11.29)–(4.11.33) and $|z| - |u| \le |z - u|$ in (4.11.36) and assuming that the solution $u(x, y)$ of (4.11.27)–(4.11.28) is bounded by N_0, where $N_0 > 0$ is a constant, we have

$$|z(x, y) - u(x, y)| \le a(x, y) + \int_{x_0}^{x} c(s, y)|z(s, y) - u(s, y)|\, ds$$

$$+ M_0 \int_{x_0}^{x} \int_{y_0}^{y} c(s, t)|z(s, t) - u(s, t)|\, ds\, dt$$

$$+ M_0 \int_{x_0}^{x} \int_{y_0}^{y} g(s, t) \left(\int_{x_0}^{s} \int_{y_0}^{t} c(\xi, \eta) \right.$$

$$\left. \times |z(\xi, \eta) u(\xi, \eta)|\, d\xi\, d\eta \right) ds\, dt, \qquad (4.11.37)$$

where

$$a(x, y) = \epsilon + M_0 \int_{x_0}^{x} \int_{y_0}^{y} g(s, t) \left(\int_{x_0}^{s} \int_{y_0}^{t} N_0 c(\xi, \eta)\, d\xi\, d\eta \right) ds\, dt. \qquad (4.11.38)$$

Let P_0, P, D^+, R and $v(s, t; x, y)$, $w(s, t; x, y)$ (with suitable changes) be as in Theorem 4.8.8. Now a suitable application of Theorem 4.8.8 yields

$$|z(x, y) - u(x, y)| \le f_2(x, y) + \int_{x_0}^{x} c(s, y) f_2(s, y) \exp\left(\int_{s}^{x} c(\xi, y)\, d\xi \right) ds, \qquad (4.11.39)$$

where

$$f_2(x, y) = a(x, y) + M_0 Q_2(x, y) + M_0 \int_{x_0}^{x} \int_{y_0}^{y} g(s, t) Q_2(s, t)\, ds\, dt,$$

in which

$$Q_2(x, y) = \int_{x_0}^{x} \int_{y_0}^{y} w(s, t; x, y) c(s, t) \left\{ a(s, t) + M_0 \int_{x_0}^{s} \int_{y_0}^{t} v(\xi, \eta; s, t) \right.$$

$$\left. \times a(\xi, \eta) c(\xi, \eta)\, d\xi\, d\eta \right\} ds\, dt,$$

where $v(s, t; x, y)$ and $w(s, t; x, y)$ are as in Theorem 4.8.8 with suitable changes in the values of p, q, b, r and h. If the right-hand side of (4.11.39) is bounded then we obtain the relative boundedness of the solutions of $u(x, y)$ and $z(x, y)$ of (4.11.25)–(4.11.26) and (4.11.27)–(4.11.28).

If $a(x, y)$ defined in (4.11.38) is small enough and, say, less then ϵ_0, where $\epsilon_0 > 0$ is arbitrary, then from (4.11.39) we have

$$|z(x, y) - u(x, y)| \leq \epsilon_0 \left\{ 1 + M_0 Q_2(x, y) + \int_{x_0}^{x}\int_{y_0}^{y} g(s, t) Q_2(s, t)\, ds\, dt \right.$$

$$+ \int_{x_0}^{x} c(s, y) \exp\left(\int_{s}^{x} c(\xi, y)\, d\xi \right) \left[1 + M_0 Q_2(s, y) \right.$$

$$\left. \left. + M_0 \int_{x_0}^{s}\int_{y_0}^{y} g(\xi, t) Q_2(\xi, t)\, d\xi\, dt \right] ds \right\}, \qquad (4.11.40)$$

If in (4.11.40) the expression in braces is bounded and $\epsilon_0 \to 0$, then we obtain $|z(x, y) - u(x, y)| \to 0$, which gives the equivalence between the solutions of (4.11.25)–(4.11.26) and (4.11.27)–(4.11.28).

Note that Theorem 4.8.8 can be used to study the stability boundedness and continuous dependence of the solutions of (4.11.25)–(4.11.26) and (4.11.27)–(4.11.28) by following arguments similar to those given in Section 4.11.1 (see also Pachpatte, 1979a, 1980c) with suitable modifications. Further, note that the integral inequality established in Theorem 4.8.8 can be used to study the similar problem for nonlinear non-self-adjoint partial differential and integro-differential equations of the form

$$u_{xy}(x, y) = (b_0(x, y) u(x, y))_x + b_0(x, y) F(x, y, u(x, y)), \qquad (4.11.41)$$

and

$$z_{xy}(x, y) = (b_0(x, y) z(x, y))_x + b_0(x, y) F(x, y, z(x, y))$$

$$+ H\left(x, y, \int_{x_0}^{x}\int_{y_0}^{y} k(x, y, s, t, z(s, t))\, ds\, dt \right), \qquad (4.11.42)$$

with the given boundary conditions and some suitable conditions on the functions involved in (4.11.41) and (4.11.42).

4.11.3 Perturbations of Hyperbolic Partial Differential Equations

This section deals with the variation of constants formulae established by Pachpatte (1982b) (taking into consideration the correction suggested by Beesack (1987) that the integral term involved in the function f in Pachpatte (1982b) must be absent) which gives the relationship between the solutions of a Bianchi-type partial differential equation and its perturbed partial integro-differential equation in three independent variables. Applications of these formulae investigated in Pachpatte (1982b) by using the inequality given in Theorem 4.9.2 are also given. In fact, the results given in Pachpatte (1982b) are motivated by the important role played by the variation of constants formulae in the theory of ordinary differential and integral equations; see Bernfeld and Lord (1978) and the references cited therein.

Consider the nonlinear hyperbolic parial differential equation

$$L[u] = f(x, y, z, u),$$
$$u(x_0, y, z) = u(x, y_0, z) = u(x, y, z_0) = u_0,$$
(D)

and its perturbed hyperbolic partial integro-differential equation

$$L[v] = f(x, y, z, v) + g\left(x, y, z, v, \int_{x_0}^{x}\int_{y_0}^{y}\int_{z_0}^{z} k(x, y, z, s, t, r, v)\,ds\,dt\,dr\right),$$
$$v(x_0, y, z) = v(x, y_0, z) = v(x, y, z_0) = u_0,$$
(E)

where

$$L[m] = m_{xyz} + a_1 m_{xy} + a_2 m_{yz} + a_3 m_{zx}$$
$$+ a_4 m_x + a_5 m_y + a_6 m_z + a_7 m,$$

$(x, y, z) \in \Delta = [x_0, a] \times [y_0, b] \times [z_0, c]$, $u, v \in C[\Delta, R]$, $a_i \in C[\Delta, R]$, $i = 1, 2, \ldots, 7$, $k \in C[\Delta^2 \times R, R]$, $f \in C[\Delta \times R, R]$, $g \in C[\Delta \times R \times R, R]$, u_0 is a constant, in which R denotes the set of real numbers. By a solution of equation (D) we mean a function u which satisfies (D) and is continuous in Δ along with the derivatives, $u_x, u_y, u_z, u_{xy}, u_{yz}, u_{zx}$ and u_{xyz}. Similarly, we can define a solution v of equation (E). We use $u = u(x, y, z, x_0, y_0, z_0, u_0)$

and $v = v(x, y, z, x_0, y_0, z_0, u_0)$ to denote the solutions of (D) and (E) respectively.

A useful nonlinear variation of constants formula established by Pachpatte (1982b) is embodied in the following theorem.

Theorem 4.11.1 *Suppose that equation* (D) *admits a unique solution* $u = u(x, y, z, x_0, y_0, z_0, u_0)$. *Suppose also that*

$$\phi(x, y, z, x_0, y_0, z_0, u_0) = \frac{\partial}{\partial u_0} u(x, y, z, x_0, y_0, z_0, u_0),$$

exists and is continuous for $x \geq x_0$, $y \geq y_0$, $z \geq z_0$ *and that* $\phi^{-1}(x, y, z, x_0, y_0, z_0, u_0)$ *exists for all* $x \geq x_0$, $y \geq y_0$, $z \geq z_0$. *If* $w = w(x, y, z)$ *is a solution of*

$$w_{xyz} = \phi^{-1}(x, y, z, x_0, y_0, z_0, w) \Bigg[A(x, y, z, x_0, y_0, z_0, w)$$

$$+ g\left(x, y, z, v, \int_{x_0}^{x} \int_{y_0}^{y} \int_{z_0}^{z} k(x, y, z, s, t, r, v) \, ds \, dt \, dr \right) \Bigg], \quad (4.11.43)$$

$$w(x_0, y, z) = w(x, y_0, z) = w(x, y, z_0) = u_0,$$

where

$$A(x, y, z, x_0, y_0, z_0, w) = -[a_1[u_{yw}w_x + u_{xw}w_y + u_{ww}w_xw_y + u_ww_{xy}]$$
$$+ a_2[u_{zw}w_y + u_{yw}w_z + u_{ww}w_yw_z + u_ww_{yz}]$$
$$+ a_3[u_{xw}w_z + u_{zw}w_x + u_{ww}w_zw_x + u_ww_{zx}]$$
$$+ a_4 u_w w_x + a_5 u_w w_y + a_6 u_w w_z$$
$$+ \{u_{yzw}w_x + u_{zxw}w_y + u_{xyw}w_x + u_{zww}w_xw_y$$
$$+ u_{xww}w_yw_z + u_{yww}w_zw_x + u_{zw}w_{xy}$$
$$+ u_{xw}w_{yz} + u_{yw}w_{zx}$$
$$+ (w_{yz}w_x + w_{zx}w_y + w_{xy}w_z)u_{ww}$$
$$+ u_{www}w_xw_yw_z\}], \quad (4.11.44)$$

then any solution $v = v(x, y, z, x_0, y_0, z_0, u_0)$ *of* (E) *satisfies the relation*

$$v(x, y, z, x_0, y_0, z_0, u_0)$$
$$= u\left(x, y, z, x_0, y_0, z_0, u_0 + \int_{x_0}^{x}\int_{y_0}^{y}\int_{z_0}^{z} \phi^{-1}(s, t, r, x_0, y_0, z_0, w(s, t, r))\right.$$
$$\times \left[A(s, t, r, x_0, y_0, z_0, w(s, t, r)) + g(s, t, r, v(s, t, r, x_0, y_0, z_0, u_0)),\right.$$
$$\left.\left.\int_{x_0}^{s}\int_{y_0}^{t}\int_{z_0}^{r} k(s, t, r, \xi, \eta, \rho, v(\xi, \eta, \rho, x_0, y_0, z_0, u_0))\, d\xi\, d\eta\, d\rho\right] ds\, dt\, dr\right),$$
(4.11.45)

as far as $w(x, y, z)$ exists for $x \geq x_0$, $y \geq y_0$, $z \geq z_0$.

□

Proof: Let $u(x, y, z, x_0, y_0, z_0, u_0)$ be any solution of (D) existing for $x \geq x_0$, $y \geq y_0$, $z \geq z_0$. The method of variation of parameters requires determining a function $w(x, y, z)$ so that

$$v(x, y, z, x_0, y_0, z_0, u_0) = u(x, y, z, x_0, y_0, z_0, w(x, y, z)),$$
$$w(x_0, y, z) = w(x, y_0, z) = w(x, y, z_0) = u_0,$$
(4.11.46)

is a solution of (E). Differentiating (4.11.46) partially with respect to z, y, x respectively we have

$$v_z = u_z + u_w w_z, \tag{4.11.47}$$
$$v_y = u_y + u_w w_y, \tag{4.11.48}$$
$$v_x = u_x + u_w w_x, \tag{4.11.49}$$

Differentiating (4.11.47), (4.11.48), (4.11.49) partially with respect to x, z, y respectively we have

$$v_{zx} = u_{zx} + u_{xw} w_z + u_{zw} w_x + u_{ww} w_z w_x + u_w w_{zx}, \tag{4.11.50}$$
$$v_{yz} = u_{yz} + u_{zw} w_y + u_{yw} w_z + u_{ww} w_y w_z + u_w w_{yz}, \tag{4.11.51}$$
$$v_{xy} = u_{xy} + u_{yw} w_x + u_{xw} w_y + u_{ww} w_x w_y + u_w w_{xy}. \tag{4.11.52}$$

Differentiating (4.11.52) partially with respect to z we have

$$v_{xyz} = u_{xyz} + u_{yzw}w_x + u_{zxw}w_y + u_{xyw}w_z + u_{zww}w_xw_y$$
$$+ u_{xww}w_yw_z + u_{yww}w_zw_x + u_{zw}w_{xy} + u_{xw}w_{yz} + u_{yw}w_{zx}$$
$$+ (w_{yz}w_x + w_{zx}w_y + w_{xy}w_z)u_{ww}$$
$$+ u_{www}w_xw_yw_z + u_ww_{xyz}. \qquad (4.11.53)$$

Multiplying both sides of (4.11.46), (4.11.47), (4.11.48), (4.11.49), (4.11.50), (4.11.51), (4.11.52) by $a_7, a_6, a_5, a_4, a_3, a_2, a_1$ respectively and adding in (4.11.53) we have

$$L[v] = L[u] - A(x, y, z, x_0, y_0, z_0, w) + u_w w_{xyz}, \qquad (4.11.54)$$

where $A(x, y, z, x_0, y_0, z_0, w)$ is as defined in (4.11.44). From (D), (E) and (4.11.54) we have

$$f(x, y, z, v) + g\left(x, y, z, v, \int_{x_0}^{x}\int_{y_0}^{y}\int_{z_0}^{z} k(x, y, z, s, t, r, v)\, ds\, dt\, dr\right)$$
$$= f(x, y, z, u) - A(x, y, z, x_0, y_0, z_0, w) + u_w w_{xyz},$$

which because of (4.11.46) and the fact that $\phi^{-1}(x, y, z, x_0, y_0, z_0, u_0)$ exists reduces to

$$w_{xyz} = \phi^{-1}(x, y, z, x_0, y_0, z_0, w)\left[A(x, y, z, x_0, y_0, z_0, w)\right.$$
$$+ g\left(x, y, z, v(x, y, z, x_0, y_0, z_0, u_0),\right.$$
$$\left.\left.\int_{x_0}^{x}\int_{y_0}^{y}\int_{z_0}^{z} k(x, y, z, s, t, r, v(s, t, r, x_0, y_0, z_0, u_0))\, ds\, dt\, dr\right)\right].$$
$$(4.11.55)$$

The solutions of (4.11.55) then determine $w(x, y, z)$. Consequently, if $w(x, y, z)$ is a solution of (4.11.55), then $v(x, y, z, x_0, y_0, z_0, u_0)$ given by (4.11.46) is a solution of (E). From (4.11.55), $w(x, y, z)$ must satisfy the integral equation

$$w(x, y, z) = u_0 + \int_{x_0}^{x}\int_{y_0}^{y}\int_{z_0}^{z} \phi^{-1}(s, t, r, x_0, y_0, z_0, w(s, t, r))$$

$$\times \Bigg[A(s, t, r, x_0, y_0, z_0, w(s, t, r))$$

$$+ g \Bigg(s, t, r, v(s, t, r, x_0, y_0, z_0, u_0),$$

$$\int_{x_0}^{s}\int_{y_0}^{t}\int_{z_0}^{r} k(s, t, r, \xi, \eta, \rho, v(\xi, \eta, \rho, x_0, y_0, z_0, u_0))$$

$$\times d\xi\, d\eta\, d\rho \Bigg) \Bigg]\, ds\, dt\, dr. \qquad (4.11.56)$$

Substituting (4.11.56) in (4.11.46) we obtain (4.11.45). Note that $v(x, y, z, x_0, y_0, z_0, u_0)$ exists for those values of $x \geq x_0$, $y \geq y_0$, $z \geq z_0$ for which the solution $w(x, y, z)$ of (4.11.43) exists.

∎

Another interesting and useful representation formula established by Pachpatte (1982b) is given in the following theorem.

Theorem 4.11.2 *Under the assumption of Theorem 4.11.1 the following relation is also valid.*

$$v(x, y, z, x_0, y_0, z_0, u_0) = u(x, y, z, x_0, y_0, z_0, u_0)$$

$$+ \int_{x_0}^{x}\int_{y_0}^{y}\int_{z_0}^{z} B(x, y, z, x_0, y_0, z_0, w(s, t, r))\, ds\, dt\, dr$$

$$+ \int_{x_0}^{x}\int_{y_0}^{y}\int_{z_0}^{z} \phi(x, y, z, x_0, y_0, z_0, w(s, t, r))$$

$$\times \phi^{-1}(s, t, r, x_0, y_0, z_0, w(s, t, r))$$

$$\times \left[A(s, t, r, x_0, y_0, z_0, w(s, t, r)) \right.$$

$$+ g \left(s, t, r, v(s, t, r, x_0, y_0, z_0, u_0), \right.$$

$$\left. \left. \int_{x_0}^{s} \int_{y_0}^{t} \int_{z_0}^{r} k(s, t, r, \xi, \eta, \rho, v(\xi, \eta, \rho, x_0, y_0, z_0, u_0)) \right. \right.$$

$$\left. \left. \times d\xi \, d\eta \, d\rho \right) \right] ds \, dt \, dr,$$

(4.11.57)

where $A(x, y, z, x_0, y_0, z_0, w)$ is as given in (4.11.44) and

$$B(x, y, z, x_0, y_0, z_0, w) = u_{www} w_s w_t w_r$$
$$+ (w_{tr} w_s + w_{sr} w_t + w_{st} w_r) u_{ww}. \quad (4.11.58)$$

□

Proof: For $x_0 \leq s \leq x$, $y_0 \leq t \leq y$, $z_0 \leq r \leq z$, we have

$$u_{str}(x, y, z, x_0, y_0, z_0, w(s, t, r))$$
$$= B(x, y, z, x_0, y_0, z_0, w(s, t, r))$$
$$+ \phi(x, y, z, x_0, y_0, z_0, w(s, t, r)) w_{str}(s, t, r). \quad (4.11.59)$$

Integrating (4.11.59) first with respect to r from z_0 to z, then integrating the resulting equation with respect to t from y_0 to y, and finally integrating the resulting equation with respect to s from x_0 to x we have

$$u(x, y, z, x_0, y_0, z_0, w(x, y, z))$$

$$= u(x, y, z, x_0, y_0, z_0, u_0)$$

$$+ \int_{x_0}^{x} \int_{y_0}^{y} \int_{z_0}^{z} B(x, y, z, x_0, y_0, z_0, w(s, t, r)) \, ds \, dt \, dr$$

$$+ \int_{x_0}^{x} \int_{y_0}^{y} \int_{z_0}^{z} \phi(x, y, z, x_0, y_0, z_0, w(s, t, r)) w_{str}(s, t, r) \, ds \, dt \, dr. \quad (4.11.60)$$

If $w(x, y, z)$ is any solution of (4.11.43) then the result (4.11.57) follows from (4.11.60), (4.11.46) and (4.11.43).

∎

In the following two theorems we present the applications of Theorems 4.11.1 and 4.11.2 given by Pachpatte (1982b) to study the boundedness of solutions of equation (E). A special version of the inequality given in Theorem 4.9.2 is used to obtain the required results.

We say that the solution $u(x, y, z, x_0, y_0, z_0, u_0)$ of (D) is globally uniformly stable if there exists a constant $M > 0$ such that $|u(x, y, z, x_0, y_0, z_0, u_0)| \leq M|u_0|$ for all $x \geq x_0 \geq 0$, $y \geq y_0 \geq 0$, $z \geq z_0 \geq 0$ and $|u_0| < \infty$.

Theorem 4.11.3 *Let the solution $u(x, y, z, x_0, y_0, z_0, u_0)$ of (D) be globally uniformly stable. Assume that the hypotheses of Theorem 4.11.1 hold and the functions involved in (E) satisfy*

$$|k(x, y, z, s, t, r, u(s, t, r, x_0, y_0, z_0, w(s, t, r)))|$$
$$\leq q(s, t, r)|w(s, t, r)|, \qquad (4.11.61)$$

$$|\phi^{-1}(x, y, z, x_0, y_0, z_0, w(x, y, z))[A(x, y, z, x_0, y_0, z_0, w(x, y, z))$$
$$+ g(x, y, z, u(x, y, z, x_0, y_0, z_0, w(x, y, z)), \bar{u})]|$$
$$\leq p(x, y, z)[|w(x, y, z)| + |\bar{u}|], \qquad (4.11.62)$$

where $p(x, y, z)$ and $q(x, y, z)$ are real-valued nonnegative continuous functions defined on Δ and

$$\int_{x_0}^{a}\int_{y_0}^{b}\int_{z_0}^{c} p(s, t, r)\,ds\,dt\,dr < \infty, \quad \int_{x_0}^{a}\int_{y_0}^{b}\int_{z_0}^{c} q(s, t, r)\,ds\,dt\,dr < \infty. \quad (4.11.63)$$

Then any solution $v(x, y, z, x_0, y_0, z_0, u_0)$ of (E) is bounded on Δ.

□

Proof: By Theorem 4.11.1 any solution $v(x, y, z, x_0, y_0, z_0, u_0)$ of (E) satisfies (4.11.46), where $w(x, y, z)$ given by (4.11.56) is a solution of (4.11.43). Using (4.11.56), (4.11.61) and (4.11.62) we have

$$|w(x, y, z)| \leq |u_0| + \int_{x_0}^{x}\int_{y_0}^{y}\int_{z_0}^{z} p(s, t, r) \left[|w(s, t, r)| \right.$$

$$\left. + \int_{x_0}^{s}\int_{y_0}^{t}\int_{z_0}^{r} q(\xi, \eta, \rho)|w(\xi, \eta, \rho)|\,d\xi\,d\eta\,d\rho \right] ds\,dt\,dr.$$

Now a suitable application of Theorem 4.9.2 part (ii) yields

$$|w(x, y, z)| \leq |u_0| \left[1 + \int_{x_0}^{x}\int_{y_0}^{y}\int_{z_0}^{z} p(s, t, r)\exp\left(\int_{x_0}^{s}\int_{y_0}^{t}\int_{z_0}^{r} [p(\xi, \eta, \rho) \right.\right.$$

$$\left.\left. + q(\xi, \eta, \rho)]\,d\xi\,d\eta\,d\rho \right) ds\,dt\,dr \right]. \qquad (4.11.64)$$

The right-hand side of (4.11.64) can be made sufficiently small by using (4.11.63) and assuming that $|u_0|$ is sufficiently small, i.e.

$$|w(x, y, z)| \leq \epsilon, \qquad (4.11.65)$$

where $\epsilon > 0$ is arbitrary. From (4.11.46), (4.11.65) and the global uniform stability of the solution $u(x, y, z, x_0, y_0, z_0, u_0)$ of (D) we obtain

$$|v(x, y, z, x_0, y_0, z_0, u_0)| \leq M\epsilon,$$

which implies the boundedness of the solution of (E), where $M > 0$ is a constant. The proof is complete.

∎

Theorem 4.11.4 *Assume that the hypotheses of Theorem 4.11.2 hold and the functions involved in (4.11.57) satisfy*

$$\int_{x_0}^{a}\int_{y_0}^{b}\int_{z_0}^{c} |B(x, y, z, x_0, y_0, z_0, w(s, t, r))|\,ds\,dt\,dr \leq M^*, \qquad (4.11.66)$$

$$|\phi(x, y, z, x_0, y_0, z_0, w(s, t, r))\phi^{-1}(s, t, r, x_0, y_0, z_0, w(s, t, r))| \leq N,$$

$$(4.11.67)$$

$$|k(x, y, z, s, t, r, v(s, t, r, x_0, y_0, z_0, u_0))|$$
$$\leq q(s, t, r)|v(s, t, r, x_0, y_0, z_0, u_0)|, \qquad (4.11.68)$$
$$|A(x, y, z, x_0, y_0, z_0, w(x, y, z)) + g(x, y, z, v(x, y, z, x_0, y_0, z_0, u_0), \bar{v})|$$
$$\leq p(x, y, z)[|v(x, y, z, x_0, y_0, z_0, u_0)| + |\bar{v}|], \qquad (4.11.69)$$

where M^* and N are positive constants and $p(x, y, z)$ and $q(x, y, z)$ are real-valued nonnegative continuous functions defined on Δ and

$$\int_{x_0}^{a}\int_{y_0}^{b}\int_{z_0}^{c} Np(s, t, r)\,ds\,dt\,dr < \infty, \quad \int_{x_0}^{a}\int_{y_0}^{b}\int_{z_0}^{c} q(s, t, r)\,ds\,dt\,dr < \infty.$$
$$(4.11.70)$$

Then for every bounded solution $u(x, y, z) = u(x, y, z, x_0, y_0, z_0, u_0)$ of (D) on Δ, the corresponding solution $v(x, y, z) = v(x, y, z, x_0, y_0, z_0, u_0)$ of (E) is bounded on Δ.

$\square$

Proof: Using the nonlinear variation of constants formula established in Theorem 4.11.2 the solutions of (D) and (E) are related by (4.11.57). From (4.11.57) and (4.11.66)–(4.11.69) we obtain

$$|v(x, y, z)| \leq \mu + \int_{x_0}^{x}\int_{y_0}^{y}\int_{z_0}^{z} Np(s, t, r)\left[|v(s, t, r)|\right.$$
$$\left. + \int_{x_0}^{s}\int_{y_0}^{t}\int_{z_0}^{r} q(\xi, \eta, \rho)|v(\xi, \eta, \rho)|\,d\xi\,d\eta\,d\rho\right]ds\,dt\,dr,$$

where $\mu = M + M^*$, in which $M > 0$ is the upper bound for $|u(x, y, z)|$. Now a suitable application of Theorem 4.9.2 part (ii) yields

$$|v(x, y, z)| \leq \mu\left[1 + \int_{x_0}^{x}\int_{y_0}^{y}\int_{z_0}^{z} Np(s, t, r)\exp\left(\int_{x_0}^{s}\int_{y_0}^{t}\int_{z_0}^{r}[Np(\xi, \eta, \rho)\right.\right.$$
$$\left.\left. + q(\xi, \eta, \rho)]\,d\xi\,d\eta\,d\rho\right)ds\,dt\,dr\right].$$

The above estimation in view of the assumption (4.11.70) implies the boundedness of $v(x, y, z)$ on Δ and the theorem is proved.

∎

We note that although the results on the boundedness and other properties of the solutions of certain second-order hyperbolic partial integro-differential equations in two independent variables have been studied by Pachpatte (1979a, 1980c), the results presented here are of interest because the approach to the problem is different from those of Pachpatte (1979a, 1980c).

4.11.4 Simultaneous Integral Equations in Two Variables

This section presents the applications of Theorem 4.3.2 to study the boundedness and uniqueness of the solutions of certain nonlinear simultaneous integral equations in two independent variables. For similar applications for certain hyperbolic simultaneous partial differential equations see Shinde and Pachpatte (1983, 1984).

As a first application we obtain bounds on the solutions of a system of simultaneous Volterra integral equations in two independent variables of the form

$$u(x, y) = a(x, y) + \int_0^x \int_0^y F(x, y, s, t, u(s, t), v(s, t))\,ds\,dt,$$

$$v(x, y) = b(x, y) + \int_0^x \int_0^y G(x, y, s, t, u(s, t), v(s, t))\,ds\,dt,$$

(H)

where $u, v, a, b \in C[\Delta, R]$, $F, G \in C[\Delta^2 \times R \times R, R]$, where $\Delta = R_+ \times R_+$, $R_+ = [0, \infty)$. We assume that

$$|F(x, y, s, t, u, v)| \leq h_1(s, t)|u| + \exp(\mu(s+t))h_2(s, t)|v|, \quad (4.11.71)$$

$$|G(x, y, s, t, u, v)| \leq \exp(-\mu(s+t))h_3(s, t)|u| + h_4(s, t)|v|, \quad (4.11.72)$$

$$|a(x, y)| \leq c_1, \quad |b(x, y)| \leq c_2, \quad (4.11.73)$$

where $h_i(x, y)$, $i = 1, 2, 3, 4$ are real-valued nonnegative continuous functions defined on Δ and c_1, c_2 and μ are nonnegative constants.

Let $(u(x, y), v(x, y))$ be a solution of (H) existing on Δ. Using (4.11.71)–(4.11.73) in (H) we have

$$|u(x, y)| \leq c_1 + \int_0^x \int_0^y h_1(s, t)|u(s, t)|\, ds\, dt$$

$$+ \int_0^x \int_0^y \exp(\mu(s + t))h_2(s, t)|v(s, t)|\, ds\, dt, \quad (4.11.74)$$

$$|v(x, y)| \leq c_2 + \int_0^x \int_0^y \exp(-\mu(s + t))h_3(s, t)|u(s, t)|\, ds\, dt$$

$$+ \int_0^x \int_0^y h_4(s, t)|v(s, t)|\, ds\, dt, \quad (4.11.75)$$

Now an application of Theorem 4.3.2 to (4.11.74)–(4.11.75) yields

$$|u(x, y)| \leq (c_1 + c_2)\exp\left(\mu(x + y) + \int_0^x \int_0^y h(s, t)\, ds\, dt\right), \quad (4.11.76)$$

$$|v(x, y)| \leq (c_1 + c_2)\exp\left(\int_0^x \int_0^y h(s, t)\, ds\, dt\right), \quad (4.11.77)$$

where $h(x, y) = \max\{[h_1(x, y) + h_3(x, y)], [h_2(x, y) + h_4(x, y)]\}$. The inequalities in (4.11.76), (4.11.77) gives the bounds on the solution (u, v) of (H).

As a second application, we shall discuss the uniqueness of the solutions of (H). We assume that the functions F and G in (H) satisfy

$$|F(x, y, s, t, u, v) - F(x, y, s, t, \bar{u}, \bar{v})|$$
$$\leq h_1(s, t)|u - \bar{u}| + \exp(\mu(s + t))h_2(s, t)|v - \bar{v}|, \quad (4.11.78)$$

$$|G(x, y, s, t, u, v) - G(x, y, s, t, \bar{u}, \bar{v})|$$
$$\leq \exp(-\mu(s + t))h_3(s, t)|u - \bar{u}| + h_4(s, t)|v - \bar{v}|, \quad (4.11.79)$$

where $h_i(x, y), i = 1, 2, 3, 4$ and μ are as given above. Now if (u, v) and $(\bar{u}, \bar{v})$ are the solutions of (H), then

$$|u - \bar{u}| \le \int_0^x \int_0^y |F(x, y, s, t, u, v) - F(x, y, s, t, \bar{u}, \bar{v})| \, ds \, dt, \quad (4.11.80)$$

$$|v - \bar{v}| \le \int_0^x \int_0^y |G(x, y, s, t, u, v) - G(x, y, s, t, \bar{u}, \bar{v})| \, ds \, dt, \quad (4.11.81)$$

Using (4.11.78) and (4.11.79) in (4.11.80) and (4.11.81) respectively we have

$$|u - \bar{u}| \le \int_0^x \int_0^y h_1(s, t)|u - \bar{u}| \, ds \, dt$$

$$+ \int_0^x \int_0^y \exp(\mu(s + t)) h_2(s, t)|v - \bar{v}| \, ds \, dt, \quad (4.11.82)$$

$$|v - \bar{v}| \le \int_0^x \int_0^y \exp(-\mu(s + t)) h_3(s, t)|u - \bar{u}| \, ds \, dt$$

$$+ \int_0^x \int_0^y h_4(s, t)|v - \bar{v}| \, ds \, dt, \quad (4.11.83)$$

Now an application of Theorem 4.3.2 to (4.11.82)–(4.11.83) yields $|u - \bar{u}| \le 0$, $|v - \bar{v}| \le 0$. Therefore $u = \bar{u}, v = \bar{v}$, i.e. there is at most one solution of (H).

4.12 Miscellaneous Inequalities

4.12.1

Let $u(x, y), a(x, y), b(x, y)$ and $c(x, y)$ be nonnegative continuous functions defined for $x, y \in R_+$. If

$$u(x, y) \le a(x, y) + \int_0^x \int_0^y [c(s, t)u(s, t) + b(s, t)] \, ds \, dt,$$

for $x, y \in R_+$, then

$$u(x, y) \le a(x, y) + p(x, y) \exp\left(\int_0^x \int_0^y c(s, t)\, ds\, dt\right),$$

for $x, y \in R_+$, where

$$p(x, y) = \int_0^x \int_0^y [c(s, t)a(s, t) + b(s, t)]\, ds\, dt,$$

for $x, y \in R_+$.

4.12.2 Bainov and Simeonov (1992, p. 110)

Let $u(x, y)$, $a(x, y)$ and $k(x, y)$ be nonnegative continuous functions defined for $x, y \in R_+$.

(i) Let $a(x, y)$ be nonincreasing in each variable $x, y \in R_+$. If

$$u(x, y) \le a(x, y) + \int_x^\infty \int_y^\infty k(s, t)u(s, t)\, ds\, dt,$$

and $\int_x^\infty \int_y^\infty k(s, t)\, ds\, dt < \infty$, for $x, y \in R_+$, then

$$u(x, y) \le a(x, y) \exp\left(\int_x^\infty \int_y^\infty k(s, t)\, ds\, dt\right),$$

for $x, y \in R_+$.

(ii) Let $a(x, y)$ be nondecreasing in x and nonincreasing in y. If

$$u(x, y) \le a(x, y) + \int_0^x \int_y^\infty k(s, t)u(s, t)\, ds\, dt,$$

and $\int_0^x \int_y^\infty k(s, t)\, ds\, dt < \infty$, for $x, y \in R_+$, then

$$u(x, y) \le a(x, y) \exp\left(\int_0^x \int_y^\infty k(s, t)\, ds\, dt\right),$$

for $x, y \in R_+$.

4.12.3 Pachpatte (1979b)

Let $u(x, y)$, $b(x, y)$ and $c(x, y)$ be nonnegative continuous functions defined for $x, y \in R_+$. Let $a(x, y) > 0$ for $x, y \in R_+$ be a continuous function and $a_x(x, y)$, $a_y(x, y)$, $a_{xy}(x, y)$ be nonnegative and continuous functions defined for $x, y \in R_+$ and

$$a_{xy}(x, y) \leq q(x, y)a(x, y),$$

for $x, y \in R_+$, where $q(x, y) \geq 0$ is a continuous function defined for $x, y \in R_+$.

(i) If

$$u(x, y) \leq a(x, y) + \int_0^x \int_0^y c(s, t)u(s, t)\,ds\,dt,$$

for $x, y \in R_+$, then

$$u(x, y) \leq \frac{a(x, 0)a(0, y)}{a(0, 0)} \exp\left(\int_0^x \int_0^y [q(s, t) + c(s, t)]\,ds\,dt\right),$$

for $x, y \in R_+$.

(ii) If

$$u(x, y) \leq a(x, y) + \int_0^x \int_0^y b(s, t)\left(u(s, t) + \int_0^s \int_0^t c(\sigma, \eta)u(\sigma, \eta)\,d\sigma\,d\eta\right)\,ds\,dt,$$

for $x, y \in R_+$, then

$$u(x, y) \leq a(x, 0) + a(0, y) - a(0, 0)$$
$$+ \int_0^x \int_0^y [q(s, t)a(s, t) + b(s, t)Q(s, t)]\,ds\,dt,$$

for $x, y \in R_+$, where

$$Q(x, y) = \frac{a(x, 0)a(0, y)}{a(0, 0)} \exp\left(\int_0^x \int_0^y [q(s, t) + b(s, t) + c(s, t)]\,ds\,dt\right),$$

for $x, y \in R_+$.

4.12.4 Pachpatte (unpublished manuscript)

Let $\phi(x)$, $f(t)$ and $g(t)$ be positive and continuously differentiable functions for $x \in [0, 1]$, $t \in [0, T]$, $T > 0$, and define $c(x, t) = \phi(x) + xg(t) + f(t)$ for $x \in [0, 1]$, $t \in [0, T]$. Let $u(x, t)$ and $p(x, t)$ be nonnegative continuous functions defined for $x \in [0, 1]$, $t \in [0, T]$ and

$$u(x, t) \leq c(x, t) + \int_0^x \int_0^s \int_0^t p(s_1, t_1) u(s_1, t_1) \, dt_1 \, ds_1 \, ds,$$

holds for $x \in [0, 1]$, $t \in [0, T]$.

(i) If $\phi'(x) \geq 0$, $f'(t) \geq 0$, $g'(t) \geq 0$ for $x \in [0, 1]$, $t \in [0, T]$, $\phi''(x)$ exists and nonnegative for $x \in [0, 1]$, then

$$u(x, t) \leq c(0, t) \exp \left[x \left(\frac{c_x(0, t)}{c(0, t)} \right) + \int_0^x \int_0^s \frac{\phi''(s_1)}{c(s_1, 0)} \, ds_1 \, ds \right.$$

$$\left. + \int_0^x \int_0^s \int_0^t p(s_1, t_1) \, dt_1 \, ds_1 \, ds \right],$$

for $x \in [0, 1]$, $t \in [0, T]$.

(ii) If $\phi'(x) \geq 0$, $f'(t) \geq 0$, $g'(t) \geq 0$, for $x \in [0, 1]$, $t \in [0, T]$, then

$$u(x, t) \leq c(x, 0) \exp \left[\int_0^t \frac{f'(t_1) + xg'(t_1)}{c(0, t_1)} \, dt_1 + \int_0^t \int_0^x \int_0^s p(s_1, t_1) \, ds_1 \, ds \, dt_1 \right],$$

for $x \in [0, 1]$, $t \in [0, T]$.

4.12.5 Pachpatte (unpublished manuscript)

Let $\phi(x)$, $f(t)$, $g(t)$ be positive and continuously differentiable functions for $x \in [0, 1]$, $t \in [0, T]$ and define $c(x, t) = \phi(x) + xg(t) + f(t)$, for $x \in [0, 1]$, $t \in [0, T]$. Let $u(x, t)$, $p(x, t)$ and $q(x, t)$ be nonnegative continuous functions defined for $x \in [0, 1]$, $t \in [0, T]$ and

$$u(x, t) \leq c(x, t) + \int_0^x \int_0^s \int_0^t p(s_1, t_1) \left(u(s_1, t_1) + \int_0^{s_1} \int_0^{s_2} \int_0^{t_1} q(s_3, t_3) \right.$$

$$\left. \times u(s_3, t_3) \, dt_3 \, ds_3 \, ds_2 \right) dt_1 \, ds_1 \, ds,$$

holds for $x \in [0, 1]$, $t \in [0, T]$.

(i) If $\phi'(x) \geq 0$, $f'(t) \geq 0$, $g'(t) \geq 0$ for $x \in [0, 1]$, $t \in [0, T]$, $\phi''(x)$ exists and nonnegative for $x \in [0, 1]$, then

$$u(x, t) \leq c(x, t) + x(c_x(0, t)) + \int_0^x \int_0^s \phi(s_1) \, ds_1 \, ds$$

$$+ \int_0^x \int_0^s \int_0^t p(s_1, t_1) Q_1(s_1, t_1) \, dt_1 \, ds_1 \, ds,$$

for $x \in [0, 1]$, $t \in [0, T]$, where

$$Q_1(x, t) = c(0, t) \exp\left[x\left(\frac{c_x(0, t)}{c(0, t)}\right) + \int_0^x \int_0^{s_2} \frac{\phi''(s_3)}{c(s_3, 0)} \, ds_3 \, ds_2 \right.$$

$$\left. + \int_0^x \int_0^s \int_0^t [p(s_3, t_3) + q(s_3, t_3)] \, dt_3 \, ds_3 \, ds \right],$$

for $x \in [0, 1]$, $t \in [0, T]$.

(ii) If $\phi'(x) \geq 0$, $f'(t) \geq 0$, $g'(t) \geq 0$ for $x \in [0, 1]$, $t \in [0, T]$, then

$$u(x, t) \leq c(x, 0) + \int_0^t [f'(t_1) + xg'(t_1)] \, dt_1$$

$$+ \int_0^t \int_0^x \int_0^s p(s_1, t_1) Q_2(t_1, s_1) \, ds_1 \, ds \, dt_1,$$

for $x \in [0, 1]$, $t \in [0, T]$, where

$$Q_2(x, t) = c(x, 0) \exp\left[\int_0^t \frac{f'(\sigma) + xg'(\sigma)}{c(0, \sigma)} \, d\sigma \right.$$

$$\left. + \int_0^t \int_0^x \int_0^{s_2} [p(s_3, t_3) + q(s_3, t_3)] \, ds_3 \, ds_2 \, dt_3 \right],$$

for $x \in [0, 1]$, $t \in [0, T]$.

4.12.6 Pachpatte (unpublished manuscript)

Let $\phi_i(x)$, $f_i(t)$, $g_i(t)$ for $i = 1, 2$ be positive and continuously differentiable functions for $x \in [0, 1]$, $t \in [0, T]$ and define $c_i(x, t) = \phi_i(x) + x g_i(t) + f_i(t)$, $i = 1, 2$, for $x \in [0, 1]$, $t \in [0, T]$ and $c(x, t) = c_1(x, t) + c_2(x, t)$, $\phi(x) = \phi_1(x) + \phi_2(x)$, $f(t) = f_1(t) + f_2(t)$, $g(t) = g_1(t) + g_2(t)$, for $x \in [0, 1]$, $t \in [0, T]$. Let $u(x, t)$, $v(x, t)$, $h_i(x, t)$, $i = 1, 2, 3, 4$, be nonnegative continuous functions defined for $x \in [0, 1]$, $t \in [0, T]$ and define

$$h(x, t) = \max\{[h_1(x, t) + h_3(x, t)], [h_2(x, t) + h_4(x, t)]\},$$

for $x \in [0, 1]$, $t \in [0, T]$ and

$$u(x, t) \leq c_1(x, t) + \int_0^x \int_0^s \int_0^t h_1(s_1, t_1) u(s_1, t_1) \, dt_1 \, ds_1 \, ds$$

$$+ \int_0^x \int_0^s \int_0^t h_2(s_1, t_1) \bar{v}(s_1, t_1) \, dt_1 \, ds_1 \, ds,$$

$$v(x, t) \leq c_2(x, t) + \int_0^x \int_0^s \int_0^t h_3(s_1, t_1) \bar{u}(s_1, t_1) \, dt_1 \, ds_1 \, ds$$

$$+ \int_0^x \int_0^s \int_0^t h_4(s_1, t_1) v(s_1, t_1) \, dt_1 \, ds_1 \, ds,$$

holds for $x \in [0, 1]$, $t \in [0, T]$, where $\bar{u}(x, t) = \exp(-\mu(x + t)) u(x, t)$, $\bar{v}(x, t) = \exp(\mu(x + t)) v(x, t)$ and μ is a nonnegative constant.

(i) If $\phi'(x) \geq 0$, $f'(t) \geq 0$, $g'(t) \geq 0$, for $x \in [0, 1]$, $t \in [0, T]$, $\phi''(x)$ exists and nonnegative for $x \in [0, 1]$, then

$$u(x, t) \leq c(0, t) \exp\left[\mu(x + t) + x\left(\frac{c_x(0, t)}{c(0, t)}\right) + \int_0^x \int_0^s \frac{\phi''(s_1)}{c(s_1, 0)} \, ds_1 \, ds \right.$$

$$\left. + \int_0^x \int_0^s \int_0^t h(s_1, t_1) \, dt_1 \, ds_1 \, ds \right],$$

$$v(x,t) \le c(0,t)\exp\left[x\left(\frac{c_x(0,t)}{c(0,t)}\right) + \int_0^x\int_0^s \frac{\phi''(s_1)}{c(s_1,0)}\,ds_1\,ds\right.$$
$$\left. + \int_0^x\int_0^s\int_0^t h(s_1,t_1)\,dt_1\,ds_1\,ds\right],$$

for $x \in [0,1]$, $t \in [0,T]$.

(ii) If $\phi'(x) \ge 0$, $f'(t) \ge 0$, $g'(t) \ge 0$, for $x \in [0,1]$, $t \in [0,T]$, then

$$u(x,t) \le c(x,0)\exp\left[\mu(x+t) + \int_0^t \frac{f'(\sigma) + xg'(\sigma)}{c(0,\sigma)}\,d\sigma\right.$$
$$\left. + \int_0^t\int_0^x\int_0^s h(s_1,t_1)\,ds_1\,ds\,dt_1\right],$$

$$v(x,t) \le c(x,0)\exp\left[\int_0^t \frac{f'(\sigma) + xg'(\sigma)}{c(0,\sigma)}\,d\sigma + \int_0^t\int_0^x\int_0^s h(s_1,t_1)\,ds_1\,ds\,dt_1\right],$$

for $x \in [0,1]$, $t \in [0,T]$.

4.12.7

Let $u(x,y)$, $u_x(x,y)$, $u_y(x,y)$, $u_{xy}(x,y)$ and $c(x,y)$ be nonnegative continuous functions defined for $x, y \in R_+$, and $u(x,0) = u(0,y) = 0$. If

$$u_{xy}(x,y) \le a(x) + b(y) + M\left[u(x,y) + \int_0^x\int_0^y c(s,t)u_{st}(s,t)\,ds\,dt\right],$$

for $x, y \in R_+$, where $a(x) > 0$, $b(y) > 0$, $a'(x) \ge 0$, $b'(y) \ge 0$ are continuous functions defined for $x, y \in R_+$, and $M \ge 0$ is a constant, then

$$u_{xy}(x,y) \le \frac{[a(0) + b(y)][a(x) + b(0)]}{[a(0) + b(0)]} \exp\left(M\int_0^x\int_0^y [1 + c(s,t)]\,ds\,dt\right),$$

for $x, y \in R_+$.

4.12.8 Pachpatte (1980e)

Let $u(x, y)$, $u_x(x, y)$, $u_y(x, y)$, $u_{xy}(x, y)$, $p(x, y)$ and $q(x, y)$ be nonnegative continuous functions defined for $x, y \in R_+$ and $u(x, 0) = u(0, y) = 0$. Let $a(x) > 0$, $b(y) > 0$, $a'(x) \geq 0$, $b'(y) \geq 0$ be continuous functions defined for $x, y \in R_+$.

(i) If

$$u_{xy}(x, y) \leq a(x) + b(y) + \int_0^x \int_0^y p(s, t) \left(u(s, t) + u_{st}(s, t) \right.$$

$$\left. + \int_0^s \int_0^t q(\sigma, \eta) u_{\sigma\eta}(\sigma, \eta) \, d\sigma \, d\eta \right) ds \, dt,$$

for $x, y \in R_+$ then

$$u_{xy}(x, y) \leq a(x) + b(y) + \int_0^x \int_0^y p(s, t) \left[\frac{[a(0) + b(t)][a(s) + b(0)]}{a(0) + b(0)} \right.$$

$$\left. \times \exp \left(\int_0^s \int_0^t [1 + p(\sigma, \eta) + q(\sigma, \eta)] \, d\sigma \, d\eta \right) \right] ds \, dt,$$

for $x, y \in R_+$.

(ii) Let $p(x, y) \geq 1$ and $M \geq 0$ be a constant. If

$$u_{xy}(x, y) \leq a(x) + b(y) + M \left[u(x, y) + \int_0^x \int_0^y p(s, t) \left(u(s, t) + u_{st}(s, t) \right. \right.$$

$$\left. \left. + \int_0^s \int_0^t q(\sigma, \eta) u_{\sigma\eta}(\sigma, \eta) \, d\sigma \, d\eta \right) ds \, dt \right],$$

for $x, y \in R_+$, then

$$u_{xy}(x, y) \leq a(x) + b(y) + \int_0^x \int_0^y 2M p(s, t) \left[\frac{[a(0) + b(t)][a(s) + b(0)]}{a(0) + b(0)} \right.$$

$$\left. \times \exp \left(\int_0^s \int_0^t [1 + 2M p(\sigma, \eta) + q(\sigma, \eta)] \, d\sigma \, d\eta \right) \right] ds \, dt,$$

for $x, y \in R_+$.

4.12.9 Pachpatte (1979c)

Suppose $u(x, y)$, $u_{xy}(x, y)$, $u_{xxyy}(x, y)$, $a(x, y)$, $b(x, y)$ and $c(x, y)$ are nonnegative continuous functions defined on a domain D. Let $P_0(x_0, y_0)$ and $P(x, y)$ be two points in D such that $(x - x_0)(y - y_0) \geq 0$ and let R be the rectangular region whose opposite corners are the points P_0 and P. Let $v(s, t; x, y)$ be the solution of the characteristic initial value problem

$$L[v] = v_{st} - [1 + b(s, t) + c(s, t) + b(s, t)c(s, t)]v = 0,$$

$$v(s, y) = v(x, t) = 1,$$

and let D^+ be a connected subdomain of D which contains P and on which $v \geq 0$ (see Figure 4.1). If $R \subset D^+$ and $u(x, y)$ satisfies

$$u_{xxyy}(x, y) \leq a(x, y) + b(x, y)\left[u(x, y) + u_{xy}(x, y)\right.$$

$$\left. + \int_{x_0}^{x}\int_{y_0}^{y} c(s, t)(u(s, t) + u_{st}(s, t) + u_{sstt}(s, t))\,ds\,dt\right],$$

and $u(x, y_0) = u(x_0, y) = 0$, then $u(x, y)$ also satisfies

$$u_{xxyy}(x, y) \leq a(x, y) + b(x, y)\int_{x_0}^{x}\int_{0}^{y} a(s, t)[1 + c(s, t)]v(s, t; x, y)\,ds\,dt.$$

4.12.10 Pachpatte (1979c)

Suppose $u(x, y)$, $u_{xy}(x, y)$, $u_{xxyy}(x, y)$, $a(x, y)$, $b(x, y)$, $c(x, y)$ are nonnegative continuous functions defined on a domain D and $b(x, y) \geq 1$, $c(x, y) \geq 1$ there. Let $P_0(x_0, y_0)$ and $P(x, y)$ be two points in D such that $(x - x_0)(y - y_0) \geq 0$ and let R be the rectangular region whose opposite corners are the points P_0 and P. Let $v(s, t; x, y)$ be the solution of the characteristic initial value problem

$$L[v] = v_{st} - b(s, t)[1 + 2c(s, t)]v = 0,$$

$$v(s, y) = v(x, t) = 1,$$

and let D^+ be a connected subdomain of D which contains P and on which $v \geq 0$ (see Figure 4.1). If $R \subset D^+$ and $u(x, y)$ satisfies

$$u_{xxyy}(x, y) \leq a(x, y) + b(x, y) \int_{x_0}^{x} \int_{y_0}^{y} c(s, t)(u(s, t) \\ + u_{st}(s, t) + u_{sstt}(s, t))\, ds\, dt,$$

and $u(x, y_0) = u(x_0, y) = 0$, then $u(x, y)$ also satisfies

$$u_{xxyy}(x, y) \leq a(x, y) + b(x, y) \int_{x_0}^{x} \int_{y_0}^{y} c(s, t)[a(s, t) + b(s, t)Q(s, t)]\, ds\, dt,$$

where

$$Q(x, y) = \int_{x_0}^{x} \int_{y_0}^{y} \left[a(s, t)(1 + c(s, t)) + 2b(s, t)c(s, t) \right. \\ \left. \times \int_{x_0}^{s} \int_{y_0}^{t} a(\sigma, \eta)[2 + c(\sigma, \eta)]v(\sigma, \eta; s, t)\, d\sigma\, d\eta \right] ds\, dt.$$

4.12.11 Pachpatte (1981b)

Suppose $u(x, y)$, $a(x, y)$, $b(x, y)$, $c(x, y)$ and $p(x, y)$ are nonnegative continuous functions defined on a domain D which lies in the first quadrant of (x, y) plane with $a(x, y) > 0$ there, and nondecreasing in both variables x and y. Let $P_0(x_0, y_0)$ and $P(x, y)$ be two points in D such that $(x - x_0)(y - y_0) \geq 0$ and let $R \subset D$ be the rectangular region whose opposite corners are the points P_0 and P. Let $v(s, t; x, y)$ be the solution of the characteristic initial value problem

$$L[v] = v_{st} - b(s, t)[c(s, t) + p(s, t)]v = 0,$$
$$v(s, y) = v(x, t) = 1,$$

and let D^+ be a connected subdomain of D which contains P and on which $v \geq 0$ (see Figure 4.1). If $R \subset D^+$ and $u(x, y)$ satisfies

$$u(x, y) \leq a(x, y) + b(x, y) \int_{x_0}^{x} \int_{y_0}^{y} c(s, t) \left(u(s, t) + b(s, t) \right. \\ \left. \times \int_{x_0}^{s} \int_{y_0}^{t} p(\sigma, \eta)u(\sigma, \eta)\, d\sigma\, d\eta \right) ds\, dt,$$

then $u(x, y)$ also satisfies

$$u(x, y) \le a(x, y) \left[1 + b(x, y) \int_{x_0}^{x} \int_{y_0}^{y} \left\{ c(s, t) + b(s, t)c(s, t) \right. \right.$$

$$\left. \left. \times \int_{x_0}^{s} \int_{y_0}^{t} [c(\sigma, \eta) + p(\sigma, \eta)] v(\sigma, \eta; s, t) \, d\sigma \, d\eta \right\} ds \, dt \right].$$

4.12.12 Bainov and Simeonov (1992, p. 119)

Let x, α belong to a domain $D \subset R^n$, and $\alpha \le x$. Let $u(x)$, $a(x)$, $b(x) \ge 0$, $q(x) \ge 0$ be continuous functions in D. Let $v(s; x)$ be the solution of the characteristic initial value problem

$$(-1)^n v_s(s; x) - q(s)b(s)v(s; x) = 0 \text{ in } D,$$

$$v(s, x) = 1 \text{ on } s_i = x_i, i = 1, 2, \ldots, n,$$

and let D^+ be a connected subdomain of D containing x, on which $v \ge 0$ for all $s \in D^+$. If $[\alpha, x] \subset D^+$ and

$$u(x) \le a(x) + q(x) \int_{\alpha}^{x} b(s) u(s) \, ds,$$

then

$$u(x) \le a(x) + q(x) \int_{\alpha}^{x} a(s) b(s) v(s; x) \, ds.$$

4.12.13 Bainov and Simeonov (1992, p. 113)

Let $\alpha, \beta \in R^n$, $\alpha < \beta$. Let $u(x)$, $a(x)$, $f(x)$ and $g(x)$ be nonnegative continuous functions for $x \in [\alpha, \beta]$, with $a(x)$ nondecreasing in $[\alpha, \beta]$. Then the inequality

$$u(x) \le a(x) + \int_{\alpha}^{x} f(s) \left[u(s) + \int_{\alpha}^{s} g(\sigma) u(\sigma) \, d\sigma \right] ds, \quad x \in [\alpha, \beta],$$

implies

$$u(x) \leq a(x) + \int_\alpha^x a(s)f(s)\exp\left(\int_\alpha^s [f(\sigma) + g(\sigma)]\,d\sigma\right) ds, \quad x \in [\alpha, \beta].$$

4.12.14 Chandra and Davis (1976)

Let $G(x)$, $H(x)$ be continuous, nonnegative matrices for $x^0 \leq x$. If

$$u(x) \leq a(x) + G(x)\int_{x^0}^x H(y)u(y)\,dy, \quad x^0 \leq x,$$

then

$$u(x) \leq a(x) + G(x)\int_{x^0}^x V(x,y)H(y)a(y)\,dy, \quad x^0 \leq x,$$

where $V(x, y)$ satisfies

$$V(x, y) = I + \int_y^x H(z)G(z)V(z, y)\,dz, \quad x^0 \leq y \leq x.$$

4.12.15 Thandapani and Agarwal (1982)

Let $u(x)$, $a(x)$, $b(x)$, $f_{ri}(x)$ ($i = 1, 2, \ldots, r$) be nonnegative continuous functions defined on $\Omega \subset R^n$. Let $v(s; x)$ be the solution of the characteristic initial value problem

$$(-1)^n v_s(s; x) - \sum_{r=1}^m E_s^r(s, b)v(s; x) = 0, \quad \text{in } \Omega,$$

$$v(s; x) = 1 \text{ on } s_i = x_i, \quad 1 \leq i \leq n,$$

and let D^+ be a connected subdomain of Ω containing x such that $v \geq 0$ for all $s \in D^+$. If $D \subset D^+$ and

$$u(x) \leq a(x) + b(x)\sum_{r=1}^m E^r(x, u),$$

where

$$E^r(x, u) = \int_{x^0}^{x} f_{r1}(x^1) \int_{x^0}^{x^1} f_{r2}(x^2) \ldots \int_{x^0}^{x^{r-1}} f_{rr}(x^r)u(x^r)\,dx^r \ldots dx^1,$$

then

$$u(x) \leq a(x) + b(x) \int_{x^0}^{x} \sum_{r=1}^{m} E_s^r(s, a)v(s; x)\,ds.$$

4.12.16 Singare and Pachpatte (1982)

Let $u(x)$, $a(x)$, $b(x)$, $f(x)$ and $g(x)$ be nonnegative continuous functions defined on $\Omega \subset R^n$. Let $v(s; x)$ and $w(s; x)$ be the solutions of the characteristic initial value problems

$$(-1)^n v_{s_1 \ldots s_n}(s; x) - [f(s) + g(s)b(s)]v(s; x) = 0, \text{ in } \Omega,$$

$$v(s; x) = 1 \text{ on } s_i = x_i, \, 1 \leq i \leq n,$$

and

$$(-1)^n w_{s_1 \ldots s_n}(s; x) + f(s)w(s; x) = 0, \text{ in } \Omega,$$

$$w(s; x) = 1 \text{ on } s_i = x_i, \quad 1 \leq i \leq n,$$

respectively and let D^+ be a connected subdomain of Ω containing x such that $v \geq 0$ and $w \geq 0$ for all $s \in D^+$ and suppose further that $D \subset D^+$.

(i) If

$$u(x) \leq a(x) + b(x) \int_{x^0}^{x} f(s) \left(\int_{x^0}^{s} g(\sigma)u(\sigma)\,d\sigma \right) ds,$$

for all $x \in D$, then

$$u(x) \leq a(x) + b(x) \int_{x^0}^{x} w(s; x) \left[f(s) \int_{x^0}^{s} g(\sigma)a(\sigma)v(\sigma; s)\,d\sigma \right] ds,$$

for all $x \in D$.

(ii) If $a(x)$ is positive and nondecreasing function in x and

$$u(x) \leq a(x) + b(x) \int_{x^0}^{x} f(s) \left(\int_{x^0}^{s} g(\sigma) u(\sigma) \, d\sigma \right) ds,$$

for all $x \in D$, then

$$u(x) \leq a(x) \left[1 + b(x) \int_{x^0}^{x} w(s; x) \left[f(s) \int_{x^0}^{s} g(\sigma) v(\sigma; s) \, d\sigma \right] ds \right],$$

for all $x \in D$.

4.12.17 Singare and Pachpatte (1982)

Let $u(x)$, $a(x)$, $b(x)$, $f(x)$, $g(x)$ and $h(x)$ be nonnegative continuous functions defined on $\Omega \subset R^n$. Let $v(s; x)$, $w(s; x)$, $e(s; x)$ be the solutions of the characteristic initial value problems

$$(-1)^n v_{s_1 \ldots s_n}(s; x) - [g(s) + h(s)b(s)] v(s; x) = 0, \text{ in } \Omega,$$

$$v(s; x) = 1 \text{ on } s_i = x_i, \quad 1 \leq i \leq n,$$

$$(-1)^n w_{s_1 \ldots s_n}(s; x) - [f(s) - g(s)] w(s; x) = 0, \text{ in } \Omega,$$

$$w(s; x) = 1 \text{ on } s_i = x_i, \quad 1 \leq i \leq n,$$

$$(-1)^n e_{s_1 \ldots s_n}(s; x) + f(s) e(s; x) = 0, \text{ in } \Omega,$$

$$e(s; x) = 1 \text{ on } s_i = x_i, 1 \leq i \leq n,$$

respectively and let D^+ be a connected subdomain of Ω containing x such that $v \geq 0$, $w \geq 0$, $e \geq 0$ for all $s \in D^+$, and suppose further that $D \subset D^+$.

(i) If

$$u(x) \leq a(x) + b(x) \int_{x^0}^{x} f(s) \left(\int_{x^0}^{s} g(\xi) \left(\int_{x^0}^{\xi} h(\sigma) u(\sigma) \, d\sigma \right) d\xi \right) ds,$$

for all $x \in D$, then

$$u(x) \leq a(x) + b(x) \int_{x^0}^{x} e(s; x) f(s) \left(\int_{x^0}^{s} w(\xi; s) \left[g(\xi) \int_{x^0}^{\xi} h(\sigma) a(\sigma) \right. \right.$$

$$\left. \left. \times v(\sigma; \xi) \, d\sigma \right] d\xi \right) ds,$$

for all $x \in D$.

(ii) If $a(x)$ is a positive and nondecreasing function in x and

$$u(x) \leq a(x) + b(x) \int_{x^0}^{x} f(s) \left(\int_{x^0}^{s} g(\xi) \left(\int_{x^0}^{\xi} h(\sigma) u(\sigma) \, d\sigma \right) d\xi \right) ds,$$

for all $x \in D$, then

$$u(x) \leq a(x) \left[1 + b(x) \int_{x^0}^{x} e(s; x) f(s) \left(\int_{x^0}^{s} w(\xi; s) \right. \right.$$

$$\left. \left. \times \left[g(\xi) \int_{x^0}^{\xi} h(\sigma) v(\sigma; \xi) \, d\sigma \right] d\xi \right) ds \right],$$

for all $x \in D$.

4.12.18 Shinde and Pachpatte (1983)

Let $u(x)$, $v(x)$, $a(x)$, $b(x)$, $p(x)$, $h_i(x)$ ($i = 1, 2, 3, 4$) be nonnegative continuous functions defined on $\Omega \subset R^n$. Let $w(s; x)$ be the solution of the characteristic initial value problem

$$(-1)^n w_{s_1 \ldots s_n}(s; x) - h(s) p(s) w(s; x) = 0, \quad \text{in } \Omega,$$

$$w(s; x) = 1 \text{ on } s_i = x_i, \quad 1 \leq i \leq n,$$

Let D^+ be a connected subdomain of Ω containing x such that $w \geq 0$ for all $s \in D^+$. If $D \subset D^+$ and

$$u(x) \leq a(x) + p(x) \left[\int_{x^0}^{x} h_1(s) u(s) \, ds + \int_{x^0}^{x} \exp\left(\mu \sum_{i=1}^{n} |s_i| \right) h_2(s) v(s) \, ds \right],$$

$$v(x) \leq b(x) + p(x) \left[\int_{x^0}^{x} \exp\left(-\mu \sum_{i=1}^{n} |s_i| \right) h_3(s) u(s) \, ds + \int_{x^0}^{x} h_4(s) v(s) \, ds \right],$$

for all $x \in D$, where μ is a nonnegative constant, then

$$u(x) \leq \exp\left(\mu \sum_{i=1}^{n} |x_i| \right) Q_1(x), \quad v(x) \leq Q_1(x),$$

for all $x \in D$, where

$$Q_1(x) = f(x) + p(x) \int_{x^0}^{x} w(s; x)h(s)f(s)\,ds,$$

in which

$$f(x) = \exp\left(-\mu \sum_{i=1}^{n} |x_i|\right) a(x) + b(x),$$

$$h(x) = \max\{[h_1(x) + h_3(x)], [h_2(x) + h_4(x)]\},$$

for all $x \in D$.

4.12.19 Shinde and Pachpatte (1983)

Let $u(x)$, $v(x)$, $a(x)$, $b(x)$, $p(x)$, $k(x)$, $h_i(x)$, $g_i(x)$ ($i = 1, 2, 3, 4$) be nonnegative continuous functions defined on $\Omega \subset R^n$. Let $w(s; x)$ and $e(s; x)$ be the solutions of the characteristic initial value problems

$$(-1)^n w_{s_1\ldots s_n}(s; x) - [k(s) + p(s)(h(s) + g(s))]w(s; x) = 0, \text{ in } \Omega,$$

$$w(s; x) = 1 \text{ on } s_i = x_i, \quad 1 \le i \le n,$$

and

$$(-1)^n e_{s_1\ldots s_n}(s; x) - [h(s)p(s) - k(s)]e(s; x) = 0, \text{ in } \Omega,$$

$$e(s; x) = 1 \text{ on } s_i = x_i, \quad 1 \le i \le n,$$

respectively. Let D^+ be a connected subdomain of Ω containing x such that $w \ge 0$, $e \ge 0$ for all $s \in D^+$. If $D \subset D^+$ and

$$u(x) \le a(x) + p(x) \left[\int_{x^0}^{x} h_1(s)u(s)\,ds + \int_{x^0}^{x} \exp\left(\mu \sum_{i=1}^{n} |s_i|\right) h_2(s)v(s)\,ds \right.$$

$$+ \int_{x^0}^{x} k(s) \left(\int_{x^0}^{s} g_1(\sigma)u(\sigma)\,d\sigma \right) ds$$

$$\left. + \int_{x^0}^{x} k(s) \left(\int_{x^0}^{s} \exp\left(\mu \sum_{i=1}^{n} |\sigma_i|\right) g_2(\sigma)v(\sigma)\,d\sigma \right) ds \right],$$

$$v(x) \leq b(x) + p(x) \left[\int_{x^0}^{x} \exp\left(-\mu \sum_{i=1}^{n} |s_i|\right) h_3(s) u(s) \, ds + \int_{x^0}^{x} h_4(s) v(s) \, ds \right.$$

$$+ \int_{x^0}^{x} k(s) \left(\int_{x^0}^{s} g_3(\sigma) u(\sigma) \, d\sigma \right) ds$$

$$+ \left. \int_{x^0}^{x} k(s) \left(\int_{x^0}^{s} \exp\left(-\mu \sum_{i=1}^{n} |\sigma_i|\right) g_4(\sigma) v(\sigma) \, d\sigma \right) ds \right],$$

for all $x \in D$, where μ is a nonnegative constant, then

$$u(x) \leq \exp\left(\mu \sum_{i=1}^{n} |x_i|\right) Q_2(x), \quad v(x) \leq Q_2(x),$$

for all $x \in D$, where

$$Q_2(x) = f(x) + p(x) \int_{x^0}^{x} e(s; x) \left\{ h(s) f(s) + k(s) \right.$$

$$\left. \times \int_{x^0}^{s} w(\sigma; s) f(\sigma) [h(\sigma) + g(\sigma)] \, d\sigma \right\} ds,$$

in which

$$f(x) = \exp\left(-\mu \sum_{i=1}^{n} |x_i|\right) a(x) + b(x),$$

$$h(x) = \max\{[h_1(x) + h_3(x)], [h_2(x) + h_4(x)]\},$$

$$g(x) = \max\{[g_1(x) + g_3(x)], [g_2(x) + g_4(x)]\},$$

for all $x \in D$.

4.12.20 Pachpatte (1988b)

Let $u(x)$, $a(x)$ and $h(x)$ be nonnegative continuous functions defined for $x \in R_+^n$ and let $p_j(x)$, $j = 1, 2, \ldots, n-1$ be positive continuous functions defined for $x \in R_+^n$, $a(x)$ be nondecreasing in each variable x_i of x in R_+^n. If

$$u(x) \leq a(x) + H[x, p_1, \ldots, p_{n-1}, hu],$$

for $x \in R_+^n$, where by the notation $H[x, p_1, \ldots, p_{n-1}, b]$ we mean the integral

$$\int_0^{x_1} \frac{1}{p_1(s_1, x_2, \ldots, x_n)} \cdots \int_0^{x_{n-1}} \frac{1}{p_{n-1}(s_1, \ldots, s_{n-1}, x_n)}$$
$$\times \int_0^{x_n} b(s_1, \ldots, s_{n-1}, s_n) \, ds_n \, ds_{n-1} \cdots ds_1,$$

then

$$u(x) \leq a(x) \exp(H[x, p_1, \ldots, p_{n-1}, h]),$$

for $x \in R_+^n$.

4.13 Notes

The results given in Section 4.2 deal with some fundamental inequalities which can be considered as two independent variable generalizations of the well-known Gronwall–Bellman inequality in one independent variable. Theorem 4.2.1 is due to Wendroff and is taken from Beckenbach and Bellman (1961, p. 154). The inequality given in Theorem 4.2.2 is a slight variant of the inequality given by Bondge et al. (1980).

Section 4.3 deals with some basic inequalities related to Wendroff's inequality and investigated by Pachpatte (1980b, 1995e). Theorem 4.3.1 is taken from Pachpatte (1995e) and Theorem 4.3.2 is a two independent variable generalization of the Greene's inequality given in Theorem 1.6.1 and is new to the literature. Theorems 4.3.3 and 4.3.4 are taken from Pachpatte (1980b).

The inequalities dealt with in Section 4.4 are due to Pachpatte (1980b). Theorem 4.4.1 is taken from Pachpatte (1980b), which in turn is a two independent variable generalization of the well-known inequality established by Pachpatte (1973a). Theorems 4.4.2, 4.4.3 and 4.4.4 are the two independent variable generalizations of the inequalities in Chapter 1 due to Pachpatte given in Theorems 1.7.4, 1.7.2 and 1.7.5 respectively. We believe that these inequalities are new to the literature.

Section 4.5 contains the Wendroff-like integral inequalities established by Pachpatte (1988b,c). Theorems 4.5.1, 4.5.2 and 4.5.3 are taken from

Pachpatte (1988c) and Theorem 4.5.4 is taken from Pachpatte (1988b). These results are mainly developed in order to apply them in the study of certain higher order partial differential equations.

The results in Section 4.6 are due to Snow (1971, 1972) which in turn can be considered as further generalizations of Wendroff's inequality and provides explicit bounds on the unknown function involving the Riemann function. Theorem 4.6.1 and 4.6.2 are taken from Snow (1971) and (1972) respectively. The results in Section 4.7 are further generalizations of Snow's inequality given by Ghoshal and Masood (1974a,b). Theorems 4.7.1 and 4.7.2 are taken from Ghoshal and Masood (1974a) and (1974b) respectively.

Section 4.8 is devoted to the more general inequalities in two independent variables investigated by Pachpatte (1979a, 1980c,d). Theorem 4.8.1 is taken from Pachpatte (1979a), Theorems 4.8.2 and 4.8.3 are taken from Pachpatte (1980c), Theorems 4.8.4–4.8.7 are taken from Pachpatte (1979a) and Theorem 4.8.8 is taken from Pachpatte (1980d).

Section 4.9 contains some basic integral inequalities of the Wendroff type in n independent variables investigated by Bondge and Pachpatte (1979a) and Singare and Pachpatte (1981). Theorem 4.9.1 is taken from Bondge and Pachpatte (1979a); Theorems 4.9.2 and 4.9.3 deal with more general inequalities and are new to the literature. Theorem 4.9.4 is taken from Singare and Pachpatte (1981).

The results given in Section 4.10 deal with Young's inequality (Young, 1973) in n independent variables and its further generalizations given by Bondge and Pachpatte (1980a). Theorem 4.10.1 is due to Young (1973) established in 1973. Theorems 4.10.2–4.10.4 are due to Bondge and Pachpatte (1980a).

Section 4.11 contains applications of some of the inequalities given in earlier sections to study some basic qualitative properties of the solutions of certain partial differential, integral and integro-differential equations. Section 4.12 contains some miscellaneous inequalities which can be used in certain new applications.

Chapter Five

Multidimensional Nonlinear Integral Inequalities

5.1 Introduction

In recent years considerable interest has been shown in developing various aspects of the theory of partial differential and integral equations both for their own sake and for their applications in science and technology. One of the most useful tools in the development of the qualitative theory of partial differential and integral equations is integral inequalities involving functions of many independent variables, which provide explicit bounds on the unknown functions. In the past few years, many new and useful integral inequalities involving multivalued functions and their partial derivatives have been found. This chapter gives some basic nonlinear multidimensional integral inequalities recently discovered in the literature. These inequalities can be used as ready and powerful tools in the study of various problems in the theory of certain partial differential, integral and integro-differential equations. Some applications are discussed to illustrate how these inequalities may be used to study the qualitative behaviour of solutions of certain partial differential and integro-differential equations, and miscellaneous inequalities which can be used in certain applications are also given.

5.2 Generalizations of Wendroff's Inequality

The fundamental role played by Wendroff's inequality and its generalizations and variants in the development of the theory of partial differential and integral equations is well known. In this section we present some basic nonlinear generalizations of Wendroff's inequality established by Bondge and Pachpatte (1979b, 1980a) and some new variants, which can be used as tools in the study of certain partial differential and integral equations.

Bondge and Pachpatte (1979b) proved the following useful nonlinear generalization of Wendroff's inequality.

Theorem 5.2.1 *Let $u(x, y)$ and $p(x, y)$ be nonnegative continuous functions defined for $x, y \in R_+$. Let $g(u)$ be a continuously differentiable function defined for $u \geq 0$, $g(u) > 0$ for $u > 0$ and $g'(u) \geq 0$ for $u \geq 0$. If*

$$u(x, y) \leq a(x) + b(y) + \int_0^x \int_0^y p(s, t) g(u(s, t)) \, ds \, dt, \qquad (5.2.1)$$

for $x, y \in R_+$, where $a(x) > 0$, $b(y) > 0$, $a'(x) \geq 0$, $b'(y) \geq 0$ are continuous functions defined for $x, y \in R_+$, then for $0 \leq x \leq x_1$, $0 \leq y \leq y_1$,

$$u(x, y) \leq \Omega^{-1} \left[\Omega(a(0) + b(y)) + \int_0^x \frac{a'(s)}{g(a(s) + b(0))} \, ds \right.$$
$$\left. + \int_0^x \int_0^y p(s, t) \, ds \, dt \right], \qquad (5.2.2)$$

where

$$\Omega(r) = \int_{r_0}^r \frac{ds}{g(s)}, \quad r > 0, \, r_0 > 0, \qquad (5.2.3)$$

Ω^{-1} *is the inverse function of Ω and x_1, y_1 are chosen so that*

$$\Omega(a(0) + b(y)) + \int_0^x \frac{a'(s)}{g(a(s) + b(0))} \, ds + \int_0^x \int_0^y p(s, t) \, ds \, dt \in \text{Dom}(\Omega^{-1}),$$

for all x, y lying in the subintervals $0 \leq x \leq x_1$, $0 \leq y \leq y_1$ of R_+.

□

MULTIDIMENSIONAL NONLINEAR INTEGRAL INEQUALITIES

Proof: We note that since $g'(u) \geq 0$ on R_+, the function $g(u)$ is monotonically increasing on $(0, \infty)$. Define a function $z(x, y)$ by the right-hand side of (5.2.1), then $z(x, 0) = a(x) + b(0)$, $z(0, y) = a(0) + b(y)$, and

$$z_{xy}(x, y) = p(x, y)g(u(x, y)). \tag{5.2.4}$$

Using $u(x, y) \leq z(x, y)$ in (5.2.4) and the fact that $z(x, y) > 0$, we observe that

$$\frac{z_{xy}(x, y)}{g(z(x, y))} \leq p(x, y). \tag{5.2.5}$$

From (5.2.5) and by using the facts that $z_x(x, y) \geq 0$, $z_y(x, y) \geq 0$, $z(x, y) > 0$, $g'(z(x, y)) \geq 0$ for $x, y \in R_+$, we observe that

$$\frac{z_{xy}(x, y)}{g(z(x, y))} \leq p(x, y) + \frac{z_x(x, y)g'(z(x, y))z_y(x, y)}{[g(z(x, y))]^2},$$

i.e.

$$\frac{\partial}{\partial y}\left(\frac{z_x(x, y)}{g(z(x, y))}\right) \leq p(x, y). \tag{5.2.6}$$

Keeping x fixed in (5.2.6), we set $y = t$; then, integrating with respect to t from 0 to y and using the fact that $z(x, 0) = a(x) + b(0)$, we have

$$\frac{z_x(x, y)}{g(z(x, y))} \leq \frac{a'(x)}{g(a(x) + b(0))} + \int_0^y p(x, t)\,dt. \tag{5.2.7}$$

From (5.2.3) and (5.2.7) we observe that

$$\frac{\partial}{\partial x}\Omega(z(x, y)) = \frac{z_x(x, y)}{g(z(x, y))}$$

$$\leq \frac{a'(x)}{g(a(x) + b(0))} + \int_0^y p(x, t)\,dt. \tag{5.2.8}$$

Keeping y fixed in (5.2.8), set $x = s$; then, integrating with respect to s from 0 to x and using the fact that $z(0, y) = a(0) + b(y)$ we have

$$\Omega(z(x, y)) \leq \Omega(a(0) + b(y)) + \int_0^x \frac{a'(s)}{g(a(s) + b(0))}\,ds$$

$$+ \int_0^x \int_0^y p(s, t)\,ds\,dt. \tag{5.2.9}$$

Now substituting the bound on $z(x, y)$ from (5.2.9) in $u(x, y) \leq z(x, y)$, we obtain the desired bound in (5.2.2). The subintervals for x and y are obvious.

∎

Remark 5.2.1 From the proof of Theorem 5.2.1, it is easy to observe that, in addition to (5.2.2), we can conclude that

$$u(x, y) \leq \Omega^{-1} \left[\Omega(a(x) + b(0)) + \int_0^y \frac{b'(t)}{g(a(0) + b(t))} dt \right.$$
$$\left. + \int_0^x \int_0^y p(s, t) \, ds \, dt \right], \quad (5.2.10)$$

where the expression in the square bracket on the right-hand side of (5.2.10) belongs to the domain of Ω^{-1}. We also note that the above conclusion applies to the following Theorems 5.2.3 and 5.2.4 and also some of the results given in our subsequent discussion.

△

Bondge and Pachpatte (1980a) gave the following generalization of Wendroff's inequality.

Theorem 5.2.2 *Let $u(x, y)$, $a(x, y)$, $b(x, y)$ and $c(x, y)$ be nonnegative continuous functions defined for $x, y \in R_+$. Let $g(u)$, $h(u)$ be continuously differentiable functions defined for $u \geq 0$, $g(u) > 0$, $h(u) > 0$ for $u > 0$ and $g'(u) \geq 0$, $h'(u) \geq 0$ for $u \geq 0$, and let $g(u)$ be subadditive and submultiplicative for $u \geq 0$. If*

$$u(x, y) \leq a(x, y) + b(x, y) h \left(\int_0^x \int_0^y c(s, t) g(u(s, t)) \, ds \, dt \right), \quad (5.2.11)$$

for $x, y \in R_+$, then for $0 \leq x \leq x_2$, $0 \leq y \leq y_2$,

$$u(x, y) \leq a(x, y) + b(x, y) h \left(G^{-1} \left[G(A(x, y)) \right. \right.$$
$$\left. \left. + \int_0^x \int_0^y c(s, t) g(b(s, t)) \, ds \, dt \right] \right), \quad (5.2.12)$$

where

$$A(x, y) = \int_0^x \int_0^y c(s, t)g(a(s, t)) \, ds \, dt, \qquad (5.2.13)$$

$$G(r) = \int_{r_0}^r \frac{ds}{g(h(s))}, \quad r > 0, \, r_0 > 0, \qquad (5.2.14)$$

G^{-1} is the inverse function of G and x_2, y_2 are chosen so that

$$G(A(x, y)) + \int_0^x \int_0^y c(s, t)g(b(s, t)) \, ds \, dt \in \mathrm{Dom}(G^{-1}),$$

for all x, y lying in the subintervals $0 \le x \le x_2$, $0 \le y \le y_2$ of R_+. $\square$

Proof: From the hypotheses on g and h, we note that the functions g and h are monotonically increasing on $(0, \infty)$. Define a function $z(x, y)$ by

$$z(x, y) = \int_0^x \int_0^y c(s, t)g(u(s, t)) \, ds \, dt. \qquad (5.2.15)$$

From (5.2.15) and using the fact that $u(x, y) \le a(x, y) + b(x, y)h(z(x, y))$ from (5.2.11) and the hypotheses on g we have

$$z(x, y) \le A(x, y) + \int_0^x \int_0^y c(s, t)g(b(s, t))g(h(z(s, t))) \, ds \, dt, \qquad (5.2.16)$$

for $x, y \in R_+$, where $A(x, y)$ is defined by (5.2.13). Now fix $\alpha, \beta \in R_+$ such that $0 \le x \le \alpha$, $0 \le y \le \beta$; then from (5.2.16) we observe that

$$z(x, y) \le A(\alpha, \beta) + \int_0^x \int_0^y c(s, t)g(b(s, t))g(h(z(s, t))) \, ds \, dt, \qquad (5.2.17)$$

for $0 \le x \le \alpha$, $0 \le y \le \beta$. Define a function $v(x, y)$ by the right-hand side of (5.2.17); then $v(x, 0) = v(0, y) = A(\alpha, \beta)$, $z(x, y) \le v(x, y)$ and

$$v_{xy}(x, y) = c(x, y)g(b(x, y))g(h(z(x, y)))$$

$$\le c(x, y)g(b(x, y))g(h(v(x, y))). \qquad (5.2.18)$$

Now first assume that $A(\alpha, \beta) > 0$; then from (5.2.18) we observe that

$$\frac{v_{xy}(x, y)}{g(h(v(x, y)))} \leq c(x, y)g(b(x, y)). \tag{5.2.19}$$

As in the proof of Theorem 5.2.1, from (5.2.19) we observe that

$$\frac{\partial}{\partial y}\left(\frac{v_x(x, y)}{g(h(v(x, y)))}\right) \leq c(x, y)g(b(x, y)). \tag{5.2.20}$$

By keeping x fixed in (5.2.20), setting $y = t$, and then integrating with respect to t from 0 to β we have

$$\frac{v_x(x, \beta)}{g(h(v(x, \beta)))} \leq \frac{v_x(x, 0)}{g(h(v(x, 0)))} + \int_0^\beta c(x, t)g(b(x, t))\,dt. \tag{5.2.21}$$

From (5.2.14) and (5.2.21) we observe that

$$\frac{\partial}{\partial x}G(v(x, \beta)) \leq \frac{\partial}{\partial x}G(v(x, 0)) + \int_0^\beta c(x, t)g(b(x, t))\,dt. \tag{5.2.22}$$

Now keeping y fixed in (5.2.22), setting $x = s$, and then integrating with respect to s from 0 to α we get

$$G(v(\alpha, \beta)) \leq G(A(\alpha, \beta)) + \int_0^\alpha \int_0^\beta c(s, t)g(b(s, t))\,ds\,dt, \tag{5.2.23}$$

Since $z(\alpha, \beta) \leq v(\alpha, \beta)$ and $\alpha, \beta \in R_+$ are arbitrary from (5.2.23) we have

$$z(x, y) \leq G^{-1}\left[G(A(x, y)) + \int_0^x \int_0^y c(s, t)g(b(s, t))\,ds\,dt\right], \tag{5.2.24}$$

for $0 \leq x \leq x_1$, $0 \leq y \leq y_1$. The desired bound in (5.2.12) follows by using (5.2.24) in $u(x, y) \leq a(x, y) + b(x, y)h(z(x, y))$.

If $A(\alpha, \beta)$ in (5.2.17) is nonnegative, we carry out the above procedure with $A(\alpha, \beta) + \epsilon$ instead of $A(\alpha, \beta)$, where $\epsilon > 0$ is an arbitrary small constant, and subsequently pass to the limit as $\epsilon \to 0$ to obtain (5.2.12).

∎

The following theorem provides another useful generalization of Wendroff's inequality.

Theorem 5.2.3 *Let $u(x, y)$, $c(x, y)$ and $p(x, y)$ be nonnegative continuous functions defined for $x, y \in R_+$. Let $g(u)$, $g'(u)$, $a(x)$, $a'(x)$, $b(y)$ and $b'(y)$ be as in Theorem 5.2.1 and $g(u)$ be submultiplicative on R_+. If*

$$u(x, y) \leq a(x) + b(y) + \int_0^x \int_0^y c(s, t) u(s, t) \, ds \, dt$$

$$+ \int_0^x \int_0^y p(s, t) g(u(s, t)) \, ds \, dt, \qquad (5.2.25)$$

for $x, y \in R_+$, then for $0 \leq x \leq x_3$, $0 \leq y \leq y_3$,

$$u(x, y) \leq q(x, y) \left[\Omega^{-1} \left[\Omega(a(0) + b(y)) + \int_0^x \frac{a'(s)}{g(a(s) + b(0))} \, ds \right. \right.$$

$$\left. \left. + \int_0^x \int_0^y p(s, t) g(q(s, t)) \, ds \, dt \right] \right], \qquad (5.2.26)$$

where

$$q(x, y) = \exp \left(\int_0^x \int_0^y c(s, t) \, ds \, dt \right), \qquad (5.2.27)$$

and Ω, Ω^{-1} are as defined in Theorem 5.2.1 and x_3, y_3 are chosen so that

$$\Omega(a(0) + b(y)) + \int_0^x \frac{a'(s)}{g(a(s) + b(0))} \, ds$$

$$+ \int_0^x \int_0^y p(s, t) g(q(s, t)) \, ds \, dt \in \text{Dom}(\Omega^{-1}),$$

for all x, y lying in the subintervals $0 \leq x \leq x_3$, $0 \leq y \leq y_3$ of R_+. □

Proof: We note that, since $a'(x) \geq 0$, $b'(y) \geq 0$, $g'(u) \geq 0$ for $x, y, u \in R_+$, the functions $a(x), b(y), g(u)$ are monotonically increasing on $(0, \infty)$.

Define a function $m(x, y)$ by

$$m(x, y) = a(x) + b(y) + \int_0^x \int_0^y p(s, t) g(u(s, t)) \, ds \, dt, \quad (5.2.28)$$

then $m(x, 0) = a(x) + b(0)$, $m(0, y) = a(0) + b(y)$ and (5.2.25) can be restated as

$$u(x, y) \leq m(x, y) + \int_0^x \int_0^y c(s, t) u(s, t) \, ds \, dt. \quad (5.2.29)$$

Since $m(x, y)$ is positive and nondecreasing in each variable $x, y \in R_+$, by applying Theorem 4.2.2 to (5.2.29) we have

$$u(x, y) \leq m(x, y) q(x, y), \quad (5.2.30)$$

where $q(x, y)$ is defined by (5.2.27). From (5.2.28) and (5.2.30) we have

$$m_{xy}(x, y) = p(x, y) g(u(x, y))$$
$$\leq p(x, y) g(m(x, y) q(x, y))$$
$$\leq p(x, y) g(m(x, y)) g(q(x, y)),$$

i.e.

$$\frac{m_{xy}(x, y)}{g(m(x, y))} \leq p(x, y) g(q(x, y)). \quad (5.2.31)$$

Now by following the same arguments as in the proof of Theorem 5.2.1 below the inequality (5.2.5) we get

$$m(x, y) \leq \Omega^{-1} \left[\Omega(a(0) + b(y)) + \int_0^x \frac{a'(s)}{g(a(s) + b(0))} \, ds \right.$$
$$\left. + \int_0^x \int_0^y p(s, t) g(q(s, t)) \, ds \, dt \right]. \quad (5.2.32)$$

Using (5.2.32) in (5.2.30) we get the required inequality in (5.2.26). ∎

A slight variant of Theorem 5.2.3 is given in the following theorem.

Theorem 5.2.4 Let $u(x, y)$, $c(x, y)$, $p(x, y)$, $g(u)$, $g'(u)$, $a(x)$, $a'(x)$, $b(y)$ and $b'(y)$ be as in Theorem 5.2.3. If

$$u(x, y) \leq a(x) + b(y) + \int_0^x c(s, y) u(s, y) \, ds + \int_0^x \int_0^y p(s, t) g(u(s, t)) \, ds \, dt, \tag{5.2.33}$$

for $x, y \in R_+$, then for $0 \leq x \leq x_4$, $0 \leq y \leq y_4$,

$$u(x, y) \leq F(x, y) \left[\Omega^{-1} \left[\Omega(a(0) + b(y)) + \int_0^x \frac{a'(s)}{g(a(s) + b(0))} \, ds \right. \right.$$

$$\left. \left. + \int_0^x \int_0^y p(s, t) g(F(s, t)) \, ds \, dt \right] \right], \tag{5.2.34}$$

where

$$F(x, y) = \exp \left(\int_0^x c(s, y) \, ds \right), \tag{5.2.35}$$

for $x, y \in R_+$, Ω, Ω^{-1}, are as defined in Theorem 5.2.1 and x_4, y_4 are chosen so that

$$\Omega(a(0) + b(y)) + \int_0^x \frac{a'(s)}{g(a(s) + b(0))} \, ds$$

$$+ \int_0^x \int_0^y p(s, t) g(F(s, t)) \, ds \, dt \in \text{Dom}(\Omega^{-1}),$$

for all x, y lying in the subintervals $0 \leq x \leq x_4$, $0 \leq y \leq y_4$ of R_+. □

The proof of this theorem can be completed by following the proof of Theorem 5.2.3, with suitable changes. The details are omitted here.

Remark 5.2.2 We note that, if the inequality (5.2.33) in Theorem 5.2.4 is replaced by

$$u(x, y) \leq a(x) + b(y) + \int_0^y c(x, t) u(x, t) \, dt + \int_0^x \int_0^y p(s, t) g(u(s, t)) \, ds \, dt,$$

then the bound obtained in (5.2.34) is replaced by

$$u(x, y) \leq F_0(x, y) \left[\Omega^{-1} \left[\Omega(a(0) + b(y)) + \int_0^x \frac{a'(s)}{g(a(s) + b(0))} ds \right. \right.$$
$$\left. \left. + \int_0^x \int_0^y p(s, t) g(F_0(s, t)) \, ds \, dt \right] \right],$$

where

$$F_0(x, y) = \exp \left(\int_0^y c(x, t) \, dt \right),$$

for $x, y \in R_+$, and the expression in the inner square bracket on the right-hand side belongs to the domain of Ω^{-1}.

△

5.3 Wendroff-type Inequalities

During the past twenty years or so many authors have developed extensions and variants of Wendroff's inequality and exhibited applications in partial differential and integral equations. In this section we shall deal with the Wendroff-type inequalities investigated by Pachpatte (1995e) and Bondge and Pachpatte (1979b,c) which can be used in the study of certain partial differential and integral equations.

Pachpatte (1995e) established the Wendroff-type inequalities in the following theorem.

Theorem 5.3.1 *Let $u(x, y)$, $a(x, y)$ and $b(x, y)$ be nonnegative continuous functions defined on R_+^2 and $L: R_+^3 \to R_+$ be a continuous function which satisfies the condition*

$$0 \leq L(x, y, v) - L(x, y, w) \leq M(x, y, w)(v - w), \tag{L}$$

for $x, y \in R_+$ and $v \geq w \geq 0$, where $M: R_+^3 \to R_+$ is a continuous function.
(i) *If*

$$u(x, y) \leq a(x, y) + b(x, y) \int_0^x \int_0^y L(s, t, u(s, t)) \, ds \, dt, \tag{5.3.1}$$

for $x, y \in R_+$, then

$$u(x, y) \leq a(x, y) + b(x, y)A(x, y)\exp\left(\int_0^x \int_0^y M(s, t, a(s, t))b(s, t)\,ds\,dt\right), \quad (5.3.2)$$

for $x, y \in R_+$, where

$$A(x, y) = \int_0^x \int_0^y L(s, t, a(s, t))\,ds\,dt, \quad (5.3.3)$$

for $x, y \in R_+$.

(ii) Let $F(u)$ be a continuous, strictly increasing, convex and submultiplicative function for $u > 0$, $\lim_{u \to \infty} F(u) = \infty$, F^{-1} denotes the inverse function of F, $\alpha(x, y)$, $\beta(x, y)$ be continuous and positive functions defined on R_+^2 and $\alpha(x, y) + \beta(x, y) = 1$. If

$$u(x, y) \leq a(x, y) + b(x, y)F^{-1}\left(\int_0^x \int_0^y L(s, t, F(u(s, t)))\,ds\,dt\right), \quad (5.3.4)$$

for $x, y \in R_+$, then

$$u(x, y) \leq a(x, y) + b(x, y)F^{-1}\left[\left(\int_0^x \int_0^y L(s, t, \alpha(s, t)\right.\right.$$

$$\times F(a(s, t)\alpha^{-1}(s, t)))\,ds\,dt\bigg)$$

$$\times \exp\left(\int_0^x \int_0^y M(s, t, \alpha(s, t)F(a(s, t)\alpha^{-1}(s, t)))\right.$$

$$\times \beta(s, t)F(b(s, t)\beta^{-1}(s, t))\,ds\,dt\bigg)\bigg], \quad (5.3.5)$$

for $x, y \in R_+$.

(iii) Let $g(u)$, $g'(u)$, $h(u)$, $h'(u)$ be as in Theorem 5.2.2. If

$$u(x, y) \leq a(x, y) + b(x, y) h \left(\int_0^x \int_0^y L(s, t, g(u(s, t))) \, ds \, dt \right), \quad (5.3.6)$$

for $x, y \in R_+$, then for $0 \leq x \leq x_1$, $0 \leq y \leq y_1$,

$$u(x, y) \leq a(x, y) + b(x, y) h \left(G^{-1} \left[G(B(x, y)) \right. \right.$$

$$\left. \left. + \int_0^x \int_0^y M(s, t, g(a(s, t))) g(b(s, t)) \, ds \, dt \right] \right), \quad (5.3.7)$$

where

$$B(x, y) = \int_0^x \int_0^y L(s, t, g(a(s, t))) \, ds \, dt, \quad (5.3.8)$$

G, G^{-1} be as defined in Theorem 5.2.2 and x_1, y_1 be chosen so that

$$G(B(x, y)) + \int_0^x \int_0^y M(s, t, g(a(s, t))) g(b(s, t)) \, ds \, dt \in \text{Dom}(G^{-1}),$$

for all x, y lying in the subintervals $0 \leq x \leq x_1$, $0 \leq y \leq y_1$ of R_+. □

Proof: (i) Define a function $z(x, y)$ by

$$z(x, y) = \int_0^x \int_0^y L(s, t, u(s, t)) \, ds \, dt. \quad (5.3.9)$$

From (5.3.9) and using the fact that $u(x, y) \leq a(x, y) + b(x, y) z(x, y)$ and the conditions on the function L we observe that

$$z(x, y) \leq \int_0^x \int_0^y L(s, t, a(s, t) + b(s, t) z(s, t)) \, ds \, dt$$

$$= A(x, y) + \int_0^x \int_0^y [L(s, t, a(s, t) + b(s, t) z(s, t))$$

$$- L(s, t, a(s, t))] \, ds \, dt$$

$$\leq A(x, y) + \int_0^x \int_0^y M(s, t, a(s, t))b(s, t)z(s, t)\, ds\, dt, \quad (5.3.10)$$

where $A(x, y)$ is defined by (5.3.3). Clearly $A(x, y)$ is nonnegative and nondecreasing in each variable $x, y \in R_+$. Now an application of Theorem 4.2.2 yields

$$z(x, y) \leq A(x, y) \exp\left(\int_0^x \int_0^y M(s, t, a(s, t))b(s, t)\, ds\, dt\right). \quad (5.3.11)$$

Using (5.3.11) in $u(x, y) \leq a(x, y) + b(x, y)z(x, y)$ we get the required inequality in (5.3.2).

(ii) Rewrite (5.3.4) as

$$u(x, y) \leq \alpha(x, y)a(x, y)\alpha^{-1}(x, y) + \beta(x, y)b(x, y)\beta^{-1}(x, y)$$

$$\times F^{-1}\left(\int_0^x \int_0^y L(s, t, F(u(s, t)))\, ds\, dt\right). \quad (5.3.12)$$

Since F is convex, submultiplicative and monotonic, from (5.3.12) we see that

$$F(u(x, y)) \leq \alpha(x, y)F(a(x, y)\alpha^{-1}(x, y)) + \beta(x, y)F(b(x, y)\beta^{-1}(x, y))$$

$$\times \int_0^x \int_0^y L(s, t, F(u(s, t)))\, ds\, dt. \quad (5.3.13)$$

The estimate given in (5.3.5) follows by first applying the inequality established in (i) to (5.3.13) and then applying F^{-1} to both sides of the resulting inequality.

(iii) Define a function $z(x, y)$ by

$$z(x, y) = \int_0^x \int_0^y L(s, t, g(u(s, t)))\, ds\, dt. \quad (5.3.14)$$

From (5.3.14) and using the fact that $u(x, y) \leq a(x, y) + b(x, y)h(z(x, y))$ and, as noted in the proof of Theorem 5.2.2, that the functions g and h are monotonically increasing on $(0, \infty)$, and using the conditions on the function

L we observe that

$$z(x, y) \leq \int_0^x \int_0^y L(s, t, g(a(s, t) + b(s, t)h(z(s, t)))) \, ds \, dt$$

$$\leq \int_0^x \int_0^y L(s, t, g(a(s, t)) + g(b(s, t))g(h(z(s, t)))) \, ds \, dt$$

$$= B(x, y) + \int_0^x \int_0^y [L(s, t, g(a(s, t)) + g(b(s, t))g(h(z(s, t))))$$
$$- L(s, t, g(a(s, t)))] \, ds \, dt$$

$$\leq B(x, y) + \int_0^x \int_0^y M(s, t, g(a(s, t)))g(b(s, t))$$
$$\times g(h(z(s, t))) \, ds \, dt, \qquad (5.3.15)$$

where $B(x, y)$ is defined by (5.3.8). The rest of the proof follows by the same arguments as in the proof of Theorem 5.2.2 below the inequality (5.2.16), with suitable changes. Further details are omitted here.

∎

The next two theorems proved by Bondge and Pachpatte (1979b) can be used more effectively in certain situations.

Theorem 5.3.2 Let $u(x, y)$, $u_x(x, y)$, $u_y(x, y)$ and $u_{xy}(x, y)$ be nonnegative continuous functions defined for $x, y \in R_+$, $u(x, 0) = u(0, y) = 0$, and $p(x, y) \geq 1$ be a continuous function defined for $x, y \in R_+$. Let $g(u)$, $g'(u)$, $a(x)$, $a'(x)$, $b(y)$, $b'(y)$ be as in Theorem 5.2.1. If

$$u_{xy}(x, y) \leq a(x) + b(y) + M \left[u(x, y) + \int_0^x \int_0^y p(s, t) g(u_{st}(s, t)) \, ds \, dt \right], \qquad (5.3.16)$$

for $x, y \in R_+$, where $M \geq 0$ is a constant, then for $0 \leq x \leq x_2$, $0 \leq y \leq y_2$,

$$u_{xy}(x, y) \leq H^{-1} \left[H(a(0) + b(y)) + \int_0^x \frac{a'(s)}{a(s) + b(0) + g(a(s) + b(0))} \, ds \right.$$
$$\left. + M \int_0^x \int_0^y p(s, t) \, ds \, dt \right], \qquad (5.3.17)$$

where
$$H(r) = \int_{r_0}^{r} \frac{ds}{s + g(s)}, \quad r > 0, r_0 > 0, \quad (5.3.18)$$

H^{-1} is the inverse function of H, and x_2, y_2 are chosen so that

$$H(a(0) + b(y)) + \int_{0}^{x} \frac{a'(s)}{a(s) + b(0) + g(a(s) + b(0))} ds$$

$$+ M \int_{0}^{x}\int_{0}^{y} p(s, t)\, ds\, dt \in \mathrm{Dom}(H^{-1}),$$

for all x, y lying in the subintervals $0 \leq x \leq x_2$, $0 \leq y \leq y_2$ of R_+.

□

Proof: We note that since $g'(u) \geq 0$ on R_+, the function g is monotonically increasing on $(0, \infty)$. Define a function $z(x, y)$ by the right-hand side of (5.3.16), then $z(x, 0) = a(x) + b(0)$, $z(0, y) = a(0) + b(y)$, $u_{xy}(x, y) \leq z(x, y)$ and

$$z_{xy}(x, y) = M[u_{xy}(x, y) + p(x, y)g(u_{xy}(x, y))]$$
$$\leq M p(x, y)[z(x, y) + g(z(x, y))],$$

i.e.
$$\frac{z_{xy}(x, y)}{z(x, y) + g(z(x, y))} \leq M p(x, y). \quad (5.3.19)$$

Now by following the same arguments as in the proof of Theorem 5.2.1 below the inequality (5.2.5) with suitable modifications we obtain the estimate for $z(x, y)$ such that

$$z(x, y) \leq H^{-1}\Bigg[H(a(0) + b(y)) + \int_{0}^{x} \frac{a'(s)}{a(s) + b(0) + g(a(s) + b(0))} ds$$

$$+ M \int_{0}^{x}\int_{0}^{y} p(s, t)\, ds\, dt\Bigg]. \quad (5.3.20)$$

Using (5.3.20) in $u_{xy}(x, y) \leq z(x, y)$ we obtain the required bound in (5.3.17).

■

Theorem 5.3.3 Let $u(x, y)$, $u_{xy}(x, y)$, $p(x, y)$, $g(u)$, $g'(u)$, $a(x)$, $a'(x)$, $b(y)$ and $b'(y)$ be as in Theorem 5.3.2. If

$$u_{xy}(x, y) \leq a(x) + b(y) + \int_0^x \int_0^y p(s, t) g(u(s, t) + u_{st}(s, t)) \, ds \, dt, \quad (5.3.21)$$

for $x, y \in R_+$, then for $0 \leq x \leq x_3$, $0 \leq y \leq y_3$,

$$u_{xy}(x, y) \leq a(x) + b(y) + \int_0^x \int_0^y p(s, t) g \left(H^{-1} \left[H(a(0) + b(t)) \right.\right.$$

$$+ \int_0^s \frac{a'(s_1)}{a(s_1) + b(0) + g(a(s_1) + b(0))} \, ds_1$$

$$\left.\left. + \int_0^s \int_0^t p(s_1, t_1) \, ds_1 \, dt_1 \right] \right) ds \, dt, \quad (5.3.22)$$

where H, H^{-1} are as defined in Theorem 5.3.2 and x_3, y_3 are chosen so that

$$H(a(0) + b(y)) + \int_0^x \frac{a'(s_1)}{a(s_1) + b(0) + g(a(s_1) + b(0))} \, ds_1$$

$$+ \int_0^x \int_0^y p(s_1, t_1) \, ds_1 \, dt_1 \in \text{Dom}(H^{-1}),$$

for all x, y lying in the subintervals $0 \leq x \leq x_3$, $0 \leq y \leq y_3$ of R_+.

$\square$

Proof: Since $g'(u) \geq 0$ on R_+, the function $g(u)$ is monotonically increasing on $(0, \infty)$. Define a function $z(x, y)$ by the right-hand side of (5.3.21), then $z(x, 0) = a(x) + b(0)$, $z(0, y) = a(0) + b(y)$,

$$u_{xy}(x, y) \leq z(x, y), \quad (5.3.23)$$

and

$$z_{xy}(x, y) = p(x, y) g(u(x, y) + u_{xy}(x, y)). \quad (5.3.24)$$

From (5.3.23) and using $u(x, 0) = u(0, y) = 0$, it is easy to observe that

$$u(x, y) \leq \int_0^x \int_0^y z(s, t)\, ds\, dt. \qquad (5.3.25)$$

Using (5.3.23) and (5.3.25) in (5.3.24) we get

$$z_{xy}(x, y) \leq p(x, y) g\left(z(x, y) + \int_0^x \int_0^y z(s, t)\, ds\, dt\right). \qquad (5.3.26)$$

Define a function $v(x, y)$ by

$$v(x, y) = z(x, y) + \int_0^x \int_0^y z(s, t)\, ds\, dt. \qquad (5.3.27)$$

From (5.3.27) and using the facts that $z_{xy}(x, y) \leq p(x, y) g(v(x, y))$ from (5.3.26) and $z(x, y) \leq v(x, y)$ from (5.3.27) we observe that

$$v_{xy}(x, y) = z_{xy}(x, y) + z(x, y)$$
$$\leq p(x, y)[v(x, y) + g(v(x, y))],$$

i.e.

$$\frac{v_{xy}(x, y)}{v(x, y) + g(v(x, y))} \leq p(x, y). \qquad (5.3.28)$$

Now by following the same arguments as in the proof of Theorem 5.2.1 below the inequality (5.2.5) with suitable modifications we obtain the estimate for $v(x, y)$ such that

$$v(x, y) \leq H^{-1}\left[H(a(0) + b(y)) + \int_0^x \frac{a'(s_1)}{a(s_1) + b(0) + g(a(s_1) + b(0))}\, ds_1 \right.$$
$$\left. + \int_0^x \int_0^y p(s_1, t_1)\, ds_1\, dt_1 \right]. \qquad (5.3.29)$$

Using (5.3.29) in (5.3.26), and first keeping x fixed, setting $y = t$ and integrating from 0 to y, then keeping y fixed, setting $x = s$ in the resulting inequality and integrating from 0 to x we obtain

$$z(x, y) \leq a(x) + b(y) + \int_0^x \int_0^y p(s, t) g \left(H^{-1} \left[H(a(0) + b(t)) \right. \right.$$
$$+ \int_0^s \frac{a'(s_1)}{a(s_1) + b(0) + g(a(s_1) + b(0))} \, ds_1$$
$$\left. \left. + \int_0^s \int_0^t p(s_1, t_1) \, ds_1 \, dt_1 \right] \right) ds \, dt. \qquad (5.3.30)$$

Using (5.3.30) in (5.3.23) we get the desired inequality in (5.3.22).

■

The following two theorems established by Bondge and Pachpatte (1979c) can be used in certain applications.

Theorem 5.3.4 *Let $\phi(x, y)$, $a(x, y)$, $b(x, y)$ be nonnegative continuous functions defined for $x, y \in R_+$, and $u(x, y)$ be a positive continuous function defined for $x, y \in R_+$. Let $F(u)$, $\alpha(x, y)$, $\beta(x, y)$ be as in Theorem 5.3.1 part (ii). If*

$$u(s, t) \geq \phi(x, y) - a(s, t) F^{-1} \left(\int_x^s \int_y^t b(s_1, t_1) F(\phi(s_1, t_1)) \, ds_1 \, dt_1 \right), \qquad (5.3.31)$$

for $0 \leq x \leq s < \infty$, $0 \leq y \leq t < \infty$, then

$$u(s, t) \geq \alpha(s, t) F^{-1} \left(\alpha^{-1}(s, t) F(\phi(x, y)) \exp \left(-\beta(s, t) F(a(s, t) \beta^{-1}(s, t)) \right. \right.$$
$$\left. \left. \times \int_x^s \int_y^t b(s_1, t_1) \, ds_1 \, dt_1 \right) \right), \qquad (5.3.32)$$

for $0 \leq x \leq s < \infty$, $0 \leq y \leq t < \infty$.

□

Proof: Rewrite (5.3.31) as

$$\phi(x, y) \leq \alpha(s, t)u(s, t)\alpha^{-1}(s, t) + \beta(s, t)a(s, t)\beta^{-1}(s, t)$$
$$\times F^{-1}\left(\int_x^s \int_y^t b(s_1, t_1)F(\phi(s_1, t_1))\,ds_1\,dt_1\right). \quad (5.3.33)$$

Since F is convex, submultiplicative and monotonic, from (5.3.33) we have

$$\alpha(s, t)F(u(s, t)\alpha^{-1}(s, t)) \geq F(\phi(x, y)) - \beta(s, t)F(a(s, t)\beta^{-1}(s, t))$$
$$\times \int_x^s \int_y^t b(s_1, t_1)F(\phi(s_1, t_1))\,ds_1\,dt_1. \quad (5.3.34)$$

Now an application of Theorem 4.9.4 part (i) with $n = 2$ yields the desired bound in (5.3.32).

∎

Theorem 5.3.5 *Let $u(x, y)$, $a(x, y)$ and $b(x, y)$ be as in Theorem 5.3.4. Let $g(u)$ and $g'(u)$ be as in Theorem 5.2.1. If*

$$u(s, t) \geq u(x, y) - a(s, t) \int_x^s \int_y^t b(\sigma, \eta)g(u(\sigma, \eta))\,d\sigma\,d\eta, \quad (5.3.35)$$

for $0 \leq x \leq s < \infty$, $0 \leq y \leq t < \infty$, then for $0 \leq x \leq s \leq s_1$, $0 \leq y \leq t \leq t_1$,

$$u(s, t) \geq \Omega^{-1}\left[\Omega(u(x, y)) - a(s, t)\int_x^s \int_y^t b(\sigma, \eta)\,d\sigma\,d\eta\right], \quad (5.3.36)$$

where Ω, Ω^{-1} are as defined in Theorem 5.2.1 and

$$\Omega(u(x, y)) - a(s, t)\int_x^s \int_y^t b(\sigma, \eta)\,d\sigma\,d\eta \in \mathrm{Dom}(\Omega^{-1}),$$

for $0 \leq x \leq s \leq s_1$, $0 \leq y \leq t \leq t_1$.

□

Proof: Since $g'(u) \geq 0$ on R_+, the function $g(u)$ is monotonically increasing on $(0, \infty)$. Rewrite (5.3.35) as

$$u(x, y) \leq u(s, t) + a(s, t) \int_x^s \int_y^t b(\sigma, \eta) g(u(\sigma, \eta))\, d\sigma\, d\eta. \qquad (5.3.37)$$

For fixed s and t in R_+ we define for $0 \leq x \leq s$, $0 \leq y \leq t$,

$$z(x, y) = u(s, t) + a(s, t) \int_x^s \int_y^t b(\sigma, \eta) g(u(\sigma, \eta))\, d\sigma\, d\eta, \qquad (5.3.38)$$

then $z(x, t) = z(s, y) = u(s, t)$, $u(x, y) \leq z(x, y)$ and

$$z_{xy}(x, y) = a(s, t) b(x, y) g(u(x, y))$$
$$\leq a(s, t) b(x, y) g(z(x, y)),$$

i.e.

$$\frac{z_{xy}(x, y)}{g(z(x, y))} \leq a(s, t) b(x, y). \qquad (5.3.39)$$

From (5.3.39) we observe that

$$\frac{z_{xy}(x, y)}{g(z(x, y))} \leq a(s, t) b(x, y) + \frac{g'(z(x, y)) z_y(x, y) z_x(x, y)}{[g(z(x, y))]^2}. \qquad (5.3.40)$$

Here we note that $z_x(x, y)$ and $z_y(x, y)$ are nonpositive, which implies that $z_x(x, y) z_y(x, y)$ is nonnegative and hence (5.3.40) is true. Now (5.3.40) is equivalent to

$$\frac{\partial}{\partial y}\left(\frac{z_x(x, y)}{g(z(x, y))}\right) \leq a(s, t) b(x, y). \qquad (5.3.41)$$

By keeping x fixed in (5.3.41), setting $y = \eta$ and integrating from y to t we have

$$\frac{z_x(x, t)}{g(z(x, t))} - \frac{z_x(x, y)}{g(z(x, y))} \leq a(s, t) \int_y^t b(x, \eta)\, d\eta. \qquad (5.3.42)$$

From (5.2.3) and (5.3.42) we observe that

$$\frac{\partial}{\partial x} \Omega(z(x, t)) - \frac{\partial}{\partial x} \Omega(z(x, y)) \leq a(s, t) \int_y^t b(x, \eta)\, d\eta. \qquad (5.3.43)$$

Now keeping y fixed in (5.3.43), setting $x = \sigma$ and integrating from x to s we have

$$\Omega(z(x, y)) \leq \Omega(u(s, t)) + a(s, t) \int_x^s \int_y^t b(\sigma, \eta) \, d\sigma \, d\eta,$$

which implies

$$\Omega(u(s, t)) \geq \Omega(u(x, y)) - a(s, t) \int_x^s \int_y^t b(\sigma, \eta) \, d\sigma \, d\eta. \qquad (5.3.44)$$

The desired bound in (5.3.36) follows from (5.3.44). The subintervals of s and t in R_+ are obvious.

∎

5.4 Generalizations of Pachpatte's Inequalities

Pachpatte (1974d, 1975h) investigated some nonlinear Bihari-type integral inequalities which are applicable in certain general situations. In this section we present some basic two independent variable generalizations of the certain inequalities in Pachpatte (1974d, 1975h) obtained by Bondge and Pachpatte (1979b, 1980a) which can be used as tools in the study of certain partial integro-differential and integral equations.

Bondge and Pachpatte (1979b) established the following generalization of the inequality given by Pachpatte (1974d).

Theorem 5.4.1 *Let $u(x, y)$ and $p(x, y)$ be nonnegative continuous functions defined for $x, y \in R_+$. Let $g(u)$, $g'(u)$, $a(x)$, $a'(x)$, $b(y)$ and $b'(y)$ be as in Theorem 5.2.1. If*

$$u(x, y) \leq a(x) + b(y) + \int_0^x \int_0^y p(s, t) \left(u(s, t) \right.$$
$$\left. + \int_0^s \int_0^t p(s_1, t_1) g(u(s_1, t_1)) \, ds_1 \, dt_1 \right) ds \, dt, \qquad (5.4.1)$$

for $x, y \in R_+$, then for $0 \le x \le x_1$, $0 \le y \le y_1$,

$$u(x, y) \le a(x) + b(y) + \int_0^x \int_0^y p(s, t) H^{-1} \left[H(a(0) + b(t)) \right.$$
$$+ \int_0^s \frac{a'(s_1)}{a(s_1) + b(0) + g(a(s_1) + b(0))} ds_1$$
$$\left. + \int_0^s \int_0^t p(s_1, t_1) ds_1 dt_1 \right] ds dt, \tag{5.4.2}$$

where

$$H(r) = \int_{r_0}^r \frac{ds}{s + g(s)}, \quad r > 0, r_0 > 0, \tag{5.4.3}$$

H^{-1} is the inverse function of H, and x_1, y_1 are chosen so that

$$H(a(0) + b(y)) + \int_0^x \frac{a'(s_1)}{a(s_1) + b(0) + g(a(s_1) + b(0))} ds_1$$
$$+ \int_0^x \int_0^y p(s_1, t_1) ds_1 dt_1 \in \text{Dom}(H^{-1}),$$

for all x, y lying in the subintervals $0 \le x \le x_1$, $0 \le y \le y_1$ of R_+. $\square$

Proof: Since $g'(u) \ge 0$, the function $g(u)$ is monotonically increasing on $(0, \infty)$. Define a function $z(x, y)$ by the right-hand side of (5.4.1), then $z(x, 0) = a(x) + b(0)$, $z(0, y) = a(0) + b(y)$, $u(x, y) \le z(x, y)$ and

$$z_{xy}(x, y) = p(x, y) \left(u(x, y) + \int_0^x \int_0^y p(s_1, t_1) g(u(s_1, t_1)) ds_1 dt_1 \right)$$
$$\le p(x, y) \left(z(x, y) + \int_0^x \int_0^y p(s_1, t_1) g(z(s_1, t_1)) ds_1 dt_1 \right). \tag{5.4.4}$$

If we put

$$v(x, y) = z(x, y) + \int_0^x \int_0^y p(s_1, t_1) g(z(s_1, t_1)) \, ds_1 \, dt_1,$$

then $v(x, 0) = a(x) + b(0)$, $v(0, y) = a(0) + b(y)$, $z_{xy}(x, y) \leq p(x, y) v(x, y)$, $z(x, y) \leq v(x, y)$ and

$$v_{xy}(x, y) = z_{xy}(x, y) + p(x, y) g(z(x, y))$$
$$\leq p(x, y)(v(x, y) + g(v(x, y))),$$

i.e.

$$\frac{v_{xy}(x, y)}{v(x, y) + g(v(x, y))} \leq p(x, y). \tag{5.4.5}$$

Now by following the same arguments as in the proof of Theorem 5.2.1 below (5.2.5) with suitable modifications and in view of (5.4.3) we obtain the estimate for $v(x, y)$ such that

$$v(x, y) \leq H^{-1} \Bigg[H(a(0) + b(y)) + \int_0^x \frac{a'(s_1)}{a(s_1) + b(0) + g(a(s_1) + b(0))} \, ds_1$$

$$+ \int_0^x \int_0^y p(s_1, t_1) \, ds_1 \, dt_1 \Bigg]. \tag{5.4.6}$$

Using (5.4.6) in (5.4.4), and first keeping x fixed, setting $y = t$ and integrating from 0 to y, then in the resulting inequality keeping y fixed, setting $x = s$ and integrating from 0 to x we obtain

$$z(x, y) \leq a(x) + b(y) + \int_0^x \int_0^y p(s, t) H^{-1} \Bigg[H(a(0) + b(t))$$

$$+ \int_0^s \frac{a'(s_1)}{a(s_1) + b(0) + g(a(s_1) + b(0))} \, ds_1$$

$$+ \int_0^s \int_0^t p(s_1, t_1) \, ds_1 \, dt_1 \Bigg] ds \, dt. \tag{5.4.7}$$

Now using (5.4.7) in $u(x, y) \leq z(x, y)$ we obtain the desired bound in (5.4.2). The subintervals for x and y are obvious.

∎

Bondge and Pachpatte (1980a) established the following three theorems which deal with the two independent variable generalizations of certain inequalities given by Pachpatte (1974d, 1975h).

Theorem 5.4.2 *Let $u(x, y)$, $a(x, y)$ and $b(x, y)$ be nonnegative continuous functions defined for $x, y \in R_+$. Let $g(u)$, $g'(u)$ be as in Theorem 5.2.1 and in addition $g(u)$ be subadditive on R_+. If*

$$u(x, y) \leq a(x, y) + \int_0^x \int_0^y b(s, t) \left(u(s, t) + \int_0^s \int_0^t b(s_1, t_1) \right.$$

$$\left. \times g(u(s_1, t_1)) \, ds_1 \, dt_1 \right) ds \, dt, \qquad (5.4.8)$$

for $x, y \in R_+$, then for $0 \leq x \leq x_2$, $0 \leq y \leq y_2$,

$$u(x, y) \leq a(x, y) + A(x, y) + \int_0^x \int_0^y b(s, t) \left\{ H^{-1} \left[H(A(s, t)) \right. \right.$$

$$\left. \left. + \int_0^s \int_0^t p(s_1, t_1) \, ds_1 \, dt_1 \right] \right\} ds \, dt, \qquad (5.4.9)$$

where

$$A(x, y) = \int_0^x \int_0^y b(s_1, t_1) \left(a(s_1, t_1) \right.$$

$$\left. + \int_0^{s_1} \int_0^{t_1} b(s_2, t_2) g(a(s_2, t_2)) \, ds_2 \, dt_2 \right) ds_1 \, dt_1, \qquad (5.4.10)$$

H, H^{-1} are as defined in Theorem 5.4.1 and x_2, y_2 are chosen so that

$$H(A(x, y)) + \int_0^x \int_0^y p(s_1, t_1) \, ds_1 \, dt_1 \in \text{Dom}(H^{-1}),$$

for all x, y lying in the subintervals $0 \leq x \leq x_2$, $0 \leq y \leq y_2$ of R_+.

□

Theorem 5.4.3 Let $u(x, y)$, $a(x, y)$, $b(x, y)$ and $c(x, y)$ be nonnegative continuous functions defined for $x, y \in R_+$. Let $g(u)$ and $g'(u)$ be as in Theorem 5.4.2. If

$$u(x, y) \leq a(x, y) + \int_0^x \int_0^y b(s, t) \left[u(s, t) + \int_0^s \int_0^t c(s_1, t_1) u(s_1, t_1) \, ds_1 \, dt_1 \right.$$

$$\left. + \int_0^s \int_0^t (b(s_1, t_1) + c(s_1, t_1)) g(u(s_1, t_1)) \, ds_1 \, dt_1 \right] ds \, dt, \quad (5.4.11)$$

for $x, y \in R_+$, then for $0 \leq x \leq x_3$, $0 \leq y \leq y_3$,

$$u(x, y) \leq a(x, y) + B(x, y) + \int_0^x \int_0^y b(s, t) \left\{ H^{-1} \left[H(B(x, y)) \right. \right.$$

$$\left. \left. + \int_0^s \int_0^t (b(s_1, t_1) + c(s_1, t_1)) \, ds_1 \, dt_1 \right] \right\} ds \, dt, \quad (5.4.12)$$

where

$$B(x, y) = \int_0^x \int_0^y b(s_1, t_1) \left[a(s_1, t_1) + \int_0^{s_1} \int_0^{t_1} b(s_2, t_2) c(s_2, t_2) \, ds_2 \, dt_2 \right.$$

$$\left. + \int_0^{s_1} \int_0^{t_1} (b(s_2, t_2) + c(s_2, t_2)) g(a(s_2, t_2)) \, ds_2 \, dt_2 \right] ds_1 \, dt_1, \quad (5.4.13)$$

H, H^{-1} are as defined in Theorem 5.4.1 and x_3, y_3 are chosen so that

$$H(B(x, y)) + \int_0^x \int_0^y (b(s_1, t_1) + c(s_1, t_1)) \, ds_1 \, dt_1 \in \text{Dom}(H^{-1}),$$

for all x, y lying in the subintervals $0 \leq x \leq x_3$, $0 \leq y \leq y_3$ of R_+. □

Theorem 5.4.4 Let $u(x, y)$, $a(x, y)$, $b(x, y)$, $c(x, y)$ and $k(x, y)$ be nonnegative continuous functions defined for $x, y \in R_+$. Let $g(u)$, $g'(u)$ be as defined in Theorem 5.2.2. If

$$u(x, y) \leq a(x, y) + b(x, y) \int_0^x \int_0^y c(s, t) g\left(u(s, t) + b(s, t) \right.$$

$$\left. \times \int_0^s \int_0^t k(s_1, t_1) g(u(s_1, t_1)) \, ds_1 \, dt_1 \right) ds \, dt, \quad (5.4.14)$$

for $x, y \in R_+$, then for $0 \leq x \leq x_4$, $0 \leq y \leq y_4$,

$$u(x, y) \leq a(x, y) + b(x, y) \left[L(x, y) + \int_0^x \int_0^y c(s, t) g\left(b(s, t) \right. \right.$$

$$\times \left\{ \Omega^{-1} \left[\Omega(L(x, y)) + \int_0^s \int_0^t (c(s_1, t_1) + k(s_1, t_1)) \right.\right.$$

$$\left.\left. \times g(b(s_1, t_1)) \, ds_1 \, dt_1 \right] \right\} \right) ds \, dt \right], \quad (5.4.15)$$

where

$$L(x, y) = \int_0^x \int_0^y c(s_1, t_1) g\left(a(s_1, t_1) + b(s_1, t_1) \int_0^{s_1} \int_0^{t_1} k(s_2, t_2) \right.$$

$$\left. \times g(a(s_2, t_2)) \, ds_2 \, dt_2 \right) ds_1 \, dt_1, \quad (5.4.16)$$

Ω, Ω^{-1} are as defined in Theorem 5.2.1 and x_4, y_4 are chosen so that

$$\Omega(L(x, y)) + \int_0^x \int_0^y (c(s_1, t_1) + k(s_1, t_1)) g(b(s_1, t_1)) \, ds_1 \, dt_1 \in \text{Dom}(\Omega^{-1}),$$

for all x, y lying in the subintervals $0 \leq x \leq x_4$, $0 \leq y \leq y_4$ of R_+. $\square$

The details of the proofs of Theorems 5.4.2–5.4.4 follow by arguments similar to those in the proofs of Theorems 5.2.2 and 5.4.1 in view of the results given in Pachpatte (1974d, 1975h). Here we leave the details to the reader.

MULTIDIMENSIONAL NONLINEAR INTEGRAL INEQUALITIES   485

In concluding, we note that the inequalities established in Sections 5.2–5.4 can be extended very easily to n ($n \geq 3$) independent variables. The precise formulations of these results are very close to those of the results given in Sections 5.2–5.4 with suitable modifications. It is left to the reader to fill in the details where needed.

5.5 Pachpatte's Inequalities I

In view of wider applications, Wendroff's inequality given in Beckenbach and Bellman (1961) has been generalized and extended in various directions. The present section is devoted to the Wendroff-like inequalities investigated by Pachpatte (1988c, 1993, 1996d) in order to apply them in the study of certain higher order partial differential equations. In what follows we shall use the notations and definitions as given in Section 4.5 without further mention.

Pachpatte (1988c) established the Wendroff-like inequalities in the following two theorems.

Theorem 5.5.1 *Let $u(x, y)$ and $h(x, y)$ be nonnegative continuous functions defined for $x, y \in R_+$. Let $a(x)$, $b(y)$, $p(x)$, $q(y)$ be positive and twice continuously differentiable functions defined for $x, y \in R_+$, $a'(x)$, $b'(y)$, $p'(x)$, $q'(y)$ are nonnegative for $x, y \in R_+$ and define $c(x, y) = a(x) + b(y) + yp(x) + xq(y)$, for $x, y \in R_+$. Let g be a continuously differentiable function defined on R_+ and $g(u) > 0$ on $(0, \infty)$, $g'(u) \geq 0$ on R_+ and*

$$u(x, y) \leq c(x, y) + A[x, y, h(s_1, t_1)g(u(s_1, t_1))], \quad (5.5.1)$$

holds for $x, y \in R_+$.

(i) If $a''(x)$, $p''(x)$ are nonnegative for $x \geq 0$, then for $0 \leq x \leq x_1$, $0 \leq y \leq y_1$,

$$u(x, y) \leq \Omega^{-1} \Bigg[\Omega(c(0, y)) + x \left(\frac{c_x(0, y)}{g(c(0, y))} \right) \\ + \int_0^x \int_0^s \frac{a''(s_1) + yp''(s_1)}{g(c(s_1, 0))} ds_1 \, ds \\ + A[x, y, h(s_1, t_1)] \Bigg], \quad (5.5.2)$$

where

$$\Omega(r) = \int_{r_0}^{r} \frac{ds}{g(s)}, \quad r > 0, r_0 > 0, \qquad (5.5.3)$$

Ω^{-1} is the inverse of Ω and x_1, y_1 are chosen so that

$$\Omega(c(0, y)) + x \left(\frac{c_x(0, y)}{g(c(0, y))} \right) + \int_0^x \int_0^s \frac{a''(s_1) + yp''(s_1)}{g(c(s_1, 0))} ds_1 \, ds$$

$$+ A[x, y, h(s_1, t_1)] \in \text{Dom}(\Omega^{-1}),$$

for all x, y lying in the subintervals $0 \leq x \leq x_1$, $0 \leq y \leq y_1$ of R_+.

(ii) If $b''(y)$, $q''(y)$ are nonnegative for $y \geq 0$, then for $0 \leq x \leq x_2$, $0 \leq y \leq y_2$,

$$u(x, y) \leq \Omega^{-1} \left[\Omega(c(x, 0)) + y \left(\frac{c_y(x, 0)}{g(c(x, 0))} \right) \right.$$

$$\left. + \int_0^y \int_0^t \frac{b''(t_1) + xq''(t_1)}{g(c(0, t_1))} dt_1 \, dt + A[y, x, h(s_1, t_1)] \right], \qquad (5.5.4)$$

where Ω, Ω^{-1} are as defined in (i) and x_2, y_2 are chosen so that

$$\Omega(c(x, 0)) + y \left(\frac{c_y(x, 0)}{g(c(x, 0))} \right) + \int_0^y \int_0^t \frac{b''(t_1) + xq''(t_1)}{g(c(0, t_1))} dt_1 \, dt$$

$$+ A[y, x, h(s_1, t_1)] \in \text{Dom}(\Omega^{-1}),$$

for all x, y lying in the subintervals $0 \leq x \leq x_2$, $0 \leq y \leq y_2$ of R_+.

$\square$

Theorem 5.5.2 Let $u(x, y)$, $h(x, y)$, $a(x)$, $b(y)$, $p(x)$, $q(y)$, $a'(x)$, $b'(y)$, $p'(x)$, $q'(y)$, $c(x, y)$, $g(u)$ and $g'(u)$ be as in Theorem 5.5.1 and

$$u(x, y) \leq c(x, y) + A[x, y, h(s_1, t_1)[u(s_1, t_1)$$

$$+ A[s_1, t_1, h(s_2, t_2)g(u(s_2, t_2))]]], \qquad (5.5.5)$$

holds for $x, y \in R_+$.

(i) If $a''(x)$ and $p''(x)$ are nonnegative for $x \in R_+$, then for $0 \le x \le x_3$, $0 \le y \le y_3$,

$$u(x, y) \le c(0, y) + xc_x(0, y) + \int_0^x \int_0^s [a''(s_1) + yp''(s_1)] \, ds_1 \, ds$$
$$+ A[x, y, h(s_1, t_1)Q_1(s_1, t_1)], \qquad (5.5.6)$$

in which

$$Q_1(x, y) = H^{-1}\left[H(c(0, y)) + x\left(\frac{c_x(0, y)}{c(0, y) + g(c(0, y))}\right)\right.$$
$$\left. + \int_0^x \int_0^{s_2} \frac{a''(s_3) + yp''(s_3)}{c(s_3, 0) + g(c(s_3, 0))} \, ds_3 \, ds_2 + A[x, y, h(s_3, t_3)]\right], \quad (5.5.7)$$

where

$$H(r) = \int_{r_0}^r \frac{ds}{s + g(s)}, \quad r > 0, r_0 > 0, \qquad (5.5.8)$$

H^{-1} is the inverse function of H and x_3, y_3 are chosen so that

$$H(c(0, y)) + x\left(\frac{c_x(0, y)}{c(0, y) + g(c(0, y))}\right) + \int_0^x \int_0^{s_2} \frac{a''(s_3) + yp''(s_3)}{c(s_3, 0) + g(c(s_3, 0))} \, ds_3 \, ds_2$$
$$+ A[x, y, h(s_3, t_3)] \in \text{Dom}(H^{-1}),$$

for all x, y lying in the subintervals $0 \le x \le x_3, 0 \le y \le y_3$ of R_+.

(ii) If $b''(y), q''(y)$ are nonnegative for $y \in R_+$ then for $0 \le x \le x_4$, $0 \le y \le y_4$,

$$u(x, y) \le c(x, 0) + yc_y(x, 0) + \int_0^y \int_0^t [b''(t_1) + xq''(t_1)] \, dt_1 \, dt$$
$$+ A[y, x, h(s_1, t_1)Q_2(s_1, t_1)], \qquad (5.5.9)$$

in which

$$Q_2(x, y) = H^{-1}\left[H(c(x, 0)) + y\left(\frac{c_y(x, 0)}{c(x, 0) + g(c(x, 0))}\right)\right.$$
$$+ \int_0^y \int_0^{t_2} \frac{b''(t_3) + xq''(t_3)}{c(0, t_3) + g(c(0, t_3))} dt_3 dt_2$$
$$\left. + A[y, x, h(s_3, t_3)]\right], \qquad (5.5.10)$$

where H, H^{-1} are as defined in (i) and x_4, y_4 are chosen so that

$$H(c(x, 0)) + y\left(\frac{c_y(x, 0)}{c(x, 0) + g(c(x, 0))}\right) + \int_0^y \int_0^{t_2} \frac{b''(t_3) + xq''(t_3)}{c(0, t_3) + g(c(0, t_3))} dt_3 dt_2$$
$$+ A[y, x, h(s_3, t_3)] \in \mathrm{Dom}(H^{-1}),$$

for all x, y lying in the subintervals $0 \leq x \leq x_4$, $0 \leq y \leq y_4$ of R_+. □

Proof of Theorem 5.5.1: Since $g'(u) \geq 0$, the function $g(u)$ is monotonically increasing on $(0, \infty)$. Define a function $z(x, y)$ by

$$z(x, y) = c(x, y) + A[x, y, h(s_1, t_1)g(u(s_1, t_1))]. \qquad (5.5.11)$$

From (5.5.11) it is easy to observe that

$$z(0, y) = c(0, y) = a(0) + b(y) + yp(0), \qquad (5.5.12)$$
$$z(x, 0) = c(x, 0) = a(x) + b(0) + xq(0), \qquad (5.5.13)$$
$$z_x(0, y) = c_x(0, y) = a'(0) + yp'(0) + q(y), \qquad (5.5.14)$$
$$z_{xx}(x, 0) = c_{xx}(x, 0) = a''(x), \qquad (5.5.15)$$
$$z_{xxy}(x, 0) = c_{xxy}(x, 0) = p''(x), \qquad (5.5.16)$$

and

$$z_{xxyy}(x, y) = h(x, y)g(u(x, y)). \qquad (5.5.17)$$

Using $u(x, y) \leq z(x, y)$ in (5.5.17) we have

$$z_{xxyy}(x, y) \leq h(x, y) g(z(x, y)). \tag{5.5.18}$$

From (5.5.18) and using the facts that $z(x, y) > 0$, $z_y(x, y) \geq 0$, $z_{xxy}(x, y) \geq 0$, $g'(u(x, y)) \geq 0$ for $x, y \in R_+$, we observe that

$$\frac{z_{xxyy}(x, y)}{g(z(x, y))} \leq h(x, y) + \frac{z_{xxy}(x, y) g'(z(x, y)) z_y(x, y)}{[g(z(x, y))]^2},$$

i.e.

$$\frac{\partial}{\partial y}\left(\frac{z_{xxy}(x, y)}{g(z(x, y))}\right) \leq h(x, y). \tag{5.5.19}$$

By keeping x fixed in (5.5.19), we set $y = t_1$ and then integrating with respect to t_1 from 0 to y and using (5.5.13) and (5.5.16) we have

$$\frac{z_{xxy}(x, y)}{g(z(x, y))} \leq \frac{p''(x)}{g(c(x, 0))} + \int_0^y h(x, t_1) \, dt_1. \tag{5.5.20}$$

Again as above, from (5.5.20) and using the facts that $z(x, y) > 0$, $z_y(x, y) \geq 0$, $z_{xx}(x, y) \geq 0$ for $x, y \in R_+$, we observe that

$$\frac{\partial}{\partial y}\left(\frac{z_{xx}(x, y)}{g(z(x, y))}\right) \leq \frac{p''(x)}{g(c(x, 0))} + \int_0^y h(x, t_1) \, dt_1. \tag{5.5.21}$$

By keeping x fixed in (5.5.21), setting $y = t$ and then integrating with respect to t from 0 to y and using (5.5.13) and (5.5.15) we have

$$\frac{z_{xx}(x, y)}{g(z(x, y))} \leq \frac{a''(x) + y p''(x)}{g(c(x, 0))} + \int_0^y \int_0^t h(x, t_1) \, dt_1 \, dt. \tag{5.5.22}$$

As above, from (5.5.22) and using the facts that $z(x, y) > 0$, $z_x(x, y) \geq 0$ for $x, y \in R_+$, we observe that

$$\frac{\partial}{\partial x}\left(\frac{z_x(x, y)}{g(z(x, y))}\right) \leq \frac{a''(x) + y p''(x)}{g(c(x, 0))} + \int_0^y \int_0^t h(x, t_1) \, dt_1 \, dt. \tag{5.5.23}$$

Now keeping y fixed in (5.5.23), setting $x = s_1$ and then integrating with respect to s_1 from 0 to x and using (5.5.12) and (5.5.14) we have

$$\frac{z_x(x, y)}{g(z(x, y))} \leq \frac{c_x(0, y)}{g(c(0, y))} + \int_0^x \frac{a''(s_1) + yp''(s_1)}{g(c(s_1, 0))} ds_1$$

$$+ \int_0^x \int_0^y \int_0^t h(s_1, t_1) dt_1 \, dt \, ds_1. \tag{5.5.24}$$

From (5.5.3) and (5.5.24) we have

$$\frac{\partial}{\partial x} \Omega(z(x, y)) = \frac{z_x(x, y)}{g(z(x, y))} \leq \frac{c_x(0, y)}{g(c(0, y))} + \int_0^x \frac{a''(s_1) + yp''(s_1)}{g(c(s_1, 0))} ds_1$$

$$+ \int_0^x \int_0^y \int_0^t h(s_1, t_1) dt_1 \, dt \, ds_1. \tag{5.5.25}$$

Now keeping y fixed in (5.5.25), setting $x = s$ and then integrating with respect to s from 0 to x and using (5.5.12) we obtain

$$\Omega(z(x, y)) \leq \Omega(c(0, y)) + x \left(\frac{c_x(0, y)}{g(c(0, y))} \right)$$

$$+ \int_0^x \int_0^s \frac{a''(s_1) + yp''(s_1)}{g(c(s_1, 0))} ds_1 \, ds$$

$$+ A[x, y, h(s_1, t_1)]. \tag{5.5.26}$$

The desired bound in (5.5.2) now follows by substituting the bound on $z(x, y)$ from (5.5.26) in $u(x, y) \leq z(x, y)$. The subdomains for x, y are obvious.

Rewriting (5.5.11) in the form

$$z(x, y) = c(x, y) + A[y, x, h(s_1, t_1) g(u(s_1, t_1))],$$

since $A[x, y, h(s_1, t_1) g(u(s_1, t_1))] = A[y, x, h(s_1, t_1) g(u(s_1, t_1))]$ and by following the same arguments as in the proof of the inequality (5.5.2) given above with suitable modifications we obtain the required inequality in (5.5.4).

The proof of Theorem 5.5.2 follows by an argument similar to that in the proof of Theorem 5.5.1 and by closely looking at the proof of Theorem 4.5.2, with suitable modifications. The details are omitted here.

We next establish the following inequality which can be used in the study of certain partial differential and integral equations.

Theorem 5.5.3 *Let $u(x, y)$, $f(x, y)$ and $p(x, y)$ be nonnegative continuous functions defined for $x, y \in R_+$. Let $g(u)$, $g'(u)$ be as in Theorem 5.5.1 and $g(u)$ be submultiplicative on R_+. If*

$$u(x, y) \leq c + \int_0^x \int_0^s f(s_1, y) u(s_1, y) \, ds_1 \, ds + A[x, y, p(s_1, t_1) g(u(s_1, t_1))], \tag{5.5.27}$$

for $x, y \in R_+$, where $c \geq 0$ is a constant, then for $0 \leq x \leq x_5$, $0 \leq y \leq y_5$,

$$u(x, y) \leq Q(x, y)\{\Omega^{-1}[\Omega(c) + A[x, y, p(s_1, t_1) g(Q(s_1, t_1))]]\}, \tag{5.5.28}$$

where

$$Q(x, y) = \exp\left(\int_0^x \int_0^s f(s_1, y) \, ds_1 \, ds\right), \tag{5.5.29}$$

Ω, Ω^{-1} are as defined in Theorem 5.5.1 and x_5, y_5 are chosen so that

$$\Omega(c) + A[x, y, p(s_1, t_1) g(Q(s_1, t_1))] \in \mathrm{Dom}(\Omega^{-1}),$$

for all x, y lying in the subintervals $0 \leq x \leq x_5$, $0 \leq y \leq y_5$ of R_+. □

Proof: Since $g'(u) \geq 0$ on R_+, the function $g(u)$ is monotonically increasing on $(0, \infty)$. We assume that $c > 0$ and that the standard limiting argument can be used to treat the remaining case. Define a function $m(x, y)$ by

$$m(x, y) = c + A[x, y, p(s_1, t_1) g(u(s_1, t_1))]. \tag{5.5.30}$$

Then (5.5.27) can be restated as

$$u(x, y) \leq m(x, y) + \int_0^x \int_0^s f(s_1, y) u(s_1, y) \, ds_1 \, ds. \tag{5.5.31}$$

Clearly $m(x, y)$ is positive and nondecreasing in both the variables $x, y \in R_+$. From (5.5.31) we observe that

$$\frac{u(x, y)}{m(x, y)} \leq 1 + \int_0^x \int_0^s f(s_1, y) \frac{u(s_1, y)}{m(s_1, y)} \, ds_1 \, ds. \tag{5.5.32}$$

Define a function $z(x, y)$ by the right-hand side of (5.5.32); then

$$z(0, y) = 1, \quad \frac{u(x, y)}{m(x, y)} \leq z(x, y)$$

and

$$z_{xx}(x, y) = f(x, y) \frac{u(x, y)}{m(x, y)} \leq f(x, y) z(x, y),$$

i.e.

$$\frac{z_{xx}(x, y)}{z(x, y)} \leq f(x, y). \tag{5.5.33}$$

Now by following the proof of Theorem 5.5.1 we obtain

$$z(x, y) \leq \exp\left(\int_0^x \int_0^s f(s_1, y) \, ds_1 \, ds\right) = Q(x, y). \tag{5.5.34}$$

Using (5.5.34) in (5.5.32) we have

$$u(x, y) \leq Q(x, y) m(x, y). \tag{5.5.35}$$

From (5.5.30) and (5.5.35) we have

$$m_{xxyy}(x, y) = p(x, y) g(u(x, y))$$
$$\leq p(x, y) g(Q(x, y) m(x, y))$$
$$\leq p(x, y) g(Q(x, y)) g(m(x, y)). \tag{5.5.36}$$

Now by following the proof of Theorem 5.5.1 we obtain

$$m(x, y) \leq \Omega^{-1}[\Omega(c) + A[x, y, p(s_1, t_1) g(Q(s_1, t_1))]]. \tag{5.5.37}$$

By using (5.5.37) in (5.5.35) we get the required inequality in (5.5.28).

■

The inequalities given in the following theorem have been recently established by Pachpatte (1993, 1996d) and are motivated by the study of certain higher order partial differential equations.

Theorem 5.5.4 Let $u(x, y)$, $a(x, y)$ and $b(x, y)$ be nonnegative continuous functions defined for $x, y \in R_+$ and $h: R_+^3 \to R_+$ be a continuous function which satisfies the condition

$$0 \leq h(x, y, v_1) - h(x, y, v_2) \leq k(x, y, v_2)(v_1 - v_2), \quad \text{(H)}$$

for $x, y \in R_+$ and $v_1 \geq v_2 \geq 0$ where $k: R_+^3 \to R_+$ is a continuous function.
(i) If

$$u(x, y) \leq a(x, y) + b(x, y)B[x, y, h(s, t, u(s, t))], \quad (5.5.38)$$

for $x, y \in R_+$ then

$$u(x, y) \leq a(x, y) + b(x, y)p(x, y)\exp(B[x, y, k(s, t, a(s, t))b(s, t)]), \quad (5.5.39)$$

for $x, y \in R_+$, where

$$p(x, y) = B[x, y, h(s, t, a(s, t))], \quad (5.5.40)$$

for $x, y \in R_+$.

(ii) Let $F(u)$ be a continuous, strictly increasing, convex, submultiplicative function for $u > 0$, $\lim_{u \to \infty} F(u) = \infty$, F^{-1} denote the inverse function of F, and $\alpha(x, y)$, $\beta(x, y)$ be continuous and positive functions for $x, y \in R_+$ and $\alpha(x, y) + \beta(x, y) = 1$. If

$$u(x, y) \leq a(x, y) + b(x, y)F^{-1}(B[x, y, h(s, t, F(u(s, t)))]), \quad (5.5.41)$$

for $x, y \in R_+$, then

$$u(x, y) \leq a(x, y) + b(x, y)F^{-1}(B[x, y, h(s, t, \alpha(s, t)F(a(s, t)\alpha^{-1}(s, t)))]$$
$$\times \exp(B[x, y, k(s, t, \alpha(s, t)F(a(s, t)\alpha^{-1}(s, t)))$$
$$\times \beta(s, t)F(b(s, t)\beta^{-1}(s, t))])), \quad (5.5.42)$$

for $x, y \in R_+$.

(iii) Let $g(u)$ be a continuously differentiable function defined for $u \geq 0$, $g(u) > 0$ for $u > 0$ and $g'(u) \geq 0$ for $u \geq 0$ and $g(u)$ is subadditive and submultiplicative for $u \geq 0$. If

$$u(x, y) \leq a(x, y) + b(x, y)B[x, y, h(s, t, g(u(s, t)))], \quad (5.5.43)$$

for $x, y \in R_+$, then for $0 \leq x \leq x_0, 0 \leq y \leq y_0$,

$$u(x, y) \leq a(x, y) + b(x, y)\Omega^{-1}[\Omega(q(x, y)) + B[x, y, k(s, t, g(a(s, t)))g(b(s, t))]], \quad (5.5.44)$$

where

$$q(x, y) = B[x, y, h(s, t, g(a(s, t)))], \quad (5.5.45)$$

Ω, Ω^{-1} are as defined in Theorem 5.5.1 and x_0, y_0 are chosen so that

$$\Omega(q(x, y)) + B[x, y, k(s, t, g(a(s, t)))g(b(s, t))] \in \text{Dom}(\Omega^{-1}),$$

for all x, y lying in the subintervals $0 \leq x \leq x_0$, $0 \leq y \leq y_0$ of R_+. $\square$

Proof: (i) Define a function $z(x, y)$ by

$$z(x, y) = B[x, y, h(s, t, u(s, t))]. \quad (5.5.46)$$

From (5.5.46) and using $u(x, y) \leq a(x, y) + b(x, y)z(x, y)$ and the condition (H), we observe that

$$z(x, y) \leq B[x, y, h(s, t, a(s, t) + b(s, t)z(s, t))]$$
$$= p(x, y) + B[x, y, \{h(s, t, a(s, t) + b(s, t)u(s, t))$$
$$- h(s, t, a(s, t))\}]$$
$$\leq p(x, y) + B[x, y, k(s, t, a(s, t))b(s, t)z(s, t)], \quad (5.5.47)$$

where $p(x, y)$ is defined by (5.5.40). Clearly $p(x, y)$ is nonnegative for $x, y \in R_+$. It is sufficient to assume that $p(x, y)$ is positive, since the standard limiting argument can be used to treat the remaining case. Now since $p(x, y)$ is positive and monotonic nondecreasing in $x, y \in R_+$, from (5.5.47) we observe that

$$\frac{z(x, y)}{p(x, y)} \leq 1 + B\left[x, y, k(s, t, a(s, t))b(s, t)\frac{z(s, t)}{p(s, t)}\right]. \quad (5.5.48)$$

Define a function $v(x, y)$ by the right-hand side of (5.5.48), then it is easy to observe that

$$D_2^m D_1^n v(x, y) = k(x, y, a(x, y))b(x, y)\frac{z(x, y)}{p(x, y)}$$
$$\leq k(x, y, a(x, y))b(x, y)v(x, y). \quad (5.5.49)$$

The rest of the proof follows by the same arguments as in the proof of Theorem 4.5.4 with suitable changes, and hence further details are omitted.

(ii) The proof follows by the same arguments as in the proof of Theorem 5.3.1 part (ii), applying the inequality given in part (i) above. The details are omitted here.

(iii) Define a function $z(x, y)$ by

$$z(x, y) = B[x, y, h(s, t, g(u(s, t)))]. \quad (5.5.50)$$

From (5.5.50) and using $u(x, y) \leq a(x, y) + b(x, y)z(x, y)$ and following the same arguments as in the proof of inequality (5.5.47) in part (i) with suitable changes we have

$$z(x, y) \leq q(x, y) + B[x, y, k(s, t, g(a(s, t)))g(b(s, t))g(z(s, t))], \quad (5.5.51)$$

where $q(x, y)$ is defined by (5.5.45). The rest of the proof can be completed by following the proofs of Theorems 5.2.2 and 4.5.4 with suitable changes. We leave the details to the reader.

■

5.6 Pachpatte's Inequalities II

In the past few years, the inequality given by Snow (1971) has attracted much attention and a number of its generalizations and their applications have appeared in the literature. In this section we present some basic nonlinear integral inequalities in two independent variables investigated by Pachpatte (1980c,d) which can be used in the analysis of various problems in the theory of partial integro-differential and integral equations.

Pachpatte (1980c) gave the following useful inequalities in two independent variables.

Theorem 5.6.1 *Let $u(x, y)$, $a(x, y)$, $b(x, y)$, $c(x, y)$, $p(x, y)$ and $q(x, y)$ be nonnegative continuous functions defined on a domain D. Let $P_0(x_0, y_0)$ and $P(x, y)$ be two points in D such that $(x - x_0)(y - y_0) \geq 0$ and R be the rectangular region whose opposite corners are the points P_0 and P. Let $G(r)$ be a continuous, strictly increasing, convex and submultiplicative function for $r > 0$, $\lim_{r \to \infty} G(r) = \infty$ for all (x, y) in D, $\alpha(x, y)$, $\beta(x, y)$ be positive continuous functions defined on a domain D, and $\alpha(x, y) + \beta(x, y) = 1$. Let*

$v(s, t; x, y)$ and $w(s, t; x, y)$ be the solutions of the characteristic initial value problems

$$L[v] = v_{st} - [p(s, t) + \beta(s, t)G(b(s, t)\beta^{-1}(s, t))(c(s, t)$$
$$+ q(s, t))]v = 0, \qquad (5.6.1)$$
$$v(s, y) = v(x, t) = 1,$$

and

$$M[w] = w_{st} - [\beta(s, t)G(b(s, t)\beta^{-1}(s, t))c(s, t) - p(s, t)]w = 0,$$
$$w(s, y) = w(x, t) = 1, \qquad (5.6.2)$$

respectively and let D^+ be a connected subdomain of D which contains P and on which $v \geq 0$ and $w \geq 0$ (Figure 4.1). If $R \subset D^+$ and $u(x, y)$ satisfies

$$u(x, y) \leq a(x, y) + b(x, y)G^{-1}\left[\int_{x_0}^{x}\int_{y_0}^{y} c(s, t)G(u(s, t))\,ds\,dt\right.$$
$$\left. + \int_{x_0}^{x}\int_{y_0}^{y} p(s, t)\left(\int_{x_0}^{s}\int_{y_0}^{t} q(s_1, t_1)G(u(s_1, t_1))\,ds_1\,dt_1\right)ds\,dt\right], (5.6.3)$$

then $u(x, y)$ also satisfies

$$u(x, y) \leq a(x, y) + b(x, y)G^{-1}\left[\int_{0}^{x}\int_{0}^{y} w(s, t; x, y)\right.$$
$$\times \left\{\alpha(s, t)G(a(s, t)\alpha^{-1}(s, t))c(s, t)\right.$$
$$+ p(s, t)\int_{x_0}^{s}\int_{y_0}^{t} \alpha(s_1, t_1)G(a(s_1, t_1)\alpha^{-1}(s_1, t_1))$$
$$\left.\left.\times [c(s_1, t_1) + q(s_1, t_1)]v(s_1, t_1; s, t)\,ds_1\,dt_1\right\}ds\,dt\right], \qquad (5.6.4)$$

□

Proof: Rewrite (5.6.3) as

$$u(x, y) \leq \alpha(x, y)a(x, y)\alpha^{-1}(x, y) + \beta(x, y)b(x, y)\beta^{-1}(x, y)$$

$$\times G^{-1}\left[\int_{x_0}^{x}\int_{y_0}^{y} c(s, t)G(u(s, t))\,ds\,dt + \int_{x_0}^{x}\int_{y_0}^{y} p(s, t)\right.$$

$$\left.\times \left(\int_{x_0}^{s}\int_{y_0}^{t} q(s_1, t_1)G(u(s_1, t_1))\,ds_1\,dt_1\right)ds\,dt\right]. \quad (5.6.5)$$

Since G is convex, submultiplicative and monotonic, from (5.6.5) we have

$$G(u(x, y)) \leq \alpha(x, y)G(a(x, y)\alpha^{-1}(x, y)) + \beta(x, y)G(b(x, y)\beta^{-1}(x, y))$$

$$\times \left[\int_{x_0}^{x}\int_{y_0}^{y} c(s, t)G(u(s, t))\,ds\,dt + \int_{x_0}^{x}\int_{y_0}^{y} p(s, t)\right.$$

$$\left.\times \left(\int_{x_0}^{s}\int_{y_0}^{t} q(s_1, t_1)G(u(s_1, t_1))\,ds_1\,dt_1\right)ds\,dt\right]. \quad (5.6.6)$$

The estimate given in (5.6.4) follows by first applying Theorem 4.8.2 to (5.6.6) with $a(x, y) = \alpha(x, y)G(a(x, y)\alpha^{-1}(x, y))$, $b(x, y) = \beta(x, y)G(b(x, y)\beta^{-1}(x, y))$ and $u(x, y) = G(u(x, y))$ and then applying G^{-1} to both sides of the resulting inequality.

∎

Theorem 5.6.2 *Let $u(x, y)$, $a(x, y)$, $b(x, y)$, $c(x, y)$, $p(x, y)$, $q(x, y)$, $P_0(x_0, y_0)$ and $P(x, y)$ be as in Theorem 5.6.1. Let $H(r)$ be a positive, continuous, strictly increasing, subadditive and submultiplicative function for $r > 0$, $\lim_{r \to \infty} H(r) = \infty$, H^{-1} is the inverse function of H. Let $v(s, t; x, y)$ and $w(s, t; x, y)$ be the solutions of the characteristic initial value problems*

$$L[v] = v_{st} - [p(s, t) + H(b(s, t))(c(s, t) + q(s, t))]v = 0,$$
$$v(s, y) = v(x, t) = 1, \quad (5.6.7)$$

and

$$M[w] = w_{st} - [H(b(s, t))c(s, t) - p(s, t)]w = 0,$$
$$w(s, y) = w(x, t) = 1, \quad (5.6.8)$$

498   MULTIDIMENSIONAL NONLINEAR INTEGRAL INEQUALITIES

respectively and let D^+ be a connected subdomain of D which contains P and on which $v \geq 0$ and $w \geq 0$ (Figure 4.1). If $R \subset D^+$ and $u(x, y)$ satisfies

$$u(x, y) \leq a(x, y) + b(x, y) H^{-1} \left[\int_{x_0}^{x} \int_{y_0}^{y} c(s, t) H(u(s, t)) \, ds \, dt \right.$$

$$\left. + \int_{x_0}^{x} \int_{y_0}^{y} p(s, t) \left(\int_{x_0}^{s} \int_{y_0}^{t} q(s_1, t_1) H(u(s_1, t_1)) \, ds_1 \, dt_1 \right) ds \, dt \right], \quad (5.6.9)$$

then $u(x, y)$ also satisfies

$$u(x, y) \leq H^{-1} \left[H(a(x, y)) + H(b(x, y)) \left[\int_{x_0}^{x} \int_{y_0}^{y} w(s, t; x, y) \right. \right.$$

$$\times \left\{ H(a(s, t)) c(s, t) + p(s, t) \int_{x_0}^{s} \int_{y_0}^{t} H(a(s_1, t_1)) \right.$$

$$\left. \left. \times \left[c(s_1, t_1) + q(s_1, t_1) \right] v(s_1, t_1; s, t) \, ds_1 \, dt_1 \right\} ds \, dt \right] \right]. \quad (5.6.10)$$

$\square$

Proof: Since H is subadditive, submultiplicative and monotonic, from (5.6.9) we have

$$H(u(x, y)) \leq H(a(x, y)) + H(b(x, y)) \left[\int_{x_0}^{x} \int_{y_0}^{y} c(s, t) H(u(s, t)) \, ds \, dt \right.$$

$$\left. + \int_{x_0}^{x} \int_{y_0}^{y} p(s, t) \left(\int_{x_0}^{s} \int_{y_0}^{t} q(s_1, t_1) H(u(s_1, t_1)) \, ds_1 \, dt_1 \right) ds \, dt \right].$$

(5.6.11)

The desired bound in (5.6.10) follows by first applying Theorem 4.8.2 to (5.6.11) with $a(x, y) = H(a(x, y))$, $b(x, y) = H(b(x, y))$ and $u(x, y) = H(u(x, y))$ and then applying H^{-1} to both sides of the resulting inequality.

∎

Pachpatte (1980c) also gave the following two inequalities which can be used in some applications.

Theorem 5.6.3 *Let $u(x, y)$, $a(x, y)$, $b(x, y)$, $c(x, y)$, $p(x, y)$, $q(x, y)$ and $P_0(x_0, y_0)$, $P(x, y)$ be as in Theorem 5.6.1. Let $G(r)$, $\alpha(x, y)$, $\beta(x, y)$ be as in Theorem 5.6.1. Let $v(s, t; x, y)$ and $w(s, t; x, y)$ be the solutions of the characteristic initial value problems*

$$L[v] = v_{st} - \beta(s, t)G(b(s, t)\beta^{-1}(s, t))[c(s, t)$$
$$+ p(s, t) + q(s, t)]v = 0, \qquad (5.6.12)$$
$$v(s, y) = v(x, t) = 1,$$

and

$$M[w] = w_{st} - \beta(s, t)G(b(s, t)\beta^{-1}(s, t))c(s, t)w = 0,$$
$$w(s, y) = w(x, t) = 1, \qquad (5.6.13)$$

respectively and let D^+ be a connected subdomain of D which contains P and on which $v \geq 0$ and $w \geq 0$ (Figure 4.1). If $R \subset D^+$ and $u(x, y)$ satisfies

$$u(x, y) \leq a(x, y) + b(x, y)G^{-1}\left[\int_{x_0}^{x}\int_{y_0}^{y} c(s, t)G(u(s, t))\,ds\,dt\right.$$

$$+ \int_{x_0}^{x}\int_{y_0}^{y} p(s, t)\left(G(u(s, t)) + \beta(s, t)G(b(s, t)\beta^{-1}(s, t))\right.$$

$$\left.\left.\times \int_{x_0}^{s}\int_{y_0}^{t} q(s_1, t_1)G(u(s_1, t_1))\,ds_1 dt_1\right)ds\,dt\right], \qquad (5.6.14)$$

then $u(x, y)$ also satisfies

$$u(x, y) \leq a(x, y) + b(x, y)G^{-1}\left[\int_{x_0}^{x}\int_{y_0}^{y} w(s, t; x, y)\right.$$

$$\times \left\{\alpha(s, t)G(a(s, t)\alpha^{-1}(s, t))[c(s, t) + p(s, t)]\right.$$

$$+ \beta(s, t)G(b(s, t)\beta^{-1}(s, t))p(s, t)$$

$$\times \int_{x_0}^{s}\int_{y_0}^{t} \alpha(s_1, t_1) G(a(s_1, t_1)\alpha^{-1}(s_1, t_1))$$

$$\times [c(s_1, t_1) + p(s_1, t_1) + q(s_1, t_1)]$$

$$\left. \times v(s_1, t_1; s, t)\, ds_1\, dt_1 \right\} ds\, dt \Bigg]. \qquad (5.6.15)$$

$\square$

Theorem 5.6.4 *Let* $u(x, y)$, $a(x, y)$, $b(x, y)$, $c(x, y)$, $p(x, y)$, $q(x, y)$, $P_0(x_0, y_0)$ *and* $P(x, y)$ *be as in Theorem 5.6.1. Let* H *and* H^{-1} *be the same functions as defined in Theorem 5.6.2. Let* $v(s, t; x, y)$ *and* $w(s, t; x, y)$ *be the solutions of the characteristic initial value problems*

$$L[v] = v_{st} - H(b(s, t))[c(s, t) + p(s, t) + q(s, t)]v = 0, \qquad (5.6.16)$$

$$v(s, y) = v(x, t) = 1,$$

and

$$M[w] = w_{st} - H(b(s, t))c(s, t)w = 0, \qquad (5.6.17)$$

$$w(s, y) = w(x, t) = 1,$$

respectively and let D^+ *be a connected subdomain of* D *which contains* P *and on which* $v \geq 0$ *and* $w \geq 0$ *(Figure 4.1). If* $R \subset D^+$ *and* $u(x, y)$ *satisfies*

$$u(x, y) \leq a(x, y) + b(x, y) H^{-1}\Bigg[\int_{x_0}^{x}\int_{y_0}^{y} c(s, t) H(u(s, t))\, ds\, dt$$

$$+ \int_{x_0}^{x}\int_{y_0}^{y} p(s, t) \Bigg(H(u(s, t)) + G(b(s, t))$$

$$\times \int_{x_0}^{s}\int_{y_0}^{t} q(s_1, t_1) H(u(s_1, t_1))\, ds_1 dt_1 \Bigg) ds\, dt \Bigg], \qquad (5.6.18)$$

then $u(x, y)$ also satisfies

$$u(x, y) \leq H^{-1}\left[H(a(x, y)) + H(b(x, y))\left[\int_{x_0}^{x}\int_{y_0}^{y} w(s, t; x, y)\right.\right.$$

$$\times \left\{H(a(s, t))[c(s, t) + p(s, t)] + H(b(s, t))p(s, t)\right.$$

$$\times \int_{x_0}^{s}\int_{y_0}^{t} H(a(s_1, t_1))[c(s_1, t_1) + p(s_1, t_1) + q(s_1, t_1)]$$

$$\left.\left.\left.\times v(s_1, t_1; s, t)\, ds_1\, dt_1\right\} ds\, dt\right]\right]. \qquad (5.6.19)$$

$\square$

The proofs of Theorems 5.6.3 and 5.6.4 can be completed by following similar arguments to those in the proofs of Theorems 5.6.1 and 5.6.2 and applying Theorem 4.8.3. The details are omitted here.

Remark 5.6.1 Note that in the special case when $p(x, y) = q(x, y) = 0$, Theorems 5.6.1–5.6.4 reduce to the further generalizations of the integral inequality established by Snow (1971); see also the results given in Bondge and Pachpatte (1980b). In the special case when $c(x, y) = 0$, Theorems 5.6.1–5.6.4 reduce to the new inequalities which can be used in certain applications.

Δ

Pachpatte (1980d) established the following inequalities, which can be used in the analysis of a class of nonlinear non-self-adjoint hyperbolic partial integro-differential and integral equations.

Theorem 5.6.5 Let $u(x, y)$, $a(x, y)$, $b(x, y)$, $c(x, y)$, $p(x, y)$, $q(x, y)$, $r(x, y)$, $h(x, y)$ and $g(x, y)$ be nonnegative continuous functions defined on a domain D. Let $P_0(x_0, y_0)$ and $P(x, y)$ be two points in D such that $(x - x_0)(y - y_0) \geq 0$ and R is the rectangular region whose opposite corners are the points P_0 and P. Let G, G^{-1}, α, β be as in Theorem 5.6.1. Let $V(s, t; x, y)$ be the solution of the characteristic initial value problem

$$M[V] = 0, \qquad (5.6.20)$$

where M is the adjoint operator of the operator L defined by

$$L[\psi] = \psi_{st} + a_1\psi_s + a_2\psi_t + a_3\psi, \qquad (5.6.21)$$

in which $a_1 = -\beta G(b\beta^{-1})cq$, $a_2 = -\beta G(b\beta^{-1})cp$ and $a_3 = -[g + \beta G(b\beta^{-1})c(r+h)]$. Let $W(s, t; x, y)$ be the solution of the characteristic initial value problem

$$N[W] = 0, \qquad (5.6.22)$$

where N is the adjoint operator of the operator T defined by

$$T[\phi] = \phi_{st} + b_1\phi_s + b_2\phi_t + b_3\phi, \qquad (5.6.23)$$

in which $b_1 = -\beta G(b\beta^{-1})cq$, $b_2 = -\beta G(b\beta^{-1})cp$, and $b_3 = -\beta G(b\beta^{-1}) \times c(r-h)$. Let D^+ be a connected subdomain of D which contains P and on which $V \geq 0$ and $W \geq 0$ (Figure 4.5). If $R \subset D^+$ and $u(x, y)$ satisfies

$$u(x, y) \leq a(x, y) + b(x, y)G^{-1}\left[p(x, y)\int_{x_0}^{x} c(s, y)G(u(s, y))\,ds\right.$$

$$+ q(x, y)\int_{y_0}^{y} c(x, t)G(u(x, t))\,dt$$

$$+ r(x, y)\int_{x_0}^{x}\int_{y_0}^{y} c(s, t)G(u(s, t))\,ds\,dt$$

$$+ h(x, y)\int_{x_0}^{x}\int_{y_0}^{y} g(s, t)\left(\int_{x_0}^{s}\int_{y_0}^{t} c(s_1, t_1)\right.$$

$$\left.\left. \times G(u(s_1, t_1))\,ds_1\,dt_1\right)ds\,dt\right], \qquad (5.6.24)$$

then $u(x, y)$ also satisfies

$$u(x, y) \leq a(x, y) + b(x, y)G^{-1}\left[p(x, y)\int_{x_0}^{x} c(s, y)G(u(s, y))\,ds\right.$$

$$+ q(x, y)\int_{y_0}^{y} c(x, t)G(u(x, t))\,dt + r(x, y)Q_0(x, y)$$

$$\left. + h(x, y)\int_{x_0}^{x}\int_{y_0}^{y} g(s, t)Q_0(s, t)\,ds\,dt\right], \qquad (5.6.25)$$

where $Q_0(x, y)$ is defined by the right member of

$$Q(x, y) = \int_{x_0}^{x} \int_{y_0}^{y} W(s, t; x, y) c(s, t) \left\{ a(s, t) + b(s, t) h(s, t) \right.$$

$$\left. \times \left(\int_{x_0}^{s} \int_{y_0}^{t} V(s_1, t_1; s, t) a(s_1, t_1) c(s_1, t_1) \, ds_1 \, dt_1 \right) \right\} ds \, dt,$$

(5.6.26)

by replacing $a(x, y)$ by $\alpha(x, y) G(a(x, y)\alpha^{-1}(x, y))$ and $b(x, y)$ by $\beta(x, y) \times G(b(x, y)\beta^{-1}(x, y))$. Further, if $q(x, y) = 0$, then

$$u(x, y) \leq a(x, y) + b(x, y) G^{-1} \left[r(x, y) Q_0(x, y) \right.$$

$$+ h(x, y) \int_{x_0}^{x} \int_{y_0}^{y} g(s, t) Q_0(s, t) \, ds \, dt$$

$$+ p(x, y) \int_{x_0}^{x} c(s, y) f_0(s, y) \exp \left(\int_{s}^{x} c(s_1, y) \right.$$

$$\left. \left. \times \beta(s_1, y) G(b(s_1, y)\beta^{-1}(s_1, y)) p(s_1, y) \, ds_1 \right) ds \right], \quad (5.6.27)$$

where $f_0(x, y)$ is defined by the right-hand side of

$$f(x, y) = a(x, y) + b(x, y) \left[r(x, y) Q(x, y) \right.$$

$$\left. + h(x, y) \int_{x_0}^{x} \int_{y_0}^{y} g(s, t) Q(s, t) \, ds \, dt \right] \quad (5.6.28)$$

replacing $a(x, y)$ by $\alpha(x, y) G(a(x, y)\alpha^{-1}(x, y))$, $b(x, y)$ by $\beta(x, y) G(b(x, y) \times \beta^{-1}(x, y))$ and $Q(x, y)$ by $Q_0(x, y)$. Again, if $p(x, y) = 0$, then

$$u(x, y) \leq a(x, y) + b(x, y)G^{-1}\Bigg[r(x, y)Q_0(x, y)$$

$$+ h(x, y) \int_{x_0}^{x} \int_{y_0}^{y} g(s, t)Q_0(s, t)\,ds\,dt + q(x, y) \int_{y_0}^{y} c(x, t)f_0(x, t)$$

$$\times \exp\Bigg(\int_{t}^{y} c(x, t_1)\beta(x, t_1)G(b(x, t_1)\beta^{-1}(x, t_1))$$

$$\times q(x, t_1)\,dt_1\Bigg)\,dt\Bigg], \qquad (5.6.29)$$

where $Q_0(x, y)$ and $f_0(x, y)$ are as defined above. $\qquad \square$

Proof: Rewrite (5.6.24) as

$$u(x, y) \leq \alpha(x, y)a(x, y)\alpha^{-1}(x, y) + \beta(x, y)b(x, y)\beta^{-1}(x, y)$$

$$\times G^{-1}\Bigg[p(x, y) \int_{x_0}^{x} c(s, y)G(u(s, y))\,ds$$

$$+ q(x, y) \int_{y_0}^{y} c(x, t)G(u(x, t))\,dt$$

$$+ r(x, y) \int_{x_0}^{x} \int_{y_0}^{y} c(s, t)G(u(s, t))\,ds\,dt$$

$$+ h(x, y) \int_{x_0}^{x} \int_{y_0}^{y} g(s, t) \Bigg(\int_{x_0}^{s} \int_{y_0}^{t} c(s_1, t_1)$$

$$\times G(u(s_1, t_1))\,ds_1\,dt_1\Bigg)\,ds\,dt\Bigg]. \qquad (5.6.30)$$

Since G is convex, submultiplicative and monotonic we have
$$G(u(x, y)) \leq \alpha(x, y)G(a(x, y)\alpha^{-1}(x, y)) + \beta(x, y)G(b(x, y)\beta^{-1}(x, y))$$
$$\times \left[p(x, y) \int_{x_0}^{x} c(s, y)G(u(s, y)) \, ds \right.$$
$$+ q(x, y) \int_{y_0}^{y} c(x, t)G(u(x, t)) \, dt$$
$$+ r(x, y) \int_{x_0}^{x} \int_{y_0}^{y} c(s, t)G(u(s, t)) \, ds \, dt$$
$$+ h(x, y) \int_{x_0}^{x} \int_{y_0}^{y} g(s, t) \left(\int_{x_0}^{s} \int_{y_0}^{t} c(s_1, t_1) \right.$$
$$\left. \left. \times G(u(s_1, t_1)) \, ds_1 \, dt_1 \right) ds \, dt \right]. \tag{5.6.31}$$

The estimate in (5.6.25) follows by first applying Theorem 4.8.8 with $a(x, y) = \alpha(x, y)G(a(x, y)\alpha^{-1}(x, y))$, $b(x, y) = \beta(x, y)G(b(x, y)\beta^{-1}(x, y))$, and $u(x, y) = G(u(x, y))$ and then applying G^{-1} to both sides of the resulting inequality. The rest of the proof when $q(x, y) = 0$ and $p(x, y) = 0$ follows by a similar argument to that in the last part of the proof of Theorem 4.8.8, in view of the proof of the first part of this theorem, with suitable modifications. The details are omitted here.

∎

Theorem 5.6.6 Let $u(x, y)$, $a(x, y)$, $b(x, y)$, $c(x, y)$, $p(x, y)$, $q(x, y)$, $r(x, y)$, $h(x, y)$, $g(x, y)$, $P_0(x_0, y_0)$ and $P(x, y)$ be as in Theorem 5.6.5. Let H, H^{-1} be as in Theorem 5.6.2. Let $V(s, t; x, y)$ be the solution of the characteristic initial value problem (5.6.20) in which M is the adjoint operator of the operator L defined by (5.6.21) with $a_1 = -H(b)cq$, $a_2 = -H(b)cp$ and $a_3 = -[g + H(b)c(r + h)]$. Let $W(s, t; x, y)$ be the solution of the characteristic initial value problem (5.6.22) in which N is the adjoint operator of the operator T defined by (5.6.23) with $b_1 = -H(b)cq$, $b_2 = -H(b)cp$, and

506 MULTIDIMENSIONAL NONLINEAR INTEGRAL INEQUALITIES

$b_3 = -H(b)c(r-h)$. Let D^+ be a connected subdomain of D which contains P and on which $V \geq 0$ and $W \geq 0$ (Figure 4.5). If $R \subset D^+$ and $u(x, y)$ satisfies

$$u(x, y) \leq a(x, y) + b(x, y) H^{-1} \left[p(x, y) \int_{x_0}^{x} c(s, y) H(u(s, y)) \, ds \right.$$

$$+ q(x, y) \int_{y_0}^{y} c(x, t) H(u(x, t)) \, dt$$

$$+ r(x, y) \int_{x_0}^{x} \int_{y_0}^{y} c(s, t) H(u(s, t)) \, ds \, dt + h(x, y) \int_{x_0}^{x} \int_{y_0}^{y} g(s, t)$$

$$\left. \times \left(\int_{x_0}^{s} \int_{y_0}^{y} c(s_1, t_1) H(u(s_1, t_1)) \, ds_1 \, dt_1 \right) ds \, dt \right], \quad (5.6.32)$$

then $u(x, y)$ also satisfies

$$u(x, y) \leq H^{-1} \left[H(a(x, y)) + H(b(x, y)) \left[p(x, y) \int_{x_0}^{x} c(s, y) H(u(s, y)) \, ds \right. \right.$$

$$+ q(x, y) \int_{y_0}^{y} c(x, t) H(u(x, t)) \, dt + r(x, y) Q_1(x, y)$$

$$\left. \left. + h(x, y) \int_{x_0}^{x} \int_{y_0}^{y} g(s, t) Q_1(s, t) \, ds \, dt \right] \right], \quad (5.6.33)$$

where $Q_1(x, y)$ is defined by the right-hand side of (5.6.26) by replacing $a(x, y)$ by $H(a(x, y))$ and $b(x, y)$ by $H(b(x, y))$. Further, if $q(x, y) = 0$, then

$$u(x, y) \leq H^{-1} \left[f_1(x, y) + H(b(x, y)) p(x, y) \left[\int_{x_0}^{x} c(s, y) f_1(s, y) \right. \right.$$

$$\left. \left. \times \exp \left(\int_{s}^{t} c(s_1, y) H(b(s_1, y)) p(s_1, y) \, ds_1 \right) ds \right] \right], \quad (5.6.34)$$

where $f_1(x, y)$ is defined by the right-hand side of (5.6.28) by replacing $a(x, y)$ by $H(a(x, y))$, $b(x, y)$ by $H(b(x, y))$, and $Q(x, y)$ by $Q_1(x, y)$. Again, if $p(x, y) = 0$, then

$$u(x, y) \leq H^{-1}\left[f_1(x, y) + H(b(x, y))q(x, y)\left[\int_{y_0}^{y} c(x, t) f_1(x, t)\right.\right.$$

$$\left.\left.\times \exp\left(\int_{t}^{y} c(x, t_1) H(b(x, t_1)) q(x, t_1) \, dt_1\right) dt\right]\right], \quad (5.6.35)$$

where $Q_1(x, y)$ and $f_1(x, y)$ are as defined above.

$\square$

Proof: Since H is subadditive, submultiplicative and monotonic, from (5.6.32) we have

$$H(u(x, y)) \leq H(a(x, y)) + H(b(x, y))\left[p(x, y) \int_{x_0}^{x} c(s, y) H(u(s, y)) \, ds\right.$$

$$+ q(x, y) \int_{y_0}^{y} c(x, t) H(u(x, t)) \, dt$$

$$+ r(x, y) \int_{x_0}^{x} \int_{y_0}^{y} c(s, t) H(u(s, t)) \, ds \, dt$$

$$+ h(x, y) \int_{x_0}^{x} \int_{y_0}^{y} g(s, t) \left(\int_{x_0}^{s} \int_{y_0}^{t} c(s_1, t_1)\right.$$

$$\left.\left.\times H(u(s_1, t_1)) \, ds_1 \, dt_1\right) ds \, dt\right]. \quad (5.6.36)$$

The desired bound in (5.6.33) follows by first applying Theorem 4.8.8 to (5.6.36) with $a(x, y) = H(a(x, y))$, $b(x, y) = H(b(x, y))$ and $u(x, y) = H(u(x, y))$ and then applying H^{-1} to both sides of the resulting inequality. Further, by setting $q(x, y) = 0$ and $p(x, y) = 0$ in (5.6.32) and applying Theorem 4.8.8 we obtain the desired bounds in (5.6.34) and (5.6.35).

■

Remark 5.6.2 We note that the functions $V(s, t; x, y)$ and $W(s, t; x, y)$ involved in Theorems 5.6.5 and 5.6.6 are the well-known Riemann functions relative to the point $P(x, y)$. The existence and continuity of the Riemann function is well known and may be demonstrated by the method of successive approximations (Counant and Hilbert, 1962).

△

5.7 Inequalities in Many Independent Variables

The integral inequalities involving functions of many independent variables which provide explicit bounds on unknown functions play a fundamental role in the development of the theory of partial differential equations. The last few years have witnessed a great deal of research concerning such inequalities and their applications in the theory of partial differential equations. This section deals with some basic inequalities established in Pelczar (1963), Headley (1974), Beesack (1975) and Pachpatte (1981c) which provide a very useful and important device in the study of many qualitative properties of the solutions of various types of partial differential, integral and integro-differential equations.

Pelczar (1963) initiated the study of some inequalities for a broad class of operators. In order to present the main results in Pelczar (1963) we need the following definitions found there.

We call a set P *partly ordered* if for some pairs of elements $x, y \in P$ a relation $x \leq y$ is defined in such a way that: (a) for each $x \in P, x \leq x$, (b) if $x \leq y$ and $y \leq x$, then $x = y$ and (c) if $x \leq y$ and $y \leq z$, then $x \leq z$.

Let P be a partly ordered set and $Q \subset P$; we call z the *upper bound of Q in P* if $z \in P$ and if $x \in Q$, then $x \leq \hat{z}$. We call $\hat{z}$ the *supremum of the set Q* (abbreviated sup Q) if $\hat{z}$ is an upper bound of Q in P and if x is an upper bound of Q in P, then $x \leq \hat{z}$. Each partially ordered set can have at most one supremum.

The set P will be said to satisfy the condition (II) if the difference $x - y \in P$ is defined for each $x, y \in P$ in such a way that (d) if $x \leq y$, then for each $x \in P$ is $x - z \leq y - z$, (e) there exists an element $0 \in P$, such that for each $x \in P$ is $x - 0 = x$ and (f) $x = y$ if and only if $x - y = 0$. The set P will be said to satisfy the condition (II*) if, for each $x, y \in P$, there exists in P, $z = \sup\{x, y\}$.

The main result established by Pelczar (1963) is embodied in the following theorem.

Theorem 5.7.1 *Assume that:*

(a_1) *The set P is not empty, partly ordered and fulfils the conditions (II) and (II*),*
(a_2) *the functions $W(x)$ and $L(x)$ are defined in the set P, and are such that $W(P) \subset P$ and $L(P) \subset P$,*
(a_3) *if $x \leq L(x)$, then $x \leq 0$,*
(a_4) *if $x \leq y$, then $W(x) \leq W(y)$,*
(a_5) *if $0 \leq W(x) - W(y)$, then $W(x) - W(y) \leq L(x - y)$,*
(a_6) *w is a solution of the equation*

$$w = W(w), \qquad (5.7.1)$$

(a_7) *$v \in P$ is such that*

$$v \leq W(v). \qquad (5.7.2)$$

Then we have

$$v \leq w. \qquad (5.7.3)$$

□

Proof: Let z be the supremum of the set $\{w, v\}$. Then

$$w \leq z \quad \text{and} \quad v \leq z. \qquad (5.7.4)$$

From (5.7.4), condition (c) and assumption (a_4) it follows that

$$w = W(w) \leq W(z) \quad \text{and} \quad v \leq W(v) \leq W(z). \qquad (5.7.5)$$

From (5.7.5) and the definition of z as the sup $\{w, v\}$ it follows that $z \leq W(z)$. From condition (d) we have

$$z - w \leq W(z) - w = W(z) - W(w).$$

Then, from (5.7.4) and the assumption (a_5) it follows that

$$z - w \leq L(z - w). \qquad (5.7.6)$$

In view of the assumption (a_3), the inequality (5.7.6) implies that $z - w = 0$. Hence $z = w$, which means (5.7.3) holds.

∎

Remark 5.7.1 It is easy to see that we can assume that the function $L(x)$ is defined only for $0 \leq x \, (x \in P)$. It is also easy to see that under the assumptions of Theorem 5.7.1, equation (5.7.1) can have at most one solution in P.

△

As an application, consider the following equation

$$u(x) = f(x) + \int_E F(x, y, u(y)) \, dy, \tag{5.7.7}$$

where $x = (x_1, \ldots, x_n)$, $y = (y_1, \ldots, y_n)$ and E is an n-dimensional set, and the inequality

$$v(x) \leq f(x) + \int_E F(x, y, v(y)) \, dy. \tag{5.7.8}$$

By making use of Theorem 5.7.1, Pelczar (1963) proved the following important result.

Theorem 5.7.2 *Assume that*
(b_1) $F(x, y, z)$ *is defined and continuous in* $\overline{E} \times E \times R$, $R = (-\infty, \infty)$,
(b_2) $f(x)$ *is defined and continuous in* $\overline{E}$,
(b_3) *if* $z \leq \hat{z}$, *then* $F(x, y, z) \leq F(x, y, \hat{z})$,
(b_4) $|F(x, y, z) - F(x, y, \hat{z})| \leq l(x, y, |z - \hat{z}|)$, *where the function* $l(x, y, z)$ *is defined in* $\overline{E} \times E \times R$ *and such that if*

$$w(x) \leq \int_E l(x, y, w(y)) \, dy,$$

then

$$w(x) \leq 0,$$

(b_5) $u(x)$ *is a solution of equation (5.7.7) in* $\overline{E}$,
(b_6) $v(x)$ *is a continuous function defined in* $\overline{E}$ *and fulfils the inequality (5.7.8).*

Then in the set $\overline{E}$ *we have*

$$v(x) \leq u(x). \tag{5.7.9}$$

□

Proof: It is easy to see that the set P of all continuous functions $w(x)$ defined for $x \in E$ fulfils the assumptions of Theorem 5.7.1 concerning the set P. Now, for $u \in P$, we define $W(u)$ as the function which at the point u has the value at the point z defined by the formula

$$z(x) = f(x) + \int_E F(x, y, u(y)) \, dy.$$

By $L(u)$ we denote the function defined in P, which at the point $u \in P$ has the value at the point w defined by the formula

$$w(x) = \int_E l(x, y, u(y)) \, dy.$$

It is easy to see that the functions $W(u)$ and $L(u)$ fulfil all assumptions of Theorem 5.7.1. Hence for each solution w of the equation

$$w = W(w),$$

and each function, which fulfils the inequality

$$v \leq W(v),$$

we have $v \leq w$. This means that for each solution of the equation (5.7.7) and for each function $v(x)$ which fulfils the inequality (5.7.8), we have

$$v(x) \leq w(x),$$

and the proof is complete.

∎

Remark 5.7.2 In particular if we put $l(x, y, z) = kz$ and assume that the function $F(x, y, z)$ is bounded and that $k < \mu(E)$, where $\mu(E)$ is the measure of the set E, then all the assumptions of Theorem 5.7.1 are satisfied.

△

In order to establish the next theorem, given by Headly (1974), we require the following result, given in Beesack (1975, p. 88).

Theorem 5.7.3 *Let G be an open set in R^N, and for $x, y \in G$ let*

$$G(x, y) = \{z \in R^N : z_j = \lambda_j x_j + (1 - \lambda_j) y_j, \quad 0 \leq \lambda_j \leq 1, 1 \leq j \leq N\},$$

denote the rectangular parallelepiped with one diagonal joining the points x, y. Let the points x^0, $y \in G$ be such that $G_0 = G(x^0, y) \subset G$, and let the functions $a(x)$, $k(x, t, z)$ be real-valued and continuous on G_0 and on $G_T \times R$ respectively, where

$$G_T = \{(x, t): x \in G_0, t \in G(x^0, x)\}.$$

Suppose also that k is nondecreasing in z for each $(x, t) \in G_T$ and that

$$|k(x, t, z)| \le h(t) g(|z|), \qquad (5.7.10)$$

for $(x, t, z) \in G_T \times R$, where $h \in L(G_0)$ and g is continuous and nondecreasing on R_+ with $\int_1^\infty ds/g(s) = \infty$. Then the integral equation

$$u(x) = a(x) + \int_{G(x^0, x)} k(x, t, u(t)) \, dt, \qquad (5.7.11)$$

has a solution which is continuous on G_0. Moreover, if $\{\epsilon_n\}$ is a strictly decreasing sequence with $\lim \epsilon_n = 0$, and if u_n is a continuous solution on G_0 of the integral equation

$$u_n(x) = a(x) + \epsilon_n + \int_{G(x^0, x)} k(x, t, u_n(t)) \, dt, \qquad (5.7.12)$$

then $U(x) = \lim u_n(x)$ exists uniformly on G_0, and $U(x)$ is the maximal solution of (5.7.11).

□

In case $x^0 \le y$ (that is, $x_i^0 \le y_i$ for $1 \le i \le N$), so that $x \in G_0 = G(x^0, y)$ implies $x^0 \le x \le y$ and $t \in G(x^0, x)$ implies $x^0 \le t \le x$, this result is a special case of results in Walter (1970), namely Theorem II, p. 131 (with $n = 1$, $H_1(x) = G(x^0, x)$, together with Remark VI(β), pp. 136–7, with $n = 1$, and no βs) and Theorem VII, p. 139. Cases where $x^0 \not\le y$ can be reduced to this case by appropriate change of independent variables. For example, if $y_\alpha < x_\alpha^0$ for certain subscripts α while $x_\beta^0 \le y_\beta$ for the remaining subscripts β, then the change of variables $\bar{x}_\alpha^0 = -x_\alpha^0$, $\bar{x}_\beta^0 = x_\beta^0$, $\bar{y}_\alpha = -y_\alpha$, $\bar{y}_\beta = y_\beta$, $\bar{t}_\alpha = -t_\alpha$, $\bar{t}_\beta = t_\beta$, and $\bar{x}_\alpha = -x_\alpha$, $\bar{x}_\beta = x_\beta$ reduces the integral equations (5.7.11), (5.7.12) to equations over $\overline{G}_0 = G(\bar{x}^0, \bar{y})$ having $\bar{x}^0 \le \bar{y}$, to which Walter's results do apply.

MULTIDIMENSIONAL NONLINEAR INTEGRAL INEQUALITIES   513

As noted in Remark VI(β) of Walter (1970, p. 136), if hypothesis (5.7.10) is deleted, then equation (5.7.11) may have only a *local* continuous solution, that is a solution defined in $G(x^0, y^0)$ for some $y^0 \in G(x^0, y)$. In this case, if $G(x^0, y^0)$ is also a common domain of existence of solutions of (5.7.12), then the final conclusion of the theorem still holds, but only on $G(x^0, y^0)$.

Headley (1974, Theorem 1) considered the case of Theorem 5.7.3 having $k = k(t, z)$ continuous on $G_0 \times R$ and nondecreasing in z. Hypothesis (5.7.10) was overlooked in Headley (1974) and in the case $k = k(t, z)$ of the following theorem given by Headley (1974, Theorem 2).

Theorem 5.7.4 *Let G, $G_0 = G(x^0, y)$ and the functions a, k be as in Theorem 5.7.3, and let the function b be continuous on G_0 and satisfy the inequality.*

$$v(x) \leq a(x) + \int_{G(x^0, x)} k(x, t, v(t))\, dt, \quad x \in G_0. \tag{5.7.13}$$

Then

$$v(x) \leq U(x), \quad x \in G_0, \tag{5.7.14}$$

where U is the maximal solution of (5.7.11) on G_0.

□

Proof: We may assume that $x_i^0 \neq y_i$ for $1 \leq i \leq N$, since otherwise $U(x) = a(x)$ and (5.7.13) is $v(x) \leq a(x)$. For $n \geq 1$, let u_n be a continuous solution of the integral equation (5.7.12) on G_0, where $\epsilon_n = n^{-1}$ for example. Then

$$v(x) < u_n(x) \text{ for } x \in G_0, \quad n \geq 1. \tag{5.7.15}$$

For, if this were false for some n and some $x \in G_0$, then because

$$v(x^0) \leq a(x^0) < a(x^0) + \frac{1}{n} = u_n(x^0),$$

it would follow from the continuity of v and u_n that for some point $z \in G_0$ with $z \neq x^0$ we would have $v(x) < u_n(x)$ for $x \in G(x^0, z) - \{z\}$ but $v(z) = u_n(z)$. But then

$$v(z) \leq a(z) + \int_{G(x^0, z)} k(z, t, v(t))\, dt$$

$$< a(z) + \frac{1}{n} + \int_{G(x^0,z)} k(z, t, u_n(t))\,dt = u_n(z).$$

It follows that (5.7.15) must hold. Taking limits as $n \to \infty$, (5.7.14) now follows from Theorem 5.7.3.

∎

Remark 5.7.3 If hypothesis (5.7.10) is deleted, then (5.7.14) will hold on any subset $G(x^0, y^0) \subset G_0$ such that equations (5.7.11) and (5.7.12) have continuous solutions on the common domain $G(x^0, y^0)$, if such $y^0 \in G_0$ exists. This was the case considered in Headley (1974), and for $N = 2$ also by Rasmussen (1976).

△

In the further discussion, some useful integral inequalities in n independent variables given by Pachpatte (1981c) are presented, which are motivated by a well-known integral inequality due to Ważeski (1969). We use the same notation as given in Section 4.9 without further mention.

Pachpatte (1981c) gave the following general version of Ważeski's inequality given in Ważeski (1969).

Theorem 5.7.5 *Let $u(x)$ and $a(x)$ be nonnegative continuous functions defined on Ω. Let $k(x, y, z)$ and $W(x, z)$ be nonnegative continuous functions defined on $\Omega^2 \times R$ and $\Omega \times R$, respectively, and nondecreasing in the last variables, and $k(x, y, z)$ be uniformly Lipschitz in the last variable. If*

$$u(x) \leq a(x) + W\left(x, \int_{x^0}^{x} k(x, y, u(y))\,dy\right), \qquad (5.7.16)$$

then

$$u(x) \leq a(x) + W(x, r(x)), \qquad (5.7.17)$$

for $x \in \Omega$, where $r(x)$ is the solution of the equation

$$r(x) = \int_{x^0}^{x} k(x, y, a(y) + W(y, r(y)))\,dy, \qquad (5.7.18)$$

existing on Ω.

□

Proof: Define a function $z(x)$ by

$$z(x) = \int_{x^0}^{x} k(x, y, u(y)) \, dy, \qquad (5.7.19)$$

then (5.7.16) can be restated as

$$u(x) \leq a(x) + W(x, z(x)). \qquad (5.7.20)$$

Using the monotonicity assumption on k and (5.7.20) in (5.7.19) we obtain

$$z(x) \leq \int_{x^0}^{x} k(x, y, a(y) + W(y, z(y))) \, dy. \qquad (5.7.21)$$

Now a suitable application of Theorem 5.7.2 or Theorem 5.7.4 to (5.7.18) and (5.7.21) yields

$$z(x) \leq r(x), \qquad (5.7.22)$$

where $r(x)$ is the solution of equation (5.7.18). Now, using (5.7.22) in (5.7.20) we obtain the desired bound in (5.7.17).

∎

The following inequality established by Pachpatte (1981c) combines the features of two inequalities, namely, the n independent variable generalization of Wendroff's inequality (Bondge and Pachpatte, 1979a) and the integral inequality given by Headley (1974, Theorem 2), and can be used more effectively in the theory of certain integral and integro-differential equations involving n independent variables.

Theorem 5.7.6 *Let $u(x)$, $f(x)$, $g(x)$, $q(x)$, $c(x)$ be nonnegative continuous functions defined on Ω, with $f(x) > 0$ and nondecreasing in x and $q(x) \geq 1$. Let $k(x, y, z)$ and $W(x, z)$ be nonnegative continuous functions defined on $\Omega^2 \times R$ and $\Omega \times R$, respectively; $k(x, y, z)$ is nondecreasing in x and z and is uniformly Lipschitz in z and $W(x, z)$ is nondecreasing in both x and z. If*

$$u(x) \leq f(x) + q(x) \left[\int_{x^0}^{x} g(y) u(y) \, dy + \int_{x^0}^{x} g(y) q(y) \left(\int_{x^0}^{y} c(s) u(s) \, ds \right) dy \right]$$

$$+ W \left(x, \int_{x^0}^{x} k(x, y, u(y)) \, dy \right), \qquad (5.7.23)$$

for $x \in \Omega$, *then*
$$u(x) \leq E_0(x)[f(x) + W(x, r(x))], \qquad (5.7.24)$$
for $x \in \Omega$, *where*
$$E_0(x) = q(x)\left[1 + \int_{x^0}^{x} g(y)q(y)\exp\left(\int_{x^0}^{y} q(s)[g(s) + c(s)]\,ds\right)dy\right], \qquad (5.7.25)$$
and $r(x)$ *is the solution of the equation*
$$r(x) = \int_{x^0}^{x} k(x, y, E_0(y)[f(y) + W(y, r(y))])\,dy, \qquad (5.7.26)$$
existing on Ω.

□

Proof: Define a function $m(x)$ by
$$m(x) = f(x) + W\left(x, \int_{x^0}^{x} k(x, y, u(y))\,dy\right), \qquad (5.7.27)$$
then (5.7.23) can be restated as
$$u(x) \leq m(x) + q(x)\left[\int_{x^0}^{x} g(y)u(y)\,dy + \int_{x^0}^{x} g(y)q(y)\left(\int_{x^0}^{y} c(s)u(s)\,ds\right)dy\right]. \qquad (5.7.28)$$
Since $m(x)$ is positive, nondecreasing and $q(x) \geq 1$, we observe from (5.7.28) that
$$\frac{u(x)}{m(x)} \leq q(x)\left[1 + \int_{x^0}^{x} g(y)\frac{u(y)}{m(y)}\,dy + \int_{x^0}^{x} g(y)q(y)\left(\int_{x^0}^{y} c(s)\frac{u(s)}{m(s)}\,ds\right)dy\right]. \qquad (5.7.29)$$
Define a function $z(x)$ such that
$$z(x) = 1 + \int_{x^0}^{x} g(y)\frac{u(y)}{m(y)}\,dy + \int_{x^0}^{x} g(y)q(y)\left(\int_{x^0}^{y} c(s)\frac{u(s)}{m(s)}\,ds\right)dy,$$
$$z(x) = 1 \text{ on } x_j = x_j^0, \quad 1 \leq j \leq n,$$

then
$$D_1\ldots D_n z(x) = g(x)\frac{u(x)}{m(x)} + g(x)q(x)\int_{x^0}^x c(s)\frac{u(s)}{m(s)}\,ds,$$

which in view of (5.7.29) implies
$$D_1\ldots D_n z(x) \leq g(x)q(x)\left[z(x) + \int_{x^0}^x c(s)q(s)z(s)\,ds\right]. \qquad (5.7.30)$$

If we put
$$v(x) = z(x) + \int_{x^0}^x c(s)q(s)z(s)\,ds, \qquad (5.7.31)$$
$$v(x) = z(x) \text{ on } x_j = x_j^0, \quad 1 \leq j \leq n,$$

then
$$D_1\ldots D_n v(x) = D_1\ldots D_n z(x) + c(x)q(x)z(x). \qquad (5.7.32)$$

Using the facts that $D_1\ldots D_n z(x) \leq g(x)q(x)v(x)$ from (5.7.30) and $z(x) \leq v(x)$ from (5.7.31) in (5.7.32) we obtain
$$D_1\ldots D_n v(x) \leq q(x)[g(x)+c(x)]v(x). \qquad (5.7.33)$$

From (5.7.33) and using the facts that $D_n v(x) \geq 0$, $D_1\ldots D_{n-1}v(x) \geq 0$ and $v(x) > 0$, we observe that
$$\frac{D_1\ldots D_n v(x)}{v(x)} \leq q(x)[g(x)+c(x)] + \frac{(D_n v(x))(D_1\ldots D_{n-1}v(x))}{v^2(x)},$$

i.e.
$$D_n\left(\frac{D_1\ldots D_{n-1}v(x)}{v(x)}\right) \leq q(x)[g(x)+c(x)]. \qquad (5.7.34)$$

By keeping $x_1,\ldots,x_{n-1}$ fixed in (5.7.34), setting $x_n = s_n$ and then integrating with respect to s_n from x_n^0 to x_n we have
$$\frac{D_1\ldots D_{n-1}v(x)}{v(x)} \leq \int_{x^0}^x q(x_1,\ldots,x_{n-1},s_n)[g(x_1,\ldots,x_{n-1},s_n)$$
$$+ c(x_1,\ldots,x_{n-1},s_n)]\,ds_n. \qquad (5.7.35)$$

Again as above, from (5.7.35) we observe that

$$D_{n-1}\left(\frac{D_1\ldots D_{n-2}v(x)}{v(x)}\right) \leq \int_{x^0}^{x} q(x_1,\ldots,x_{n-1},s_n)[g(x_1,\ldots,x_{n-1},s_n)$$
$$+ c(x_1,\ldots,x_{n-1},s_n)]\,ds_n.$$

By keeping $x_1,\ldots,x_{n-2}$ and x_n fixed in the above inequality, setting $x_{n-1} = s_{n-1}$ and then integrating with respect to s_{n-1} from x^0_{n-1} to x_{n-1} we have

$$\frac{D_1\ldots D_{n-2}v(x)}{v(x)} \leq \int_{x^0_{n-1}}^{x_{n-1}}\int_{x^0_n}^{x_n} q(x_1,\ldots,x_{n-2},s_{n-1},s_n)$$
$$\times [g(x_1,\ldots,x_{n-2},s_{n-1},s_n)$$
$$+ c(x_1,\ldots,x_{n-2},s_{n-1},s_n)]\,ds_n\,ds_{n-1}.$$

Continuing in this way we have

$$\frac{D_1 v(x)}{v(x)} \leq \int_{x^0_2}^{x_2}\cdots\int_{x^0_n}^{x_n} q(x_1,s_2,\ldots,s_n)[g(x_1,s_2,\ldots,s_n)$$
$$+ c(x_1,s_2,\ldots,s_n)]\,ds_n\ldots ds_2.$$

Now keeping $x_2,\ldots,x_n$ fixed in the above inequality, we set $x_1 = s_1$ and then integrating with respect to s_1 from x^0_1 to x_1 we have

$$v(x) \leq \exp\left(\int_{x^0}^{x} q(s)[g(s) + c(s)]\,ds\right).$$

Substituting this bound on $v(x)$ in (5.7.30), setting $x_n = y_n$ and then integrating both sides with respect to y_n from x^0_n to x_n, then setting $x_{n-1} = y_{n-1}$ and integrating with respect to y_{n-1} from x^0_{n-1} to x_{n-1}, and continuing in this way, finally setting $x_1 = y_1$ and then integrating with respect to y_1 from x^0_1 to x_1, we obtain

$$z(x) \leq 1 + \int_{x^0}^{x} g(y)q(y)\exp\left(\int_{x^0}^{y} q(s)[g(s) + c(s)]\,ds\right) dy.$$

Substituting this bound on $z(x)$ in (5.7.29) we have

$$u(x) \le E_0(x) m(x), \quad (5.7.36)$$

where $E_0(x)$ is defined by (5.7.25). From (5.7.27) and (5.7.36) we have

$$u(x) \le E_0(x) \left[f(x) + W\left(x, \int_{x^0}^{x} k(x, y, u(y)) \, dy\right) \right]. \quad (5.7.37)$$

Now a suitable application of Theorem 5.7.5 yields the desired bound in (5.7.24).

∎

Remark 5.7.4 We note that in the special case when $c(x) = 0$, the inequality established in Theorem 5.7.6 reduces to another interesting inequality which can be used in some applications.

△

Another interesting and useful integral inequality given by Pachpatte (1981c) in n independent variables involving two nonlinear functions on the right-hand side of the inequality is embodied in the following theorem.

Theorem 5.7.7 *Let $u(x)$, $f(x)$, $g(x)$, $q(x)$, $k(x, y, z)$, $W(x, z)$ be as in Theorem 5.7.6 and furthermore $f(x) \ge 1$. Let $H: R_+ \to R_+$ be a continuously differentiable function with $H(u) > 0$ for $u > 0$, $H'(u) \ge 0$ for $u \ge 0$ and satisfies $(1/v)H(u) \le H(u/v)$ for $v \ge 1$, $u \ge 0$ and $H(u)$ is submultiplicative for $u \ge 0$. If*

$$u(x) \le f(x) + q(x) \int_{x^0}^{x} g(y) H(u(y)) \, dy + W\left(x, \int_{x^0}^{x} k(x, y, u(y)) \, dy\right), \quad (5.7.38)$$

for $x \in \Omega$, then for $x \in \Omega_1 \subset \Omega$,

$$u(x) \le E_1(x)[f(x) + W(x, r(x))], \quad (5.7.39)$$

where

$$E_1(x) = q(x) G^{-1} \left[G(1) + \int_{x^0}^{x} g(y) H(q(y)) \, dy \right], \quad (5.7.40)$$

in which
$$G(v) = \int_{v_0}^{v} \frac{ds}{H(s)}, \quad v > 0, v_0 > 0, \tag{5.7.41}$$

G^{-1} is the inverse of G and
$$G(1) + \int_{x^0}^{x} g(y)H(q(y))\,dy \in \text{Dom}(G^{-1}),$$

for all $x \in \Omega$, and $r(x)$ is a solution of the equation
$$r(x) = \int_{x^0}^{x} k(x, y, E_1(y)[f(y) + W(y, r(y))])\,dy, \tag{5.7.42}$$

existing on Ω.

□

Proof: We note that since $H'(u) \geq 0$ on R_+, the function H is monotonically increasing on $(0, \infty)$. Define a function $m(x)$ as in the proof of Theorem 5.7.6 given by (5.7.27); then (5.7.38) can be restated as

$$u(x) \leq m(x) + q(x) \int_{x^0}^{x} g(y)H(u(y))\,dy. \tag{5.7.43}$$

Since $m(x) \geq 1$ and nondecreasing and $q(x) \geq 1$, then from (5.7.43) we observe that

$$\frac{u(x)}{m(x)} \leq q(x) \left[1 + \int_{x^0}^{x} g(y)H\left(\frac{u(y)}{m(y)}\right) dy \right]. \tag{5.7.44}$$

Define a function $z(x)$ by
$$z(x) = 1 + \int_{x^0}^{x} g(y)H\left(\frac{u(y)}{m(y)}\right) dy,$$

$$z(x) = 1 \text{ on } x_j = x_j^0, \quad 1 \leq j \leq n,$$

then
$$D_1 \ldots D_n z(x) = g(x)H\left(\frac{u(x)}{m(x)}\right),$$

which in view of (5.7.44) and the submultiplicative character of H implies

$$D_1 \ldots D_n z(x) \leq g(x) H(q(x)) H(z(x)). \qquad (5.7.45)$$

From (5.7.45) and using the facts that $D_n H(z(x)) \geq 0$, $D_1 \ldots D_{n-1} z(x) \geq 0$ and $H(z(x)) > 0$, we observe that

$$\frac{D_1 \ldots D_n z(x)}{H(z(x))} \leq g(x) H(q(x)) + \frac{[D_n H(z(x))][D_1 \ldots D_{n-1} z(x)]}{[H(z(x))]^2},$$

i.e.

$$D_n \left(\frac{D_1 \ldots D_{n-1} z(x)}{H(z(x))} \right) \leq g(x) H(q(x)).$$

Now by following an argument similar to that in the proof of Theorem 5.7.6, with suitable modifications, we obtain

$$\frac{D_1 z(x)}{H(z(x))} \leq \int_{x_2^0}^{x_2} \ldots \int_{x_n^0}^{x_n} g(x_1, y_2, \ldots, y_n) H(q(x_1, y_2, \ldots, y_n)) \, dy_n \ldots dy_2.$$

$$(5.7.46)$$

From (5.7.41) and (5.7.46), we observe that

$$D_1 G(u(x)) \leq \int_{x_2^0}^{x_2} \ldots \int_{x_n^0}^{x_n} g(x_1, y_2, \ldots, y_n)$$

$$\times H(q(x_1, y_2, \ldots, y_n)) \, dy_n \ldots dy_2.$$

By keeping $x_2, \ldots, x_n$ fixed in the above inequality, setting $x_1 = y_1$ and then integrating with respect to y_1 from x_1^0 to x_1 we have

$$u(x) \leq G^{-1} \left[G(1) + \int_{x^0}^{x} g(y) H(q(y)) \, dy \right].$$

The rest of the proof is immediate by analogy with the last argument in the proof of Theorem 5.7.6. The subdomain Ω_1 of Ω is obvious.

∎

Pachpatte (1981c) gave the following inequality, which can be used in more general situations.

Theorem 5.7.8 Let $u(x)$, $f(x)$, $g(x)$, $k(x, y, z)$ and $W(x, z)$ be as in Theorem 5.7.7. Let $H: R_+ \to R_+$ be a continuously differentiable function with $H(u) > 0$ for $u > 0$, $H'(u) \geq 0$ for $u \geq 0$ and satisfying $(1/v)H(u) \leq H(u/v)$, for $v \geq 1$, $u \geq 0$. If

$$u(x) \leq f(x) + \int_{x^0}^{x} g(y) \left(u(y) + \int_{x^0}^{y} g(s) H(u(s)) \, ds \right) dy$$

$$+ W \left(x, \int_{x^0}^{x} k(x, y, u(y)) \, dy \right), \qquad (5.7.47)$$

for $x \in \Omega$, then for $x \in \Omega_2 \subset \Omega$,

$$u(x) \leq E_2(x)[f(x) + W(x, r(x))], \qquad (5.7.48)$$

where

$$E_2(x) = 1 + \int_{x^0}^{x} g(y) F^{-1} \left[F(1) + \int_{x^0}^{y} g(s) \, ds \right] dy, \qquad (5.7.49)$$

in which

$$F(\sigma) = \int_{\sigma_0}^{\sigma} \frac{ds}{s + H(s)}, \quad \sigma > 0, \sigma_0 > 0, \qquad (5.7.50)$$

F^{-1} is the inverse function of F and

$$F(1) + \int_{x^0}^{x} g(s) \, ds \in \text{Dom}(F^{-1}),$$

for all $x \in \Omega_2$, and $r(x)$ is a solution of the equation

$$r(x) = \int_{x^0}^{x} k(x, y, E_2(y)[f(y) + W(y, r(y))]) \, dy, \qquad (5.7.51)$$

existing on Ω.

$\square$

The details of the proof of this theorem follow by an argument similar to that in the proof of Theorem 5.7.7, together with the proof of Theorem 5.4.1, and the details are omitted here.

Pachpatte (1981c) has established the following generalization of the integral inequality given by Young (1973).

Theorem 5.7.9 *Let $u(x)$, $a(x)$, $b(x)$, $c(x)$ and $g(x)$ be nonnegative continuous functions defined on Ω. Let $f(x)$, $k(x, y, z)$ and $W(x, z)$ be as in Theorem 5.7.6. Let $v(y; x)$ and $e(y; x)$ be the solutions of the characteristic initial value problems*

$$(-1)^n v_{y_1 \ldots y_n}(y; x) - [a(y)b(y) + a(y)g(y) + c(y)]v(y; x) = 0 \text{ in } \Omega,$$
(5.7.52)
$$v(y; x) = 1 \text{ on } y_i = x_i, \quad 1 \leq i \leq n,$$

and

$$(-1)^n e_{y_1 \ldots y_n}(y; x) - [a(y)b(y) - c(y)]e(y; x) = 0 \text{ in } \Omega,$$
(5.7.53)
$$e(y, x) = 1 \text{ on } y_i = x_i, \quad 1 \leq i \leq n,$$

respectively, and let D^+ be a connected subdomain of Ω containing x such that $v \geq 0$, $e \geq 0$ for all $y \in D^+$. If $D \subset D^+$ and

$$u(x) \leq f(x) + a(x) \left[\int_{x^0}^{x} b(y)u(y)\,dy + \int_{x^0}^{x} c(y) \left(\int_{x^0}^{y} g(s)u(s)\,ds \right) dy \right]$$

$$+ W\left(x, \int_{x^0}^{x} k(x, y, u(y))\,dy\right), \quad (5.7.54)$$

for $x \in \Omega$, then

$$u(x) \leq E_3(x)[f(x) + W(x, r(x))], \quad (5.7.55)$$

where

$$E_3(x) = 1 + a(x) \left[\int_{x^0}^{x} e(y; x) \left\{ b(y) + c(y) \left(\int_{x^0}^{y} [b(s) + g(s)] \right.\right.\right.$$
$$\left.\left.\left. \times v(s; y)\,ds \right) \right\} dy \right], \quad (5.7.56)$$

and $r(x)$ is a solution of the equation

$$r(x) = \int_{x^0}^{x} k(x, y, E_3(y)[f(y) + W(y, r(y))]) \, dy, \qquad (5.7.57)$$

existing on Ω.

$\square$

Proof: Define a function $m(x)$ as in the proof of Theorem 5.7.6 given by (5.7.27); then (5.7.54) can be restated as

$$u(x) \leq m(x) + a(x) \left[\int_{x^0}^{x} b(y) u(y) \, dy + \int_{x^0}^{x} c(y) \left(\int_{x^0}^{y} g(s) u(s) \, ds \right) dy \right]. \qquad (5.7.58)$$

Since $m(x)$ is positive nondecreasing, we observe from (5.7.58) that

$$\frac{u(x)}{m(x)} \leq 1 + a(x) \left[\int_{x^0}^{x} b(y) \frac{u(y)}{m(y)} \, dy + \int_{x^0}^{x} c(y) \left(\int_{x^0}^{y} g(s) \frac{u(s)}{m(s)} \, ds \right) dy \right]. \qquad (5.7.59)$$

Define a function $z(x)$ such that

$$z(x) = \int_{x^0}^{x} b(y) \frac{u(y)}{m(y)} \, dy + \int_{x^0}^{x} c(y) \left(\int_{x^0}^{y} g(s) \frac{u(s)}{m(s)} \, ds \right) dy,$$

$$z(x) = 0 \text{ on } x_i = x_i^0, \quad 1 \leq i \leq n,$$

then we obtain

$$D_1 \ldots D_n z(x) = b(x) \frac{u(x)}{m(x)} + c(x) \left(\int_{x^0}^{x} g(s) \frac{u(s)}{m(s)} \, ds \right),$$

which in view of (5.7.59) implies

$$D_1 \ldots D_n z(x) \leq b(x)[1 + a(x) z(x)] + c(x) \left(\int_{x^0}^{x} g(s)[1 + a(s) z(s)] \, ds \right). \qquad (5.7.60)$$

Adding $c(x) z(x)$ to both sides of the above inequality we have

$$D_1 \ldots D_n z(x) + c(x) z(x) \leq b(x)[1 + a(x) z(x)]$$

$$+ c(x) \left[z(x) + \int_{x^0}^{x} g(s)[1 + a(s) z(s)] \, ds \right]. \qquad (5.7.61)$$

If we put

$$\psi(x) = z(x) + \int_{x^0}^{x} g(s)[1 + a(s)z(s)]\,ds, \qquad (5.7.62)$$

$$\psi(x) = z(x) = 0 \text{ on } x_i = x_i^0, \quad 1 \le i \le n,$$

then we obtain

$$D_1 \ldots D_n \psi(x) = D_1 \ldots D_n z(x) + g(x)[1 + a(x)z(x)]. \qquad (5.7.63)$$

Using $D_1 \ldots D_n z(x) \le b(x)[1 + a(x)z(x)] + c(x)\psi(x)$ from (5.7.61) and $z(x) \le \psi(x)$ from (5.7.62) in (5.7.63) we have

$$L[\psi] = D_1 \ldots D_n \psi(x) - [a(x)b(x) + a(x)g(x) + c(x)]\psi(x)$$
$$\le [b(x) + g(x)]. \qquad (5.7.64)$$

Furthermore, all pure mixed derivatives of ψ with respect to $x_1, \ldots, x_{i-1}, x_{i+1}, \ldots, x_n$ up to order $n-1$ vanish on $x_i = x_i^0$, $1 \le i \le n$. If w is a function which is n times continuously differentiable in D, then

$$wL\psi - \psi Mw = \sum_{k=1}^{n} (-1)^{k-1} D_k[(D_0 D_1 \ldots D_{k-1} w)(D_{k+1} \ldots D_n D_{n+1} \psi)], \qquad (5.7.65)$$

where

$$Mw = (-1)^n D_1 \ldots D_n w(x) - [a(x)b(x) + a(x)g(x) + c(x)]w(x),$$

with $D_0 \equiv D_{n+1} = I$ the identity. By integrating (5.7.65) over D, using s as the variable of integration, and noting that ψ vanishes together with all its mixed derivatives up to order $n-1$ on $s_k = x_k^0$, $1 \le k \le n$, we then obtain

$$\int_D (wL\psi - \psi Mw)\,ds = \sum_{k=1}^{n} (-1)^{k-1} \int_{s_k = x_k} (D_1 \ldots D_{k-1} w)(D_{k+1} \ldots D_n \psi)\,ds', \qquad (5.7.66)$$

where $ds' = ds_1 \ldots ds_{k-1}\,ds_{k+1} \ldots ds_n$.

Now let w be chosen as the function v satisfying (5.7.52). Since $v = 1$ on $s_k = x_k$, $1 \le k \le n$, it follows that $D_1 \ldots D_{k-1} v(s; x) = 0$ on $s_k = x_k$ for

$2 \leq k \leq n$. Thus (5.7.66) becomes

$$\int_D v(s;x)L\psi(s)\,ds = \int_{s_1=x_1} v(s;x)D_2\ldots D_n\psi(y)\,dy' = \psi(x). \qquad (5.7.67)$$

By the continuity of v and the fact that $v=1$ on $s=x$, there is a domain D^+ containing x on which $v \geq 0$. Now multiplying (5.7.64) throughout by v and using (5.7.62) and (5.7.67), we obtain

$$\psi(x) \leq \int_{x^0}^{x} [b(s)+g(s)]v(s;x)\,ds.$$

Now substituting this bound on $\psi(x)$ in (5.7.61) we obtain

$$Lz = D_1\ldots D_n z(x) - [a(x)b(x)-c(x)]z(x)$$

$$\leq b(x)+c(x)\left(\int_{x^0}^{x}[b(s)+g(s)]v(s;x)\,ds\right).$$

Again, by following the same argument as above we obtain the estimate for $z(x)$ such that

$$z(x) \leq \int_{x^0}^{x} e(y;x)\left\{b(y)+c(y)\left(\int_{x^0}^{y}[b(s)+g(s)]v(s;y)\,ds\right)\right\}dy.$$

Now substituting this bound on $z(x)$ in (5.7.59) we obtain

$$u(x) \leq E_3(x)m(x), \qquad (5.7.68)$$

where $E_3(x)$ is defined by (5.7.56). From the definition of $m(x)$ and (5.7.68) we have

$$u(x) \leq E_3(x)\left[f(x)+W\left(x,\int_{x^0}^{x} k(x,y,u(y))\,dy\right)\right].$$

Now a suitable application of Theorem 5.7.5 yields the desired bound in (5.7.55).

■

In concluding this section we note that there is no essential difficulty in establishing the further generalizations of the integral inequalities established by Pachpatte (1979a, 1980b,c,d) and many other theorems given in earlier sections in the set-up of Theorem 5.7.9. Since the details of these results are very close to those given in the proof of Theorem 5.7.9 with suitable modifications, they are left to the reader to fill in where needed.

5.8 Pachpatte's Inequalities III

Pachpatte (1996c, 1997, in press d, unpublished manuscript) investigated some useful integral inequalities in two independent variables which claim their origin in the inequalities given in Theorems 3.4.1 and 3.4.3 respectively by Ou-Iang (1957) and Dafermos (1979). In this section we shall give some basic inequalities investigated in Pachpatte (1996c, 1997, in press d, unpublished manuscript) which can play a vital role in the study of certain new classes of partial differential and integral equations. Here we shall use the notation and definitions as given in Section 4.5 without further mention.

The following inequality established by Pachpatte (in press d) deals with the two independent variable generalization of the inequality given by Ou-Iang in Theorem 4.3.1.

Theorem 5.8.1 *Let $u(x, y)$ and $p(x, y)$ be nonnegative continuous functions defined for $x, y \in R_+$. If*

$$u^2(x, y) \leq c^2 + 2 \int_0^x \int_0^y p(s, t) u(s, t) \, ds \, dt, \quad (5.8.1)$$

for $x, y \in R_+$, where $c \geq 0$ is a constant, then

$$u(x, y) \leq c + \int_0^x \int_0^y p(s, t) \, dt \, ds, \quad (5.8.2)$$

for $x, y \in R_+$.

□

Proof: It is sufficient to assume that c is positive, since the standard limiting argument can be used to treat the remaining case. Let $c > 0$ and define a function $z(x, y)$ by the right-hand side of (5.8.1), then

528 MULTIDIMENSIONAL NONLINEAR INTEGRAL INEQUALITIES

$z(x, 0) = z(0, y) = c^2$, $u(x, y) \leq \sqrt{z(x, y)}$ and

$$\frac{D_2 D_1 z(x, y)}{\sqrt{z(x, y)}} \leq 2p(x, y). \tag{5.8.3}$$

From (5.8.3) and the facts that $\sqrt{z(x, y)} > 0$, $D_1 z(x, y) \geq 0$, $D_2 z(x, y) \geq 0$ for $x, y \in R_+$, we observe that

$$\frac{D_2 D_1 z(x, y)}{\sqrt{z(x, y)}} \leq 2p(x, y) + \frac{[D_1 z(x, y)][D_2 \sqrt{z(x, y)}]}{[\sqrt{z(x, y)}]^2},$$

i.e.

$$D_2 \left(\frac{D_1 z(x, y)}{\sqrt{z(x, y)}} \right) \leq 2p(x, y). \tag{5.8.4}$$

By keeping x fixed in (5.8.4), setting $y = t$ and integrating with respect to t from 0 to y, and using the fact that $D_1 z(x, 0) = 0$, we have

$$\frac{D_1 z(x, y)}{\sqrt{z(x, y)}} \leq 2 \int_0^y p(x, t) \, dt. \tag{5.8.5}$$

Now keeping y fixed in (5.8.5) and setting $x = s$ and integrating with respect to s from 0 to x we have

$$\sqrt{z(x, y)} \leq c + \int_0^x \int_0^y p(s, t) \, dt \, ds. \tag{5.8.6}$$

Using (5.8.6) in $u(x, y) \leq \sqrt{z(x, y)}$, we get the required inequality in (5.8.2).

∎

Another useful integral inequality involving functions of two independent variables established by Pachpatte (in press d) is given in the following theorem.

Theorem 5.8.2 *Let $u(x, y)$, $p(x, y)$ and c be as in Theorem 5.8.1. Let $g(u)$ be a continuously differentiable function defined on R_+ and $g(u) > 0$ for $u > 0$, $g'(u) \geq 0$ for $u \geq 0$. If*

$$u^2(x, y) \leq c^2 + 2 \int_0^x \int_0^y p(s, t) u(s, t) g(u(s, t)) \, dt \, ds, \tag{5.8.7}$$

for $x, y \in R_+$, then for $0 \leq x \leq x_1$, $0 \leq y \leq y_1$,

$$u(x, y) \leq \Omega^{-1}\left[\Omega(c) + \int_0^x \int_0^y p(s, t)\, dt\, ds\right], \qquad (5.8.8)$$

where

$$\Omega(r) = \int_{r_0}^r \frac{ds}{g(s)}, \qquad r > 0, r_0 > 0, \qquad (5.8.9)$$

Ω^{-1} is the inverse of Ω and x_1, y_1 are chosen so that

$$\Omega(c) + \int_0^x \int_0^y p(s, t)\, dt\, ds \in \text{Dom}(\Omega^{-1}),$$

for all x, y lying in the subintervals $0 \leq x \leq x_1$, $0 \leq y \leq y_1$ of R_+.

$\square$

Proof: Since $g'(u) \geq 0$ on R_+, the function g is monotonically increasing on $(0, \infty)$. As in the proof of Theorem 5.8.1, we assume that $c > 0$ and define a function $z(x, y)$ by the right-hand side of (5.8.7); then $z(x, 0) = z(0, y) = c^2$, $u(x, y) \leq \sqrt{z(x, y)}$ and it is easy to observe that

$$\frac{D_2 D_1 z(x, y)}{\sqrt{z(x, y)}} \leq 2 p(x, y) g(\sqrt{z(x, y)}). \qquad (5.8.10)$$

Now by following the same arguments as in the proof of Theorem 5.8.1 below (5.8.3) up to (5.8.6), from (5.8.10) we get

$$\sqrt{z(x, y)} \leq c + \int_0^x \int_0^y p(s, t) g(\sqrt{z(s, t)})\, dt\, ds. \qquad (5.8.11)$$

Now a suitable application of Theorem 5.2.1 to (5.8.11) yields

$$\sqrt{z(x, y)} \leq \Omega^{-1}\left[\Omega(c) + \int_0^x \int_0^y p(s, t)\, dt\, ds\right]. \qquad (5.8.12)$$

Using (5.8.12) in $u(x, y) \leq \sqrt{z(x, y)}$, we get the required inequality in (5.8.8). The subintervals for x and y are obvious.

∎

The following theorem deals with fairly general inequalities established by Pachpatte (1997) which can be used in some applications.

Theorem 5.8.3 *Let $u(x, y)$, $f(x, y)$, $g(x, y)$ and $h(x, y)$ be nonnegative continuous functions defined for $x, y \in R_+$ and c be a nonnegative constant.*

(a_1) *If*

$$u^2(x, y) \leq c^2 + 2 \int_0^x \int_0^y \left[f(s, t)u(s, t) \left(u(s, t) + \int_0^s \int_0^t g(s_1, t_1) \right. \right.$$

$$\left. \left. \times u(s_1, t_1) \, dt_1 \, ds_1 \right) + h(s, t)u(s, t) \right] dt \, ds, \quad (5.8.13)$$

for $x, y \in R_+$, then

$$u(x, y) \leq p(x, y) \left[1 + \int_0^x \int_0^y \left[f(s, t) \exp \left(\int_0^s \int_0^t [f(s_1, t_1) \right. \right. \right.$$

$$\left. \left. \left. + g(s_1, t_1)] \right) dt_1 \, ds_1 \right] dt \, ds \right], \quad (5.8.14)$$

for $x, y \in R_+$, where

$$p(x, y) = c + \int_0^x \int_0^y h(s, t) \, dt \, ds, \quad (5.8.15)$$

for $x, y \in R_+$.

(a_2) *If*

$$u^2(x, y) \leq c^2 + 2 \int_0^x \int_0^y \left[f(s, t)u(s, t) \left(\int_0^s \int_0^t g(s_1, t_1)u(s_1, t_1) \, dt_1 \, ds_1 \right) \right.$$

$$\left. + h(s, t)u(s, t) \right] dt \, ds, \quad (5.8.16)$$

for $x, y \in R_+$, then

$$u(x, y) \le p(x, y) \exp\left(\int_0^x \int_0^y f(s, t) \left(\int_0^s \int_0^t g(s_1, t_1) \, dt_1 ds_1\right) dt \, ds\right),$$
(5.8.17)

for $x, y \in R_+$, where $p(x, y)$ is defined by (5.8.15).

□

Proof: (a_1) We first assume that $c > 0$ and define a function $z(x, y)$ by the right-hand side of (5.8.13). Then $z(x, 0) = z(0, y) = c^2$, $u(x, y) \le \sqrt{z(x, y)}$ and

$$D_2 D_1 z(x, y) \le 2\sqrt{z(x, y)} \left[f(x, y) \left(\sqrt{z(x, y)} + \int_0^x \int_0^y g(s_1, t_1) \sqrt{z(s_1, t_1)} \, dt_1 \, ds_1 \right) + h(x, y) \right]. \quad (5.8.18)$$

From (5.8.18) and using the facts that $\sqrt{z(x, y)} > 0$, $D_1 z(x, y) \ge 0$, $D_2 z(x, y) \ge 0$ for $x, y \in R_+$, we observe that

$$D_2\left(\frac{D_1 z(x, y)}{\sqrt{z(x, y)}}\right) \le 2 \left[f(x, y) \left(\sqrt{z(x, y)} + \int_0^x \int_0^y g(s_1, t_1) \sqrt{z(s_1, t_1)} \, dt_1 \, ds_1 \right) + h(x, y) \right]. \quad (5.8.19)$$

Now by following the same arguments as in the proof of Theorem 5.8.1 below (5.8.3) up to (5.8.6), from (5.8.19) we get

$$\sqrt{z(x, y)} \le p(x, y) + \int_0^x \int_0^y f(s, t) \left(\sqrt{z(s, t)} + \int_0^s \int_0^t g(s_1, t_1) \sqrt{z(s_1, t_1)} \, dt_1 \, ds_1 \right) dt \, ds, \quad (5.8.20)$$

where $p(x, y)$ is defined by (5.8.15). Since $p(x, y)$ is positive and monotonic nondecreasing in $x, y \in R_+$, by applying Theorem 4.4.2, we get

$$\sqrt{z(x, y)} \leq p(x, y) \left[1 + \int_0^x \int_0^y f(s, t) \exp\left(\int_0^s \int_0^t [f(s_1, t_1) + g(s_1, t_1)] \, dt_1 \, ds_1 \right) dt \, ds \right]. \quad (5.8.21)$$

Now by using (5.8.21) in $u(x, y) \leq \sqrt{z(x, y)}$ we get the required inequality in (5.8.14).

If $c \geq 0$ we carry out the above procedure with $c + \epsilon$ instead of c, where $\epsilon > 0$ is an arbitrary small constant, and subsequently pass to the limit as $\epsilon \to 0$ to obtain (5.8.14).

(a_2) The proof follows by the same arguments as those given in the proof of (a_1) above with suitable modifications, and hence the details are omitted here. ∎

A generalization of Theorem 5.8.1 in the other direction given by Pachpatte (unpublished manuscript) is contained in the following theorem.

Theorem 5.8.4 *Let $u(x, y)$, $f(x, y)$ and $g(x, y)$ be nonnegative continuous functions defined for $x, y \in R_+$ and c be a nonnegative constant. Let $L: R_+^3 \to R_+$ be a continuous function which satisfies the condition*

$$0 \leq L(x, y, v) - L(x, y, w) \leq k(x, y, w)(v - w), \quad (L)$$

for $x, y \in R_+$ and $v \geq w \geq 0$, where $k: R_+^3 \to R_+$ is a continuous function. If

$$u^2(x, y) \leq c^2 + 2 \int_0^x \int_0^y [f(s, t)u(s, t)L(s, t, u(s, t)) + g(s, t)u(s, t)] \, dt \, ds \quad (5.8.22)$$

for $x, y \in R_+$, then

$$u(x, y) \leq p(x, y) + q(x, y) \exp\left(\int_0^x \int_0^y f(s, t)k(s, t, p(s, t)) \, dt \, ds \right), \quad (5.8.23)$$

for $x, y \in R_+$, where

$$p(x, y) = c + \int_0^x \int_0^y g(s, t) \, dt \, ds, \qquad (5.8.24)$$

$$q(x, y) = \int_0^x \int_0^y f(s, t) L(s, t, p(s, t)) \, dt \, ds, \qquad (5.8.25)$$

for $x, y \in R_+$.

$\square$

Proof: It is sufficient to assume that c is positive, since the standard argument can be used to treat the remaining case. Let $c > 0$ and define a function $z(x, y)$ by the right-hand side of (5.8.22), then $z(x, 0) = z(0, y) = c^2$, $u(x, y) \leq \sqrt{z(x, y)}$ and

$$\frac{D_2 D_1 z(x, y)}{\sqrt{z(x, y)}} \leq [f(x, y) L(x, y, \sqrt{z(x, y)}) + g(x, y)]. \qquad (5.8.26)$$

Now by following the same arguments as in the proof of Theorem 5.8.1 below (5.8.3) up to (5.8.6) we get

$$\sqrt{z(x, y)} \leq p(x, y) + \int_0^x \int_0^y f(s, t) L(s, t, \sqrt{z(s, t)}) \, dt \, ds. \qquad (5.8.27)$$

Define $v(x, y)$ by

$$v(x, y) = \int_0^x \int_0^y f(s, t) L(s, t, \sqrt{z(s, t)}) \, dt \, ds. \qquad (5.8.28)$$

From (5.8.28) and using the fact that $\sqrt{z(x, y)} \leq p(x, y) + v(x, y)$ and the hypothesis (L) we observe that

$$v(x, y) \leq \int_0^x \int_0^y f(s, t) L(s, t, p(s, t) + v(s, t)) \, dt \, ds$$

$$= q(x, y) + \int_0^x \int_0^y f(s, t) \{L(s, t, p(s, t)$$

$$+ v(s, t)) - L(s, t, p(s, t))\} \, dt \, ds$$

$$\leq q(x, y) + \int_0^x \int_0^y f(s, t) k(s, t, p(s, t)) v(s, t) \, dt \, ds. \quad (5.8.29)$$

Clearly $q(x, y)$ is nonnegative and monotonic nondecreasing for $x, y \in R_+$. Now an application of Theorem 4.2.2 to (5.8.29) yields

$$v(x, y) \leq q(x, y) \exp\left(\int_0^x \int_0^y f(s, t) k(s, t, p(s, t)) \, dt \, ds\right). \quad (5.8.30)$$

The desired inequality in (5.8.23) now follows by using (5.8.30) in (5.8.27) and the fact that $u(x, y) \leq \sqrt{z(x, y)}$.

∎

Pachpatte (1996c) has established the inequalities in the following theorem which can be used more effectively in certain applications.

Theorem 5.8.5 *Let $u(x, y)$, $f(x, y)$ and $h(x, y)$ be nonnegative continuous functions defined for $x, y \in R_+$ and c be a nonnegative constant.*
(b_1) *If*

$$u^2(x, y) \leq c^2 + 2B[x, y, f(s, t)u^2(s, t) + h(s, t)u(s, t)]. \quad (5.8.31)$$

for $x, y \in R_+$, then

$$u(x, y) \leq p(x, y) \exp(B[x, y, f(s, t)]), \quad (5.8.32)$$

for $x, y \in R_+$, where

$$p(x, y) = c + B[x, y, h(s, t)], \quad (5.8.33)$$

for $x, y \in R_+$.

(b_2) *Let $g(u)$ be a continuously differentiable function defined on R_+ and $g(u) > 0$ for $u > 0$, $g'(u) \geq 0$ for $u \geq 0$. If*

$$u^2(x, y) \leq c^2 + 2B[x, y, f(s, t)u(s, t)g(u(s, t))$$
$$+ h(s, t)u(s, t)], \quad (5.8.34)$$

for $x, y \in R_+$, then for $0 \leq x \leq x_1$, $0 \leq y \leq y_1$,

$$u(x, y) \leq \Omega^{-1}[\Omega(p(x, y)) + B[x, y, f(s, t)]], \quad (5.8.35)$$

where $p(x, y)$ is defined by (5.8.33) and Ω, Ω^{-1} are as defined in Theorem 5.8.2 and x_1, y_1 are chosen so that

$$\Omega(p(x, y)) + B[x, y, f(s, t)] \in \text{Dom}(\Omega^{-1}),$$

for all x, y lying in the subintervals $0 \leq x \leq x_1$, $0 \leq y \leq y_1$ of R_+.

(b$_3$) Let L and k be as in Theorem 5.8.4. If

$$u^2(x, y) \leq c^2 + 2B[x, y, f(s, t)u(s, t)L(s, t, u(s, t)) + h(s, t)u(s, t)], \tag{5.8.36}$$

for $x, y \in R_+$, then

$$u(x, y) \leq p(x, y) + q(x, y)\exp(B[x, y, f(s, t)k(s, t, p(s, t))]), \tag{5.8.37}$$

for $x, y \in R_+$, where $p(x, y)$ is defined by (5.8.33) and

$$q(x, y) = B[x, y, f(s, t)L(s, t, p(s, t))], \tag{5.8.38}$$

for $x, y \in R_+$. $\square$

Proof: (b$_1$) Assume that $c > 0$ and define a function $z(x, y)$ by the right-hand side of (5.8.31); then $u(x, y) \leq \sqrt{z(x, y)}$ and

$$\frac{D_2^m D_1^n z(x, y)}{\sqrt{z(x, y)}} \leq 2[f(x, y)\sqrt{z(x, y)} + h(x, y)]. \tag{5.8.39}$$

From (5.8.39) and using the facts that $z(x, y) > 0$, $D_2 z(x, y) \geq 0$, $D_2^{m-1} D_1^n z(x, y) \geq 0$, for $x, y \in R_+$, we observe that (see Theorem 4.5.4)

$$D_2\left(\frac{D_2^{m-1} D_1^n z(x, y)}{\sqrt{z(x, y)}}\right) \leq 2[f(x, y)\sqrt{z(x, y)} + h(x, y)]. \tag{5.8.40}$$

Now by following the same arguments as in the proof of Theorem 4.5.4, below the inequality (4.5.48) up to (4.5.53), with suitable modifications, we get

$$\frac{D_1 z(x, y)}{\sqrt{z(x, y)}} \leq 2 \int_0^x \int_0^{s_{n-2}} \cdots \int_0^{s_1} \int_0^y \int_0^{t_{m-1}} \cdots \int_0^{t_1} [f(s, t)\sqrt{z(s, t)} + h(s, t)] \, dt \, dt_1 \ldots dt_{m-1} \, ds \, ds_1 \ldots ds_{n-2}, \tag{5.8.41}$$

Now keeping y fixed in (5.8.41), setting $x = s_{n-1}$ and then integrating with respect to s_{n-1} from 0 to x and using the fact that $\sqrt{z(0, y)} = c$, we have

$$\sqrt{z(x, y)} \leq p(x, y) + B[x, y, f(s, t)\sqrt{z(s, t)}], \quad (5.8.42)$$

where $p(x, y)$ is defined by (5.8.33). Clearly $p(x, y)$ is positive and monotonic nondecreasing in $x, y \in R_+$. By applying Theorem 4.5.4 part (i) we get

$$\sqrt{z(x, y)} \leq p(x, y) \exp(B[x, y, f(s, t)]). \quad (5.8.43)$$

Now using (5.8.43) in $u(x, y) \leq \sqrt{z(x, y)}$, we get the required inequality in (5.8.32). If $c \geq 0$, we carry out the above procedure with $c + \epsilon$ instead of c, where $\epsilon > 0$ is an arbitrary small constant, and subsequently pass to the limit as $\epsilon \to 0$ to obtain (5.8.32).

(b_2) Assume that $c > 0$ and define a function $z(x, y)$ by the right-hand side of (5.8.34), then $u(x, y) \leq \sqrt{z(x, y)}$ and

$$\frac{D_2^m D_1^n z(x, y)}{\sqrt{z(x, y)}} \leq 2[f(x, y)g(\sqrt{z(x, y)}) + h(x, y)]. \quad (5.8.44)$$

Now by following the same steps as in the proof of part (b_1) up to (5.8.42) we get

$$\sqrt{z(x, y)} \leq p(x, y) + B[x, y, f(s, t)g(\sqrt{z(s, t)})], \quad (5.8.45)$$

where $p(x, y)$ is defined by (5.8.33). The rest of the proof can be completed by following the ideas used in the proof of Theorem 5.2.2 and Theorem 4.5.4 with suitable changes. Here we omit the further details.

(b_3) Assume that $c > 0$ and define a function $z(x, y)$ by the right-hand side of (5.8.36), then $u(x, y) \leq \sqrt{z(x, y)}$ and

$$\frac{D_2^m D_1^n z(x, y)}{\sqrt{z(x, y)}} \leq 2[f(x, y)L(x, y, \sqrt{z(x, y)}) + h(x, y)]. \quad (5.8.46)$$

Now by following the same steps as in the proof of part (b_1) up to (5.8.42) we get

$$\sqrt{z(x, y)} \leq p(x, y) + B[x, y, f(s, t)L(s, t, \sqrt{z(s, t)})], \quad (5.8.47)$$

where $p(x, y)$ is defined by (5.8.33). Define a function $v(x, y)$ by

$$v(x, y) = B[x, y, f(s, t)L(s, t, \sqrt{z(s, t)})]. \quad (5.8.48)$$

From (5.8.48) and using the fact that $\sqrt{z(x,y)} \leq p(x,y) + v(x,y)$, and the hypothesis (L) we have

$$v(x,y) \leq B[x, y, f(s,t)L(s,t, p(s,t) + v(s,t))]$$
$$= q(x,y) + B[x, y, f(s,t)\{L(s,t, p(s,t) + v(s,t))$$
$$- L(s,t, p(s,t))\}]$$
$$\leq q(x,y) + B[x, y, f(s,t)k(s,t, p(s,t))v(s,t)], \quad (5.8.49)$$

where $q(x,y)$ is defined by (5.8.38). Clearly $q(x,y)$ is nonnegative and monotonic nondecreasing in each variable $x, y \in R_+$. Now an application of Theorem 4.5.4 part (i) yields

$$v(x,y) \leq q(x,y) \exp(B[x, y, f(s,t)k(s,t, p(s,t))]). \quad (5.8.50)$$

Using (5.8.50) in (5.8.47) and then $u(x,y) \leq \sqrt{z(x,y)}$, we get the desired inequality in (5.8.37). The proof of the case when c is nonnegative can be completed as mentioned in the proof of part (b_1). ∎

5.9 Pachpatte's Inequalities IV

Inspired by the study of certain new classes of differential and integral equations, Pachpatte (1994a,b) established some useful integral inequalities in one and more than one independent variables. It seems that the inequalities given in Pachpatte (1994a,b) will quite likely be a useful source for future work. In this section, we offer some basic inequalities given in Pachpatte (1994a,b) which claims their origin to be in the inequality given by Ou-Iang (1957). This discussion uses the notation and definitions as given in Section 4.5 and also the following notation and definitions without futher mention.

Let R_+^n denote the product $R_+ \times \cdots \times R_+$ (n times). A point $(x_1, \ldots, x_n)$ in R_+^n is denoted by x and we denote $dx_n \ldots dx_1$ by dx. We define the operators I_j recursively by $I_0 u(x) = u(x)$, $I_j u(x) = b_j(x) D_j u(x)$, $j = 1, 2, \ldots, n$ with $b_n(x) = 1$, where $u(x)$ and $b_j(x)$ are some functions defined for $x \in R_+^n$ and $D_j = \partial/\partial x_j$. For $x \in R_+^n$ and some functions $b_j(x) > 0$, $j = 1, 2, \ldots, n-1$, $n \geq 2$ an integer, and $k(x)$ defined for $x \in R_+^n$, we set

$$M[x, b, k(y)] = M[x_1, \ldots, x_n, b_1, \ldots, b_{n-1}, k(y)]$$

$$= \int_0^{x_1} \frac{1}{b_1(y_1, x_2, \ldots, x_n)} \cdots \int_0^{x_{n-1}} \frac{1}{b_{n-1}(y_1, \ldots, y_{n-1}, x_n)}$$

$$\times \int_0^{x_n} k(y)\,dy.$$

Pachpatte (1994a) investigated the following useful inequality.

Theorem 5.9.1 *Let $F(x, y)$ and $g(x, y)$ be nonnegative continuous functions defined for $x, y \in R_+$ and let $p > 1$ be a constant. If*

$$F^p(x, y) \leq c + B[x, y, g(s, t)F(s, t)], \tag{5.9.1}$$

for $x, y \in R_+$, where $c \geq 0$ is a constant, then

$$F(x, y) \leq [c^{(p-1)/p} + ((p-1)/p)B[x, y, g(s, t)]]^{1/(p-1)}, \tag{5.9.2}$$

for $x, y \in R_+$.

□

Proof: We first assume that $c > 0$ and define a function $z(x, y)$ by the right-hand side of (5.9.1), then $F(x, y) \leq \sqrt[p]{z(x, y)}$ and

$$D_2^m D_1^n z(x, y) \leq g(x, y) \sqrt[p]{z(x, y)}. \tag{5.9.3}$$

From (5.9.3) and the facts that $z(x, y)$ is positive and $D_2(\sqrt[p]{z(x, y)})$ and $D_2^{m-1}D_1^n z(x, y)$ are nonnegative for $x, y \in R_+$, we observe that

$$\frac{D_2^m D_1^n z(x, y)}{\sqrt[p]{z(x, y)}} \leq g(x, y) + \frac{[D_2(\sqrt[p]{z(x, y)})][D_2^{m-1}D_1^n z(x, y)]}{[\sqrt[p]{z(x, y)}]^2},$$

i.e.

$$D_2\left(\frac{D_2^{m-1}D_1^n z(x, y)}{\sqrt[p]{z(x, y)}}\right) \leq g(x, y). \tag{5.9.4}$$

By keeping x fixed in (5.9.4), we set $y = t$ and then, by integrating with respect to t from 0 to y and using the fact that $D_2^{m-1}D_1^n z(x, 0) = 0$, $m = 2, 3, \ldots$, we have

$$\frac{D_2^{m-1}D_1^n z(x, y)}{\sqrt[p]{z(x, y)}} \leq \int_0^y g(x, t)\,dt. \tag{5.9.5}$$

MULTIDIMENSIONAL NONLINEAR INTEGRAL INEQUALITIES 539

Again as above, from (5.9.5) and the facts that $z(x, y)$ is positive and $D_2(\sqrt[p]{z(x, y)})$ and $D_2^{m-2}D_1^n z(x, y)$ are nonnegative for $x, y \in R_+$, we observe that

$$D_2\left(\frac{D_2^{m-2}D_1^n z(x, y)}{\sqrt[p]{z(x, y)}}\right) \leq \int_0^y g(x, t)\, dt. \tag{5.9.6}$$

By keeping x fixed in (5.9.6), setting $y = t_1$, then integrating with respect to t_1 from 0 to y and using the fact that $D_2^{m-2}D_1^n z(x, 0) = 0$, $m = 3, 4, \ldots$, we have

$$\frac{D_2^{m-2}D_1^n z(x, y)}{\sqrt[p]{z(x, y)}} \leq \int_0^y \int_0^{t_1} g(x, t)\, dt\, dt_1.$$

Continuing in this way we obtain

$$\frac{D_1^n z(x, y)}{\sqrt[p]{z(x, y)}} \leq \int_0^y \int_0^{t_{m-1}} \cdots \int_0^{t_1} g(x, t)\, dt\, dt_1 \cdots dt_{m-1}. \tag{5.9.7}$$

From (5.9.7) and the facts that $z(x, y)$ is positive and $D_1(\sqrt[p]{z(x, y)})$ and $D_1^{n-1}z(x, y)$ are nonnegative for $x, y \in R_+$, we observe that

$$\frac{D_1^n z(x, y)}{\sqrt[p]{z(x, y)}} \leq \int_0^y \int_0^{t_{m-1}} \cdots \int_0^{t_1} g(x, t)\, dt\, dt_1 \cdots dt_{m-1}$$

$$+ \frac{[D_1(\sqrt[p]{z(x, y)})][D_1^{n-1}z(x, y)]}{[\sqrt[p]{z(x, y)}]^2},$$

i.e.

$$D_1\left(\frac{D_1^{n-1}z(x, y)}{\sqrt[p]{z(x, y)}}\right) \leq \int_0^y \int_0^{t_{m-1}} \cdots \int_0^{t_1} g(x, t)\, dt\, dt_1 \cdots dt_{m-1}. \tag{5.9.8}$$

By keeping y fixed in (5.9.8), setting $x = s$, then integrating with respect to s from 0 to x and using the fact that $D_1^{n-1}z(0, y) = 0$, for $n = 2, 3, \ldots$, we have

$$\frac{D_1^{n-1}z(x, y)}{\sqrt[p]{z(x, y)}} \leq \int_0^x \int_0^y \int_0^{t_{m-1}} \cdots \int_0^{t_1} g(x, t)\, dt\, dt_1 \cdots dt_{m-1}\, ds.$$

Continuing in this way, we obtain

$$\frac{D_1 z(x, y)}{\sqrt[p]{z(x, y)}} \le \int_0^x \int_0^{s_{n-2}} \cdots \int_0^{s_1} \int_0^y \int_0^{t_{m-1}} \cdots \int_0^{t_1} g(s, t) \, dt \, dt_1 \ldots dt_{m-1} \, ds \, ds_1 \ldots ds_{n-2}.$$
(5.9.9)

Now by keeping y fixed in (5.9.9), setting $x = s_{n-1}$, then integrating with respect to s_{n-1} from 0 to x and using the fact that $z(0, y) = c$, we have

$$(\sqrt[p]{z(x, y)})^{p-1} - (\sqrt[p]{c})^{p-1} \le ((p-1)/p) B[x, y, g(s, t)]. \quad (5.9.10)$$

From (5.9.10) and using the fact $F(x, y) \le \sqrt[p]{z(x, y)}$, we observe that

$$F(x, y) \le [c^{(p-1)/p} + ((p-1)/p) B[x, y, g(s, t)]]^{1/(p-1)}.$$

This is the required inequality in (5.9.2). If $c \ge 0$, we carry out the above procedure with $c + \epsilon$ instead of c, where $\epsilon > 0$ is an arbitrary small constant, and subsequently pass to the limit $\epsilon \to 0$ to obtain (5.9.2).

∎

Another useful inequality established by Pachpatte (1994a) is given in the following theorem.

Theorem 5.9.2 *Let* $u(x, y)$, $v(x, y)$, $h_i(x, y)$ *for* $i = 1, 2, 3, 4$ *be nonnegative continuous functions defined for* $x, y \in R_+$ *and let* $p > 1$ *be a constant. If* c_1, c_2 *and* μ *are nonnegative constants such that*

$$u^p(x, y) \le c_1 + B[x, y, h_1(s, t) u(s, t)] + B[x, y, h_2(s, t) \bar{v}(s, t)],$$
(5.9.11)

$$v^p(x, y) \le c_2 + B[x, y, h_3(s, t) \bar{u}(s, t)] + B[x, y, h_4(s, t) v(s, t)],$$
(5.9.12)

for $x, y \in R_+$, *where* $\bar{u}(x, y) = \exp(-p\mu(x + y)) u(x, y)$ *and* $\bar{v}(x, y) = \exp(p\mu(x + y)) v(x, y)$ *for* $x, y \in R_+$, *then*

$$u(x, y) \le \exp(\mu(x + y)) [(2^{p-1}(c_1 + c_1))^{(p-1)/p} + 2^{p-1}((p-1)/p)$$
$$\times B[x, y, h(s, t)]]^{1/(p-1)}, \quad (5.9.13)$$

$$v(x, y) \le [(2^{p-1}(c_1 + c_2))^{(p-1)/p} + 2^{p-1}((p-1)/p)$$
$$\times B[x, y, h(s, t)]]^{1/(p-1)}, \quad (5.9.14)$$

for $x, y \in R_+$, where

$$h(x, y) = \max\{[h_1(x, y) + h_3(x, y)], [h_2(x, y) + h_4(x, y)]\}, \quad (5.9.15)$$

for $x, y \in R_+$.

$\square$

Proof: We first multiply (5.9.11) by $\exp(-p\mu(x + y))$ and observe that

$$\exp(-p\mu(x + y))u^p(x, y) \leq c_1 + B[x, y, h_1(s, t)\overline{u}(s, t)]$$
$$+ B[x, y, h_2(s, t)v(s, t)]. \quad (5.9.16)$$

Define

$$F(x, y) = \exp(-\mu(x + y))u(x, y) + v(x, y). \quad (5.9.17)$$

By taking the pth power on both sides of (5.9.17) and using the elementary inequality

$$(a_1 + a_2)^r \leq 2^{r-1}(a_1^r + a_2^r),$$

(where $a_1, a_2 \geq 0$ are reals and $r > 1$) and (5.9.16), (5.9.12) we observe that

$$F^p(x, y) \leq 2^{p-1}[\exp(-p\mu(x + y))u^p(x, y) + v^p(x, y)]$$
$$\leq 2^{p-1}[c_1 + c_2 + B[x, y, [h_1(s, t) + h_3(s, t)]\overline{u}(s, t)]$$
$$+ B[x, y, [h_2(s, t) + h_4(s, t)]v(s, t)]]. \quad (5.9.18)$$

Now using the fact that $\exp(-p\mu(x + y)) \leq \exp(-\mu(x + y))$ and (5.9.15) in (5.9.18) we observe that

$$F^p(x, y) \leq 2^{p-1}(c_1 + c_2) + B[x, y, 2^{p-1}h(s, t)F(s, t)]. \quad (5.9.19)$$

The bounds in (5.9.13) and (5.9.14) follow from an application of Theorem 5.9.1 to (5.9.19) and splitting.

∎

The inequalities in the following two theorems involving functions of n independent variables given by Pachpatte (1994a) can be used in the study of certain new classes of partial differential and integral equations.

Theorem 5.9.3 *Let $F(x) \geq 0$, $k(x) \geq 0$ and $b_i(x) > 0$ for $i = 1, 2, \ldots, n-1$, be continuous functions defined for $x \in R_+^n$ and let $p > 1$ be a constant. If*

$$F^p(x) \leq c + M[x, b, k(y)F(y)], \quad (5.9.20)$$

for $x \in R_+^n$, where $c \geq 0$ is a constant, then

$$F(x) \leq [c^{(p-1)/p} + ((p-1)/p)M[x, b, k(y)]]^{1/(p-1)}, \qquad (5.9.21)$$

for $x \in R_+^n$.

□

Theorem 5.9.4 Let $u(x) \geq 0$, $v(x) \geq 0$, $b_i(x) > 0$, $i = 1, 2, \ldots, n-1$, and $h_j(x) \geq 0$, $i = 1, 2, 3, 4$, be continuous functions defined for $x \in R_+^n$ and let $p > 1$ be a constant. If c_1, c_2 and μ are nonnegative constants such that

$$u^p(x) \leq c_1 + M[x, b, h_1(y)u(y)] + M[x, b, h_2(y)\bar{v}(y)], \qquad (5.9.22)$$

$$v^p(x) \leq c_2 + M[x, b, h_3(y)\bar{u}(y)] + M[x, b, h_4(y)v(y)], \qquad (5.9.23)$$

for $x \in R_+^n$, where $\bar{u}(x) = \exp(-p\mu \sum_{i=1}^n x_i)u(x)$ for $x \in R_+^n$ and $\bar{v}(x) = \exp(p\mu \sum_{i=1}^n x_i)v(x)$ for $x \in R_+^n$, then

$$u(x) \leq \exp\left(\mu \sum_{i=1}^n x_i\right)[(2^{p-1}(c_1 + c_2))^{(p-1)/p}$$
$$+ 2^{p-1}((p-1)/p)M[x, b, h(y)]]^{1/(p-1)}. \qquad (5.9.24)$$

$$v(x) \leq [(2^{p-1}(c_1 + c_2))^{(p-1)/p}$$
$$+ 2^{p-1}((p-1)/p)M[x, b, h(y)]]^{1/(p-1)}, \qquad (5.9.25)$$

for $x \in R_+^n$, where

$$h(x) = \max\{[h_1(x) + h_3(x)], \quad [h_2(x) + h_4(x)]\}, \qquad (5.9.26)$$

for $x \in R_+^n$.

□

The proofs of Theorems 5.9.3 and 5.9.4 proceed much as the proofs of the theorems given above and follow by closely looking at the proofs of the main results given in Sections 5.7 and 3.10. The details are left to the reader.

The following theorem, which is a slight variant of the results established by Pachpatte (1994b) involving functions of n independent variables, can be proved by using similar arguments to those mentioned above.

Theorem 5.9.5 Let $u(x)$, $f(x)$ and $h(x)$ be nonnegative continuous functions defined for $x \in R_+^n$ and c be a nonnegative constant. Let $b_j(x) > 0$, $j = 1, 2, \ldots, n-1$, $n \geq 2$ an integer, be continuous functions defined for $x \in R_+^n$.

(i) If
$$u^2(x) \le c^2 + 2M[x, b, f(y)u^2(y) + h(y)u(y)], \qquad (5.9.27)$$
for $x \in R_+^n$, then
$$u(x) \le p(x)\exp(M[x, b, f(y)]), \qquad (5.9.28)$$
for $x \in R_+^n$, where
$$p(x) = c + M[x, b, h(y)], \qquad (5.9.29)$$
for $x \in R_+^n$.

(ii) Let $g(u)$, $g'(u)$ be as in Theorem 5.8.2. If
$$u^2(x) \le c^2 + 2M[x, b, f(y)u(y)g(u(y)) + h(y)u(y)], \qquad (5.9.30)$$
for $x \in R_+^n$, then for $0 \le x_i \le x_i^*$, $i = 1, 2, \ldots, n$,
$$u(x) \le \Omega^{-1}[\Omega(p(x)) + M[x, b, f(y)]], \qquad (5.9.31)$$
where $p(x)$ is defined by (5.9.29), Ω, Ω^{-1} are as defined in Theorem 5.8.2 and $x \in R_+^n$ be chosen so that
$$\Omega(p(x)) + M[x, b, f(y)] \in \mathrm{Dom}(\Omega^{-1}),$$
for all x_i lying in the subintervals $0 \le x_i \le x_i^*$, $i = 1, 2, \ldots, n$ of R_+.

(iii) Let $L: R_+^n \times R_+ \to R_+$ be a continuous function satisfying the condition
$$0 \le L(x, v) - L(x, w) \le k(x, w)(v - w), \qquad (5.9.32)$$
for $x \in R_+^n$ and $v \ge w \ge 0$, where $k: R_+^n \times R_+ \to R_+$ is a continuous function. If
$$u^2(x) \le c^2 + 2M[x, b, f(y)u(y)L(y, u(y)) + h(y)u(y)], \qquad (5.9.33)$$
for $x \in R_+^n$, then
$$u(x) \le p(x) + q(x)\exp(M[x, b, f(y)k(y, p(y))]), \qquad (5.9.34)$$
for $x \in R_+^n$, where $p(x)$ is defined in (5.9.29) and
$$q(x) = M[x, b, f(y)L(y, p(y))], \qquad (5.9.35)$$
for $x \in R_+^n$.

□

5.10 Pachpatte's Inequalities V

Pachpatte (1991, 1992, 1994d) investigated a number of new inequalities involving functions of two independent variables which claim their origins in the inequalities used by Haraux (1981) and Engler (1989). This section presents some basic inequalities given in Pachpatte (1991, 1992, 1994d) which can be used as tools in the study of certain new classes of partial differential and integral equations. We shall use the same notation and definitions as given in Section 4.5 without further mention.

Pachpatte (1994d) investigated the inequalities given in the following two theorems.

Theorem 5.10.1 *Let $u: E \to R_1$, $p: E \to R_+$ be continuous functions and $c \geq 1$ be a constant, where $E = [0, \alpha] \times [0, \beta]$, $\alpha > 0$, $\beta > 0$, $R_1 = [1, \infty)$ and $R_+ = [0, \infty)$. If*

$$u(x, y) \leq c + \int_0^x \int_0^y p(s, t) u(s, t) \log u(s, t) \, dt \, ds, \tag{5.10.1}$$

for $(x, y) \in E$, then

$$u(x, y) \leq c^{Q(x, y)}, \tag{5.10.2}$$

for $(x, y) \in E$, where

$$Q(x, y) = \exp\left(\int_0^x \int_0^y p(s, t) \, dt \, ds\right), \tag{5.10.3}$$

for $(x, y) \in E$.

□

Proof: Define a function $v(x, y)$ by the right-hand side of (5.10.1), then $u(x, y) \leq v(x, y)$ and

$$D_2 D_1 v(x, y) = p(x, y) u(x, y) \log u(x, y)$$

$$\leq p(x, y) v(x, y) \log v(x, y). \tag{5.10.4}$$

From (5.10.4) and the facts that $v(x, y) > 0$, $D_1 v(x, y) \geq 0$, $D_2 v(x, y) \geq 0$ for $(x, y) \in E$, we observe that

$$\frac{D_2 D_1 v(x, y)}{v(x, y)} \leq p(x, y) \log v(x, y) + \frac{D_1 v(x, y) D_2 v(x, y)}{[v(x, y)]^2},$$

i.e.
$$D_2\left(\frac{D_1 v(x, y)}{v(x, y)}\right) \leq p(x, y) \log v(x, y). \tag{5.10.5}$$

By keeping x fixed in (5.10.5), setting $y = t$, integrating from 0 to y and using the fact that $D_1 v(x, 0) = 0$, we get

$$\frac{D_1 v(x, y)}{v(x, y)} \leq \int_0^y p(x, t) \log v(x, t) \, dt, \tag{5.10.6}$$

Keeping y fixed in (5.10.6) and setting $x = s$, integrating from 0 to x and using the fact that $v(0, y) = c$, we obtain

$$\log v(x, y) \leq \log c + \int_0^x \int_0^y p(s, t) \log v(s, t) \, dt \, ds. \tag{5.10.7}$$

Now a suitable application of the inequality given in Theorem 4.2.1 (see Remark 4.2.1) yields

$$\log v(x, y) \leq (\log c) \exp\left(\int_0^x \int_0^y p(s, t) \, dt \, ds\right)$$

$$= \log c^{Q(x, y)}. \tag{5.10.8}$$

From (5.10.8) we observe that

$$v(x, y) \leq c^{Q(x, y)}. \tag{5.10.9}$$

Now using (5.10.9) in $u(x, y) \leq v(x, y)$ we get the required inequality in (5.10.2).

∎

Theorem 5.10.2 *Let* $u: E \to R_1$, $p: E \to R_+$ *be continuous functions and* $c \geq 1$ *be a constant, where* E *and* R_1 *are as defined in Theorem 5.10.1. Let* $g(u)$ *be a continuously differentiable function defined on* R_+ *and* $g(u) > 0$ *on* $(0, \infty)$ *and* $g'(u) \geq 0$ *on* R_+. *If*

$$u(x, y) \leq c + \int_0^x \int_0^y p(s, t) u(s, t) g(\log u(s, t)) \, dt \, ds, \tag{5.10.10}$$

for $(x, y) \in E$, then for $(x, y) \in E_1 \subset E$,

$$u(x, y) \leq \exp\left(\Omega^{-1}\left[\Omega(\log c) + \int_0^x \int_0^y p(s, t)\, dt\, ds\right]\right), \quad (5.10.11)$$

where

$$\Omega(r) = \int_{r_0}^r \frac{ds}{g(s)}, \quad r > 0,\ r_0 > 0, \quad (5.10.12)$$

Ω^{-1} is the inverse of Ω and $(x, y) \in E_1 \subset E$ is chosen so that

$$\Omega(\log c) + \int_0^x \int_0^y p(s, t)\, dt\, ds \in \text{Dom}(\Omega^{-1}),$$

for $(x, y) \in E_1 \subset E$.

□

Proof: Since $g'(u) \geq 0$ on R_+, the function $g(u)$ is monotonically increasing on $(0, \infty)$. Define a function $v(x, y)$ by the right-hand side of (5.10.10). By following the arguments as in the proof of Theorem 5.10.1 up to the inequality (5.10.7) we obtain

$$\log v(x, y) \leq \log c + \int_0^x \int_0^y p(s, t) g(\log v(s, t))\, dt\, ds. \quad (5.10.13)$$

Now a suitable application of Theorem 5.2.1 yields

$$\log v(x, y) \leq \Omega^{-1}\left[\Omega(\log c) + \int_0^x \int_0^y p(s, t)\, dt\, ds\right]. \quad (5.10.14)$$

From (5.10.14) we observe that

$$v(x, y) \leq \exp\left(\Omega^{-1}\left[\Omega(\log c) + \int_0^x \int_0^y p(s, t)\, dt\, ds\right]\right). \quad (5.10.15)$$

The desired inequality in (5.10.11) now follows by using (5.10.15) in $u(x, y) \leq v(x, y)$. The subdomain E_1 of (x, y) in E is obvious.

■

Another interesting and useful inequality is embodied in the following theorem.

Theorem 5.10.3 *Let* $u\colon E \to R_1$, $p, q\colon E \to R_+$ *be continuous functions and* $c \geq 1$ *be a constant, where* E *and* R_1 *are as defined in Theorem 5.10.1. Let* $g(u)$, $g'(u)$ *be as defined in Theorem 5.10.2 and furthermore we assume that* $g(u)$ *is submultiplicative on* R_+. *If*

$$u(x, y) \leq c + \int_0^x \int_0^y u(s, t) \left[p(s, t) \log u(s, t) \right.$$

$$\left. + q(s, t) g(\log u(s, t)) \right] dt\, ds, \qquad (5.10.16)$$

for $(x, y) \in E$, *then for* $(x, y) \in E_2 \subset E$,

$$u(x, y) \leq \exp\left[Q(x, y) \left\{ \Omega^{-1} \left[\Omega(\log c) + \int_0^x \int_0^y q(s, t) g(Q(s, t))\, dt\, ds \right] \right\} \right],$$
$$(5.10.17)$$

where $Q(x, y)$ *is defined by (5.10.3),* Ω, Ω^{-1} *are as defined in Theorem 5.10.2 and* $(x, y) \in E_2 \subset E$ *is chosen so that*

$$\Omega(\log c) + \int_0^x \int_0^y q(s, t) g(Q(s, t))\, dt\, ds \in \mathrm{Dom}(\Omega^{-1}),$$

for $(x, y) \in E_2 \subset E$.

□

Proof: Since $g'(u) \geq 0$ on R_+, the function $g(u)$ is monotonically increasing on $(0, \infty)$. Define a function $v(x, y)$ by the right-hand side of (5.10.16). Then by following the arguments as in the proof of Theorem 5.10.1 up to the inequality (5.10.7) we have

$$\log v(x, y) \leq \log c + \int_0^x \int_0^y p(s, t) \log v(s, t)\, dt\, ds$$

$$+ \int_0^x \int_0^y q(s, t) g(\log v(s, t))\, dt\, ds. \qquad (5.10.18)$$

Define a function $m(x, y)$ by

$$m(x, y) = \log c + \int_0^x \int_0^y q(s, t) g(\log v(s, t)) \, dt \, ds. \qquad (5.10.19)$$

Then (5.10.16) can be restated as

$$\log v(x, y) \leq m(x, y) + \int_0^x \int_0^y p(s, t) \log v(s, t) \, dt \, ds. \qquad (5.10.20)$$

Since $m(x, y)$ is positive and nondecreasing in both the variables x and y, by applying Theorem 4.2.2 to (5.10.20) we have

$$\log v(x, y) \leq Q(x, y) m(x, y), \qquad (5.10.21)$$

where $Q(x, y)$ is defined by (5.10.3). From (5.10.19) and (5.10.21) we observe that

$$m(x, y) \leq \log c + \int_0^x \int_0^y q(s, t) g(Q(s, t) m(s, t)) \, dt \, ds$$

$$\leq \log c + \int_0^x \int_0^y q(s, t) g(Q(s, t)) g(m(s, t)) \, dt \, ds. \qquad (5.10.22)$$

Now a suitable application of Theorem 5.2.1 to (5.10.22) yields

$$m(x, y) \leq \Omega^{-1} \left[\Omega(\log c) + \int_0^x \int_0^y q(s, t) g(Q(s, t)) \, dt \, ds \right]. \qquad (5.10.23)$$

Using (5.10.23) in (5.10.21) we get

$$\log v(x, y) \leq Q(x, y) \left\{ \Omega^{-1} \left[\Omega(\log c) + \int_0^x \int_0^y q(s, t) g(Q(s, t)) \, dt \, ds \right] \right\}. \qquad (5.10.24)$$

From (5.10.24) we get

$$v(x, y) \leq \exp\left[Q(x, y)\left\{\Omega^{-1}\left[\Omega(\log c)\right.\right.\right.$$
$$\left.\left.\left.+ \int_0^x \int_0^y q(s, t) g(Q(s, t))\, dt\, ds\right]\right\}\right]. \quad (5.10.25)$$

Using (5.10.25) in $u(x, y) \leq v(x, y)$ we get the desired inequality in (5.10.17). The subdomain E_2 of (x, y) in E is obvious.

∎

The inequalities in the following theorems are also established by Pachpatte (1994d) and can be used in some applications.

Theorem 5.10.4 *Let $u: E \to R_1$, $p, q: E \to R_+$ be continuous functions and $c \geq 1$ be a constant, where E and R_1 are as defined in Theorem 5.10.1. If*

$$u(x, y) \leq c + \int_0^x \int_0^y p(s, t) u(s, t) \left[\log u(s, t) + \int_0^s \int_0^t q(s_1, t_1)\right.$$
$$\left. \times \log u(s_1, t_1)\, dt_1\, ds_1\right] dt\, ds, \quad (5.10.26)$$

for $(x, y) \in E$, then

$$u(x, y) \leq c^{Q_0(x, y)}, \quad (5.10.27)$$

for $(x, y) \in E$, where

$$Q_0(x, y) = 1 + \int_0^x \int_0^y p(s, t) \exp\left(\int_0^s \int_0^t [p(s_1, t_1) + q(s_1, t_1)]\, dt_1\, ds_1\right) dt\, ds, \quad (5.10.28)$$

for $(x, y) \in E$.

□

Theorem 5.10.5 *Let $u(x, y)$, $p(x, y)$, c, $g(u)$ and $g'(u)$ be as in Theorem 5.10.2. If*

$$u(x, y) \leq c + \int_0^x \int_0^y p(s, t) u(s, t) \left[\log u(s, t) + \int_0^s \int_0^t p(s_1, t_1) \right.$$

$$\left. \times g(\log u(s_1, t_1)) \, dt_1 \, ds_1 \right] dt \, ds, \qquad (5.10.29)$$

for $(x, y) \in E$, then for $(x, y) \in E_3 \subset E$,

$$u(x, y) \leq c \exp \left[\int_0^x \int_0^y p(s, t) G^{-1} \left[G(\log c) \right. \right.$$

$$\left. \left. + \int_0^s \int_0^t p(s_1, t_1) \, dt_1 \, ds_1 \right] dt \, ds \right], \qquad (5.10.30)$$

where

$$G(r) = \int_{r_0}^r \frac{ds}{s + g(s)}, \quad r > 0, \, r_0 > 0, \qquad (5.10.31)$$

G^{-1} is the inverse function of G and $(x, y) \in E_3 \subset E$ is chosen so that

$$G(\log c) + \int_0^x \int_0^y p(s_1, t_1) \, dt_1 \, ds_1 \in \text{Dom}(G^{-1}),$$

for $(x, y) \in E_3 \subset E$.

□

The proofs of Theorems 5.10.4 and 5.10.5 follow by the same arguments as in the proofs of Theorems 5.10.1 and 5.10.2, using Theorems 4.4.1 and 5.4.1 respectively, with suitable modifications. The details are omitted.

The following theorems established by Pachpatte (1991) deals with the two independent variable generalizations of the inequalities given in Theorems 3.9.4 and 3.9.6.

Theorem 5.10.6 *Let $u: E \to R_1$, $p: E \to (0, \infty)$, $q: E \to R_+$ be continuous functions and $c \geq 1$ be a constant, where E and R_1 are as defined in*

Theorem 5.10.1. *If*

$$u(x, y) \le c + \int_0^x \int_0^y p(s, t) u(s, t)$$
$$\times \left(\int_0^s \int_0^t q(s_1, t_1) \log u(s_1, t_1) \, dt_1 \, ds_1 \right) dt \, ds \quad (5.10.32)$$

for $(x, y) \in E$, *then*

$$u(x, y) \le c^{Q_1(x, y)}, \quad (5.10.33)$$

for $(x, y) \in E$, *where*

$$Q_1(x, y) = \exp\left(\int_0^x \int_0^y p(s, t) \left(\int_0^s \int_0^t q(s_1, t_1) \, dt_1 \, ds_1 \right) dt \, ds \right), \quad (5.10.34)$$

for $(x, y) \in E$.

□

Theorem 5.10.7 *Let* $u(x, y), p(x, y), q(x, y)$ *and* c *be as in Theorem 5.10.6. Let* $g(u), g'(u)$ *be as in Theorem 5.10.2. If*

$$u(x, y) \le c + \int_0^x \int_0^y p(s, t) u(s, t)$$
$$\times \left(\int_0^s \int_0^t q(s_1, t_1) g(\log u(s_1, t_1)) \, dt_1 \, ds_1 \right) dt \, ds, \quad (5.10.35)$$

for $(x, y) \in E$, *then for* $(x, y) \in E_4 \subset E$,

$$u(x, y) \le \exp\left[\Omega^{-1} \left[\Omega(\log c) \right.\right.$$
$$\left.\left. + \int_0^x \int_0^y p(s, t) \left(\int_0^s \int_0^t q(s_1, t_1) \, dt_1 \, ds_1 \right) dt \, ds \right] \right], \quad (5.10.36)$$

where Ω, Ω^{-1} are as defined in Theorem 5.10.2 and $(x, y) \in E_4 \subset E$ is chosen so that

$$\Omega(\log c) + \int_0^x \int_0^y p(s, t) \left(\int_0^s \int_0^t q(s_1, t_1) \, dt_1 \, ds_1 \right) dt \, ds \in \mathrm{Dom}(\Omega^{-1}),$$

for $(x, y) \in E_4 \subset E$.

□

Proofs of Theorems 5.10.6 and 5.10.7: The details of the proof of Theorem 5.10.7 only are given; the proof of Theorem 5.10.6 can be completed similarly.

Since $g'(u) \geq 0$ on R_+, the function $g(u)$ is monotonically increasing on $(0, \infty)$. Define a function $z(x, y)$ by the right-hand side of (5.10.35). Then $u(x, y) \leq z(x, y)$ and

$$D_2 D_1 z(x, y) \leq p(x, y) z(x, y)$$

$$\times \left(\int_0^x \int_0^y q(s_1, t_1) g(\log z(s_1, t_1)) \, dt_1 \, ds_1 \right). \quad (5.10.37)$$

From (5.10.37) and using the facts that $z(x, y) > 0, D_1 z(x, y) \geq 0$, $D_2 z(x, y) \geq 0$, we observe that

$$D_2 \left(\frac{D_1 z(x, y)}{z(x, y)} \right) \leq p(x, y) \left(\int_0^x \int_0^y q(s_1, t_1) g(\log z(s_1, t_1)) \, dt_1 \, ds_1 \right).$$

(5.10.38)

By keeping x fixed in (5.10.38), setting $y = t$, integrating from 0 to y and using the fact that $D_1 z(x, 0) = 0$ we get

$$\frac{D_1 z(x, y)}{z(x, y)} \leq \int_0^y p(x, t) \left(\int_0^x \int_0^t q(s_1, t_1) g(\log z(s_1, t_1)) \, dt_1 \, ds_1 \right) dt.$$

(5.10.39)

Now keeping y fixed in (5.10.39), setting $x = s$, integrating from 0 to x and using the fact that $z(0, y) = c$, we obtain

$$\log z(x, y) \leq \log c + \int_0^x \int_0^y p(s, t)$$

$$\times \left(\int_0^s \int_0^t q(s_1, t_1) g(\log z(s_1, t_1)) \, dt_1 \, ds_1 \right) dt \, ds. \quad (5.10.40)$$

Define a function $v(x, y)$ by the right-hand side of (5.10.40). Then it is easy to observe that

$$D_2 D_1 \left(\frac{1}{p(x, y)} D_2 D_1 v(x, y) \right) = q(x, y) g(\log z(x, y))$$

$$\leq q(x, y) g(v(x, y)). \quad (5.10.41)$$

From (5.10.41) and using the facts that $g'(u) \geq 0$, $D_2 v(x, y) \geq 0$, $D_1((1/p(x, y))D_2 D_1 v(x, y)) \geq 0$, we observe that

$$D_2 \left(\frac{D_1((1/p(x, y))D_2 D_1 v(x, y))}{g(v(x, y))} \right) \leq q(x, y). \quad (5.10.42)$$

By keeping x fixed in (5.10.42), setting $y = t_1$, integrating from 0 to y and using the fact that $D_1((1/p(x, 0))D_2 D_1 v(x, 0)) = 0$, we obtain

$$\frac{D_1((1/p(x, y))D_2 D_1 v(x, y))}{g(v(x, y))} \leq \int_0^y q(x, t_1) \, dt_1. \quad (5.10.43)$$

As above from (5.10.43) we observe that

$$D_1 \left(\frac{(1/p(x, y))D_2 D_1 v(x, y)}{g(v(x, y))} \right) \leq \int_0^y q(x, t_1) \, dt_1. \quad (5.10.44)$$

By keeping y fixed in (5.10.44), setting $x = s_1$, integrating from 0 to x and using the fact that $(1/p(0, y))D_2 D_1 v(0, y) = 0$, we have

$$\frac{D_2 D_1 v(x, y)}{g(v(x, y))} \leq p(x, y) \left(\int_0^x \int_0^y q(s_1, t_1) \, dt_1 \, ds_1 \right). \quad (5.10.45)$$

Again as above from (5.10.45) we observe that

$$D_2\left(\frac{D_1 v(x,y)}{g(v(x,y))}\right) \leq p(x,y)\left(\int_0^x \int_0^y q(s_1,t_1)\,dt_1\,ds_1\right). \quad (5.10.46)$$

Keeping x fixed in (5.10.46), setting $y = t$, integrating from 0 to y and using the fact that $D_1 v(x, 0) = 0$, we have

$$\frac{D_1 v(x,y)}{g(v(x,y))} \leq \int_0^y p(x,t)\left(\int_0^x \int_0^t q(s_1,t_1)\,dt_1\,ds_1\right)dt. \quad (5.10.47)$$

From (5.10.12) and (5.10.47) we observe that

$$\frac{\partial}{\partial x}\Omega(v(x,y)) = \frac{D_1 v(x,y)}{g(v(x,y))} \leq \int_0^y p(x,t)\left(\int_0^x \int_0^t q(s_1,t_1)\,dt_1\,ds_1\right)dt. \quad (5.10.48)$$

Now keeping y fixed in (5.10.48), setting $x = s$, integrating from 0 to x and using the fact that $v(0, y) = \log c$, we have

$$\Omega(v(x,y)) \leq \Omega(\log c) + \int_0^x \int_0^y p(s,t)\left(\int_0^s \int_0^t q(s_1,t_1)\,dt_1\,ds_1\right)dt\,ds. \quad (5.10.49)$$

Using the bound on $v(x, y)$ from (5.10.49) in (5.10.40) we have

$$\log z(x,y) \leq \Omega^{-1}\left[\Omega(\log c) + \int_0^x \int_0^y p(s,t)\left(\int_0^s \int_0^t q(s_1,t_1)\,dt_1\,ds_1\right)dt\,ds\right]. \quad (5.10.50)$$

From (5.10.50) we get

$$z(x,y) \leq \exp\left[\Omega^{-1}\left[\Omega(\log c) + \int_0^x \int_0^y p(s,t)\left(\int_0^s \int_0^t q(s_1,t_1)\,dt_1\,ds_1\right)dt\,ds\right]\right]. \quad (5.10.51)$$

The desired inequality in (5.10.36) now follows by using (5.10.51) in $u(x, y) \leq z(x, y)$. The subdomain E_4 of (x, y) in E is obvious.

∎

Pachpatte (1992) established the inequalities given in the following theorems, which can be used in certain applications.

Theorem 5.10.8 *Let* $u: E \to R_1, h: E \to R_+$ *be continuous functions and* $c \geq 1$ *be a constant, where* E *and* R_1 *are as defined in Theorem 5.10.1. If*

$$u(x, y) \leq c + B[x, y, h(s, t)u(s, t) \log u(s, t)], \qquad (5.10.52)$$

for $(x, y) \in E$, *then*

$$u(x, y) \leq c^{\exp(B[x, y, h(s,t)])}, \qquad (5.10.53)$$

for $(x, y) \in E$.

□

Theorem 5.10.9 *Let* $u: E \to R_1, h: E \to R_+$ *be continuous functions and* $c \geq 1$ *be a constant, where* E *and* R_1 *are as defined in Theorem 5.10.1. Let* $g(u), g'(u)$ *be as in Theorem 5.10.2. If*

$$u(x, y) \leq c + B[x, y, h(s, t)u(s, t)g(\log u(s, t))]. \qquad (5.10.54)$$

for $(x, y) \in E$, *then for* $(x, y) \in E_0 \subset E$,

$$u(x, y) \leq \exp(\Omega^{-1}[\Omega(\log c) + B[x, y, h(s, t)]]), \qquad (5.10.55)$$

where Ω, Ω^{-1} *are as defined in Theorem 5.10.2 and* $(x, y) \in E_0 \subset E$ *is chosen so that*

$$\Omega(\log c) + B[x, y, h(s, t)] \in \text{Dom}(\Omega^{-1}),$$

for $(x, y) \in E_0 \subset E$.

□

The proofs of Theorems 5.10.8 and 5.10.9 can be completed by following the ideas used in the proof of Theorem 5.10.7 and Theorem 5.8.5 and looking closely at the proof of Theorem 4.5.4, with suitable changes. The details are omitted here.

In concluding this section, we note that the inequalities established in Theorems 5.10.1–5.10.9 can be extended very easily to $n \geq 3$ independent variables. The precise formulations of these results are very close to those of the results mentioned above, with suitable modifications. It is left to the reader to fill in the details where needed.

We note that the inequalities given in Chapters 4 and 5 are recently established and still admit various generalizations and extensions in different directions. For various other inequalities related to the inequalities given in Chapters 4 and 5 and their applications, the interested reader is referred to Apartsin and Men (1979), Beesack (1985), Cheung (1993), Conlan and Daiz (1963), Corduneanu (1982, 1983, 1987), DeFranco (1976), Fink (1981), Gutowski (1978), Westphal (1949), Yang (1984, 1985) and Young (1982).

5.11 Applications

In this section we present applications of some of the inequalities given in earlier sections to study the qualitative behaviour of the solutions of certain partial differential and integro-differential equations. Most of the inequalities given here are recently investigated and can be used as tools in the study of certain new classes of partial differential, integral and integro-differential equations.

5.11.1 Hyperbolic Partial Differential Equations

In this section we first present an application of Theorem 5.2.4 to obtain bounds on the solutions of a nonlinear hyperbolic partial differential equation of the form

$$u_{xy}(x, y) = (b(x, y)u(x, y))_y + f(x, y, u(x, y)), \qquad (5.11.1)$$

with the boundary conditions

$$u(x, 0) = \sigma(x), \quad u(0, y) = \tau(y), \quad \sigma(0) = \tau(0), \qquad (5.11.2)$$

where $\sigma, \tau: R_+ \to R$, $b: R_+^2 \to R$, $f: R_+^2 \times R \to R$ are continuous functions and

$$|b(x, y)| \le c(x, y), \qquad (5.11.3)$$

$$|f(x, y, u)| \le p(x, y)g(|u|), \qquad (5.11.4)$$

where $c(x, y)$, $p(x, y)$, $g(r)$ are as defined in Theorem 5.2.4. It is easy to observe that the problem (5.11.1)–(5.11.2) is equivalent to the integral equation

$$u(x, y) = a(x, y) + \int_0^x b(s, y)u(s, y)\,ds$$
$$+ \int_0^x \int_0^y f(s, t, u(s, t))\,ds\,dt, \qquad (5.11.5)$$

where $u(x, y)$ is a solution of (5.11.1) with (5.11.2),

$$a(x, y) = \sigma(x) + \tau(y) - \sigma(0) - \int_0^x b(s, 0)\sigma(s)\,ds.$$

We assume that

$$|a(x, y)| \le k, \qquad (5.11.6)$$

where $k \ge 0$ is a constant. Using (5.11.3), (5.11.4) and (5.11.6) in (5.11.5) we have

$$|u(x, y)| \le k + \int_0^x c(s, y)|u(s, y)|\,ds + \int_0^x \int_0^y p(s, t)g(|u(s, t)|)\,ds\,dt. \qquad (5.11.7)$$

Now a suitable application of Theorem 5.2.4 to (5.11.7) yields

$$|u(x, y)| \le F(x, y)\left[\Omega^{-1}\left[\Omega(k) + \int_0^x \int_0^y p(s, t)g(F(s, t))\,ds\,dt\right]\right], \qquad (5.11.8)$$

where $F(x, y)$, Ω, Ω^{-1} are as defined in Theorem 5.2.4. If the right-hand side in (5.11.8) is bounded then we obtain the boundedness of the solutions of (5.11.1)–(5.11.2).

We next apply Theorem 5.5.3 to obtain the bound on the solution of a nonlinear fourth-order partial differential equation of the form

$$u_{xxyy}(x, y) + (a(x, y)u(x, y))_{yy} = F(x, y, u(x, y)), \qquad (5.11.9)$$

with the given boundary conditions

$$u(x, 0) = \phi_0(x), \quad u_y(x, 0) = \phi_1(x),$$
$$u(0, y) = \psi_0(y), \quad u_x(0, y) = \psi_1(y), \quad (5.11.10)$$

where $a: R_+^2 \to R$, $F: R_+^2 \times R \to R$ are continuous functions and $\phi_0(x)$, $\psi_0(y)$, $\phi_1(x)$, $\psi_1(y)$ are real-valued twice continuously differentiable functions for $x, y \in R_+$ and

$$\phi_0(0) = \psi_0(0), \quad \phi_1(0) = \psi_0'(0), \quad \phi_0'(0) = \psi_1(0), \quad \phi_1'(0) = \psi_1'(0). \quad (5.11.11)$$

It is easy to observe that the problem (5.11.9)–(5.11.11) is equivalent to the integral equation

$$u(x, y) = k(x, y) - \int_0^x \int_0^s a(s_1, y) u(s_1, y) \, ds_1 \, ds$$
$$+ A[x, y, F(s_1, t_1, u(s_1, t_1))], \quad (5.11.12)$$

where $u(x, y)$ is a solution of (5.11.9) with (5.11.10)–(5.11.11) and the notation $A[x, y, F]$ is defined as in Theorem 5.5.3,

$$k(x, y) = \phi_0(x) + \psi_0(y) + x\psi_1(y) + y\phi_1(x) - \phi_0(0) - \phi_0'(0)x$$
$$- y\phi_1(0) - xy\phi_1'(0) + \int_0^x \int_0^s a(s_1, 0)\phi_0(s_1) \, ds_1 \, ds. \quad (5.11.13)$$

We assume that

$$|k(x, y)| \leq c, \quad (5.11.14)$$
$$|a(x, y)| \leq f(x, y), \quad (5.11.15)$$
$$|F(x, y, u)| \leq p(x, y) g(|u|), \quad (5.11.16)$$

where c, $f(x, y)$, $p(x, y)$, $g(r)$ are as defined in Theorem 5.5.3. Using (5.11.14)–(5.11.16) in (5.11.12) we have

$$|u(x, y)| \leq c + \int_0^x \int_0^s f(s_1, y) |u(s_1, y)| \, ds_1 \, ds$$
$$+ A[x, y, p(s_1, t_1) g(|u(s_1, t_1)|)]. \quad (5.11.17)$$

Now an application of Theorem 5.5.3 yields

$$|u(x, y)| \leq Q(x, y)\{\Omega^{-1}[\Omega(c) + A[x, y, p(s_1, t_1)g(Q(s_1, t_1))]]\}, \quad (5.11.18)$$

where $Q(x, y)$, Ω, Ω^{-1} are as defined in Theorem 5.5.3. If the right-hand side of (5.11.18) is bounded, then we obtain the bound on the solutions of (5.11.9)–(5.11.11). For similar applications, see Pachpatte (1979a, 1980d, 1988c)

5.11.2 Hyperbolic Partial Integro-differential Equations

This section presents an application of Theorem 5.4.2 to obtain bounds on the solutions of a nonlinear hyperbolic partial integro-differential equation of the form

$$u_{xy}(x, y) = f\left(x, y, u(x, y), \int_0^x \int_0^y k(x, y, s, t, u(s, t)) \, ds \, dt\right), \quad (5.11.19)$$

with the given boundary conditions

$$u(x, 0) = \sigma(x), \quad u(0, y) = \tau(y), \quad u(0, 0) = 0, \quad (5.11.20)$$

where $\sigma, \tau: R_+ \to R$, $k: R_+^4 \times R \to R$, $f: R_+^2 \times R \times R \to R$ are continuous functions. The problem (5.11.19)–(5.11.20) is equivalent to the integral equation

$$u(x, y) = \sigma(x) + \tau(y) + \int_0^x \int_0^y f\left(s, t, u(s, t),\right.$$

$$\left.\int_0^s \int_0^t k(s, t, s_1, t_1, u(s_1, t_1)) \, ds_1 \, dt_1\right) ds \, dt, \quad (5.11.21)$$

where $u(x, y)$ is a solution of (5.11.19) with (5.11.20). We assume that

$$|\sigma(x) + \tau(y)| \leq a(x, y), \quad (5.11.22)$$

$$|k(x, y, s, t, u)| \leq b(s, t)g(|u|), \quad (5.11.23)$$

$$|f(x, y, u, v)| \leq b(x, y)[|u| + |v|], \quad (5.11.24)$$

where $a(x, y)$, $b(x, y)$, $g(r)$ are as defined in Theorem 5.4.2. Using (5.11.22)–(5.11.24) in (5.11.21) we have

$$|u(x, y)| \leq a(x, y) + \int_0^x \int_0^y b(s, t) \left(|u(s, t)| \right.$$
$$\left. + \int_0^s \int_0^t b(s_1, t_1) g(|u(s_1, t_1)|) \, ds_1 \, dt_1 \right) ds \, dt. \quad (5.11.25)$$

Now an application of Theorem 5.4.2 to (5.11.25) yields

$$|u(x, y)| \leq a(x, y) + A(x, y) + \int_0^x \int_0^y b(s, t) \left\{ H^{-1} \left[H(A(s, t)) \right. \right.$$
$$\left. \left. + \int_0^s \int_0^t b(s_1, t_1) \, ds_1 \, dt_1 \right] \right\} ds \, dt, \quad (5.11.26)$$

where $A(x, y)$, H, H^{-1} are as defined in Theorem 5.4.2. The right-hand side of (5.11.26) gives the bound on the solution $u(x, y)$ of (5.11.19)–(5.11.20).

The inequalities established in Theorems 5.5.1 and 5.5.2 can be used to study the problems of boundedness of the solutions of nonlinear fourth-order partial differential and integro-differential equations of the forms

$$u_{xxyy}(x, y) = f(x, y, u(x, y)), \quad (5.11.27)$$

and

$$u_{xxyy}(x, y) = f\left(x, y, u(x, y), \right.$$
$$\left. \int_0^x \int_0^{s_2} \int_0^y \int_0^{t_2} k(x, y, s_3, t_3, u(s_3, t_3)) \, dt_3 \, dt_2 \, ds_3 \, ds_2 \right) \quad (5.11.28)$$

respectively with the given boundary conditions (5.11.10)–(5.11.11), and under some suitable conditions on the functions involved in (5.11.27) and (5.11.28). For more details, the reader is referred to Pachpatte (1979c, 1988c).

5.11.3 Higher Order Hyperbolic Partial Differential Equations

In this section we present an application of Theorem 5.10.8 given by Pachpatte (1992), to obtain the bound on the solution of the following higher order hyperbolic partial differential equation of the form

$$D_2^m D_1^n u(x, y) = h(x, y) u(x, y) \log |u(x, y)|, \tag{5.11.29}$$

with the given boundary conditions

$$\frac{\partial^j u(x, 0)}{\partial y^j} = \alpha_j(x), \quad 0 \le j \le m - 1, \tag{5.11.30}$$

$$\frac{\partial^i u(0, y)}{\partial x^i} = \beta_i(y), \quad 0 \le i \le n - 1, \tag{5.11.31}$$

where $h \in C[R_+^2, R]$, $\alpha_j \in C^{(n)}[R_+, R]$, $\beta_i \in C^{(m)}[R_+, R]$ and

$$\alpha_j^{(i)}(0) = \beta_i^{(j)}, \quad 0 \le j \le m - 1, \quad 0 \le i \le n - 1. \tag{5.11.32}$$

It is easy to observe that the problem (5.11.29)–(5.11.32) is equivalent to the integral equation

$$u(x, y) = q(x, y) + B[x, y, h(s, t) u(s, t) \log |u(s, t)|], \tag{5.11.33}$$

where $u(x, y)$ is a solution of (5.11.29) with (5.11.30)–(5.11.32), the notation $B[x, y, f]$ is defined as in Theorem 5.10.8 and

$$q(x, y) = \sum_{i=1}^{n} \frac{x^{i-1}}{(i-1)!} \beta_{i-1}(y) + \sum_{j=1}^{m} \frac{y^{j-1}}{(j-1)!} \alpha_{j-1}(x)$$

$$- \sum_{i=1}^{n} \frac{x^{i-1}}{(i-1)!} \sum_{j=1}^{m} \frac{y^{j-1}}{(j-1)!} \alpha_{j-1}^{(i-1)}(0). \tag{5.11.34}$$

From (5.11.33) we observe that

$$|u(x, y)| \le |q(x, y)| + B[x, y, |h(s, t)||u(s, t)||\log |u(s, t)||]. \tag{5.11.35}$$

From (5.11.35) we observe that

$$1 + |u(x, y)| \le 1 + |q(x, y)| + B[x, y, |h(s, t)|(1 + |u(s, t)|)$$
$$\times \log(1 + |u(s, t)|)]. \tag{5.11.36}$$

We assume that
$$1 + |q(x, y)| \leq c, \tag{5.11.37}$$
where $c \geq 1$ is a constant. Using (5.11.37) in (5.11.36) we have
$$1 + |u(x, y)| \leq c + B[x, y, |h(s, t)|(1 + |u(s, t)|)$$
$$\times \log(1 + |u(s, t)|)]. \tag{5.11.38}$$

Now an application of Theorem 5.10.8 to (5.11.38) yields
$$1 + |u(x, y)| \leq c^{\exp(B[x, y, |h(s,t)|])}. \tag{5.11.39}$$

From (5.11.39) we observe that
$$|u(x, y)| \leq [c^{\exp(B[x, y, |h(s,t)|])} - 1], \tag{5.11.40}$$
for $(x, y) \in E$, where E is defined as in Theorem 5.10.8. The inequality (5.11.40) gives the bound on the solution $u(x, y)$ of (5.11.29)–(5.11.32) in terms of the known functions.

5.11.4 Multivariate Hyperbolic Partial Integro-differential Equations

In this section we present applications of Theorem 5.7.6 to the boundedness and behavioural relationships of the solutions of some nonlinear hyperbolic partial integro-differential equations given by Pachpatte (1981c).

As a first application, we obtain a bound on the solution of a nonlinear hyperbolic partial integro-differential equation of the form

$$D_1 \ldots D_n u(x) = A\left(x, u(x), \int_{x^0}^{x} B(x, y, u(y)) dy\right) + F(x, u(x)), \tag{5.11.41}$$

with the conditions prescribed on $x_i = x_i^0$, $1 \leq i \leq n$, where $F: \Omega \times R \to R$, $B: \Omega^2 \times R \to R$, $A: \Omega \times R^2 \to R$ are continuous functions. We assume that

$$|B(x, y, u)| \leq c(y)|u|, \tag{5.11.42}$$

$$|A(x, u, v)| \leq g(x)(|u| + |v|), \tag{5.11.43}$$

$$|F(x, u)| \leq k(x, |u|), \tag{5.11.44}$$

MULTIDIMENSIONAL NONLINEAR INTEGRAL INEQUALITIES 563

where $c(y), g(x), k(x, y, \phi) = k(y, \phi)$ are as defined in Theorem 5.7.6. Let the boundary conditions be such that the given equation (5.11.41) is equivalent to the integral equation

$$u(x) = h(x) + \int_{x^0}^{x} A\left(y, u(y), \int_{x^0}^{y} B(y, s, u(s))\, ds\right) dy + \int_{x^0}^{x} F(y, u(y))\, dy, \tag{5.11.45}$$

where $h(x)$ depends on the given boundary conditions. We assume that

$$|h(x)| \leq f(x), \tag{5.11.46}$$

were $f(x)$ is defined as in Theorem 5.7.6. Using (5.11.42)–(5.11.44) and (5.11.46) in (5.11.45) we have

$$|u(x)| \leq f(x) + \int_{x^0}^{x} g(y)|u(y)|\, dy + \int_{x^0}^{x} g(y)\left(\int_{x^0}^{y} c(s)|u(s)|\, ds\right) dy$$

$$+ \int_{x^0}^{x} k(y, |u(y)|)\, dy. \tag{5.11.47}$$

Now a suitable application of Theorem 5.7.6 with $q(x) = 1$, $w(x, \phi) = \phi$ and $k(x, y, \phi) = k(y, \phi)$ yields

$$|u(x)| \leq E_0^*(x)[f(x) + r(x)], \tag{5.11.48}$$

where

$$E_0^*(x) = 1 + \int_{x^0}^{x} g(y) \exp\left(\int_{x^0}^{y} [g(s) + c(s)]\, ds\right) dy, \tag{5.11.49}$$

and $r(x)$ is a solution of the equation

$$r(x) = \int_{x^0}^{x} k(y, E_0^*(y)[f(y) + r(y)])\, dy.$$

If the right-hand side in (5.11.48) is bounded then we obtain the boundedness of the solution $u(x)$ of (5.11.41).

Our second application deals with the behavioural relationship between the solutions of (5.11.41) with the conditions prescribed on $x_i = x_i^0$,

$1 \leq i \leq n$, and the nonlinear hyperbolic partial integro-differential equation of the form

$$D_1 \ldots D_n v(x) = A_0 \left(x, v(x), \int_{x^0}^{x} B_0(x, y, v(y)) \, dy \right), \qquad (5.11.50)$$

with the conditions prescribed on $x_i = x_i^0$, $1 \leq i \leq n$, where $B: \Omega^2 \times R \to R$, $A_0: \Omega \times R^2 \to R$ are continuous functions. We assume that

$$|B(x, y, u) - B_0(x, y, v)| \leq c(y)|u - v|, \qquad (5.11.51)$$

$$|A(x, u, \bar{u}) - A_0(x, v, \bar{v})| \leq g(x)[|u - v| + |\bar{u} - \bar{v}|] \qquad (5.11.52)$$

$$|F(x, u)| \leq k(x, |u|), \qquad (5.11.53)$$

where $c(y)$, $g(x)$ and $k(x, \phi)$ are as explained above. Equations (5.11.41) and (5.11.50) are equivalent to the integral equation (5.11.45) and

$$v(x) = \bar{h}(x) + \int_{x^0}^{x} A_0 \left(y, v(y), \int_{x^0}^{y} B_0(y, s, v(s)) \, ds \right) dy, \qquad (5.11.54)$$

where $\bar{h}(x)$ depends on the given boundary conditions. From (5.11.45) and (5.11.54) we have

$$u - v = h(x) - \bar{h}(x) + \int_{x^0}^{x} \left\{ A \left(y, u(y), \int_{x^0}^{y} B(y, s, u(s)) \, ds \right) \right.$$

$$\left. - A_0 \left(y, v(y), \int_{x^0}^{y} B_0(y, s, v(s)) \, ds \right) \right\} dy + \int_{x^0}^{x} F(y, u(y)) \, dy.$$

$$(5.11.55)$$

Using (5.11.51)–(5.11.53) and $||u| - |v|| \leq |u - v|$ and assuming that $|h(x) - \bar{h}(x)| \leq f(x)$, and that the solution $v(x)$ of (5.11.50) is bounded by a constant $M \geq 0$ in (5.11.55), where $f(x)$ is as defined in Theorem 5.7.6, we have

$$|u - v| \leq f(x) + \int_{x^0}^{x} g(y) \left(|u - v| + \int_{x^0}^{y} c(s)|u(s) - v(s)| \, ds \right) dy$$

$$+ \int_{x^0}^{x} k(y, M + |u(y) - v(y)|) \, dy. \qquad (5.11.56)$$

Now a suitable application of Theorem 5.7.6 yields

$$|u - v| \leq E_0^*(x)[f(x) + r(x)], \tag{5.11.57}$$

where $E^*(x)$ is defined by (5.11.49) and $r(x)$ is a solution of the integral equation

$$r(x) = \int_{x^0}^{x} k(y, M + E_0^*(y)[f(y) + r(y)]) \, dy. \tag{5.11.58}$$

If the right-hand side of (5.11.57) is bounded then we obtain the relative boundedness of the solution $u(x)$ of (5.11.41) and the solution $v(x)$ of (5.11.50). If $f(x)$ in (5.11.57) is small enough and, say, less than ϵ, where $\epsilon > 0$ is arbitrary, if equation (5.11.58) admits only an identically zero solution, and if $E_0^*(x)$ in (5.11.57) is bounded and $\epsilon \to 0$, then we obtain $|u(x) - v(x)| \to 0$, which gives the equivalence between the solutions of (5.11.41) and (5.11.50).

5.12 Miscellaneous Inequalities

5.12.1 Dragomir and Ionescu (1987, 1989)

Let u, a, b, be nonnegative continuous functions defined on R_+^2 and $L: R_+^3 \to R_+$ be a continuous function which satisfies the condition

$$0 \leq L(x, y, v) - L(x, y, w) \leq M(x, y, w)(v - w), \tag{L}$$

for $x, y \in R_+$ and $v \geq w \geq 0$, where M is a nonnegative continuous function defined on R_+^3.

(i) If

$$u(x, y) \leq a(x, y) + b(x, y) \int_0^x \int_0^y L(s, t, u(s, t)) \, dt \, ds, \tag{*}$$

for $x, y \in R_+$, then

$$u(x, y) \leq a(x, y) + b(x, y) \left[\exp\left(\int_0^x \int_0^y P(s, t) \, dt \, ds \right) - 1 \right],$$

for $x, y \in R_+$, where $P(x, y)$ is given by

$$P(x, y) = [\{L(x, y, a(x, y))\}^2 + \{M(x, y, a(x, y))\}^2 b^2(x, y)]^{1/2},$$

for $x, y \in R_+$.

(ii) If $u(x, y)$ satisfies the inequality (*) above, then

$$u(x, y) \leq a(x, y) + b(x, y)Q(x, y)\int_0^x \int_0^y L(s, t, a(s, t))\, dt\, ds,$$

for $x, y \in R_+$, where

$$Q(x, y) = 1 + \int_0^x \int_0^y M(s, t, a(s, t))b(s, t)$$

$$\times \exp\left(\int_0^s \int_0^t M(s_1, t_1, a(s_1, t_1))b(s_1, t_1)\, dt_1\, ds_1\right) dt\, ds,$$

for $x, y \in R_+$.

5.12.2 Pachpatte (in press k)

Let $u(x, y)$, $p(x, y)$, $q(x, y)$ and $r(x, y)$ be nonnegative continuous functions defined for $x, y \in R_+$. Let $k(x, y, s, t)$ and its partial derivatives $k_1(x, y, s, t) = (\partial/\partial x)k(x, y, s, t)$, $k_2(x, y, s, t) = (\partial/\partial y)k(x, y, s, t)$, $k_{12}(x, y, s, t) = (\partial^2/\partial x \partial y)k(x, y, s, t)$ be nonnegative continuous functions for $0 \leq s \leq x < \infty$, $0 \leq t \leq y < \infty$ and $L: R_+^3 \to R_+$ be a continuous function satisfying the condition

$$0 \leq L(x, y, v) - L(x, y, w) \leq M(x, y, w)(v - w),$$

for $x, y \in R_+$, $v \geq w \geq 0$, where $M: R_+^3 \to R_+$ is continuous.

(i) If

$$u(x, y) \leq p(x, y) + q(x, y)\int_0^x \int_0^y k(x, y, s, t)[r(s, t)u(s, t)$$

$$+ L(s, t, u(s, t))]\, dt\, ds,$$

for $x, y \in R_+$, then

$$u(x, y) \leq p(x, y) + q(x, y) \left(\int_0^x \int_0^y A(s_1, t_1) \, dt_1 \, ds_1 \right)$$

$$\times \exp\left(\int_0^x \int_0^y B(s_1, t_1) \, dt_1 \, ds_1 \right),$$

for $x, y \in R_+$, where

$$A(x, y) = k(x, y, x, y)[r(x, y)p(x, y) + L(x, y, p(x, y))]$$

$$+ \int_0^x k_1(x, y, s, y)[r(s, y)p(s, y) + L(s, y, p(s, y))] \, ds$$

$$+ \int_0^y k_2(x, y, x, t)[r(x, t)p(x, t) + L(x, t, p(x, t))] \, dt$$

$$+ \int_0^x \int_0^y k_{12}(x, y, s, t)[r(s, t)p(s, t) + L(s, t, p(s, t))] \, dt \, ds,$$

$$B(x, y) = k(x, y, x, y)[r(x, y) + M(x, y, p(x, y))]q(x, y)$$

$$+ \int_0^x k_1(x, y, s, y)[r(s, y) + M(s, y, p(s, y))]q(s, y) \, ds$$

$$+ \int_0^y k_2(x, y, x, t)[r(x, t) + M(x, t, p(x, t))]q(x, t) \, dt$$

$$+ \int_0^x \int_0^y k_{12}(x, y, s, t)[r(s, t) + M(s, t, p(s, t))]q(s, t) \, dt \, ds,$$

for $x, y \in R_+$.

(ii) Let F, F^{-1}, α, β be as in Theorem 5.3.1, part (ii). If

$$u(x, y) \leq p(x, y) + q(x, y)F^{-1}\left(\int_0^x \int_0^y k(x, y, s, t)[r(s, t)F(u(s, t))\right.$$

$$\left. + L(s, t, F(u(s, t)))] \, dt \, ds \right),$$

for $x, y \in R_+$, then

$$u(x, y) \le p(x, y) + q(x, y) F^{-1} \left[\left(\int_0^x \int_0^y \overline{A}(s_1, t_1) \, dt_1 \, ds_1 \right) \right.$$

$$\left. \times \exp \left(\int_0^x \int_0^y \overline{B}(s_1, t_1) \, dt_1 \, ds_1 \right) \right],$$

for $x, y \in R_+$, where $\overline{A}(x, y)$ and $\overline{B}(x, y)$ are defined by the right-hand sides of $A(x, y)$ and $B(x, y)$ defined above by replacing $p(x, y)$ by $\alpha(x, y) F(p(x, y)\alpha^{-1}(x, y))$ and $q(x, y)$ by $\beta(x, y) F(q(x, y)\beta^{-1}(x, y))$ for $x, y \in R_+$.

5.12.3 Pachpatte (1979b)

Let $u(x, y)$, $b(x, y)$ be nonnegative continuous functions defined for $x, y \in R_+$ and $a(x, y) > 0$ for $x, y \in R_+$ be a continuous function. Let $g(u)$ be a continuously differentiable function defined for $u \ge 0$, $g(u) > 0$ for $u > 0$ and $g'(u) \ge 0$ for $u \ge 0$.

(i) If $a_x(x, y)$, $a_y(x, y)$, $a_{xy}(x, y)$ exist and are nonnegative continuous functions defined for $x, y \in R_+$ and

$$a_{xy}(x, y) \le q(x, y) g(a(x, y)),$$

for $x, y \in R_+$, where $q(x, y)$ is a nonnegative continuous function defined for $x, y \in R_+$, and suppose further that

$$u(x, y) \le a(x, y) + \int_0^x \int_0^y b(s, t) g(u(s, t)) \, dt \, ds,$$

for $x, y \in R_+$; then for $0 \le x \le x_1, 0 \le y \le y_1$,

$$u(x, y) \le \Omega^{-1} \left[\Omega(a(0, y)) + \int_0^x \frac{a_s(s, 0)}{g(a(s, 0))} \, ds \right.$$

$$\left. + \int_0^x \int_0^y [q(s, t) + b(s, t)] \, dt \, ds \right],$$

where Ω, Ω^{-1} are as defined in Theorem 5.2.1 and x_1, y_1 are chosen so that

$$\Omega(a(0, y)) + \int_0^x \frac{a_s(s, 0)}{g(a(s, 0))} \, ds + \int_0^x \int_0^y [q(s, t) + b(s, t)] \, dt \, ds \in \mathrm{Dom}(\Omega^{-1}),$$

for all x, y lying in the subintervals $0 \leq x \leq x_1$, $0 \leq y \leq y_1$ of R_+.

(ii) If $a_x(x, y)$, $a_y(x, y)$, $a_{xy}(x, y)$ exist and are nonnegative continuous functions defined for $x, y \in R_+$ and

$$a_{xy}(x, y) \leq q(x, y)[a(x, y) + g(a(x, y))],$$

for $x, y \in R_+$, where $q(x, y)$ is nonnegative continuous function defined for $x, y \in R_+$ and suppose further that

$$u(x, y) \leq a(x, y) + \int_0^x \int_0^y b(s, t) \left(u(s, t) \right.$$

$$\left. + \int_0^s \int_0^t b(s_1, t_1) g(u(s_1, t_1)) \, dt_1 \, ds_1 \right) dt \, ds,$$

for $x, y \in R_+$, then for $0 \leq x \leq x_2$, $0 \leq y \leq y_2$,

$$u(x, y) \leq a(x, 0) + a(0, y) - a(0, 0) + \int_0^x \int_0^y q(s, t)[a(s, t) + g(a(s, t))] \, dt \, ds$$

$$+ \int_0^x \int_0^y b(s, t) H^{-1} \left[H(a(0, y)) + \int_0^s \frac{a_{s_1}(s_1, 0)}{a(s_1, 0) + g(a(s_1, 0))} \, ds_1 \right.$$

$$\left. + \int_0^s \int_0^t [q(s_1, t_1) + b(s_1, t_1)] \, dt_1 \, ds_1 \right] dt \, ds,$$

where H, H^{-1} are as defined in Theorem 5.4.1 and x_2, y_2 are chosen so that

$$H(a(0, y)) + \int_0^x \frac{a_{s_1}(s_1, 0)}{a(s_1, 0) + g(a(s_1, 0))} \, ds_1$$

$$+ \int_0^x \int_0^y [q(s_1, t_1) + b(s_1, t_1)] \, dt_1 \, ds_1 \in \mathrm{Dom}(H^{-1}),$$

for all x, y lying in the subintervals $0 \leq x \leq x_2$, $0 \leq y \leq y_2$ of R_+.

5.12.4 Pachpatte (unpublished manuscript)

Let $u(x,t)$ and $p(x,t)$ be nonnegative continuous functions defined for $x \in I_1 = [0,1]$, $t \in I_T = [0,T]$. Let $\phi(x)$, $h(t)$, $f(t)$ be positive and continuous functions defined for $x \in I_1$, $t \in I_T$ and define $c(x,t) = \phi(x) + xh(t) + f(t)$ for $x \in I_1$, $t \in I_T$ and let $g(u)$, $g'(u)$ be as in Theorem 5.5.1 and

$$u(x,t) \le c(x,t) + \int_0^x \int_0^s \int_0^t p(s_1,t_1) g(u(s_1,t_1))\, dt_1\, ds_1\, ds,$$

holds for $x \in I_1$, $t \in I_T$.

(i) If $\phi(x)$ is twice continuously differentiable function such that $\phi'(x) \ge 0$, $\phi''(x) \ge 0$ for $x \in I_1$, and $f(t)$, $h(t)$ be continuously differentiable functions such that $f'(t) \ge 0$, $h'(t) \ge 0$ for $t \in I_T$, then for $0 \le x \le x_1$, $0 \le t \le t_1$,

$$u(x,t) \le \Omega^{-1}\left[\Omega(c(0,t)) + x\left(\frac{c_x(0,t)}{g(c(0,t))}\right) + \int_0^x \int_0^s \frac{\phi''(s_1)}{g(c(s_1,0))}\, ds_1\, ds \right.$$

$$\left. + \int_0^x \int_0^s \int_0^t p(s_1,t_1)\, dt_1\, ds_1\, ds\right],$$

where Ω, Ω^{-1} are as defined in Theorem 5.5.1 and x_1, t_1 are chosen so that

$$\Omega(c(0,t)) + x\left(\frac{c_x(0,t)}{g(c(0,t))}\right) + \int_0^x \int_0^s \frac{\phi''(s_1)}{g(c(s_1,0))}\, ds_1\, ds$$

$$+ \int_0^x \int_0^s \int_0^t p(s_1,t_1)\, dt_1\, ds_1\, ds \in \mathrm{Dom}(\Omega^{-1}),$$

for all x,t lying in the subintervals $0 \le x \le x_1$ of I_1 and $0 \le t \le t_1$ of I_T.

(ii) If $\phi(x)$, $f(t)$, $h(t)$ be continuously differentiable functions such that $\phi'(x) \ge 0$, $f'(t) \ge 0$, $h'(t) \ge 0$ for $x \in I_1$, $t \in I_T$, then for $0 \le x \le x_2$, $0 \le t \le t_2$,

$$u(x,t) \le \Omega^{-1}\left[\Omega(c(x,0)) + \int_0^t \frac{[f'(t_1) + xh'(t_1)]}{g(c(0,t_1))}\, dt_1 \right.$$

$$\left. + \int_0^t \int_0^x \int_0^s p(s_1,t_1)\, ds_1\, ds\, dt_1\right],$$

where Ω, Ω^{-1} are as defined in Theorem 5.5.1 and x_2, t_2 are chosen so that

$$\Omega(c(x,0)) + \int_0^t \frac{[f'(t_1) + xh'(t_1)]}{g(c(0,t_1))} dt_1$$

$$+ \int_0^t \int_0^x \int_0^s p(s_1, t_1) ds_1 ds dt_1 \in \text{Dom}(\Omega^{-1}),$$

for all x, t lying in the subintervals $0 \le x \le x_2$ of I_1 and $0 \le t \le t_2$ of I_T.

5.12.5 Pachpatte (unpublished manuscript)

Let $u(x,t)$ and $p(x,t)$ be nonnegative continuous functions defined for $x \in I_1 = [0,1]$, $t \in I_T = [0,T]$. Let $\phi(x)$, $h(t)$, $f(t)$ be positive and continuous functions defined for $x \in I_1$, $t \in I_T$ and define $c(x,t) = \phi(x) + xh(t) + f(t)$ for $x \in I_1$, $t \in I_T$. Let $g(u)$, $g'(u)$ be as in Theorem 5.5.1 and

$$u(x,t) \le c(x,t) + \int_0^x \int_0^s \int_0^t p(s_1,t_1) \left(u(s_1,t_1) + \int_0^{s_1} \int_0^{s_2} \int_0^{t_1} p(s_3,t_3) \right.$$

$$\left. \times g(u(s_3,t_3)) dt_3 ds_3 ds_2 \right) dt_1 ds_1 ds,$$

holds for $x \in I_1$, $t \in I_T$.

(i) If $\phi(x)$ is a twice continuously differentiable function such that $\phi'(x) \ge 0$, $\phi''(x) \ge 0$ for $x \in I_1$ and $f(t)$, $h(t)$ are continuously differentiable functions such that $f'(t) \ge 0$, $h'(t) \ge 0$ for $t \in I_T$, then for $0 \le x \le x_3$, $0 \le t \le t_3$,

$$u(x,t) \le c(0,t) + xc_x(0,t) + \int_0^x \int_0^s \phi''(s_1) ds_1 ds$$

$$+ \int_0^x \int_0^s \int_0^t p(s_1,t_1) Q_1(s_1,t_1) dt_1 ds_1 ds,$$

in which

$$Q_1(x,t) = H^{-1}\left[H(c(0,t)) + x\left(\frac{c_x(0,t)}{c(0,t)+g(c(0,t))}\right)\right.$$

$$+ \int_0^x\int_0^{s_2}\frac{\phi''(s_3)}{c(s_3,0)+g(c(s_3,0))}\,ds_3\,ds_2$$

$$\left.+ \int_0^x\int_0^s\int_0^t p(s_3,t_3)\,dt_3\,ds_3\,ds\right],$$

H, H^{-1} are as defined in Theorem 5.5.2 and x_3, t_3 are chosen so that

$$H(c(0,t)) + x\left(\frac{c_x(0,t)}{c(0,t)+g(c(0,t))}\right)$$

$$+ \int_0^x\int_0^{s_2}\frac{\phi''(s_3)}{c(s_3,0)+g(c(s_3,0))}\,ds_3\,ds_2$$

$$+ \int_0^x\int_0^s\int_0^t p(s_3,t_3)\,dt_3\,ds_3\,ds \in \text{Dom}(H^{-1}),$$

for all x, t lying in the subintervals $0 \le x \le x_3$ of I_1 and $0 \le t \le t_3$ of I_T.

(ii) If $\phi(x)$, $f(t)$, $h(t)$ are continuously differentiable functions such that $\phi'(x) \ge 0$, $f'(t) \ge 0$, $h'(t) \ge 0$ for $x \in I_1$, $t \in I_T$, then for $0 \le x \le x_4$, $0 \le t \le t_4$,

$$u(x,t) \le c(x,0) + \int_0^t [f'(t_1) + xh'(t_1)]\,dt_1$$

$$+ \int_0^t\int_0^x\int_0^s p(s_1,t_1)Q_2(s_1,t_1)\,ds_1\,ds\,dt_1,$$

in which

$$Q_2(x,t) = H^{-1}\left[H(c(x,0)) + \int_0^t \frac{[f'(t_2)+xh'(t_2)]}{c(0,t_2)+g(c(0,t_2))}\,dt_2\right.$$

$$\left.+ \int_0^t\int_0^x\int_0^s p(s_3,t_3)\,ds_3\,ds\,dt_3\right],$$

H, H^{-1} are as defined in Theorem 5.5.2 and x_4, t_4 are chosen so that

$$H(c(x,0)) + \int_0^t \frac{[f'(t_2) + xh'(t_2)]}{c(0,t_2) + g(c(0,t_2))} dt_2$$

$$+ \int_0^t \int_0^x \int_0^s p(s_3, t_3) \, ds_3 \, ds \, dt_3 \in \text{Dom}(H^{-1}),$$

for all x, t lying in the subintervals $0 \leq x \leq x_4$ of I_1 and $0 \leq t \leq t_4$ of I_T.

5.12.6 Rasmussen (1976)

Let $P_0(x_0, y_0)$ and $P_1(x_1, y_1)$ be points in the domain D such that $(x_1 - x_0)(y_1 - y_0) \geq 0$ and such that the closed rectangle R_1 with opposite vertices P_0 and P_1 is contained in D. Let $\phi(x, y)$, $g(x, y)$ and $k(x, y, u)$ be functions, with the former two continuous on D and with the latter continuous on $D \times R$ nondecreasing in u, and satisfying the Lipschitz condition $|k(x, y, u) - k(x, y, v)| \leq L|u - v|$. If for all (x, y) in R_1,

$$\phi(x, y) \leq g(x, y) + \int_{x_0}^x \int_{y_0}^y k(t, s, \phi(s, t)) \, ds \, dt,$$

then $\phi(x, y) \leq U(x, y)$ on R_1, where $U(x, y)$ is a maximal solution of the equation

$$u(x, y) = g(x, y) + \int_{x_0}^x \int_{y_0}^y k(t, s, u(s, t)) \, ds \, dt.$$

5.12.7 Pachpatte (unpublished manuscript)

Let $u(x, y)$, $b(x, y)$, $c(x, y)$ and $p(x, y)$ be nonnegative continuous functions defined for $x, y \in R_+$. Let $a(x, y) > 0$ for $x, y \in R_+$ be a continuous and nondecreasing function in x.

(i) Let g, g', h, h' be as in Theorem 5.2.2. If

$$u(x, y) \leq a(x, y) + \int_0^x b(s, y) \left(u(s, y) + \int_0^s p(s_1, y) u(s_1, y) \, ds_1 \right) ds$$

$$+ h \left(\int_0^x \int_0^y c(s, t) g(u(s, t)) \, dt \, ds \right),$$

for $x, y \in R_+$, then for $0 \le x \le x_1, 0 \le y \le y_1$,

$$u(x, y) \le \psi(x, y)\left[a(x, y) + h\left(G^{-1}\left[G\left(\int_0^x\int_0^y c(s, t)g(a(s, t)\psi(s, t))\,dt\,ds\right)\right.\right.\right.$$
$$\left.\left.\left. + \int_0^x\int_0^y c(s, t)g(\psi(s, t))\,dt\,ds\right]\right)\right],$$

where

$$\psi(x, y) = 1 + \int_0^x b(s, y)\exp\left(\int_0^s [b(s_1, y) + p(s_1, y)]\,ds_1\right)ds,$$

G, G^{-1} are as defined in Theorem 5.2.2 and x_1, y_1 are chosen so that

$$G\left(\int_0^x\int_0^y c(s, t)g(a(s, t)\psi(s, t))\,dt\,ds\right)$$
$$+ \int_0^x\int_0^y c(s, t)g(\psi(s, t))\,dt\,ds \in \mathrm{Dom}(G^{-1})$$

for all x, y lying in the subintervals $0 \le x \le x_1, 0 \le y \le y_1$ of R_+.

(ii) Let $k(x, y, u)$ be a nonnegative continuous function defined on $R_+^2 \times R$, which is nondecreasing in z and is uniformly Lipschitz in z. If

$$u(x, y) \le a(x, y) + \int_0^x b(s, y)\left(u(s, y) + \int_0^s p(s_1, y)u(s_1, y)\,ds_1\right)ds$$
$$+ \int_0^x\int_0^y k(s, t, u(s, t))\,dt\,ds,$$

for $x, y \in R_+$, then

$$u(x, y) \le \psi(x, y)[a(x, y) + r(x, y)],$$

for $x, y \in R_+$, where $\psi(x, y)$ is as defined above and $r(x, y)$ is a solution of the integral equation

$$r(x, y) = \int_0^x \int_0^y k(s, t, \psi(s, t)[a(s, t) + r(s, t)]) \, dt \, ds,$$

existing for $x, y \in R_+$.

5.12.8 Pachpatte (unpublished manuscript)

Let $u(x, y), b(x, y)$ and $c(x, y)$ be nonnegative continuous functions defined for $x, y \in R_+$. Let $a(x, y) \geq 1$ for $x, y \in R_+$ be a continuous and nondecreasing function in x.

(i) Let g, g', h, h' be as in Theorem 5.2.2 and let H, H' be as in Theorem 5.7.8. If

$$u(x, y) \leq a(x, y) + \int_0^x b(s, y) \left(u(s, y) + \int_0^s b(s_1, y) H(u(s_1, y)) \, dy \right) ds$$

$$+ h \left(\int_0^x \int_0^y c(s, t) g(u(s, t)) \, dt \, ds \right),$$

for $x, y \in R_+$, then for $0 \leq x \leq x_1, 0 \leq y \leq y_1$,

$$u(x, y) \leq \psi_1(x, y) \left[a(x, y) + h \left(G^{-1} \left[G \left(\int_0^x \int_0^y c(s, t) \right. \right. \right. \right.$$

$$\left. \left. \left. \times g(a(s, t) \psi_1(s, t)) \, dt \, ds \right) + \int_0^x \int_0^y c(s, t) g(\psi_1(s, t)) \, dt \, ds \right] \right) \right],$$

where G, G^{-1} are as defined in Theorem 5.2.2,

$$\psi_1(x, y) = 1 + \int_0^x b(s, y) F^{-1} \left[F(1) + \int_0^s b(s_1, y) \, ds_1 \right] ds,$$

in which F, F^{-1} are as defined in Theorem 5.7.8 and x_1, y_1 are chosen so that

$$F(1) + \int_0^x b(s_1, y) \, ds_1 \in \text{Dom}(F^{-1}),$$

and

$$G\left(\int_0^x \int_0^y c(s,t)g(a(s,t)\psi_1(s,t))\,dt\,ds\right)$$
$$+ \int_0^x \int_0^y c(s,t)g(\psi_1(s,t))\,dt\,ds \in \mathrm{Dom}(G^{-1}),$$

for all x, y lying in the subintervals $0 \le x \le x_1, 0 \le y \le y_1$ of R_+.

(ii) Let $k(x, y, u)$ be a nonnegative continuous function defined on $R_+^2 \times R$ which is nondecreasing in u and is uniformly Lipschitz in z. If

$$u(x,y) \le a(x,y) + \int_0^x b(s,y)\left(u(s,y) + \int_0^s b(s_1, y)H(u(s_1, y))\,ds_1\right)ds$$
$$+ \int_0^x \int_0^y k(s, t, u(s,t))\,dt\,ds,$$

for $x, y \in R_+$, then for $0 \le x \le x_2, 0 \le y \le y_2$,

$$u(x,y) \le \psi_1(x,y)[a(x,y) + r(x,y)],$$

where $\psi_1(x, y)$ is as defined above and $r(x, y)$ is a solution of the integral equation

$$r(x,y) = \int_0^x \int_0^y k(s, t, \psi_1(s,t)[a(s,t) + r(s,t)])\,dt\,ds,$$

existing for $x, y \in R_+$.

5.12.9 Pachpatte (1981b)

Let $u(x, y), b(x, y), c(x, y), p(x, y)$ and $q(x, y)$ be nonnegative continuous functions defined on a domain D. Let g, g' be as in Theorem 5.2.3 and $k > 0$ is a constant. Let $P_0(x_0, y_0)$ and $P(x, y)$ be two points in D such that $(x - x_0)(y - y_0) \ge 0$ and let $R \subset D$ be a rectangular region whose opposite corners are the points P_0 and P. Let $v(s, t; x, y)$ be the solution of the characteristic initial value problem

$$L[v] = v_{st} - b(s,t)[c(s,t) + p(s,t)]v = 0, \quad v(s,y) = v(x,t) = 1,$$

and let D^+ be a connected subdomain of D which contains P and on which $v \geq 0$ (Figure 4.1). If $R \subset D^+$ and $u(x,y)$ satisfies

$$u(x,y) \leq k + b(x,y)\left(\int_{x_0}^{x}\int_{y_0}^{y} c(s,t)u(s,t)\,ds\,dt\right.$$

$$+ \int_0^x\int_0^y c(s,t)b(s,t)\left(\int_0^s\int_0^t p(s_1,t_1)u(s_1,t_1)\,ds_1\,dt_1\right)ds\,dt\right)$$

$$+ \int_{x_0}^{x}\int_{y_0}^{y} q(s,t)g(u(s,t))\,ds\,dt,$$

for $(x,y) \in R$, then for $(x,y) \in R_1 \subset R$,

$$u(x,y) \leq Q(x,y)\Omega^{-1}\left[\Omega(k) + \int_{x_0}^{x}\int_{y_0}^{y} q(s,t)g(Q(s,t))\,ds\,dt\right],$$

where

$$Q(x,y) = 1 + b(x,y)\left(\int_{x_0}^{x}\int_{y_0}^{y}\left\{c(s_1,t_1) + b(s_1,t_1)c(s_1,t_1)\right.\right.$$

$$\times \left(\int_{x_0}^{s_1}\int_{y_0}^{t_1}[c(s_2,t_2) + p(s_2,t_2)]v(s_2,t_2;s_1,t_1)\,ds_2\,dt_2\right)\right\}ds_1\,dt_1\right),$$

for $(x,y) \in R$, Ω, Ω^{-1} are as defined in Theorem 5.2.1 and x_1, y_1 are chosen so that

$$\Omega(k) + \int_{x_0}^{x}\int_{y_0}^{y} q(s,t)g(Q(s,t))\,ds\,dt \in \mathrm{Dom}(\Omega^{-1}),$$

for all $(x,y) \in R_1 \subset R$.

5.12.10 Bondge and Pachpatte (1979a)

Let $u(x)$ and $p(x)$ be nonnegative continuous functions defined on Ω, where Ω is defined as in Section 4.9. Let $a_i(x_i) > 0$, $a'_i(x_i) \geq 0$ for $1 \leq i \leq n$ are continuous functions defined for $x_i \geq x_i^0$. Let $H : R_+ \to R_+$ be continuously differentiable function with $H(u) > 0$ for $u > 0$ and $H'(u) \geq 0$ for $u \geq 0$.

(i) If

$$u(x) \leq \sum_{i=1}^{n} a_i(x_i) + \int_{x^0}^{x} p(s) H(u(s))\, ds,$$

for $x \in \Omega$, then for $x^0 \leq x \leq x^*$,

$$u(x) \leq G^{-1}\left[G\left(\sum_{i=2}^{n} a_i(x_i) + a_1(x_1^0) \right) \right.$$

$$+ \int_{x_1^0}^{x_1} \frac{a'_1(s_1)}{H(\sum_{i=3}^{n} a_i(x_i) + a_2(x_2^0) + a_1(s_1))}\, ds_1$$

$$\left. + \int_{x_0}^{x} p(s)\, ds \right],$$

where

$$G(r) = \int_{r_0}^{r} \frac{ds}{H(s)}, \quad r > 0,\, r_0 > 0,$$

and G^{-1} is the inverse of G, and x^* be chosen so that

$$G\left(\sum_{i=2}^{n} a_i(x_i) + a_1(x_1^0) \right) + \int_{x_1^0}^{x_1} \frac{a'_1(s_1)}{H(\sum_{i=3}^{n} a_i(x_i) + a_2(x_2^0) + a_1(s_1))}\, ds_1$$

$$+ \int_{x^0}^{x} p(s)\, ds \in \mathrm{Dom}(G^{-1}),$$

for all x lying in the parallelepiped $x^0 \leq x \leq x^*$ in Ω.

(ii) If

$$u(x) \le \sum_{i=1}^{n} a_i(x_i) + \int_{x^0}^{x} p(s)\left(u(s) + \int_{x^0}^{s} p(y)H(u(y))\,dy\right)ds,$$

for $x \in \Omega$, then for $x^0 \le x \le x^{**}$,

$$u(x) \le \sum_{i=1}^{n} a_i(x_i) + \int_{x^0}^{x} p(s)W^{-1}\left[W\left(\sum_{i=1}^{n} a_i(s_i) + a_1(x_1^0)\right)\right.$$

$$+ \int_{x_1^0}^{s_1} \frac{a_1'(y_1)}{(\sum_{i=3}^{n} a_i(s_i) + a_2(x_2^0) + a_1(y_1)) + H(\sum_{i=3}^{n} a_i(s_i) + a_2(x_2^0) + a_1(y_1))}$$

$$\left.+ \int_{x^0}^{x} p(y)\,dy\right]ds,$$

where

$$W(r) = \int_{r_0}^{r} \frac{dt}{t + H(t)}, \quad r > 0, r_0 > 0,$$

and W^{-1} is the inverse function of W, and x^{**} is chosen so that

$$W\left(\sum_{i=1}^{n} a_i(x_i) + a_1(x_1^0)\right)$$

$$+ \int_{x_1^0}^{x_1} \frac{a_1'(s_1)}{(\sum_{i=3}^{n} a_i(x_i) + a_2(x_2^0) + a_1(s_1)) + H(\sum_{i=3}^{n} a_i(x_i) + a_2(x_2^0) + a_1(s_1))}\,ds_1$$

$$+ \int_{x^0}^{x} p(s)\,ds \in \text{Dom}(W^{-1}),$$

for all $x \in \Omega$ lying in the parallelepiped $x^0 \le x \le x^{**}$ in Ω.

5.12.11 Singare and Pachpatte (1981)

Let $\phi(x), a(x), b(x)$ and $c(x)$ be nonnegative continuous functions defined on Ω, and let $u(x)$ be a positive continuous function defined on Ω.

(i) Let $H(r)$ be a positive, continuous, strictly increasing, convex and submultiplicative function for $r > 0$, $\lim_{r \to \infty} H(r) = \infty$. Let $\alpha(x), \beta(x)$ be positive continuous functions defined on Ω with $\alpha(x) + \beta(x) = 1$. If

$$u(s) \geq \phi(x) - a(s)H^{-1}\left[\int_x^s h(\sigma)H(\phi(\sigma))\,d\sigma \right.$$

$$\left. + \int_x^s b(\sigma)\left(\int_\sigma^s c(\xi)H(\phi(\xi))\,d\xi\right) d\sigma\right],$$

is satisfied for $x \leq s$; $x, s \in \Omega$, then

$$u(s) \geq \alpha(s)H^{-1}\left[\alpha^{-1}(s)H(\phi(x))\left\{1 + \beta(s)H(a(s)\beta^{-1}(s))\int_x^s b(\sigma)\right.\right.$$

$$\left.\left.\times \exp\left(\int_\sigma^s [\beta(s)H(a(s)\beta^{-1}(s))b(\xi) + c(\xi)]\,d\xi\right) d\sigma\right\}^{-1}\right],$$

for $x \leq s$; $x, s \in \Omega$.

(ii) Let $G(r)$ be a positive, continuous, strictly increasing, subadditive and submultiplicative function for $r > 0$ with $\lim_{r \to \infty} G(r) = \infty$, and let G^{-1} denote the inverse function of G. If

$$u(s) \geq \phi(x) - a(s)G^{-1}\left[\int_x^s b(\sigma)G(\phi(\sigma))\,d\sigma \right.$$

$$\left. + \int_x^s b(\sigma)\left(\int_\sigma^s c(\xi)G(\phi(\xi))\,d\xi\right) d\sigma\right],$$

for $x \leq s$; $x, s \in \Omega$, then

$$u(s) \geq G^{-1}\left[G(\phi(x))\left\{1 + G(a(s))\int_x^s b(\sigma)\right.\right.$$

$$\left.\left.\times \exp\left(\int_\sigma^s [b(\xi)G(a(s)) + c(\xi)]\,d\xi\right)\right\}^{-1}\right],$$

for $x \leq s$; $x, s \in \Omega$.

5.12.12 Pachpatte (unpublished manuscript)

Let $u(x), b(x)$ and $c(x)$ be nonnegative continuous functions defined on Ω. Let $a(x) \geq 1$ for $x \in \Omega$ be a continuous and nondecreasing function in x. Let $H(u)$ and $H'(u)$ be as in Theorem 5.7.8. If

$$u(x) \leq a(x) + \int_{x^0}^{x} b(y) H\left(u(y) + \int_{x^0}^{y} c(s) H(u(s)) \, ds \right) dy,$$

for $x \in \Omega$, then for $x \in \Omega_1 \subset \Omega$,

$$u(x) \leq a(x) Q(x),$$

where

$$Q(x) = 1 + \int_{x^0}^{x} b(y) H\left(G^{-1}\left[G(1) + \int_{x^0}^{y} [b(s) + c(s)] \, ds \right] \right) dy, \quad (**)$$

$$G(r) = \int_{r_0}^{r} \frac{ds}{H(s)}, \quad r > 0, r_0 > 0,$$

G^{-1} is the inverse function of G and

$$G(1) + \int_{x^0}^{x} [b(s) + c(s)] \, ds \in \mathrm{Dom}(G^{-1}),$$

for all $x \in \Omega_1 \subset \Omega$.

5.12.13 Pachpatte (unpublished manuscript)

Let $u(x), b(x), c(x)$ and $q(x)$ be nonnegative continuous functions defined on Ω and $k \geq 1$ is a constant. Let $H(u), H'(u)$ be as in Theorem 5.7.8. Let $g(u)$ be a continuously differentiable function defined for $u \geq 0, g(u) > 0$ for $u > 0$ and $g'(u) \geq 0$ for $u \geq 0$ and $g(u)$ is submultiplicative. If

$$u(x) \leq k + \int_{x^0}^{x} b(y) H\left(u(y) + \int_{x^0}^{y} c(s) H(u(s)) \, ds \right) dy$$

$$+ \int_{x^0}^{x} q(y) g(u(y)) \, dy,$$

for $x \in \Omega$, then for $x \in \Omega_1 \subset \Omega$,

$$u(x) \leq Q(x)E^{-1}\left[E(k) + \int_{x^0}^{x} q(y)g(Q(y))\,dy\right],$$

where $Q(x)$ is defined in (**) with G, G^{-1} as involved therein,

$$E(r) = \int_{r_0}^{r} \frac{ds}{g(s)}, \quad r > 0, r_0 > 0,$$

E^{-1} is the inverse function of E, and

$$G(1) + \int_{x^0}^{x} [b(s) + c(s)]\,ds \in \mathrm{Dom}(G^{-1}),$$

$$E(k) + \int_{x^0}^{x} q(y)g(Q(y))\,dy \in \mathrm{Dom}(E^{-1}),$$

for $x \in \Omega_1 \subset \Omega$.

5.12.14 Pachpatte (unpublished manuscript)

Let $u(x), b(x)$ and $c(x)$ be nonnegative continuous functions defined for $x \in \Omega$. Let $a(x) \geq 1$ for $x \in \Omega$ be a continuous and nondecreasing function in x. Let $H(u)$ and $H'(u)$ be as in Theorem 5.7.8 and $k(x, y, u), W(x, u)$ be as in Theorem 5.7.6. If

$$u(x) \leq a(x) + \int_{x^0}^{x} b(y)H\left(u(y) + \int_{x^0}^{y} c(s)H(u(s))\,ds\right)dy$$

$$+ W\left(x, \int_{x^0}^{x} k(x, y, u(y))\,dy\right),$$

for $x \in \Omega$, then for $x \in \Omega_1 \subset \Omega$,

$$u(x) \leq Q(x)[a(x) + W(x, r(x))],$$

where $Q(x)$ is defined as in $(**)$ with G, G^{-1} involved therein and $r(x)$ is a solution of the integral equation

$$r(x) = \int_{x^0}^{x} k(x, y, Q(x)[a(y) + W(y, r(y))]) \, dy,$$

existing on Ω.

5.12.15 Pachpatte (unpublished manuscript)

Let $u(x)$ and $b(x)$ be nonnegative continuous functions defined on Ω. Let $a(x) \geq 1$ for $x \in \Omega$ be a continuous and nondecreasing function in x. Let $H(u)$ and $H'(u)$ be as in Theorem 5.7.8. If

$$u(x) \leq a(x) + \int_{x^0}^{x} b(y) \left[u(y) + \int_{x^0}^{y} b(s) \left(\int_{x^0}^{s} b(\sigma) H(u(\sigma)) \, d\sigma \right) ds \right] dy,$$

for $x \in \Omega$, then for $x \in \Omega_1 \subset \Omega$,

$$u(x) \leq a(x) Q_0(x),$$

where

$$Q_0(x) = 1 + \int_{x^0}^{x} b(y) \left[1 + \int_{x^0}^{y} b(s) G^{-1} \left[G(1) + \int_{x^0}^{s} b(\sigma) \, d\sigma \right] ds \right] dy, \quad (***)$$

$$G(r) = \int_{r_0}^{r} \frac{ds}{s + H(s)}, \quad r > 0, r_0 > 0,$$

G^{-1} is the inverse of G and

$$G(1) + \int_{x^0}^{x} b(\sigma) \, d\sigma \in \text{Dom}(G^{-1}),$$

for all $x \in \Omega_1 \subset \Omega$.

5.12.16 Pachpatte (unpublished manuscript)

Let $u(x)$, $b(x)$ and $q(x)$ be nonnegative continuous functions defined on Ω and $k \geq 1$ be a constant. Let $H(u)$ and $H'(u)$ be as in Theorem 5.7.8. Let

$g(u)$ and $g'(u)$ be as in Theorem 5.2.3. If

$$u(x) \le k + \int_{x^0}^{x} b(y) \left[u(y) + \int_{x^0}^{y} b(s) \left(\int_{x^0}^{s} b(\sigma) H(u(\sigma)) \, d\sigma \right) ds \right] dy$$

$$+ \int_{x^0}^{x} q(y) g(u(y)) \, dy,$$

for $x \in \Omega$, then for $x \in \Omega_1 \subset \Omega$,

$$u(x) \le Q_0(x) E^{-1} \left[E(k) + \int_{x^0}^{x} q(y) g(Q_0(y)) \, dy \right],$$

where $Q_0(x)$ is defined as in $(***)$ with G, G^{-1} as involved therein,

$$E(r) = \int_{r_0}^{r} \frac{ds}{g(s)}, \quad r > 0, r_0 > 0,$$

E^{-1} is the inverse function of E, and

$$G(1) + \int_{x^0}^{x} b(\sigma) \, d\sigma \in \text{Dom}(G^{-1}),$$

$$E(k) + \int_{x^0}^{x} q(y) g(Q_0(y)) \, dy \in \text{Dom}(E^{-1}),$$

for all $x \in \Omega_1 \subset \Omega$.

5.12.17 Pachpatte (unpublished manuscript)

Let $u(x)$, and $b(x)$ be nonnegative continuous functions defined for $x \in \Omega$. Let $a(x) \ge 1$ for $x \in \Omega$ be a continuous and nondecreasing function in x. Let $H(u)$ and $H'(u)$ be as in Theorem 5.7.8 and $k(x, y, u)$ and $W(x, u)$ be as in Theorem 5.7.6. If

$$u(x) \le a(x) + \int_{x^0}^{x} b(y) \left[u(y) + \int_{x^0}^{y} b(s) \left(\int_{x^0}^{s} b(\sigma) H(u(\sigma)) \, d\sigma \right) ds \right] dy$$

$$+ W \left(x, \int_{x^0}^{x} k(x, y, u(y)) \, dy \right),$$

for $x \in \Omega$, then for $x \in \Omega_1 \subset \Omega$,

$$u(x) \leq Q_0(x)[a(x) + W(x, r(x))],$$

where $Q_0(x)$ is defined as in $(***)$ with G, G^{-1} as involved therein, and $r(x)$ is a solution of the integral equation

$$r(x) = \int_{x^0}^{x} k(x, y, Q_0(y)[a(y) + W(y, r(y))]) \, dy,$$

existing on Ω.

5.12.18 Singare and Pachpatte (1982)

Let $u(x)$, $a(x)$, $b(x)$, $f(x)$ and $g(x)$ be nonnegative continuous functions defined on Ω, $G(u)$ be a continuous strictly increasing convex and submultiplicative function for $u > 0$, $\lim_{u \to \infty} G(u) = \infty$, for all $x \in \Omega$; let $\alpha(x)$, $\beta(x)$ be positive continuous functions defined on a domain Ω, and $\alpha(x) + \beta(x) = 1$. Let $v(s; x)$ and $e(s; x)$ be the solutions of the characteristic initial value problems

$$(-1)^n v_{s_1 \ldots s_n}(s; x) - [f(s) + g(s)\beta(s)G(b(s)\beta^{-1}(s))]v(s; x) = 0 \text{ in } \Omega,$$
$$v(s; x) = 1 \text{ on } s_i = x_i, \quad 1 \leq i \leq n,$$

and

$$(-1)^n e_{s_1 \ldots s_n}(s; x) + f(s)e(s; x) = 0 \text{ in } \Omega,$$
$$e(s; x) = 1 \text{ on } s_i = x_i, \quad 1 \leq i \leq n,$$

respectively and let D^+ be a connected subdomain of Ω containing x such that $v \geq 0$, $e \geq 0$ for all $s \in D^+$. If $D \subset D^+$ and

$$u(x) \leq a(x) + b(x)G^{-1}\left(\int_{x^0}^{x} f(s) \left(\int_{x^0}^{s} g(\sigma)G(u(\sigma)) \, d\sigma\right) ds\right),$$

for all $x \in D$, then

$$u(x) \leq a(x) + b(x) G^{-1} \left[\int_{x^0}^{x} e(s; x) \left(f(s) \left(\int_{x^0}^{s} \alpha(\sigma) G(a(\sigma)\alpha^{-1}(\sigma)) g(\sigma) \right. \right. \right.$$
$$\left. \left. \left. \times v(\sigma; s) \, d\sigma \right) ds \right) \right],$$

for all $x \in D$.

5.12.19 Shinde and Pachpatte (1983)

Let $u(x)$, $v(x)$, $a(x)$, $b(x)$, $p(x)$, $h_i(x)$ ($i = 1, 2, 3, 4$) be nonnegative continuous functions defined on Ω, and $h(x) = \max\{[h_1(x) + h_3(x)], [h_2(x) + h_4(x)]\}$. Let $G(r)$ be a continuous, strictly increasing convex and submultiplicative function for $r > 0$, $\lim_{r \to \infty} G(r) = \infty$ for all x in Ω, G^{-1} be the inverse function of G; $\alpha(x)$, $\beta(x)$ are positive continuous functions defined on Ω and $\alpha(x) + \beta(x) = 1$.

Let $w(s; x)$ be the solution of the characteristic initial value problem

$$(-1)^n w_{s_1 \ldots s_n}(s; x) - \beta(s) G(p(s)\beta^{-1}(s)) h(s) w(s; x) = 0 \text{ in } \Omega,$$
$$w(s; x) = 1 \text{ on } s_i = x_i, 1 \leq i \leq n.$$

Let D^+ be a connected subdomain of Ω containing x such that $w \geq 0$ for all $s \in D^+$. If $D \subset D^+$ and

$$u(x) \leq a(x) + p(x) G^{-1} \left[\int_{x^0}^{x} h_1(s) G(u(s)) \, ds \right.$$
$$\left. + \int_{x^0}^{x} \exp\left(\mu \sum_{i=1}^{n} |s_i|\right) h_2(s) G(v(s)) \, ds \right],$$

$$v(x) \leq b(x) + p(x) G^{-1} \left[\int_{x^0}^{x} \exp\left(-\mu \sum_{i=1}^{n} |s_i|\right) h_3(s) G(u(s)) \, ds \right.$$
$$\left. + \int_{x^0}^{x} h_4(s) G(v(s)) \, ds \right],$$

for $x \in D$, where μ is a nonnegative constant, then

$$u(x) \le G^{-1}\left[\exp\left(\mu \sum_{i=1}^{n} |x_i|\right) Q(x)\right], \quad v(x) \le G^{-1}[Q(x)],$$

for $x \in D$, where

$$Q(x) = f(x) + \beta(x)G(p(x)\beta^{-1}(x))\left(\int_{x^0}^{x} w(s;x)h(s)f(s)\,ds\right),$$

in which

$$f(x) = \exp\left(-\mu \sum_{i=1}^{n} |x_i|\right) \alpha(x)G(a(x)\alpha^{-1}(x)) + \alpha(x)G(b(x)\alpha^{-1}(x)),$$

for $x \in D$.

5.12.20 Pachpatte (1993, 1996d)

Let $u(x) \ge 0$, $a(x) \ge 0$, $b(x) \ge 0$, $p_i(x) > 0$ for $i = 1, 2, \ldots, n-1$, be continuous functions defined for $x \in R_+^n$ and $q: R_+^n \times R_+ \to R_+$ be a continuous function which satisfies the condition

$$0 \le q(x, v_1) - q(x, v_2) \le k(x, v_2)(v_1 - v_2),$$

for $x \in R_+^n$ and $v_1 \ge v_2 \ge 0$, where $k: R_+^n \times R_+ \to R_+$ is a continuous function. Let the notation $M[x, p, q]$ be as defined in Section 5.9.

(i) If

$$u(x) \le a(x) + b(x)M[x, p, q(y, u(y))],$$

for all $x \in R_+^n$, then

$$u(x) \le a(x) + b(x)M[x, p, q(y, a(y))]\exp(M[x, p, k(y, a(y))b(y)]),$$

for all $x \in R_+^n$.

(ii) Let $F(u)$ be a continuous, strictly increasing, convex, submultiplicative function for $u > 0$, $\lim_{u \to \infty} F(u) = \infty$, F^{-1} denotes the inverse function of F and $\alpha(x)$, $\beta(x)$ be continuous and positive functions defined for $x \in R_+^n$ and $\alpha(x) + \beta(x) = 1$. If

$$u(x) \le a(x) + b(x)F^{-1}(M[x, p, q(y, F(u(y)))]),$$

for $x \in R_+^n$, then

$$u(x) \leq a(x) + b(x)F^{-1}(M[x, p, q(y, \alpha(y)F(a(y)\alpha^{-1}(y)))]$$
$$\times \exp(M[x, p, k(y, \alpha(y)F(a(y)\alpha^{-1}(y)))\beta(y)F(b(y)\beta^{-1}(y))])),$$

for $x \in R_+^n$.

(iii) Let $g(u), g'(u), \Omega, \Omega^{-1}$ be as in Theorem 5.2.1. If

$$u(x) \leq a(x) + b(x)M[x, p, q(y, g(u(y))],$$

for $x \in R_+^n$, then for $0 \leq x \leq x^*$,

$$u(x) \leq a(x) + b(x)\Omega^{-1}[\Omega(c(x)) + M[x, p, k(y, g(a(y)))g(b(y))]],$$

where

$$c(x) = M[x, p, q(y, g(a(y)))],$$

and $x^* \in R_+^n$ is chosen so that

$$\Omega(c(x)) + M[x, p, k(y, g(a(y)))g(b(y))] \in \text{Dom}(\Omega^{-1}),$$

for all $x \in R_+^n$ such that $0 \leq x \leq x^*$.

5.13 Notes

The results given in Section 5.2 deal with some basic nonlinear generalizations of Wendroff's inequality given by Bondge and Pachpatte (1979b, 1980a). Theorems 5.2.1 and 5.2.2 are taken from Bondge and Pachpatte (1979b) and (1980a) respectively. Theorems 5.2.3 and 5.2.4 are new and motivated by certain applications.

The inequalities dealt with in Section 5.3 are due to Pachpatte (1995e) and Bondge and Pachpatte (1979b,c). Theorem 5.3.1 was recently established by Pachpatte (1995e). Theorems 5.3.2 and 5.3.3 are given by Bondge and Pachpatte (1979b) and Theorems 5.3.4 and 5.3.5 are given by Bondge and Pachpatte (1979c).

Section 5.4 contains some useful generalizations of the inequalities given by Pachpatte (1974d, 1975h). Theorem 5.4.1 is due to Bondge and Pachpatte (1979b). Theorems 5.4.2, 5.4.3 and 5.4.4 are due to Bondge and Pachpatte (1980a).

Sections 5.5 and 5.6 are devoted to the inequalities recently investigated by Pachpatte (1980c,d, 1988c, 1993, 1996d). Theorems 5.5.1 and 5.5.2 are taken from Pachpatte (1988c). Theorem 5.5.3 is new and motivated by certain applications. Theorem 5.5.4 is taken from Pachpatte (1993, 1996d). Theorems 5.6.1 to 5.6.4 are taken from Pachpatte (1980c) and Theorems 5.6.5 and 5.6.6 are taken from Pachpatte (1980d).

The results given in Section 5.7 deal with the integral inequalities in many independent variables established by various investigators (Beesack, 1975; Headley, 1974; Pachpatte, 1981c; Pelczar, 1963). Theorems 5.7.1 and 5.7.2 are due to Pelczar (1963), Theorem 5.7.3 is taken from Beesack (1975) and Theorem 5.7.4 is taken from Headley (1974). Theorems 5.7.5–5.7.9 are due to Pachpatte (1981c).

The inequalities given in Sections 5.8–5.10 were recently discovered by Pachpatte (1991, 1992, 1994a,b,d, 1996c, 1997, in press d, unpublished manuscript). Theorems 5.8.1 and 5.8.2 are taken from Pachpatte (in press d). Theorems 5.8.3, 5.8.4, 5.8.5 are taken from Pachpatte (1997), (unpublished manuscript) and (1996c), respectively. Theorems 5.9.1 to 5.9.4 are taken from Pachpatte (1994a), while Theorem 5.9.5 is taken from Pachpatte (1994b). The results in Theorems 5.10.1 and 5.10.2 are taken from Pachpatte (1994d) and Theorem 5.10.3 is new. Theorems 5.10.4 and 5.10.5 are taken from Pachpatte (1994d). Theorems 5.10.6 and 5.10.7 are taken from Pachpatte (1991). Theorems 5.10.8 and 5.10.9 are taken from Pachpatte (1992).

Section 5.11 is devoted to the applications of some of the inequalities given in earlier sections to the various properties of the solutions of certain partial differential, integral and integro-differential equations. Section 5.12 contains some miscellaneous inequalities which can also be used in some new applications.

References

1. Abramovich, J., "On Gronwall and Wendroff type inequalities", *Proc. Am. Math. Soc.*, vol. 87, 481–486, 1983.
2. Agarwal, R. P. and P. Y. H. Pang, *Opial Inequalities With Applications in Differential and Difference Equations*, New York, Kluwer Academic Publishers, 1995.
3. Agarwal, R. P., "On an integral inequality in n independent variables", *J. Math. Anal. Appl.*, vol. 85, 192–196, 1982.
4. Agarwal, R. P., "On integrodifferential inequalities in two independent variables", *J. Math. Anal. Appl.*, vol. 89, 581–597, 1982.
5. Agarwal, R. P., "Inequalities involving partial derivatives", *J. Math. Anal. Appl.*, vol. 89, 628–638, 1982.
6. Agarwal, R. P. and E. Thandapani, "Remarks on generalizations of Gronwall's inequality", *Chinese J. Math.*, vol. 9, 1–22, 1981.
7. Akinyele, O., "On Gronwall-Bellman-Bihari type integral inequalities in several variables with retardation", *J. Math. Anal. Appl.*, vol. 104, 1–24, 1984.
8. Antosiewicz, H. A., "An inequality for approximate solutions of ordinary differential equations", *Math. Z.*, vol. 78, 44–52, 1962.
9. Apartsin, A. S. and Y. T. Men, "Nonimprovable bounds for solutions of certain integral inequalities", *Sibirskii Mat. Z.*, vol. 20, 192–195, 1979.
10. Arino, O. and I. Györi, "Stability results based on Gronwall type inequalities for some functional-differential systems, *Coll. Math. Soc. J. Bolyai 47 Differential Equations, Qualitative Theory*, 37–59, Szeged, Hungary, 1984.
11. Asirov, S. and S. Atdaev, "Ob odnom obobscenii neravnenstva Bellman", *Izv. Akad. Nauk. Turkmen SSR Ser. Fiz-Tehn.Him. Geol. Nauk.*, No. 3, 102–104, 1973.

12. Azbelev, N. B. and Z. B. Tsalyuk, "On integral inequalities I", *Mat. Sb.*, vol. 56, 325–342, 1962 (in Russian).
13. Babkin, B. N., "On a generalization of a theorem of Academician S. A. Chaplygin on differential inequality", *Molotov Gas. Univ.*, vol. 8, 3–6, 1953 (in Russian).
14. Bailey, N. T. J., *The Mathematical Theory of Infectious Diseases and its Applications*, New York, Hafner, 1975.
15. Bainov, D. and P. Simeonov, *Integral Inequalities and Applications*, New York, Kluwer Academic Publishers, 1992.
16. Barbu, V., *Differential Equations* (in Romanian), Iasi, Ed. Junimea, 1985.
17. Beckenbach, E. F. and R. Bellman, *Inequalities*, Springer-Verlag, Berlin, New York, 1961.
18. Beesack, P. R., "Comparison theorems and integral inequalities for Volterra integral equations", *Proc. Am. Math. Soc.*, vol. 20, 61–66, 1969.
19. Beesack, P .R., "Gronwall inequalities", *Carleton Mathematical Lecture Notes*, No. 11, 1975.
20. Beesack, P. R., "On integral inequalities of Bihari type", *Acta Math. Acad. Sci. Hungar.*, vol. 28, 81–88, 1976.
21. Beesack, P. R., "On Lakshmikantham's comparison method for Gronwall inequalities", *Ann. Polon. Mat.*, vol. 35, 187–222, 1977.
22. Beesack, P. R., "On some Gronwall-type integral inequalities in n independent variables", *J. Math. Anal. Appl.*, vol. 100, 393–408, 1984b.
23. Beesack, P. R., "Asymptotic behaviour of solutions of some general nonlinear differential equations and integral inequalities", *Proc. Royal Soc. Edinburgh*, vol. 98A, 49–67, 1984a.
24. Beesack, P. R., "Systems of multidimensional Volterra integral equations and inequalities", *Nonlinear Analysis TMA*, vol. 9, 1451–1486, 1985.
25. Beesack, P. R., "On some variation of parameter methods for integro-differential, integral, and quasilinear partial integro-differential equations", *Appl. Math. Comp.*, vol. 22, 189–215, 1987.
26. Bellman, R., *Stability Theory of Differential Equations*, New York, McGraw-Hill, 1953.
27. Bellman, R., *Dynamic Programming*, Princeton NJ, Princeton University Press, 1957.
28. Bellman, R. and K. L. Cooke, *Differential-Difference Equations*, New York, Academic Press, 1963.
29. Bellman, R., "The stability of solutions of linear differential equations", *Duke Math. J.*, vol. 10, 643–647, 1943.
30. Bellman, R., "Asymptotic series for the solutions of linear differential-difference equations", *Rendiconti del circolo Matematica Di Palermo*, vol. 7, 1–9, 1958.
31. Bellman, R., "Upper and lower bounds for solutions of the matrix Riccati equation", *J. Math. Anal. Appl.*, vol. 17, 373–379, 1967.

32. Bernfeld, S. R. and M. E. Lord, "A nonlinear variation of constants method for integro-differential and integral equations", *Appl. Math. Comp.*, vol. 4, 1-14, 1978.
33. Bihari, I., "A generalization of a lemma of Bellman and its application to uniqueness problems of differential equations", *Acta Math. Acad. Sci. Hungar.* vol. 7, 71-94, 1956.
34. Bihari, I., "Researches of the boundedness and stability of the solutions of nonlinear differential equations", *Acta Math. Acad. Sci. Hungar.*, vol. 8, 261-278, 1957.
35. Bobisud, L. E., "Existence of solutions to some nonlinear diffusion problems", *J. Math. Anal. Appl.*, vol. 168, 413-424, 1992.
36. Bobrowski, D., J. Popenda and J. Werbowski, "On the system of integral inequalities with delay of Gronwall-Bellman type", *Fasc. Math.*, vol. 10, 97-104, 1978.
37. Bondge, B. K. and B. G. Pachpatte, "On Wendroff type integral inequalities in n independent variables", *Chinese J. Math.*, vol. 7, 37-46, 1979a.
38. Bondge, B. K. and B. G. Pachpatte, "On nonlinear integral inequalities of the Wendroff type", *J. Math. Anal. Appl.*, vol. 70, 161-169, 1979b.
39. Bondge, B. K. and B. G. Pachpatte, "On some fundamental integral inequalities in two independent variables", *J. Math. Anal. Appl.*, vol. 72, 533-544, 1979c.
40. Bondge, B. K. and B. G. Pachpatte, "On two independent variable generalizations of certain integral inequalities", *Tamkang J. Math.*, vol. 11, 37-47, 1980a.
41. Bondge, B. K. and B. G. Pachpatte, "On some partial integral inequalities in two independent variables", *Funkcialaj Ekvacioj*, vol. 23, 327-334, 1980b.
42. Bondge, B. K. and B. G. Pachpatte, "On some fundamental integral inequalities in n independent variables", *Bull. Inst. Math. Acad. Sinica*, vol. 8, 553-560, 1980a.
43. Bondge, B. K., B. G. Pachpatte and W. Walter, "On generalized Wendroff type inequalities and their applications", *Nonlinear Analysis TMA*, vol. 4, 491-495, 1980.
44. Bownds, J. M. and J. M. Cushing, "Some stability theorems for systems of Volterra integral equations", *Appl. Anal.*, vol. 5, 65-77, 1975.
45. Brauer, F., "Bounds for solutions of ordinary differential equations", *Proc. Am. Math. Soc.*, vol. 14, 36-43, 1963.
46. Brauer, F., "Nonlinear differential equations with forcing terms", *Proc. Am. Math. Soc.*, vol. 15, 758-765, 1964.
47. Brauer, F. and J. S. W. Wong, "On asymptotic behavior of perturbed linear systems", *J. Differential Equations*, vol. 6, 142-153, 1969.
48. Brezis, H., *Operateure Maximaux Monotones et Semigroups de Contractions dans les Espaces de Hilbert*, Amsterdam, North-Holland, 1973.
49. Burton, T. A., *Volterra Integral and Differential Equations,* New York, Academic Press, 1983.

50. Butler, G. and T. Rogers, "A generalization of a lemma of Bihari and applications to pointwise estimates for integral equations", *J. Math. Anal. Appl.*, vol. 33, 77-81, 1971a.
51. Butler, G. and T. Rogers, "Pointwise bounds on derivatives of solutions to ordinary differential equations", *SIAM J. Math. Anal.*, vol. 2, 521-528, 1971b.
52. Chandirov, G. I., "On a generalization of Gronwall's inequality and its applications", *Uchen. Zap. Azerb. Univ. Ser. Fiz. Mat. Nauk*, vol. 6, 3-11, 1956 (in Russian).
53. Chandra, J. and B .A. Fleishman, "On a generalization of the Gronwall-Bellman lemma in partially order Banach spaces", *J. Math. Anal. Appl.*, vol. 31, 668-681, 1970.
54. Chandra, J. and V. Lakshmikantham, "Some pointwise estimates for solutions of a class of nonlinear functional-integral inequalities", *Ann. Mat. Pura Appl.*, vol. 94, 63-74, 1972.
55. Chandra, J. and P. W. Davis, "Linear generalizations of Gronwall's inequality", *Proc. Am. Math. Soc.*, vol. 60, 157-160, 1976.
56. Chaplygin, S. A., *Collected Papers on Mechanics and Mathematics,* vol. 1 (in Russian), Moscow, Gostekhizdat, 1954.
57. Cheung, W. S., "On some new integro-differential inequalities of the Gronwall and Wendroff type", *J. Math. Anal. Appl.*, vol. 178, 438-449, 1993.
58. Chu, S. C. and F. T. Metcalf, "On Gronwall's inequality", *Proc. Am. Math. Soc.*, vol. 18, 439-440, 1967.
59. Coddington, E. A. and N. Levinson, *Theory of Ordinary Differential Equations*, New York, McGraw-Hill, 1955.
60. Cohen, D. S., "The asymptotic behavior of a class of nonlinear differential equations", *Proc. Am. Math. Soc.* .vol. 18, 607-609, 1967.
61. Conlan, J. and J. B. Daiz, "Existence of solutions of nth order hyperbolic partial differential equation", *Contr. Diff. Eqs.* vol. 2, 277-289, 1963.
62. Constantin, A., "A Gronwall-like inequality and its applications", *Rend. Mat. Acc. Lincei*, vol. 9, 111-115, 1990a.
63. Constantin, A., "On some integro-differential and integral inequalities and applications", *Ann. Univ. Din Timisoara Facult. de Mat. Infor.*, vol. 30, 1-21, 1990b.
64. Constantin, A., "Some observations on Conti's result", *Rend. Math. Acc. Lincei*, vol. 9, 137-145, 1991.
65. Constantin, A., "On pointwise estimates for solutions of Volterra integral equations", *Bollettino U.M.I.*, vol. 6-A, 215-225, 1992.
66. Constantin, A., "On the asymptotic behavior of second order nonlinear differential equations", *Rendiconti di Matematica, Serie VII*, vol. 13, 627-634, 1993.
67. Coppel, W. A., *Stability and Asymptotic Behavior of Differential Equations,* Boston, Heath, 1965.
68. Copson, E. T., "On the Riemann-Green function", *Arch. Rat. Mech. Anal.*, vol. 1, 324-348, 1958.

REFERENCES

69. Corduneanu, A., "A note on Gronwall inequality in two independent variables", *J. Integral Equations,* vol. 4, 261–276, 1982.
70. Corduneanu, A., "An integral inequality in two independent variables", *An. Sti. Univ. "Al. I. Cuza" Iasi Sect. I. Mat.,* vol. 39, 69–74, 1983.
71. Corduneanu, A., "Integral inequalities in two independent variables", *Rev. Roumaines Math. Pures Appl.,* vol. 82, 331–341, 1987.
72. Corduneanu, C., *Principles of Differential and Integral Equations,* New York, Chelsea Publishing Co., 1977.
73. Corduneanu, C., *Integral Equations and Stability of Feedback Systems,* New York, Academic Press, 1973.
74. Corduneanu, C., *Integral Equations and Applications,* Cambridge, Cambridge University Press, 1991.
75. Courant, R. and D. Hilbert, *Partial Differential Equations: Methods of Mathematical Physics, Vol. II,* New York, Wiley Interscience, 1962.
76. Dafermos, C. M., "The second law of thermodynamics and stability", *Arch. Rat. Mech. Anal.,* vol. 70, 167–179, 1979.
77. Dannan, F. M., "Integral inequalities of Gronwall–Bellman–Bihari type and asymptotic behavior of certain second order nonlinear differential equations", *J. Math. Anal. Appl.,* vol. 108, 151–164, 1985.
78. Das, K. M., "A note on an inequality due to Greene", *Proc. Am. Math. Soc.,* vol. 77, 424–425, 1979.
79. Davis, W. P. and J. A. Chatfield, "Concerning product integrals and exponentials", *Proc. Am. Math. Soc.,* vol. 25, 743–747, 1970.
80. DeFranco, R. J., "Gronwall's inequality for systems of multiple Volterra integral equations", *Funkcial. Ekvac.,* vol. 19, 1–9, 1976.
81. Deo, S. G. and M. G. Murdeshwar, "A note on Gronwall's inequality", *Bull. London Math. Soc.,* vol. 3, 34–36, 1971.
82. Dhakne, M. B. and B. G. Pachpatte, "On perturbed abstract functional integrodifferential equation", *Acta Math. Sci.,* vol. 8, 263–282, 1988a.
83. Dhakne, M. B. and B. G. Pachpatte, "On a general class of abstract functional integro-differential equations", *Indian J. Pure Appl. Math.,* vol. 19, 728–746, 1988b.
84. Dhakne, M. B. and B. G. Pachpatte, "On some abstract functional integrodifferential equations", *Indian J. Pure Appl. Math.,* vol. 22, 109–134, 1991.
85. Dhakne, M. B. and B. G. Pachpatte, "On a quasilinear functional integrodifferential equation in a Banach space", *Indian J. Pure Appl. Math.,* vol. 25, 275–297, 1994.
86. Dhongade, U. D. and S. G. Deo, "Some generalizations of Bellman–Bihari integral inequalities", *J. Math. Anal. Appl.,* vol. 44, 218–226, 1973.
87. Dhongade, U. D. and S. G. Deo, "A nonlinear generalizations of Bihari's inequality", *Proc. Am. Math. Soc.,* vol. 54, 211–216, 1976.
88. Diekmann, O., "Run for your life. A note on the asymptotic speed of propagation of an epidemic", *J. Differential Equations,* vol. 33, 58–73, 1979.

89. Distefano, N., "A Volterra integral equation in the stability of some linear hereditary phenomena", *J. Math. Anal. Appl.*, vol. 23, 365-383, 1968.
90. Dragomir, S. S., "The Gronwall type lemmas and applications, *Monografii Matematice"*, Univ. Timisoara, No. 29, 1987a.
91. Dragomir, S. S., "On Volterra integral equations with kernels of (L)-type", *Ann. Univ. Timisoara Facult de Mat. Infor.*, vol. 25, 21-41, 1987b.
92. Dragomir, S. S., "On some Gronwall type lemmas", *Studia Univ. Babes-Bolyai, Mathematica*, vol. 33, 29-36, 1988.
93. Dragomir, S. S., "On some nonlinear generalizations of Gronwall's inequality", *Coll. Sci. Fac. Sci. Kraqujevac, sec. Math. Tech. Sci.*, vol. 13, 23-28, 1992.
94. Dragomir, S. S. and N. M. Ionescu, "The Wendroff type lemma and applications (I)", *Itin. Seminar on functional equations, approximation and convexity*, 135-142, Cluj-Napoca, 1987.
95. Dragomir, S. S. and N. M. Ionescu, "On nonlinear integral inequalities in two independent variables", *Studia Univ. Babes-Bolyai, Mathematica*, vol. 34, 11-17, 1989.
96. Engler, H., "Global regular solutions for the dynamic antiplane shear problem in nonlinear viscoelasticity", *Math. Z.* vol. 202, 251-259, 1989.
97. Erbe, L. H. and Qingkai Kong, "Stieltjes integral inequalities of Gronwall type and applications", *Ann. di Mat. Pura Appl.*, vol. CLVII, 77-97, 1990.
98. Filatov, A. N. and L. V. Sharova, *Integral Inequalities and the Theory of Nonlinear Oscillations*, Moscow, Nauka, 1976 (in Russian).
99. Fink, A. M., "Wendroff's inequalities", *Nonlinear Analysis TMA*, vol. 5, 873-874, 1981.
100. Gamidov, Sh. T., "Some integral inequalities for boundary value problems for differential equations", *Diff. Eqs.* vol. 5, 463-472, 1969.
101. Garbedian, P., *Partial Differential Equations*, New York, John Wiley and Sons, Inc., 1964.
102. Ghoshal, S. K. and M. A. Masood, "Generalized Gronwall's inequality and its applications to a class of non-self-adjoint linear and nonlinear hyperbolic partial differential equations, to establish their uniqueness, continuous dependence and comparison theorems", *Indian J. Pure Appl. Math.*, vol. 5, 297-310, 1974a.
103. Ghoshal, S. K. and M. A. Masood, "Gronwall's vector inequality and its applications to a class of non-self-adjoint linear and nonlinear hyperbolic partial differential equations", *J. Indian Math. Soc.*, vol. 38, 383-394, 1974b.
104. Gollwitzer, H. E., "A note on a functional inequality", *Proc. Amer. Math. Soc.*, vol. 23, 642-647, 1969.
105. Grace, S. R. and B. S. Lalli, "Asymptotic behavior of certain second order integro-differential equations", *J. Math. Anal. Appl.*, vol. 76, 84-90, 1980a.
106. Gracc, S. R. and B. S. Lalli, "On boundedness and asymptotic behavior of certain second order integro-differential equations", *J. Math. Physical Sci.*, vol. 14, 191-203, 1980b.

107. Graef, J. R. and P. W. Spikes, "On the nonoscillation, convergence to zero, and integrability of solutions of a second order differential equation", *Math. Nachr.*, vol. 130, 139-149, 1987.
108. Greene, D. E., "An inequality for a class of integral systems", *Proc. Am. Math. Soc.*, vol. 62, 101-104, 1977.
109. Gripenberg, G., S. O. Londen and O. Staffans, *Nonlinear Volterra and Integral Equations*, Cambridge, Cambridge University Press, 1990.
110. Gripenberg, G., "Periodic solutions of an epidemic model", *J. Math. Biol.*, vol. 10, 271-280, 1980.
111. Gripenberg, G., "On some epidemic models", *Quart. Appl. Math.*, vol. 39, 317-327, 1981.
112. Gronwall, T. H., "Note on the derivatives with respect to a parameter of the solutions of a system of differential equations", *Ann. Math.*, vol. 20, 292-296, 1919.
113. Gutowski, R., "Etude d'une inégalité intégrale non linéaire en deux variables", *Ann. Polon. Mat.*, vol. 25, 247-252, 1978.
114. Halanay, A., *Differential Equations: Stability, Oscillations, Time-Lag*, New York, Academic Press, 1966.
115. Hale, J. K., *Theory of Functional Differential Equations*, Berlin, Springer-Verlag, 1977.
116. Hallam, T. G., "Asymptotic expansions for the solutions of a class of nonhomogenous differential equations", *Arch. Rat. Mech. Anal.* vol. 33, 139-154, 1969.
117. Hallam, T. G., "On stability and L^p solutions of ordinary differential equations", *Ann. Math. Pura Appl.* vol. LXXXV, 307-326, 1970.
118. Haraux, A., *Nonlinear Evolution Equations. Global Behavior of Solutions*, Lecture Notes in Math. No. 847, Berlin, New York, Springer-Verlag, 1981.
119. Hardy, G. H., J. E. Littlewood and G. Pólya, *Inequalities*, Cambridge, Cambridge University Press, 1934.
120. Hartman, P., *Ordinary Differential Equations*, New York, Wiley, 1964.
121. Hastings, S. P., "On the asymptotic growth of solutions to nonlinear equation", *Proc. Am. Math. Soc.*, vol. 17, 40-47, 1966.
122. Headley, V. B., "A multidimensional nonlinear Gronwall inequlity", *J. Math. Anal. Appl.* vol. 47, 250-255, 1974.
123. Helton, B. W., "A product integral representation for a Gronwall inequality", *Proc. Am. Math. Soc.* vol. 23, 493-500, 1969.
124. Helton, J. C., "Two generalizations of the Gronwall inequality by product integration", *Rocky Mountain J. Math.* vol. 7, 733-749, 1977.
125. Herod, J. V., "A Gronwall inequality for linear Stieltjes integrals", *Proc. Am. Math. Soc.* vol. 23, 34-36, 1969.
126. Hille, E., *Lectures on Ordinary Differential Equations*, New York, Addison-Wesley, 1969.

127. Ikehata, R. and N. Okazawa, "Yosida approximation and nonlinear hyperbolic equations", *Nonlinear Analysis TMA*, Vol. 15, 479–495, 1990.
128. Jones, G. S., "Fundamental inequalities for discrete and discontinuous functional equations", *J. Soc. Ind. Appl. Math.* vol. 12, 43–57, 1964.
129. Joseph, D. D. and E. M. Sparrow, "Nonlinear diffusion induced by nonlinear sources", *Quart. Appl. Math.*, vol. 28, 327–342, 1970.
130. Kamke, E., *Differentialgleichungen reeller Funktionen*, 2. Aufl, Leipzig, Akademische Verlagsgesellschaft, 1945.
131. Kong, Q. and B. Zhang, "Some generalizations of Gronwall–Bihari integral inequalities and their applications", *China Ann. Math.* vol. 10B, 371–385, 1989.
132. Krasnoselskii, A. M., *Positive Solutions of Operator Equation*, Groningen, Noordhoff, 1964.
133. Kusano, T. and H. Onose, "Nonlinear oscillation of a sublinear delay equation of arbitrary order", *Proc. Am. Math. Soc.* vol. 40, 219–224, 1973.
134. Ladas, G., "Oscillation and asymptotic behavior of solutions of differential equations with retarded argument", *J. Diff. Eq.* vol. 10, 281–290, 1971.
135. Lakshmikantham, V., "A variation of constants formula and Bellman–Gronwall–Reid inequalities", *J. Math. Anal. Appl.* vol. 41, 199–204, 1973.
136. Lakshmikantham, V. and S Leela, *Differential and Integral Inequalities*, Vol. I, New York, Academic Press, 1969.
137. Lalli, B. S., "Boundedness and asymptotic behavior of certain nth order integro-differential equations", *J. Nigerian Math. Soc.* vol. 3, 71–82, 1984.
138. Langenhop, C. E., "Bounds on the norm of a solution of a general differential equation", *Proc. Am. Math. Soc.* vol. 11, 795–799, 1960.
139. LaSalle, J. P., "Uniqueness theorems and successive approximations", *Ann. Math.* vol. 50, 722–730, 1949.
140. Lasota, A., A. Strauss, and W. Walter, "Infinite systems of differential inequalities defined recursively", *J. Diff. Eq.* vol. 9, 93–107, 1971.
141. Li, Y. S., "The bound, stability and error estimates for the solution of nonlinear differential equations", *Chinese Math. Acta*, vol. 3, 34–41, 1963.
142. Lin, C. T., "On some new integro-differential inequalities", *Soochow J. Math.* vol. 7, 125–134, 1981.
143. Lin, C. T., "On Pachpatte type integral inequalities", *Chinese J. Math.* vol. 18, 305–313, 1990.
144. Lin, C. T. and C. M. Wu, "A note on an inequality ascribed to Pachpatte", *Tamkang J. Math.* vol. 20, 25–29, 1989.
145. Losonczi, L., "A generalization of the Gronwall–Bellman lemma and its applications", *J. Math. Anal. Appl.*, vol. 44, 701–709, 1973.
146. Mahmudov, A. P. and V. M. Musaev, "On the theory of solutions of nonlinear Volterra Urysohn integral equations", *Akad. Nauk. Azerbaidzan SSR Dokl.* vol. 25, 3–6, 1969.

147. Mamedov, Ya. D., S. Ashirov and S. Atdaev, *Theorems on Inequalities*, Ashkabad, Ilim, 1980 (in Russian).
148. Mao, X. R., "Lebesgue-Stieltjes integral inequalities and stochastic stabilities", *Quart. J. Math. Oxford*, vol. 40, 301-311, 1989.
149. Maroni, P., "Une généralisation non linéaire de l'inégualité de Gronwall", *Bull. Acad. Polon. Sci. Ser. Sci. Math. Astron. et. Phys.* vol. 16, 703-709, 1968.
150. Martynyuk, A. A. and R. Gutowski, *Integral Inequalities and Stability of Motion*, Kiev, Nauk. Dumka, 1979 (in Russian).
151. Medveď, M., "Bihari type inequalities with multiple integral and delay", *Periodica Math. Hungarica*, vol. 27, 207-212, 1993.
152. Mikhailov, V. P., *Partial Differential Equations*, Moscow, Mir Publications, English translation, 1978.
153. Miller, R. K., *Nonlinear Volterra Integral Equations*, Menlo Park CA, W. A. Benjamin, 1971.
154. Mingarelli, A. B., "On a Stieltjes version of Gronwall's inequality", *Proc. Am. Math. Soc.* vol. 82, 249-252, 1981.
155. Mitrinović, D. S., *Analytic Inequalities*, Berlin, New York, Springer-Verlag, 1970.
156. Mitrinović, D. S. and J. E. Pečarić, *Differential and Integral Inequalities*, Belgrade, Naucna Knjiga, 1988.
157. Mitrinović, D. S., J. E. Pečarić and A. M. Fink, *Inequalities Involving Functions and Their Integrals and Derivatives*, New York, Kluwer Academic Publishers, 1991.
158. Mitrinović, D. S., J. E. Pečarić and A. M. Fink, *Classical and New Inequalities in Analysis*, New York, Kluwer Academic Publishers, 1992.
159. Morchalo, J., "A functional differential inequality for linear integro-differential equations with retardation", *Ann. Soc. Math. Poln. Ser. I: Comment Math.* vol. 21, 177-186, 1979.
160. Morchalo, J., "On the applications of differential inequalities to stability theory", *Demonstratio Mathematica*, vol. 17, 955-974, 1984.
161. Morro, A., "A Gronwall-like inequality and its applications to continuum thermodynamics", *Boll. U. M. I.* vol. 6, 553-562, 1982.
162. Movlyankulov, Kh. and A. N. Filatov, "On an approximate method for constructing solutions of integral equations", *Trudy Inst. Kibernet. Akad. Nauk Uzb SSR.* vol. 2, 11-18, 1972 (in Russian).
163. Muldowney, J. S. and J S. W. Wong, "Bounds for solutions of nonlinear integrodifferential equations", *J. Math. Anal. Appl.* vol. 23, 487-499, 1968.
164. Murge, M. G. and B. G. Pachpatte, "Existence and uniqueness of solutions of nonlinear Itô type stochastic integral equations", *Acta Math. Sci.* vol. 7, 207-216, 1987.
165. Murge, M. G. and B. G. Pachpatte, "On averaging method for second order Itô type stochastic equations", *Indian J. Pure Appl. Math.* vol. 23, 25-37, 1992.

166. Norbury, J. and A. M. Stuart, "Volterra integral equations and a new Gronwall inequality. Part I: the linear case", *Proc. Royal Soc. Edinburgh*, vol. 106A, 361–373, 1987.
167. Okrasinski, W., "On a nonlinear convolution equation occurring in the theory of water precolation", *Ann. Polon. Math.* vol. 37, 223–229, 1980.
168. Olech, C., "On a system of integral inequalities", *Colloq. Math.* vol. 16, 137–139, 1967.
169. Olekhnik, S. N., "Boundedness and unboundedness of solutions of some systems of ordinary differential equations", *Vestnik Moskov. Univ. Ser. I Mat.* vol. 27, 34–44, 1972.
170. Opial, Z., "Sur une système d'inégualités intégrales", *Ann. Polon. Math.* vol. 7, 247–254, 1960.
171. Ou-Iang, L., "The boundedness of solutions of linear differential equations $y'' + A(t)y = 0$", *Shuxue Jinzhan*, vol 3, 409–415, 1957.
172. Pachpatte, B. G., "A note on Gronwall–Bellman inequality", *J. Math. Anal. Appl.* vol. 44, 758–762, 1973a.
173. Pachpatte, B. G., "Integral perturbations of nonlinear systems of differential equations", *Bull. Soc. Math. Gréce*, vol. 14, 92–97, 1973b.
174. Pachpatte, B. G., "Stability and asymptotic behavior of perturbed nonlinear systems, *J. Diff. Eq.* vol. 16, 14–25, 1974b.
175. Pachpatte, B. G., "An integral inequality similar to Bellman–Bihari inequality", *Bull. Soc. Math. Gréce.* vol. 15, 7–12, 1974d.
176. Pachpatte, B. G., "Integral inequalities of Gronwall–Bellman type and their applications", *J. Math. Physical Sci.* vol. 8, 309–318, 1974a.
177. Pachpatte, B. G., "A note on some integral inequalities", *The Math. Student*, vol. XLII, 409–411, 1974c.
178. Pachpatte, B. G., "A note on integral inequalities of the Bellman–Bihari type", *J. Math. Anal. Appl.* vol. 49, 295–301, 1975h.
179. Pachpatte, B. G., "On some integral inequalities similar to Bellman–Bihari inequalities", *J. Math. Anal. Appl.* vol. 49, 794–802, 1975b.
180. Pachpatte, B. G., "On some generalizations of Bellman's Iemma", *J. Math. Anal. Appl.* vol. 51, 141–150, 1975a.
181. Pachpatte, B. G., "Perturbations of nonlinear systems of differential equations", *J. Math. Anal. Appl.* vol. 51, 550–556, 1975f.
182. Pachpatte, B. G., "On some integrodifferential equations in Banach spaces", *Bull. Austral. Math. Soc.* vol. 12, 337–350, 1975i.
183. Pachpatte, B. G., "A note on an integral inequality", *J. Math. Physical Sci.* vol. 9, 11–14, 1975c.
184. Pachpatte, B. G., "Boundedness and asymptotic behavior of solutions of second order integrodifferential equations", *J. Math. Physical Sci.* vol. 9, 171–175, 1975e.
185. Pachpatte, B. G., "On an integral inequality of Gronwall–Bellman", *J. Math. Physical Sci*, vol. 9, 405–416, 1975g.

186. Pachpatte, B. G., "On some integral inequalities of the Gronwall-Bellman type", *Indian J. Pure Appl. Math.* vol. 6, 769-772, 1975d.
187. Pachpatte, B. G., "On preserving stability of Volterra integral equations", *Bull. Math. de la Soc. Sci. Math. de la R. S. de Romaine*, vol. 20, 165-172, 1976a.
188. Pachpatte, B. G., "A new integral inequality for differential and integral equations", *Proc. Nat. Acad. Sci. India.* vol. 46(A), 21-26, 1976f.
189. Pachpatte, B. G., "On some new integral inequalities for differential and integral equations", *J. Math. Physical Sci.* vol. 10, 101-116, 1976b.
190. Pachpatte, B. G., "On the asymptotic behavior of functional integrodifferential equations", *J. Math. Physical Sci.* vol. 10, 11-16, 1976d.
191. Pachpatte, B. G., "On some nonlinear generalizations of Gronwall's inequality", *Proc. Indian Acad. Sci.* vol. 84A, 1-9. 1976g.
192. Pachpatte, B. G., "On a class of nonlinear nth order integrodifferential equations", *J. M. A. C. T.* vol. 9, 37-45, 1976e.
193. Pachpatte, B. G., "On a class of integral inequalities of Bellman-Bihari type", *Indian J. Pure Appl. Math.* vol. 7, 1166-1177, 1976c.
194. Pachpatte, B. G., "A note on Gronwall type integral and integrodifferential inequalities", *Tamkang J. Math.* vol. 8, 53-59, 1977c.
195. Pachpatte, B. G., "On some new generalizations of Gronwall's inequality and their discrete analogues", *J. M. A. C. T.*, vol. 10, 55-69, 1977d.
196. Pachpatte, B. G., "On some integral inequalities and their applications to integrodifferential equations", *Indian J. Pure Appl. Math.* vol. 8, 1157-1175, 1977a.
197. Pachpatte, B. G., "On some fundamental integrodifferential and integral inequalities", *An. Sti. Univ. Al. I. Cuza" Iasi*, vol. 23, 77-86, 1977b.
198. Pachpatte, B. G., "On some fundamental integrodifferential inequalities for differential equations", *Chinese J. Math.* vol. 6, 17-23, 1978c.
199. Pachpatte, B. G., "On some fundamental integral inequalities with applications", *Bul. Sti. Inst. Politch Traian Vuia Timisoara*, vol. 24, 25-33, 1978b.
200. Pachpatte, B. G., "On some fundamental integrodifferential inequalities and their discrete analogues", *Proc. Indian Acad. Sci.* vol. 87A, 201-207, 1978a.
201. Pachpatte, B. G., "On some new integral and integrodifferential inequalities in two independent variables and their applications", *J. Diff. Eq.* vol. 33, 249-272, 1979a.
202. Pachpatte, B. G., "On Wendroff type partial integral inequalities", *Tamkang J. Math.* vol. 10, 141-150, 1979b.
203. Pachpatte, B. G., "On some integrodifferential inequalities similar to Wendroff's inequality", *An. Sti. Univ. "Al. I. Cuza" Iasi*, vol. 25, 127-136, 1979c.
204. Pachpatte, B. G., "A note on some new partial integral inequalities", *Chinese J. Math.* vol 8, 13-17, 1980e.
205. Pachpatte, B. G., "On some fundamental partial integral inequalities", *J. Math. Anal. Appl.* vol. 73, 238-251, 1980c.

206. Pachpatte, B. G., "On some new integrodifferential inequalities of the Wendroff type", *J. Math. Anal. Appl.* vol. 73, 491–500, 1980b.
207. Pachpatte, B. G., "On some new integral inequalities for non-self-adjoint hyperbolic partial integrodifferential equations", *J. Math. Anal. Appl.* vol. 76, 58–71, 1980d.
208. Pachpatte, B. G., "A note on second order integrodifferential inequalities of the Gronwall-Bellman type", *Univ. Beograd Publ. Elek. Fak. Ser. Mat. Fiz.* No. 678-No. 715, 35–41, 1980a.
209. Pachpatte, B. G., "On some partial integral inequalities in n independent variables", *J. Math. Anal. Appl.* vol. 79, 256–272, 1981c.
210. Pachpatte, B. G., "On certain integral inequalities for partial differential and integral equations", *An. Sti. Univ. Al. I. Cuza" Iasi* vol. 27, 365–374, 1981b.
211. Pachpatte, B. G., Some inequalities between functions and their derivatives, *Acta Ciencia Indica*, vol. 7, 223–234, 1981a.
212. Pachpatte, B. G., "A note on some fundamental integrodifferential inequalities", *Tamkang. J. Math.* vol. 13, 63–67, 1982a.
213. Pachpatte, B. G., "Perturbations of Bianchi type partial integrodifferential equation", *Bull. Inst. Math. Acad. Sinica*, vol. 10, 347–358, 1982b.
214. Pachpatte, B. G., "A note on Greene's inequality", *Tamkang J. Math*, vol. 15, 49–54, 1984.
215. Pachpatte, B. G., "On Wendroff like integral inequalities", *Chinese J. Math*, vol. 16, 137–155, 1988c.
216. Pachpatte, B. G., "On certain multidimensional integral inequalities and their applications", *Chung Yuan J.*, vol. 17, 14–21, 1988b.
217. Pachpatte, B. G., "On some integral inequalities and their applications to higher order differential equations", *Soochow J. Math.* vol. 14, 221–230, 1988a.
218. Pachpatte, B. G., "On a generalized Volterra integral equation", *Bul. Univ. Din Brasov*, vol. 32, 5–12, 1990.
219. Pachpatte, B. G., "On Bihari like integral and discrete inequalities", *Soochow J. Math.* vol. 17, 213–232, 1991.
220. Pachpatte, B. G., "On Gronwall like integral inequalities", *Indian J. Pure Appl. Math.* vol. 23, 131–140, 1992.
221. Pachpatte, B. G., "On certain nonlinear integral inequalities and their discrete analogues", *Facta Univ. (NIS), Ser. Math. Inform.* vol. 8, 21–34, 1993.
222. Pachpatte, B. G., "On a certain inequality arising in the theory of differential equations", *J. Math. Anal. Appl.*, vol. 182, 143–157, 1994a.
223. Pachpatte, B. G., "On some fundamental integral inequalities arising in the theory of differential equations", *Chinese J. Math.*, vol. 22, 261–273, 1994b.
224. Pachpatte, B. G., "Inequalities related to a certain integral inequality arising in the theory of differential equations", *Studia Univ. Babes-Bolyai, Mathematica*, vol. 39, 33–50, 1994d.

225. Pachpatte, B. G., "On a new inequality suggested by the study of a certain nonlinear convolution equation", *An. Univ. din Timisoara, Seria Math.*, vol. 32, 93–102, 1994c.
226. Pachpatte, B. G., "On some new inequalities related to certain inequalities in the theory of differential equations", *J. Math. Anal. Appl.*, vol. 189, 128–144, 1995a.
227. Pachpatte, B. G., "On a new inequality suggested by the study of certain epidemic models", *J. Math. Anal. Appl.*, vol. 195, 638–644, 1995d.
228. Pachpatte, B. G., "A note on certain integral inequalities with delay", *Periodica Math. Hungar.*, vol. 31, 229–234, 1995b.
229. Pachpatte, B. G., "On some inequalities useful in the theory of certain higher order differential and difference equations", *Studia Univ. Babes-Bolyai, Mathematica*, vol. 40, 15–34, 1995c.
230. Pachpatte, B.G., "Inequalities similar to a certain inequality used in the theory of differential equations", *Chinese J. Math.*, vol. 24, 55–68, 1996a.
231. Pachpatte, B. G., "On certain nonlinear integral inequalities and their discrete analogues II", *Recent Progress in Inequalities, Memorial conference vol. I, Scientific Review*, No. 21–22, 99–110, 1996d.
232. Pachpatte, B. G., "Comparison theorems related to a certain inequality used in the theory of differential equations", *Soochow J. Math.*, vol. 22, 383–394, 1996b.
233. Pachpatte, B. G., "Inequalities related to a certain inequality used in the theory of differential equations", *Tamkang J. Math.*, in press a.
234. Pachpatte, B. G., "On Gronwall type inequalities occuring in the theory of differential equations", *Fasciculi Mathematici*, No. 26, 111–123, 1996c.
235. Pachpatte, B. G., "On nonlinear integral and discrete inequalities in two independent variables", *Bul. Sti. Tech. Inst. Politehn. Timisoara*, vol. 40, 29–38, 1995e.
236. Pachpatte, B. G., "A note on a certain inequality in the theory of differential equations", *Octogon*, vol. 5, 44–48, 1997.
237. Pachpatte, B. G., "On certain new integral inequalities and their discrete analogues", *Radovi Mat.*, in press d.
238. Pachpatte, B. G., "Generalizations of certain inequalities used in the theory of differential equations", *An. Sti. Univ.,, Al. I. Cuza" Iasi*, in press e.
239. Pachpatte, B. G., "On nonlinear Volterra-type integral and discrete inequalities", *An. Univ. Din Timisoara, Ser. Math.*, in press k.
240. Pachpatte, B. G., "A note on certain nonlinear functional inequalities", *An. Univ. Bucuresti, Ser. Math.*, in press l.
241. Pachpatte, B. G., "On some inequalities suggested by a certain inequality used in the theory of differential equations", *An. Sti. Univ. ,,Al. I. Cuza" Iasi*, in press m.
242. Pachpatte, B. G., "Generalizations of a certain inequality used in the theory of differential equations", *Ann. Univ. Bucuresti, Ser. Math.*, in press f.

243. Pachpatte, B. G., "A note on a certain integral inequality", *Octogon*, vol. 4, 43–45, 1996e.
244. Pachpatte, B. G., "A note on certain integral inequalities and their discrete analogues", *Octogon*, vol. 4, 18–21, 1996d.
245. Pachpatte, B. G., "On an integral inequality used in the theory of differential equations", *Bul. Sti. Tech. Inst., Politehn, Timisora*, in press h.
246. Pachpatte, B. G., "A note on certain inequalities arising in the theory of differential and difference equations", *Octogon*, in press i.
247. Pachpatte, B. G., "Inequalities for certain integrodifferential equations and their discrete analogues", *Soochow J. Math.*, in press j.
248. Pachpatte, B. G., "Inequalities for partial differential and integral equations", unpublished manuscript.
249. Peano, G., "Sull'integrabilità delle equazioni differenziali del primo ordine", *Atti. R. Accad. Sc. Toriao*, vol. 21, 677–685, 1885–86.
250. Pelczar, A., "On some inequalities", *Zeszyty Nauk. Univ. Jag. Prace Math.*, vol. LXXVII, 77–80, 1963.
251. Pinto, M., "Integral inequalities of Bihari-type and applications", *Funckcial. Ekv.*, vol. 33, 387–404, 1990.
252. Popenda, J., "A note on Gronwall-Bellman inequality", *Fasciculi Math.*, vol. 16, 29–41, 1986.
253. Ráb, M., "Linear integral inequalities", *Arch. Math. I. Scripta Fac. Sci. Nat. Ujep Brunensis*, vol. 15, 37–46, 1979.
254. Rasmussen, D. L., "Gronwall's inequality for functions of two independent variables", *J. Math. Anal. Appl.*, vol. 55, 407–417, 1976.
255. Redheffer, R. M., "Differential and integral inequalities", *Proc. Am. Math. Soc.*, vol. 15, 715–716, 1964.
256. Reid, W. T., "Properties of solutions of an infinite system of ordinary linear differential equations of first order with auxiliary boundary conditions", *Trans. Am. Math. Soc.*, vol. 32, 284–318, 1930.
257. Reid, W. T., *Ordinary Differential Equations*, New York, London, Springer-Verlag, 1971.
258. Rodrigues, H. M., "On growth and decay of solutions of perturbed retarded linear equations", *Tôhoku Math. J.*, vol. 32, 593–605, 1980.
259. Rogers, T., "A functional integral inequality and bounds for nonlinear differential equations", *Bull. Acad. Polon. des. Sci. Ser. Sci. Math. Astr et. Phys.*, vol. 22, 269–272, 1974.
260. Rudakov, V. P., "On a two-sided estimate of the norms of solutions of a class of nonlinear systems", *Diff. Eq.*, vol. 6, 371–372, 1970 (in Russian).
261. Sansone, G. and R. Conti, *Nonlinear Differential Equations*, New York, Macmillan, 1964.
262. Sardarly, S. M., "Investigation of the solutions of nonlinear integral equations on the halfline with respect to a parameter", *Uchen. Zap. Azerb. Univ.*, vol. 2, 24–30, 1965 (in Russian).

263. Sato, T., "Sur l'équation intégrale non linéaire de Volterra", *Comp. Math.*, vol. 11, 271-289, 1953.
264. Schmaedeke, W. W. and G. R. Sell, "The Gronwall inequality for modified Stieltjes integrals", *Proc. Amer. Math. Soc.*, vol. 19, 1217-1222, 1968.
265. Schröder, J., *Operator inequalities*, New York, Academic Press, 1980.
266. Shih, M. H. and C. C. Yeh, "Some integral inequalities in n independent variables", *J. Math. Anal. Appl.*, vol. 84, 569-593, 1981.
267. Shinde, B. B. and B. G. Pachpatte, "On multidimensional simultaneous integral inequalities", *J. Math. Physical Sci.*, vol. 17, 219-243, 1983.
268. Shinde, B. B. and B. G. Pachpatte, "On some fundamental simultaneous integral inequalities in n independent variables", *Ganit J. Bangladesh Math. Soc.*, vol. 4, 54-67, 1984.
269. Singare, V. M. and B. G. Pachpatte, Lower bounds on some integral inequalities in n independent variables", *Indian J. Pure Appl. Math.*, vol. 12, 318-331, 1981.
270. Singare V. M. and B. G. Pachpatte, "On some fundamental multidimensional integral inequalities", *J. Math. Physical Sci.*, vol. 16, 431-452, 1982.
271. Sneddon, I. N., *Elements of Partial Differential Equations*, New York, McGraw-Hill, 1957.
272. Snow, D. R., "A two independent variable Gronwall type inequality", in *Inequalities III* (ed. O. Shisha), 330-340, New York, Academic Press, 1971.
273. Snow, D. R., "Gronwall's inequality for systems of partial differential equations in two independent variables", *Proc. Am. Math. Soc.*, vol. 33, 46-54, 1972.
274. Sobolev, S. L., *Partial Differential Equations of Mathematical Physics*, Oxford, Pergamon Press, 1964.
275. Stachurska, B., "On a nonlinear integral inequality", *Zeszyty Nauk. Univ. Jag, Prace Mat.*, vol. 15, 151-157, 1971.
276. Strauss, A. and J. Yorke, "Perturbation theorems for ordinary differential equations", *J. Diff. Eq.*, vol. 3, 15-30, 1967.
277. Sugiyama, S., "On comparison theorems of nonlinear Volterra integral equations", *Kōdai Math. Sem. Rep.*, vol. 27, 147-154, 1976.
278. Szarski, J., *Differential Inequalities*, Warsaw, PWN, 1965.
279. Talpalaru, P., "Some results concerning the asymptotic equivalence of integro-differential equations", *Ann. Sti. Univ. „Al. I. Cuza" Iasi*, vol. 19, 117-131, 1973.
280. Thandapani, E. and R. P. Agarwal, "On some new integro-differential inequalities: theory and applications", *Tamkang J. Math.*, vol. 11, 169-184, 1980.
281. Thandapani, E. and R. P. Agarwal, "On some new inequalities in n independent variables", *J. Math. Anal. Appl.*, vol. 86, 542-561, 1982.
282. Thompson, D. E., "Some theorems on a Volterra equation of the second kind", *Can. Math. Bull.*, vol. 14, 584-589, 1971.
283. Tong, J., "The asymptotic behavior of a class of nonlinear differential equations of second order", *Proc. Am. Math. Soc.*, vol. 84, 235-236, 1982.

284. Trench, W. F., On the asymptotic behavior of solutions of second order linear differential equations, *Proc. Am. Math. Soc.*, vol. 14, 12-14, 1963.
285. Tricomi, F. G., *Integral Equations*, New York, Interscience, 1957.
286. Tsamatos, P. Ch. and S.K. Ntouyas, "On a Bellman-Bihari type integral inequality with delay", *Periodica Math. Hungarica*, vol. 23, 91-94, 1991.
287. Tsutsumi, M. and I. Fukuda, "On solutions of the derivative nonlinear Schrödinger equation: existence and uniqueness theorem", *Funkcialaj Ekvaciej*, vol. 23, 259-277, 1980.
288. Turinici, M., "A class of operator inequalities on ordered linear spaces", *Demonstratio Mathematica*, vol. 15, 145-153, 1982.
289. Turinici, M., "Abstract Gronwall-Bellman inequalities in ordered metrizable uniform spaces", *J. Integral Eq.*, vol. 6, 105-117, 1984.
290. Turinici, M., "Abstract comparison principle and multivariate Gronwall-Bellman inequalities", *J. Math. Anal. Appl.*, vol. 117, 100-127, 1986.
291. Ved', Yu. A., "On perturbations of linear homogeneous differential equations with variable coefficients", in *Issled. Integro-Diffents. Uravn. Kirghizia 3*, Ilim, Frunze, 1965 (in Russian).
292. Venckova, M., "On the boundedness of solutions of higher order differential equations", *Arch. Math. (Brno)*, vol. 13, 235-242, 1977.
293. Viswanathan, B., "A generalization of Bellman's lemma", *Proc. Amer. Math. Soc.*, vol. 14, 15-18, 1963.
294. Walter, W., *Differential and Integral Inequalities*, Berlin, New York, Springer-Verlag, 1970.
295. Wang, C. L., "A short proof of a Greene theorem", *Proc. Am. Math. Soc.*, vol. 69, 357-358, 1978.
296. Waltman, P., "On the asymptotic behavior of a nonlinear equation", *Proc. Am. Math. Soc.*, vol. 15, 918-923, 1964.
297. Waltman, P., "On the asymptotic behavior of solutions of an n-th order equation", *Montasch Math.*, vol. 69, 427-430, 1965.
298. Ważeski, T., "Sur un systéme des inéqualities integrales ordinaires non linéares", *Bull. Acad. Polon. Sci. Sér. Sci. Math. Astronom. Phys.*, vol. 17, 226-229, 1969.
299. Werbowski, J., "An estimation of the solution of the system of nonlinear equations of Volterra type with delay", *Fasciculi Math.*, vol. 9, 67-75, 1975.
300. Westphal, H., "Zur Abschätzung der Lösungen nichtlinearer parabolischer Differentialgleichungen", *Math. Z.*, vol. 51, 690-695, 1949.
301. Willett, D., "Nonlinear vector integral equations as contraction mappings", *Arch. Rational Mech. Anal.*, vol. 15, 79-86, 1964.
302. Willett, D., "A linear generalization of Gronwall's inequality", *Proc. Am. Math. Soc.*, vol. 16, 774-778, 1965.
303. Willett, D. and J. S. W. Wong, "On the discrete analogues of some generalizations of Gronwall's inequality", *Monatsh. Math.*, vol. 69, 362-367, 1965.

REFERENCES

304. Wong, J. S. W., "On an integral inequality of Gronwall", *Rev. Romaine Math. Pures Appl.*, vol. 12, 1519-1522, 1967.
305. Wright, F. M., M. L. Klasi and D. R. Kennenbeck, "The Gronwall inequality for wighted integrals", *Proc. Am. Math. Soc.*, vol. 30, 504-510, 1971.
306. Yang, C. C., "On some nonlinear generalizations of Gronwall-Bellman's inequality", *Bull. Inst. Math. Acad. Sinica*, vol. 7, 15-19, 1979.
307. Yang, E. H., "On asymptotic behavior of certain integro-differential equations", *Proc. Am. Math. Soc.*, vol. 90, 271-276, 1984.
308. Yang, E. H., "On the most general form of Bellman-type linear inequalities involving multiple-fold integral functionals", *J. Math. Anal. Appl.*, vol. 103, 184-197, 1984.
309. Yang, E. H., "On some new integral inequalities in n independent variables", *J. Math. Anal. Appl.*, vol. 109, 171-181, 1985.
310. Young, E. C., "Gronwall's inequality in n independent variables", *Proc. Am. Math. Soc.*, vol. 41, 241-244, 1973.
311. Young, E. C., "On Bellman-Bihari integral inequalities", *Int. J. Math. Math. Sci.*, vol. 5, 97-103, 1982.
312. Young, E. C., "Functional integral inequalities in N variables", *Chinese J. Math.*, vol. 10, 1-7, 1982.
313. Young, E. C., "On integral inequalities of Gronwall-Bellman type", *Proc. Am. Math. Soc.*, vol. 94, 636-640, 1985.
314. Zakai, M., "Some moment inequalities for stochastic integrals and for solutions of stothastic differential equations", *Israel J. Math.*, vol. 5, 170-176, 1967.
315. Zettl, A., "Square integrable solutions of $Ly = f(t, y)$", *Proc. Am. Math. Soc.*, vol. 26, 635-639, 1970.
316. Ziebur, A. D., "On the Gronwall-Bellman lemma", *J. Math. Anal. Appl.*, vol. 22, 92-95, 1968.

Index

A

Agarwal and Thandapani inequality, 92–3

B

Bainov and Simeonov inequalities, 440, 449–50
Bellman inequality, 9–12
Bellman's lemma or inequality, 12
Bihari inequalities, 107–17
Bondge and Pachpatte inequalities, 578–9

C

Chandra and Davis inequality, 450
Comparison theorem, 102
Coppel inequality, 89

D

Differential equations, 297–303
    estimates of solutions, 203–6
    higher order hyperbolic partial, 561–2
    hyperbolic partial, 428–37, 556–9
    non-self-adjoint hyperbolic, 424–7
Dragomir inequalities, 221–7, 303–4
Dragomir and Ionescu inequalities, 565–6

E

Epidemic models, 294–7

G

Gamidov inequality, 25–9, 88–9
Gronwall-Bellman inequality, 12, 12–16, 107

Gronwall-Bellman-Bihari inequalities, 117–26
Gronwall inequality, 9–12

H

Haraux and Engler inequalities, 267–9
Hastings inequality, 91
Higher order hyperbolic partial differential equations, 561–2
Higher order integro-differential equations, 81–4, 198–203
Hyperbolic partial differential equations, 556–9
    higher order, 561–2
    perturbations, 428–37
Hyperbolic partial integro-differential equations, 417–24, 559–60
    multivariate, 562–5

I

Integral equations, involving product integrals, 84–6
Integro-differential equations, 301–3
    higher order, 81–4, 198–203
    hyperbolic partial, 417–24, 559–60
    multivariate hyperbolic partial, 562–5
    non-self-adjoint hyperbolic, 424–7
    perturbed, 192–8
    second-order, 75–8
Integro-differential inequalities, 44–52, 171–80
Iterated integrals
    inequalities with, 181–6
    nonlinear integral inequalities involving, 276–91

INDEX

J
Jones inequality, 95–6

L
Langenhop inequalities, 107–17
Lebesgue integrability, 138
Linear integral inequalities, 9–97
  Agarwal and Thandapani inequality, 92–3
  applications, 75–86
  Bellman inequality, 9–12
  Coppel inequality, 89
  Gamidov inequality, 25–9, 88–9
  Gronwall inequality, 9–12
  Gronwall–Bellman inequality, 12–16
  Hastings inequality, 91
  integro-differential inequalities, 44–52
  involving product integrals, 62–75
  Jones inequality, 95–6
  Mahmudov and Musaev inequality, 90–1
  Pachpatte's inequalities, 32–44, 88, 92–4
  Rodrigues inequality, 25–9
  Rudakov inequality, 87
  Sansone and Conti inequality, 86–7
  Sardarly inequality, 87
  Thandapani and Agarwal inequality, 94
  Ved' inequality, 92
  Volterra-type integral inequalities, 17–25
  with several iterated integrals, 52–62
  Young inequality, 95

M
Mahmudov and Musaev inequality, 90–1
Minkowski inequality, 137
Multidimensional linear integral inequalities, 323–457
  applications, 417–39
  Bainov and Simeonov inequalities, 440, 449–50
  Chandra and Davis inequalities, 450
  Pachpatte's inequalities, 327–54, 374–96, 441–9, 455–6
  several variables, 396–409
  Shinde and Pachpatte inequality, 453–5
  Singare and Pachpatte inequality, 451–3
  Snow's inequalities, 354–74
  Thandapani and Agarwal inequalities, 450–1
  Wendroff inequalities, 323–7, 468–79
  Young's inequality, 409–17
Multidimensional nonlinear integral inequalities, 459–589
  applications, 556–65
  Bondge and Pachpatte, 578–9
  Dragomir and Ionescu inequalities, 565–6
  many independent variables, 508–27
  Pachpatte's inequalities, 479–508, 527–77, 581–5, 587–8
  Rasmussen inequality, 573
  Shinde and Pachpatte inequality, 586–7
  Singare and Pachpatte inequalities, 579–80, 585–6
  Wendroff's inequality, 460–8
Multivariate hyperbolic partial integro-differential equations, 562–5

N
Non-self-adjoint hyperbolic differential and integro-differential equations, 424–7
Nonlinear differential equations, 99
  second order, 187–91
Nonlinear integral equations, 297–300
Nonlinear integral inequalities, 99–321
  applications, 186–206, 291

INDEX

Bihari and Langenhop inequalities, 107–17
Dragomir inequalities, 221–7, 303–4
epidemic models, 294–7
Gronwall–Bellman–Bihari inequalities, 117–26
Haraux and Engler inequalities, 267–9
integral, 135–48
integro-differential inequalities, 171–80
involving comparison, 100–7
iterated integrals, 276
Ou-Iang and Dafermos inequalites, 233–5
Pachpatte's inequalities, 148–71, 206–18, 227–33, 236–67, 270–6, 304–20
Rogers' inequality, 206
Volterra-type kernels, 126–35
Willett and Wong inequality, 207
with iterated integrals, 181–6

O

Ou-lang and Dafermos inequalites, 233–5

P

Pachpatte's inequalities, 32–44, 88, 92–4, 148–71, 206–18, 227–33, 236–67, 270–6, 304–20, 327–54, 374–96, 441–9, 455–6, 479–508, 527–56, 566–77, 581–5, 587–8
Perturbation of Volterra integral equations, 78–81
Perturbed integro-differential equations, 192–8
Product integrals
  inequalities involving, 62–75
  integral equations involving, 84–6

R

Rasmussen inequality, 573
Rodrigues inequality, 25–9
Rogers' inequality, 206
Rudakov inequality, 87

S

Sansone and Conti inequality, 86–7
Sardarly inequality, 87
Second order integro-differential equations, 75–8
Second order nonlinear differential equations, 187–91
Shinde and Pachpatte inequalities, 453–5, 586–7
Simultaneous inequalities, 30–2
Simultaneous integral equations in two variables, 437–9
Singare and Pachpatte inequalities, 451–3, 579–80, 585–6
Snow's inequalities, 354–74

T

Thandapani and Agarwal inequalities, 94, 450–1

V

Ved' inequality, 92
Volterra-type integral equations, 78–81, 291–4
Volterra-type integral inequality, 17–25
Volterra-type kernels, 126–35

W

Wendroff's inequalities, 323–7, 460–79
Willett and Wong inequality, 207

Y

Young's inequality, 95, 409–17